NONLINEAR MODELING
AND FORECASTING

NONLINEAR MODELING AND FORECASTING

PROCEEDINGS OF THE WORKSHOP
ON NONLINEAR MODELING
AND FORECASTING
HELD SEPTEMBER, 1990
IN SANTA FE, NEW MEXICO

Editors

Martin Casdagli
Santa Fe Institute

Stephen Eubank
*Los Alamos National Laboratory and
Santa Fe Institute*

Proceedings Volume XII

SANTA FE INSTITUTE
STUDIES IN THE SCIENCES OF COMPLEXITY

Addison-Wesley Publishing Company
The Advanced Book Program
Redwood City, California • Menlo Park, California
Reading, Massachusetts • New York • Don Mills,Ontario
Wokingham, United Kingdom • Amsterdam • Bonn
Sydney • Singapore • Tokyo • Madrid • San Juan

Editor: *Barbara Holland*
Production Manager: *Pam Suwinsky*
Production Assistant: *Karl Matsumoto*

Director of Publications, Santa Fe Institute: *Ronda K. Butler-Villa*
Publications Assistant, Santa Fe Institute: *Della L. Ulibarri*

This volume was typeset using T$_E$Xtures on a Macintosh II computer. Camera-ready output from a NEC Silentwriter 2.

Library of Congress Cataloging-in-Publication Data

Nonlinear modeling and forecasting / [edited by] Martin Casdagli,
    Stephen Eubank.
        p.   cm.— (Santa Fe Institute studies in the sciences of complexity.
    Lecture notes)
    Papers from a Workshop on "Nonlinear Modeling and Forecasting"
    held at Santa Fe, N.M., Sept. 17-21, 1991, and sponsored by the
    Santa Fe Institute and NATO.
    Includes bibliographical references and index.
    1. Time-series analysis—Congresses.
    2. Nonlinear theories—Congresses.   3. Forecasting—Congresses
    I. Casdagli, Martin.   II. Eubank, Stephen.   III. Santa Fe Institute
    (Santa Fe, N.M.) IV. North Atlantic Treaty Organization.
    V. Workshop on "Nonlinear Modeling and Forecasting"
    (1991: Santa Fe, N.M.) VI. Series.
    QA280.N66 1992   519.5'5'01175—dc20   91-40068
    ISBN 0-201-52764-2 (hb).—ISBN 0-201-58788-2 (pb)

1 2 3 4 5 6 7 8 9 10-MA-95 94 93 92

About the Santa Fe Institute

The *Santa Fe Institute* (SFI) is a multidisciplinary graduate research and teaching institution formed to nurture research on complex systems and their simpler elements. A private, independent institution, SFI was founded in 1984. Its primary concern is to focus the tools of traditional scientific disciplines and emerging new computer resources on the problems and opportunities that are involved in the multidisciplinary study of complex systems—those fundamental processes that shape almost every aspect of human life. Understanding complex systems is critical to realizing the full potential of science, and may be expected to yield enormous intellectual and practical benefits.

All titles from the *Santa Fe Institute Studies in the Sciences of Complexity* series will carry this imprint which is based on a Mimbres pottery design (circa A.D. 950–1150), drawn by Betsy Jones.

PROCEEDINGS VOLUMES

Volume	Editor	Title
I	David Pines	Emerging Syntheses in Science, 1987
II	Alan S. Perelson	Theoretical Immunology, Part One, 1988
III	Alan S. Perelson	Theoretical Immunology, Part Two, 1988
IV	Gary D. Doolen et al.	Lattice Gas Methods for Partial Differential Equations, 1989
V	Philip W. Anderson, Kenneth Arrow, & David Pines	The Economy as an Evolving Complex System, 1988
VI	Christopher G. Langton	Artificial Life: Proceedings of an Interdisciplinary Workshop on the Synthesis and Simulation of Living Systems, 1988
VII	George I. Bell & Thomas G. Marr	Computers and DNA, 1989
VIII	Wojciech H. Zurek	Complexity, Entropy, and the Physics of Information, 1990
IX	Alan S. Perelson & Stuart A. Kauffman	Molecular Evolution on Rugged Landscapes: Proteins, RNA and the Immune System, 1990
X	Christopher Langton et al.	Artificial Life II, 1991
XI	John A. Hawkins & Murray Gell-Mann	Evolution of Human Languages, 1992
XII	Martin Casdagli & Stephen Eubank	Nonlinear Modeling and Forecasting, 1992

LECTURES VOLUMES

Volume	Editor	Title
I	Daniel L. Stein	Lectures in the Sciences of Complexity, 1989
II	Erica Jen	1989 Lectures in Complex Systems, 1990
III	Lynn Nadel & Daniel L. Stein	1990 Lectures in Complex Systems, 1991

LECTURE NOTES VOLUMES

Volume	Author	Title
I	John Hertz, Anders Krogh, & Richard Palmer	Introduction to the Theory of Neural Computation, 1990
II	Gérard Weisbuch	Complex Systems Dynamics, 1990

Contributors to This Volume

Henry D. I. Abarbanel, Scripps Institution of Oceanography, University of California

Christopher G. Atkeson, Massachusetts Institute of Technology

C. W. Barnes, Los Alamos National Laboratory

Aviv Bergman, SRI International Artificial Intelligence Center and Stanford University

W. A. Brock, University of Wisconsin, Madison

Martin Casdagli, Santa Fe Institute

James P. Crutchfield, University of California, Berkeley

Stephen Eubank, Santa Fe Institute and Los Alamos National Laboratory

J. Doyne Farmer, Los Alamos National Laboratory and the Santa Fe Institute

G. W. Flake, University of Maryland

Bryan Galdrikian, Los Alamos National Laboratory and the Santa Fe Institute

Clive W. J. Granger, University of California, San Diego

Peter Grassberger, University of Wuppertal

Bernardo A. Huberman, Xerox PARC

Norman F. Hunter, Jr., Los Alamos National Laboratory

R. D. Jones, Los Alamos National Laboratory

Matthew Koebbe, University of California, Santa Cruz

Eric J. Kostelich, Arizona State University, Tempe

Wallace E. Larimore, Adaptics, Inc.

Daniel P. Lathrop, University of Texas, Austin

Blake LeBaron, University of Wisconsin, Madison

L. A. Lee, Los Alamos National Laboratory

Y. C. Lee, University of Maryland

André Longtin, Los Alamos National Laboratory and the Santa Fe Institute

Gottfried Mayer-Kress, University of California, Santa Cruz

W. C. Mead, Los Alamos National Laboratory

Alistair I. Mees, The University of Western Australia

Thomas P. Meyer, CCSR, University of Illinois at Urbana-Champaign

M. K. O'Rourke, Los Alamos National Laboratory

Norman H. Packard, Beckman Institute, University of Illinois at Urbana-Champaign

S. M. Potter, University of California, Los Angeles

T. Subba Rao, University of Manchester Institute of Science and Technology

David E. Rumelhart, Stanford University

Richard L. Smith, University of North Carolina, Chapel Hill

Sara A. Solla, AT&T Bell Laboratories

L. Stokbro, Neils Bohr Institute

William W. Taylor, The RTA Corporation

Timo Teräsvirta, University of California, San Diego

James Theiler, Los Alamos National Laboratory and the Santa Fe Institute

Contributors to This Volume

Brent Townshend, TCT
D. K. Umberger, Nordic Institute for Theoretical Physics
Grace Wahba, University of Wisconsin
Andreas S. Weigend, Stanford University

Preface

INTRODUCTION

The Santa Fe Institute and NATO sponsored a workshop on "Nonlinear Modeling and Forecasting" from September 17-21 at the Inn at Loretto in Santa Fe, New Mexico. The purpose of the workshop was to bring together researchers working on different aspects of nonlinear modeling from a wide range of disciplines, who would not normally meet. The central theme of the conference was the construction of nonlinear models from time-series data. Approaches to this problem have drawn from the disciplines of multivariate function approximation and neural nets, dynamical systems and chaos, statistics, information theory, and control theory. Applications have been made to economics, mechanical engineering, meteorology, speech processing, biology, and fluid dynamics. The interdisciplinary field of nonlinear modeling has grown rapidly over the last decade. This is largely due to the increasing availability of computer resources, which both allows for the collection of increasingly large data sets, and the analysis of the data sets with numerically intensive algorithms. The field has also grown with the increasing recognition of the ubiquity and importance of the effects of nonlinear dynamics in the natural and social sciences. The organizers and sponsors of the workshop felt the time was ripe to bring together researchers from this rapidly expanding, interdisciplinary field in order to stimulate cross-fertilization of ideas and search for unifying themes. One of the products of the workshop is this volume, which consists of papers mostly contributed from the invited speakers. The contents and connection between these papers will be briefly summarized below. Another product of the workshop is a time-series competition to be held at the Santa Fe Institute in the Fall of 1991, which we also describe below.

There are a variety of books, papers, and review articles relevant to nonlinear modeling and forecasting that the interested reader might like to consult. For a statistical approach to nonlinear time-series analysis, see the book by Tong.[45] See also the books by Box and Jenkins,[2] Priestly[34] and Brockwell and Davis[6] for a background in linear time-series analysis. There are several recent papers and review articles on nonlinear time-series modeling motivated by deterministic chaos. For papers see Farmer,[20] Crutchfield,[13] Casdagli,[7] Mees,[31] Cremers and Hübler,[12] Sugihara and May,[41] Sayers,[40] and Brock.[5] For review articles see Eubank and Farmer,[19] Grassberger et al.,[23] Casdagli et al.,[11] and Casdagli.[8] For a background in dynamical systems and chaos, see the books by Guckenheimer and Holmes,[24] Shaw,[37] and Glass and Mackey,[22] the review article by Eckmann and Ruelle,[16] and the proceedings volumes by Mayer-Kress[30] and Abraham et al.[1] The editors have also found the book on neural nets by Hertz et al.,[26] the book on splines by Wahba,[46] and the book on bootstrapping statistics by Efron[17] to be useful reading.

THE PROCEEDINGS

For organizational purposes, we have somewhat arbitrarily divided the papers submitted to the proceedings into four sections: function approximation, statistics, dynamical systems, and applications. Some minor changes were made to the contents of the papers on the recommendation of the editors.

The first five papers discuss various approaches to nonlinear multivariate function approximation. Mees describes a local linear approach based on tesselations, which may be useful for investigating the topology of attractors as well as forecasting. Solla discusses the learning capability of neural nets. Neural nets may be regarded as a special form of nonlinear multivariate function approximation, and have received much attention over recent years. See also the paper by Weigend et al.[47] The papers by Stockbro and by Mead et al. describe hybrid approaches which are in some sense intermediate between neural nets and spline techniques. The paper by Wahba on splines is also included, although she was unable to attend the conference.

Some form of nonlinear multivariate function approximation is implicit in any nonlinear modeling and forecasting algorithm, and indeed several of the other papers in this proceedings propose their own approaches to this central problem. One of the questions debated in the workshop was which function approximation technique to use in which circumstances. It was generally agreed that the answer probably depends on a number of factors, including the computer resources available, the length of the time series, and the noise level or complexity of the source of the time-series data. It was also generally agreed that the situation would be clarified by performing comparison tests of the wide variety of algorithms on a collection of time-series data. These discussions stimulated interest in a time-series competition (see below).

The next five papers address statistical issues in time-series analysis. Smith discusses statistical problems associated with estimating the correlation dimension of a fractal or chaotic attractor from limited data. Brock and Potter describe a nonparametric form of hypothesis testing based on correlation dimension ideas. Theiler et al. discuss various hypothesis-testing strategies based on generating "surrogate data" under a null hypothesis. Granger and Terasvirta summarize results of a comparative study of various function approximation schemes on time series simulated at a variety of noise levels. Subba-Rao reviews the "bispectrum" which is a nonlinear generalization of the spectral approach to linear time-series analysis. It was agreed that there are highly non-trivial statistical problems involved in estimating nonlinear models from data and assessing their adequacy. A reading of these papers will apprise researchers of some of the statistical techniques which are currently available for addressing such problems.

The next seven papers discuss a variety of different approaches to nonlinear modeling and forecasting, and are loosely categorized under "dynamical systems." Abarbanel discusses invariants associated with chaotic attractors, in particular local Liapunov exponents and divergence rates, which limit the predictive accuracies of any forecasting model in making multi-step predictions (see also the paper by Kostelich and Lathrop which discusses the estimation of invariants from a time series using unstable periodic orbits). Meyer and Packard describe a genetic algorithm approach to identifying special intervals of predictability in a time series (see also the papers by LeBaron and by Weigend et al. which focus attention on special intervals of a time series). Casdagli describes an approach to the nonlinear modeling of "input-ouput" time series (see also the paper by Hunter for applications of this approach). Larimore presents a canonical variate analysis approach to state-space reconstruction and nonlinear modeling, and provides a promising alternative to modeling with delay coordinates. Bergman et al. describe an approach to the modeling and coding of symbol sequences using neural nets and an information theoretic criterion. Crutchfield also develops the modeling of symbol sequences, and draws on results from computation theory and statistical mechanics. Mayer-Kress and Koebbe describe an exploratory graphical technique for time-series analysis known as a "recurrence plot," and is included although they were unable to attend the conference. These papers should give the reader a sense of the wide variety of approaches that are possible in analyzing time-series data. Even from a practical point of view, there is much more to nonlinear modeling and forecasting than function approximation and statistics: an understanding of the wide variety of phenomena possible in nonlinear dynamical systems should also be taken into account.

The last seven papers are about applications to a variety of time-series data, though we would like to emphasize that all of these applications use nonlinear modeling and forecasting ideas developed by the authors themselves. LeBaron presents evidence for forecastability in the Standard and Poors index during intervals of low volatility. Weigend et al. present results from forecasting sunspots and exchange rates using a form of neural-net algorithm which avoids over-fitting to data. Townshend presents evidence for nonlinear forecastability in speech time series using a combination of filtering and piecewise-linear modeling techniques. Taylor describes a technique for separating speech time series from background contaminating time series of a different dynamical form. Hunter presents results on the nonlinear modeling and forecasting of mechanical vibrations and ice-age time series, based on an input-output approach to modeling. Kostelich and Lathrop describe results on characterizing attractors in fluid dynamics experiments using unstable periodic orbits. Atkeson presents results on modeling robot arms using piecewise-polynomial modeling.

At present, the number of different techniques available for nonlinear modeling and forecasting appears to outnumber the applications to time-series data that have been made. It would be desirable for this trend to reverse. Perhaps this might be achieved by making widely available both nonlinear modeling and forecasting algorithms and time-series data bases (see time-series competition below).

OTHER TOPICS

Several topics were discussed at the workshop but are not included in these proceedings. We now summarize these topics, and give existing references in the literature.

An important issue in the nonlinear modeling of time-series data is the construction of a state space. This step is a prerequisite to the steps of function approximation and statistical evaluation, and also poses interesting theoretical problems. Sauer discussed results on the embedding of chaotic attractors in the absence of noise; see Sauer et al.[35] These results provide extensions to a theorem of Takens.[42] Farmer discussed results on state-space reconstruction in the presence of low levels of observational noise; see Casdagli et al.[10] Scargle discussed some advantages of filtering a time series before modeling; see Scargle.[36] Nonlinear modeling and forecasting may also be applied to the noise reduction of time-series data. Noise-reduction algorithms were discussed by Kostelich, Hammel, and Sidorowich; see Kostelich and Yorke,[29] Hammel,[25] and Farmer and Sidorowich.[21] Techniques of nonlinear modeling applicable to the control of dynamical systems were discussed by Hübler and by Delchamps; see Hübler[28] and Delchamps.[15] Lapedes discussed approaches to the nonlinear modeling of DNA sequences using information theoretic ideas; see Stolorz.[39] Tishby presented a technique for the analysis of speech time series; see Tishby.[44] Poggio described a regularized radial basis-function technique designed for applications to noisy data; see Poggio.[33] Breeden presented results on estimating parameters in hidden variable models from time-series data; see Breeden and Hübler.[3]

Some recent papers on nonlinear modeling and forecasting that have appeared since the conference include the following. Several authors have constructed a range of models from the nonlinear deterministic extreme to the linear stochastic extreme, with the intention of investigating whether a time series displays evidence for deterministic chaos as opposed to nonlinear stochastic behavior; see Briggs,[4] Hseih,[27] Casdagli et al.,[11] and Casdagli.[9] Ellner has pointed out some of the pitfalls in the identification of low-dimensional chaos from the results of the multi-step forecasting of time series.[18] Smith has analyzed effects of noise contamination on dimension calculations from a statistical point of view.[38] Theiler has analyzed the effects of performing dimension calculations on "long memory" time series, dominated by low frequencies.[43] Grassberger et al. have proposed a new noise-reduction algorithm.[23] Ott et al. have devised an algorithm for the control of chaos[32]; this has been applied to a physical system by Ditto et al.[14]

TIME-SERIES COMPETITION

A time-series competition will be held at SFI in the Fall of 1991, to be organized by Andreas Weigend and Neil Gershenfeld. The intention is to compare the forecasting algorithms of the competitors on some carefully selected time series from a variety of sources. The complete set of time series used in the competition, as well as several other time series, will be collected in an archive on the SFI computer network, and made available at the end of the competition.

ACKNOWLEDGMENTS

Funding for workshop came from NATO and SFI, including core funding from the National Science Foundation (PHY-8714918), the U.S. Department of Energy (ER-FG05-88ER25-54), and the John D. and Catherine T. MacArthur Foundation. We would like to thank David Campbell, director of the CNLS, for encouraging us to apply for NATO funding, and Doyne Farmer for his advice on whom to invite and how to organize the workshop. It also gives great pleasure to thank Andi Sutherland and Ginger Richardson for their help in overcoming the administrative difficulties in organizing the workshop. Finally, we would like to express our gratitude to Ronda Butler-Villa and Della Ulibarri for their excellent work in processing the submitted papers into the format of this proceedings volume.

Martin Casdagli
Santa Fe Institute
1660 Old Pecos Trail
Santa Fe, NM 87501

Stephen Eubank
Santa Fe Institute
1660 Old Pecos Trail
Santa Fe, NM 87501

November 15, 1991

REFERENCES

1. Abraham, N. B., A. M. Albano, A. Passamante, and P. E. Rapp, eds. *Measures of Complexity and Chaos NATO Advanced Science Institute Series*, Vol. 208. New York: Plenum, 1989.
2. Box, G. E. P., and G. M. Jenkins. *Time Series Analysis Forecasting and Control*. San Francisco: Holden-Day, 1970.
3. Breeden, J. L., and A. Hübler. "Reconstructing Equations of Motion from Experimental Data with Hidden Variables." Preprint, University of Illinois, 1990.
4. Briggs, K. "Improved Methods for the Analysis of Chaotic Time Series." Mathematics Research Paper 90-2, La Trobe University, Melbourne, Australia, 1990.
5. Brock, W. "Causality, Chaos, Explanation and Prediction in Economics and Finance." In *Beyond Belief: Randomness, Prediction and Explanation in Science*, edited by J. Casti and A. Karlqvist. 1990.
6. Brockwell, P. J., and R. A. Davis. *Time Series: Theory and Methods*. New York: Springer, 1987.
7. Casdagli, M. "Nonlinear Prediction of Chaotic Time Series." *Physica D* **35** (1989).
8. Casdagli, M. "Nonlinear Forecasting, Chaos and Statistics." In *Modeling Complex Phenomena*, edited by L. Lam and V. Naroditsky. New York: Springer-Verlag, to appear.
9. Casdagli, M. "Chaos and Deterministic Versus Stochastic Nonlinear Modeling." *J. Roy. Stat. Soc*, to appear.
10. Casdagli, M., S. Eubank, J. D. Farmer, and J. Gibson. *Physica D*, to appear.
11. Casdagli, M., D. Des Jardins, S. Eubank, J. D. Farmer, J. Gibson, N. Hunter, and J. Theiler. "Nonlinear Modeling of Chaotic Time Series: Theory and Applications." In *EPRI Workshop on Applications of Chaos*, edited by J. Kim and J. Stringer. To appear.
12. Cremers, J., and A. Hübler. "Construction of Differential Equations from Experimental Data." *Z. Naturforsch.* **42a** (1987): 797–802.

13. Crutchfield, J. P., and B. S. McNamara. "Equations of Motion from a Data Series." *Complex Systems* **1** (1987): 417–452.

14. Ditto, W. L., S. N. Rauseo, and M. L. Spano. "Experimental Control of Chaos." *Phys. Rev. Lett.* **65** (1990): 3211–3213.

15. Delchamps, D. "Feedback Systems with Measurement Quantization: Chaos, Invariant Measures and Statistical Stability." Cornell University preprint, 1989.

16. Eckmann, J. P., and D. Ruelle. "Ergodic Theory of Chaos and Strange Attractors." *Rev. Mod. Phys* **57** (1985): 617.

17. Efron, B. *The Jackknife, the Bootstrap and Other Resampling Plans.* CBMS-NSF Regional Conference Series in Applied Mathematics, Vol 38. SIAM, 1982.

18. Ellner, S. "Detecting Low-Dimensional Chaos in Population Dynamics Data: A Critical Review." In *Does Chaos Exist in Ecological Systems*, edited by J. Logan and F. Hain. University of Virginia Press, to appear.

19. Eubank, S., and J. D. Farmer. "An Introduction to Chaos and Randomness." In *1989 Lectures in Complex Systems*, edited by E. Jen, 75–190. Santa Fe Institute Studies in the Science of Complexity, Lect. Vol II. Redwood City, CA: Addison-Wesley, 1989.

20. Farmer, J. D., and J. J. Sidorowich. "Predicting Chaotic Time Series." *Phys. Rev. Lett.* **59(8)** (1987): 845–848.

21. Farmer, J. D., and J. J. Sidorowich. "Optimal Shadowing and Noise Reduction." *Physica D* **47** (1991): 373.

22. Glass, L., and M. C.Mackey. *From Clocks to Chaos: The Rhythms of Life.* Princeton University Press, 1988.

23. Grassberger, P., T. Schreiber and C. Schaffrath. "Nonlinear Time Sequence Analysis." Preprint, University of Wuppertal, 1991.

24. Guckenheimer, J., and P. J. Holmes. *Nonlinear Oscillations, Dynamical Systems, and Bifurcations of Vector Fields.* New York: Springer-Verlag, 1983.

25. Hammel, S. M. "Noise Reduction for Chaotic Systems." Preprint, Naval Surface Warfare Center, Silver Spring Maryland, 1989.

26. Hertz, J., A. Krogh, and R. G. Palmer. *Introduction to the Theory of Neural Computation.* Studies in the Sciences of Complexity, Lect. Notes Vol. I. Redwood City, CA: Addison-Wesley, 1991.

27. Hseih, D. "Chaos and Nonlinear Dynamics: Application to Financial Markets." *J. Finance*, to appear.

28. Hübler, A., and E. Lusher. "Resonant Stimulation and Control of Nonlinear Oscillators." *Naturwissenschaften* **76** (1989): 67.

29. Kostelich, E. J., and J. A. Yorke. "Noise Reduction in Dynamical Systems." *Phys. Rev. A* **38(3)** (1988): 1649.

30. Mayer-Kress, G., ed. *Dimensions and Entropies in Chaotic Systems—Quantification of Complex Behavior.* Springer Series in Synergetics, Vol. 32. Berlin: Springer-Verlag, 1986.

31. Mees, A. I. "Modelling Complex Systems." In *Dynamics of Complex Inter-connected Biological Systems*, edited by T. Vincent et al. Boston: Birkhauser, 1990.

32. Ott, E., C. Grebogi, and J. A. Yorke. "Controlling Chaos." *Phys. Rev. Lett.* **64** (1990): 1196–1199.

33. Poggio, T., and F. Girosi. "Regularization Algorithms for Learning that are Equivalent to Multilayer Networks." *Science* **247** (1990): 978-982.

34. Priestly, M. B. *Spectral Analysis and Time Series*, Vols. I and II. London: Academic Press, 1981.

35. Sauer, T., J. A. Yorke, and M. Casdagli. "Embedology." *J. Stat. Phys.*, to appear.

36. Scargle, J. D. "Modeling Chaotic and Random Processes." *J. Astrophys.*, to appear.

37. Shaw, R. *The Dripping Faucet as a Model Dynamical System*. Santa Cruz, CA: Aerial Press, 1984.

38. Smith, R. L. "Estimating Dimension in Noisy Chaotic Time Series." Preprint, University of North Carolina, 1991.

39. Stolorz, P., A. Lapedes, and X. Yuan. "Predicting Protein Secondary Structure Using Neural Nets and Statistics." *J. Mol. Bio.*, submitted.

40. Sayers, C. "Chaos and the Business Cycle." In *The Ubiquity of Chaos*, edited by S. Krasner. Publication No. 89-15S, AAAS, 1333 "H" St. Washington, DC, 20005, 1990.

41. Sugihara, G., and R. May. "Nonlinear Forecasting as a Way of Distinguishing Chaos from Measurement Error in a Data Series." *Nature* **344** (1990): 734–741.

42. Takens, F. "Detecting Strange Attractors in Fluid Turbulence." *Dynamical Systems and Turbulence*, edited by D. Rand and L.-S. Young. Berlin, Springer-Verlag, 1981.

43. Theiler, J. "Some Comments on the Correlation Dimension of $1/f^\alpha$ Noise." *Phys. Lett. A* **155** (1991): 480–493.

44. Tishby, T. "A Dynamical Systems Approach to Speech Processing." *International Conference on Acoustics Speech and Signal Processing*, 1990.

45. Tong, H. *Nonlinear Time Series Analysis: A Dynamical Systems Approach*. Oxford University Press, 1990.

46. Wahba, G. *Spline Models for Observational Data*. CBMS-NSF Regional Conference Series in Applied Mathematics, Vol 59. SIAM, 1990.

47. Weigend, A. "Predicting Sunspots and Exchange Rates with Connectionist Networks." This volume.

Contents

Preface
Martin Casdagli and Stephen Eubank   xiii

Section 1. Function Approximation   1

Tesselations and Dynamical Systems
Alistair I. Mees   3

Supervised Learning: A Theoretical Framework
Sara A. Solla   25

Prediction of Chaotic Time Series using CNLS-Net–
Example: The Mackey-Glass Equation
*W. C. Mead, R. D. Jones, Y. C. Lee, C. W. Barnes,
G. W. Flake, L. A. Lee, and M. K. O'Rourke*   39

Forecasting with Weighted Maps
L. Stokbro and D. K. Umberger   73

Multivariate Function and Operator Estimation, Based on
Smoothing Splines and Reproducing Kernels
Grace Wahba   95

Section 2. Statistics   113

Optimal Estimation of Fractal Dimension
Richard L. Smith   115

Diagnostic Testing for Nonlinearity, Chaos, and General
Dependence in Time-Series Data
W. A. Brock and S. M. Potter   137

Using Surrogate Data to Detect Nonlinearity in Time
Series
James Theiler, Bryan Galdrikian, André Longtin,
Stephen Eubank, and J. Doyne Farmer   **163**

Experiments in Modeling Nonlinear Relationships Between
Time Series
Clive W. J. Granger and Timo Teräsvirta   **189**

Analysis of Nonlinear Time Series (and Chaos) by Bispectral Methods
T. Subba Rao   **199**

Section 3. Dynamical Systems **227**

Local and Global Lyapunov Exponents on a Strange
Attractor
Henry D. I. Abarbanel   **229**

Local Forecasting of High-Dimensional Chaotic Dynamics
Thomas P. Meyer and Norman H. Packard   **249**

A Dynamical Systems Approach to Modeling Input-Output
Systems
Martin Casdagli   **265**

Identification and Filtering of Nonlinear Systems Using
Canonical Variate Analysis
Wallace E. Larimore   **283**

Forecasting Probabilities with Neural Networks
Aviv Bergman, Peter Grassberger, and Thomas P. Meyer   **305**

Semantics and Thermodynamics
James P. Crutchfield   **317**

Use of Recurrence Plots in the Analysis of Time-Series
Data
Matthew Koebbe and Gottfried Mayer-Kress   **361**

Section 4. Applications **379**

Nonlinear Forecasts for the S&P Stock Index
Blake LeBaron   **381**

Predicting Sunspots and Exchange Rates with Connectionist Networks

   Andreas S. Weigend, Bernardo A. Huberman, and
   David E. Rumelhart   395

Nonlinear Prediction of Speech Signals
   Brent Townshend   433

Application of Nonlinear Prediction to Signal Separation
   William W. Taylor   455

Application of Nonlinear Time-Series Models to Driven
Systems
   Norman F. Hunter, Jr.   467

Periodic Saddle Orbits in Experimental Strange
Attractors
   Daniel P. Lathrop and Eric J. Kostelich   493

Memory-Based Approaches to Approximating Continuous
Functions
   Christopher G. Atkeson   521

Index
                                                                        523

Section I: Function Approximation

Alistair I. Mees
Mathematics Department, The University of Western Australia, Nedlands, Perth, Western
Australia 6009

Tesselations and Dynamical Systems

Time-series data from a dynamical system may be used to obtain geometric and dynamical information about the system. Geometric information includes dimension of a manifold containing the state space, as well as information about folds, branches, and other chaos indicators. Dynamical information comes from map reconstruction, resulting in a model with stated smoothness properties which may be tested in the way that any scientific theory may be tested: it allows falsifiable predictions to be made.

1. INTRODUCTION

This paper concerns the central theme of this conference: reconstructing dynamical systems directly from data. It adopts a strongly geometric approach, and tries to get qualitative as well as quantitative information. The methods apply to all kinds of systems for which a deterministic model is appropriate including chaotic systems. There has been criticism that much of the work claiming that real-world systems are chaotic is unscientific; the methods of this paper can be used in several ways to do a better job.

The basic hypothesis is that the data comes from a deterministic dynamical system. If it is incorrect, this will be apparent when predictions are examined or other tests are performed. If it is correct, and if the approach of this paper is appropriate to the particular case, we will gain at least some predictive power, and possibly qualitative insights as well, arising from the geometric methodology used.

There is no assumption that all of the data comes from a single orbit, or that it does not contain transients; indeed, transients and multiple orbits will generally make the job a good deal easier, by giving information about parts of the state space that might otherwise not be available. For example, it is difficult to obtain information about the stable manifold of an attractor without having some parts of orbits off the attractor.

Any modeling errors will be ascribed to noise. In the present paper we do not try to account for noise in the dynamics or errors in the data, for three reasons:

1. The case where the data is known accurately and the system is not subject to significant disturbances occurs frequently and is interesting in its own right.
2. The methodology for the case of noisy dynamics or data is still under development.
3. The methodology for the noise-free case is a prerequisite for the noisy case and involves non-trivial mathematics and computation.

Even without the contribution of this conference, there is already a literature on the problem of inducing dynamical maps from data.[5,8,9,15,16,31] The key idea is that state-space data reconstructed by embedding one or more time series contains dynamical information, since the image of each point is known unless it is the last on a sample orbit. Interpolating on the points with known images allows estimation of the images of other points and so gives an approximation to the dynamical map. Prediction, modeling, filtering, control, and so on all become possible once such an approximation is available.

The approach described here is somewhat related to the piecewise linear fitting ideas of Farmer and Sidorowich,[9] although it was initially developed without knowledge of their work. It is not restricted to piecewise linear approximations, and has certain possible advantages such as guaranteed continuity or smoothness, as well as the ability to give geometric insights. Its main disadvantages compared to the Farmer and Sidorowich method, or to radial basis functions,[15] is that it becomes slow in high dimensions. It is, however, worth remembering that high-dimensional systems are likely to require huge amounts of data to model, and it seems likely that any approach will run into difficulty.

We assume that the data has come from a deterministic dynamical system and has been reconstructed by embedding if required, so that we are given a sequence of points $\{x_t \in \mathbf{R}^k\}$, $t = 1, \ldots, n$, with k sufficiently large that x_t is "as good as" the true system state. Call this time series "the data." There has been so much written on embedding[2,8,15,18,20,29] that we take the reconstructed series as given. Note, though, that the concept of a good embedding dimension k is not well defined and, in fact, the size of the *window* (which takes into account embedding lag as well as embedding dimension) appears to be more important than the dimension.[2] The tesselation method can sometimes give information about good choices of dimension, as we shall see.

2. GEOMETRY

Sometimes, the state space of the dynamical system is an unknown submanifold of the Euclidean space. The only information available is a finite number of points, which in effect form a cloud in the space, together with their ordering in the time series. Can one identify such a submanifold from the point cloud? Even if the statespace is the entire containing Euclidean space, there may sometimes be useful geometric information to be had, such as evidence that the action of the dynamics folds a region onto itself in a horseshoe-like manner. How can we extract this kind of information from the data?

The natural way to represent, for example, a two-dimensional submanifold M^2 of $\mathbf{R}^3$, using only a finite set of points on the manifold, is to triangulate it. If the points are dense enough, each triangle, defined by neighboring data points, will be an approximation to a piece of the tangent manifold TM^2. For higher-dimensional manifolds M^n, the triangles are replaced by simplices of the appropriate dimension.

2.1 TESSELATION AND TRIANGULATION

Let us begin by considering Euclidean space rather than manifolds. There are many ways to triangulate a given set of points, but in Euclidean space there is a natural one, called the *Delaunay triangulation* by Sibson and Green.[11] It is best introduced via its dual, the *Dirichlet tesselation*.

First we need a couple of notions from convex analysis. The *convex hull* $\mathrm{co}S$ of a set S is the set of all convex combinations of collections of points of S; that is,

$$\mathrm{co}S = \left\{ z \ : \ z = \sum_{i=1}^{m} \lambda_i x_i \right\}$$

for some $\{x_i\}_{i=1}^{m} \subset S$ and $\lambda_i \geq 0$, $\sum_{i=1}^{m} \lambda_i = 1$. (Actually, it suffices by Caratheodory's theorem[24] to take $m = k + 1$ if S is k-dimensional.) The *closure* of a set S is denoted by $\overline{S}$. For the case where S contains finitely many points, it can be shown[24] that the convex hull is the polytope which is the intersection of a finite number of closed half-spaces containing S.

Given data points $\{x_t\}$, we associate with each x_t a territory V_t consisting of all points which are closer to x_t than to other data points. That is, the *tile* (or *Voronoi polygon*) belonging to x_t is

$$V_t = \{z \ : \ |z - x_t| < |z - x_s|, \quad s = 1, \ldots, n, \quad s \neq t\}.$$

The closure of the union of all such tiles is $\mathbf{R}^k$; being intersections of open half-spaces, the tiles are the interiors of polytopes. Data points strictly inside the convex hull co$\{x_s\}$ have bounded tiles, while those on the boundary have unbounded tiles.

If two tiles share a face (i.e., $\overline{V}_s \cap \overline{V}_t \neq \emptyset$), we say x_s and x_t are *neighbors*. Connecting neighbors by line segments which then define simplices results in a triangulation of the convex hull called the Delaunay triangulation. It has the optimality property that its simplices are most nearly regular with respect to local boundary changes. For example, in $\mathbf{R}^2$ they are as nearly as possible equilateral triangles.

The present reasons for selecting this particular triangulation are that the optimality property implies that the simplices will be of small diameter, which is important for minimizing the approximation error, and that there are reasonably efficient methods for building the triangulation.[4] Later we shall discover a stronger reason for the choice.

Constructing either one of the Dirichlet tesselation or the Delaunay triangulation is sufficient, because the other one can then be derived easily. The method used for all of the examples in this paper was Bowyer's, which successively inserts points into an existing triangulation, calculating the neighbors of each new point and adjusting the existing neighborhood structure as required to fit the optimality property.

For n points in $\mathbf{R}^k$, Bowyer's method takes a time of order $n^{1+1/k}$ as $n \to \infty$ for given k, and this can in principle be improved to $O(n \log n)$. Unfortunately,[19] the number of neighbors of a point grows exponentially fast with k and so both storage and time requirements become onerous in high dimensions. The worst case is where each of the n points is a neighbor of every other, in which case storage of order n^2 and time of order $n^{2+1/k}$ will be required. Even though this is not prohibitive for many problems, the methods here are probably most useful for $k = 2, 3, 4$. For higher-dimensional state spaces, other approaches may be more appropriate, especially given that geometric information of the kind we are pursuing is probably less useful in high dimensions.

2.2 MANIFOLDS

In many cases the natural state space is some p-dimensional manifold M^p embedded in $\mathbf{R}^k$ where $k > p$. It would be interesting to triangulate M, not $\mathbf{R}^k$. A successful triangulation would allow calculation of the manifold's topological dimension p, its genus, and other information useful in identifying it. In low dimensions, triangulation would assist in visualization. The simplices would be local approximations to the tangent space and so would be useful in answering dynamical questions.

Sometimes M is known and has a simple form, such as a 2-sphere, in which case it may be possible to find a special purpose method for triangulating it.[23] For other cases we have to find some general method for triangulating M rather than $\mathbf{R}^k$, even when the dimension of M is not known *a priori*.

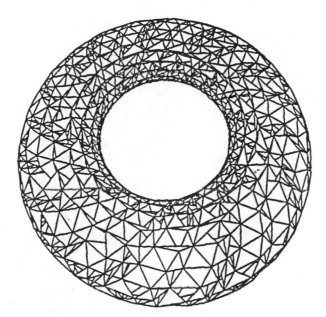

FIGURE 1   A trajectory on a 2-torus, triangulated as described in the text. A simplified hidden-line removal algorithm has been used, which results in an incorrect representation in places; nevertheless, some of the holes in the torus are genuine and result from incorrect identification of tangent triangles.

Provided the embedding is good and there are enough data points, points which are close in $\mathbf{R}^k$ will usually be close on M. (In the limit of infinitely many points, this is always true by the definition of embedding.) We can therefore identify p and triangulate M simultaneously, by first triangulating the points as a subset of $\mathbf{R}^k$, then examining the simplices that result.

One method is to look only at the shape of each simplex. If $p < k$, then probably a face of the simplex represents a part of the tangent manifold. The other faces arise from the fact that at least one point of the simplex comes from a different part of M, and can be removed. This point can be identified by the fact that (at least in the large data limit) it is usually further from the centroid than the others. By removing one point from every simplex, then repeating the whole process as long as most simplices are strongly asymmetrical, we reduce the dimension of the triangulation in steps. If the process is taken too far, the tendency will be to disconnect the simplices from one another, but when all goes well, the end result will be a cover of all or part of M with p-dimensional simplices.

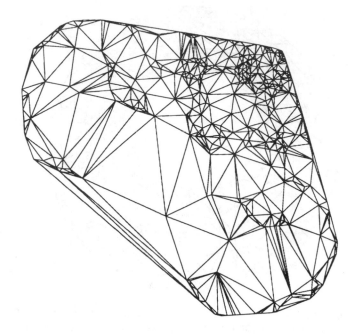

FIGURE 2 A set of points in the plane and their Delaunay triangulation.

For example, triangulating a set of points in $\mathbf{R}^3$ which lie on a torus $S^1 \times S^1$ results in tetrahedra, each of which has one face F that is an approximation to the local tangent space $T(S^1 \times S^1)$. The other three faces result from joining the points defining F to a point Q on some other part of $S^1 \times S^1$. Typically, Q will be much further from the points of F than are the points of F from one another. By examining all of the tetrahedra, and discarding from each the point furthest from the centroid, we are left with triangles the union of which should be an approximation to $S^1 \times S^1$. Figure 1 shows this in operation; the points used were from a trajectory on the torus, and the trajectory can even be seen in the edges of the triangles. There were too few points to give a perfect triangulation: examination will reveal that occasionally a triangle is produced which does not well approximate the tangent space.

Notice that the torus is much clearer to the eye in the triangulated representation of Figure 1 than it would be in the point-cloud representation. We would probably be confident both that $p = 2$ is the manifold dimension, and that this particular two-dimensional manifold can be embedded in three dimensions.

The author has shown elsewhere[15] that applying the process to points from trajectories of the Lorenz system[27] suggests that the attractor lies on a branched two-dimensional manifold. The latter is a well-known simplified model of the Lorenz attractor, providing a good approximation to the dynamics.[12,27] This calculation might therefore have provided the insight to make such a model had it not already been known. It should be emphasized here that the tesselation computer program knows nothing about the Lorenz dynamics, nor anything about manifolds except that they are locally Euclidean.

Research is currently directed to developing improved methods for identifying the manifold, and to investigating direct methods which do not require an initial triangulation in $\mathbf{R}^k$.

2.3 IMAGES OF THE TRIANGULATION

Triangulation may be helpful in identifying a branched manifold. This is one of the classic chaos generators; an even more important one is folding. It is sometimes possible to identify folding by using triangulation.

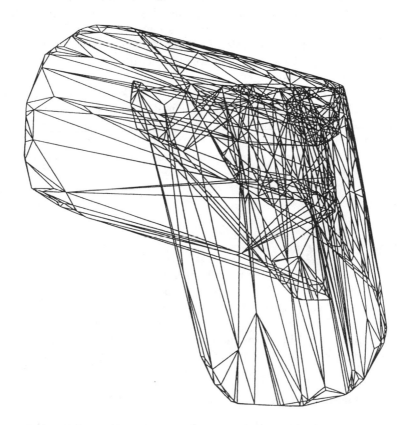

FIGURE 3   The image of the triangulation from Figure 2 under the approximation described in the text. The part of the plane in question is folded over like a piece of origami.

Continuity of f ensures that, as long as there is sufficient data, points which were neighbors in the tesselation will remain close after one step of the mapping. Moreover, the edges of the small triangles (or the facets of the simplices) will, to a first approximation, be mapped into straight lines (or portions of affine submanifolds). Consider pairs of points which are neighbors in the triangulation. Find where they are mapped to, then draw the line segments joining these images. The resulting set of simplices—not necessarily a triangulation, since they may overlap—gives an approximation to how the part of the manifold covered by the data is affected by the map.

We have shown elsewhere[15] how applying the process to a few points from a Poincaré section of a piecewise linear approximation to the forced Duffing equation reveals a double bend, highly suggestive of what the map actually does to a region of the plane.

Figure 2 shows a triangulation of some points in the plane on an attractor of a certain model of interacting neurons,[1] and Figure 3 shows the image of the triangulation. The folding effect of the map is clear: it is non-invertible, and its operation can be regarded as a piece of origami, making the folded paper model suggested by Figure 3. This insight enabled a detailed study of this map by a different but related method.[14]

In more than three dimensions, visualization of the kind we have used to identify branches and folds is no longer available. Nor is it clear how one can identify folds automatically. It appears to be necessary to resort to more conventional means such as calculation of Lyapunov exponents, fractal dimensions, or other indicators of the form of the attractors. Before we can discuss this, it is necessary to consider approximations in more detail.

3. APPROXIMATION

Until now we have been concerned mainly with qualitative features of the dynamics of our time series. When we considered the image of a triangulation, however, we were in effect constructing a predictor. We turn now to a more detailed examination of map approximation via tesselation, in preparation for building predictors. Of course, the map is not uniquely defined by a finite set of points and their images. We shall consider two different map definitions based on tesselations.

3.1 LOCAL COORDINATES

A natural way to approximate the dynamics is to produce a map that interpolates on nearby data points in the embedding (or state) space. To formalize this, we use Sibson's *local coordinates*,[26] which are a generalization of the well-known notion of barycentric coordinates. The following is a condensed description; proofs and further details are given elsewhere.[17]

DEFINITION $\lambda_i(z)$ $(i \in N(z))$ are *local coordinates* for z with respect to $\{x_t, t = 1, \ldots, n\}$ if

$$z = \sum_{i \in N(z)} \lambda_i(z) x_i \tag{3.1.1}$$

and

$$\sum_{i \in N(z)} \lambda_i(z) = 1. \tag{3.1.2}$$

The coordinates need not at present be unique, or non-negative, so Eq. (3.1.1) does not necessarily represent a convex combination. The constraint (3.1.2) will be shown to ensure that affine functions are fitted exactly by a natural approximation.

Suppose images f_t are known, where $f_t = f(x_t)$ for an otherwise unknown C^1 function f. We try to approximate f by $\hat{f}$ where

$$\hat{f}(z) = \sum_{i \in N(z)} \lambda_i(z) f_i \tag{3.1.3}$$

for some defined neighborhood scheme N.

LEMMA If $f \in C^1$, then

$$\hat{f}(z) = f(z) + o(\epsilon) \tag{3.1.4}$$

where

$$\epsilon = \text{diameter} \left(\{z\} \cup \{x_i : i \in N(z)\} \right). \tag{3.1.5}$$

Note that if $z \in \text{co}\{x_t\}$, it will be possible to define neighborhoods so that $z \in \text{co}\{x_t : t \in N(z)\}$, and then the formula for ϵ simplifies to

$$\epsilon = \text{diameter}\{x_t : t \in N(z)\}. \tag{3.1.6}$$

The approximation is in fact exact if the original map is linear or biased linear:

LEMMA If f is affine, then $\hat{f} = f$.

3.2 PIECEWISE LINEAR APPROXIMATION

If $N(z)$ contains $k + 1$ indices, then Eqs. (3.1.1–3.1.2) provide $k + 1$ equations for the $k + 1$ unknowns $\lambda_i, i \in N(z)$. The $k + 1$ points define a simplex and λ_i are the barycentric coordinates of z with respect to the vertices. If $z \in \text{co}\{x_t\}$ and we choose $N(z)$ to label the simplex containing z (which is unique for generic z), then $\lambda_i(z) \geq 0$ for all i and z is a convex combination of $x_i, i \in N(z)$.

In this case, Eq. (3.1.3) defines a function $\hat{f}$ that is continuous in z:

LEMMA If $x \in \mathrm{co}\{x_t\}$ and $N(z)$ indexes the simplex containing z (with arbitrary choice between available simplices in case of degeneracy), then

$$\hat{f}(z) = \sum_{i \in N(z)} \lambda_i(z) f_i$$

is continuous in z.

Note that this approximation is exactly the one used in section 2 where we assumed that the image of a simplex is a simplex: its success there is already apparent, and more examples will be given later.

3.3 ALMOST SMOOTH APPROXIMATION

If, instead of using the simplices of the triangulation to define neighborhoods, we use the tiles of the tesselation, we obtain a different interpretation of the basic approximation (3.1.3). Consider inserting a new point z into a tesselation induced by $\{x_t, t = 1, \ldots, n\}$, as discussed in section 2. Then z obtains a tile $T(z)$ which has been built by removing parts of other tiles T_i. If $z \in \mathrm{co}\{x_t\}$, then $T(z)$ is finite, and we shall assume this to be the case. Let the indices of all such depleted tiles comprise $N(z)$, as in Figure 4. That is, $N(z)$ is just the set of neighbors of z in the augmented tesselation.

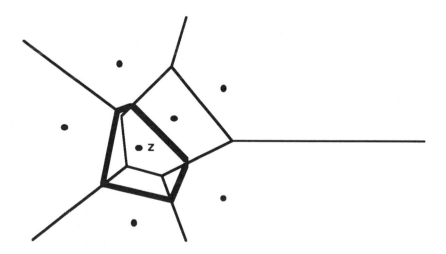

FIGURE 4 Insertion of a new tile into a tesselation. The existing tesselation is defined by points $\{x_i\}$, and insertion of z results in parts of tiles being lost to the tile of z. The points which have lost tile area are the neighbors of z in the new tesselation.

Sibson[26] discusses various possible choices of weights $\lambda_i(z)$. We shall consider only his *subtile weights*. Let μ_k be Lesbegue measure on $\mathbf{R}^k$, and define $T_i(z)$ to be the part of $T(z)$ taken from tile T_i. Then define

$$\lambda_i(z) = \frac{\mu_k(T_i(z))}{\sum_{j \in N(z)} \mu_k(T_j(x))} \tag{3.3.1}$$

so that for all i, $\lambda_i(z) \geq 0$ and $\sum_{i \in N(z)} \lambda_i(z) = 1$.

LEMMA $\lambda_i(z)$ defined by Eq. (3.3.1) is a piecewise rational function in the components of z, and is C^1 in z for $z \in \mathrm{co}\{x_t\} \setminus \bigcup\{x_t\}$.

That is, the weights are smooth functions everywhere on the convex hull of the data except at the data points themselves. They are continuous but not in general differentiable at the data points.

The effect of the lemma is that we are no longer dealing with piecewise linear approximations, but with smooth approximations: the rational functions merge smoothly across the boundaries, except perhaps at the data points themselves, where they are nevertheless continuous.

Since the basic approximation (3.1.3) is still only accurate to $o(\epsilon)$, or to $O(\epsilon^2)$ if f is C^2, the new approximation is not necessarily more accurate than the piecewise linear one. (Recall that $o(\epsilon)$ is a function that goes to zero faster than ϵ, i.e., $o(\epsilon)/\epsilon \to 0$ as $\epsilon \to 0$. In the present case, if f is a C^2 function, then we can replace $o(\epsilon)$ by $O(\epsilon^2)$, where $O(\epsilon^2)$ is a function that goes to zero *as fast as* ϵ^2, i.e., $O(\epsilon^2)/\epsilon^2 \not\to \infty$ as $\epsilon \to 0$.) Higher-order approximations are possible, though it is not possible to define them uniquely in general. Sibson discusses "spherical quadratic" approximations which extend the exact fit of affine functions to exact fit of radially quadratic functions, and Barnhill and Little[3] discuss a number of other interpolants, specifically restricted to two and three dimensions.

On the other hand, even without additional accuracy, the tesselation method is likely to be be an improvement over the simplex method in the presence of noise because the noise is averaged over more data points.

The main disadvantage of using tesselations is that $\mu_k(T_i(z))$ may be expensive to calculate. It is a sum of simplex measures, and as such is a sum of determinants. The terms in the sum can be obtained from the neighbor information already available[17] but the number of simplices taking part in the sum grows exponentially with k, as has already been discussed.

3.4 ASYMPTOTIC EXACTNESS OF THE APPROXIMATIONS

Consider a set $\Lambda \subseteq \mathbf{R}^k$, perhaps an attractor or a manifold, and a set D of data points $x_i \in \Lambda$. If the number of data points $n = |\Lambda|$ increases in such a way that Λ is well represented, both the piecewise linear and the almost smooth approximations become asymptotically exact.

PROPOSITION Let r_n be the greatest radius of a circumsphere in the triangulation of $\{x_i\}_{i=1}^n$. If $r_n \to 0$ as $n \to \infty$, then $\hat{f} \to f$, in the sense that

$$\sup_{z \in \text{co}\{x_j\}} |f(z) - \hat{f}(z)| \to 0,$$

where $\hat{f}$ is defined either as in section 3.2 or as in section 3.3.

PROOF Since f is C^1, Taylor's theorem shows that for any fixed n, the error term

$$\sup_{z \in \text{co}\{x_j\}} |f(z) - \hat{f}(z)| = o(r_n).$$

By definition of $o(r_n)$, the error goes to zero as $n \to \infty$ since $r_n \to 0$ as $n \to \infty$.

---

4. DYNAMICS

Using the approximations defined above makes it possible to approximate the dynamics of the unknown map using only the data. The work described here is a development of the author's earlier work[15,16] and the piecewise linear version has some features in common with work by Farmer and Sidorowich[9] and by May and Sugihara.[28] These publications, and others in the literature,[5,6,7,31] discuss in some detail the philosophy of nonlinear prediction, and give many applications. This makes it possible to be brief here. We simply recall that the provision of *falsifiable predictions* is the principal function of any scientific model. At the same time, we draw attention to the fact that *only* providing quantitative results is limiting and this is the reason for our emphasis until now on the geometrical approach and the possibility that it may lead to insights—that is, qualitative information.

4.1 TRAINING AND PREDICTION

The commonest way to recreate dynamics is to use a fixed set of embedded data points to build the model—in our case the tesselation—and then use the resulting $\hat{f}$ to give the image of any required point. That is, there is a "training set" of data which is quite distinct from sets requiring prediction, interpolation, and so on.

Once an approximation $\hat{f}$ is available, new orbits can be found, including, for example, a prediction from the final training point x_n to $x_{n+1}, x_{n+2}, \ldots$. Reverse time prediction is also possible, but the original map may not have been invertible so additional precautions must be taken.

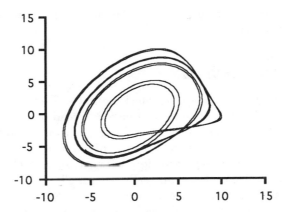

FIGURE 5   A trajectory of the Rössler attractor, with 1000 sample points.

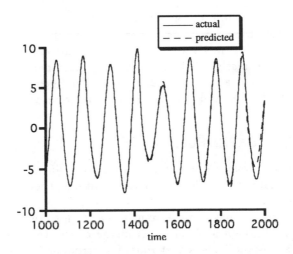

FIGURE 6   The next 1000 points of the attractor of Figure 5 shown as a time series (solid line) with the tesselation predictions shown dashed. The prediction and the actual trajectory coincide to the limits of plotting resolution in the earlier stages.

Figure 5 shows a trajectory of 1000 points from the Rössler attractor. The trajectory was used to construct a tesselation and the tesselation version of $\hat{f}$ was used to predict the next 1000 points. In Figure 6 it can be seen that the prediction is almost perfect: even after 10 periods (1000 time steps), the approximate and true orbits are very close.

4.2 RECURSIVE FORECASTING

In control and information theory, so-called recursive filtering has proved to be remarkably robust and the linear version is in widespread use as the Kalman-Bucy filter.[25] The idea is to build the model in real time, constantly updating it as more data arrives, but always being prepared to make predictions with the data available so far. In principle, the accuracy keeps improving as more and more data comes in, though the error will not in general tend to zero. The approach is particularly useful when there are external inputs.

Recursive prediction is easy to do with the approach of this paper, since the tesselation and triangulation are built by successive addition of points and at any stage the model is precisely the tesselation built so far. Thus given data $x_0, \ldots, x_t$, we build $\hat{f}^{(t)}$, the model available at time t, and make any required predictions. When x_{t+1} becomes available, we add it to the tesselation, giving $\hat{f}^{(t+1)}$.

Since a tesselation is a fairly large data structure, we may also attempt to construct a more economical model by defining some acceptable error δ and only adding x_{t+1} if either (1) $x_{t+1} \notin \text{co}\{x_1, \ldots, x_t\}$, or (2) $|\hat{f}^{(t)}(x_t) - x_{t+1}| > \delta$.

An additional possibility is to control the size of the model by choosing to delete old points from the tesselation as new ones come in: this approach is especially suitable if we believe that the data comes from a dynamical system which is non-stationary, but is sufficiently slowly varying for a stationary model over a short enough time interval to be a reasonable approximation. Such an approach is similar to one suggested by Farmer and Sidorowich[9] in a different context.

4.3 OUTSIDE THE CONVEX HULL

The approximations developed in section 3 are only defined for points within the convex hull of the data. While this is properly cautious, there are occasions when it is necessary to approximate the images of points outside the convex hull. For example, rounding errors may cause a point to be mapped outside the convex hull during prediction, although a possible advantage of the tesselation method over some other forecasting methods is that rounding errors are the *only* way in which this can happen. This could be regarded as a useful robustness property: tesselation models appear to be better than some others in producing attractors.

If transients have been made available in the modeling step, then the problem is greatly reduced. Typically, this will give enough information away from the attractor to make it rare for a point which should be on the attractor to appear outside the convex hull, even if it is mapped off the attractor.

The solution is simply to project onto the nearest point of the boundary of the convex hull, and take the image of that as the prediction. There is a deeper issue, however. When transient information has not been given, it is very difficult, if not impossible, to determine the stable manifold. Projecting onto the convex hull in this way can be regarded as a choice of stable manifold which is at least as good as any other.

5. APPLICATIONS TO REAL-WORLD DATA

There is insufficient space to give full details of applications. The method has been applied successfully to many artificial data sets as well as to several natural ones such as sunspot numbers. Some of these are discussed elsewhere.[15,17]

5.1 ELECTROENCEPHALOGRAPHIC DATA

There have been many attempts to show that EEG data may sometimes have deterministic characteristics, from the early identification of alpha, delta, and other rhythms, to more recent measures of fractal dimension.[13,21,22] A direct modeling approach may be of some value.

Figure 7 shows a training set based on a segment of an EEG taken from a patient in a relaxed state. This was embedded in three dimensions with a lag of 10, and used to build a tesselation. The resulting dynamical model was used to predict one time step ahead for the EEG shown as a solid line in Figure 8. (This is a later segment of the same EEG used for training.) The dashed line shows the result: prediction is rather poor in some places, mainly when there are large excursions going outside the range of the training set, but in other places, such as towards the end of the segment, it is very good.

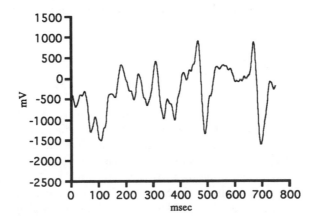

FIGURE 7 A resting EEG.

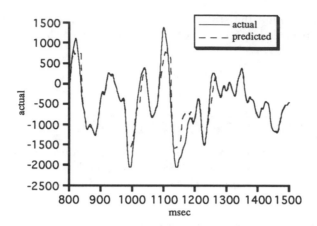

FIGURE 8   Continuation of the EEG of Figure 7 (solid line) and its one-step prediction by the tesselation method (dashed line).

In fact, the correlation coefficient between predicted and actual waveform over the last segment (times 563 to 750) is 99.8%, as against 93.7% in the middle section (times 189 to 563). More informative is the correlation between the delta waveforms $\{y_{t+1} - y_t\}$ and $\{\hat{y}_{t+1} - y_t\}$, where $\hat{y}$ is the prediction. This is -9% over the middle part but +66% over the last part.

If any conclusion is to be drawn from this casual experiment, it is perhaps that the dynamics of the last part are more like the dynamics of the training set, and that there is some evidence that determinism exists. This is unastounding given that even a simple spectral analysis reveals some periodicity, but it is evidence that simple phenomenological models may have some use in this problem. Further work need to be done before any conclusions can be reached about the scientific value of the approach.

5.2 MEASLES EPIDEMICS

Sugihara and May[28] have studied measles epidemic data using a method very similar to that of section 3.2. Let us use the method of sections 3.3 for forecasting, presenting the results in manner that facilitates comparison with Sugihara and May's results. The purpose here is not to argue the case for determinism in this data, or to show that this approach is better than the standard statistical approach, since both of these have been done by Sugihara and May.

The solid line in Figure 9 shows the Sugihara-May data as a time series: it was obtained by differencing the original epidemic data. Exactly as they did, we use the

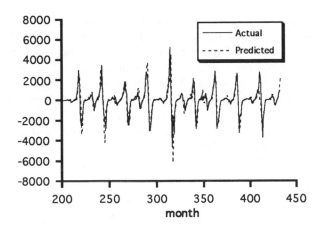

FIGURE 9   Measles epidemic differences (solid line) and one-step predictions (dashed line) by a tesselation model built from an earlier segment of the same data set.

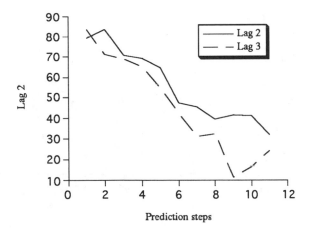

FIGURE 10   Correlation coefficient ρ for measles data as a function of number of prediction steps, using the tesselation predictor with embedding dimension three and the stated lags. The reduction in predictability as time goes on is consistent with chaos (positive Lyapunov exponent).

first 216 points as a training set, and then predict forward separately from each of the later points, for times 1 up to 12 units; the results for prediction time 1 are shown dashed in Figure 9. Figure 10 shows the correlation coefficient ρ between predicted and observed results as a function of number of prediction steps for the stated lags. These results accord well with those of Sugihara and May (their Figure 4d). Our one-step predictions with a three-dimensional embedding and a lag of 3 appear to be as good as their best effort, which had a six-dimensional embedding with a lag of 1. Note the fall off in prediction quality with prediction time, evidence for determinism as argued by Sugihara and May.

5.3 LYNX PELTS

A traditional data set for time series analysts is the lynx pelt data. See, for example, Tong.[30] Usually logarithms of pelt numbers are taken, because of the large variations from year to year. Using the recursive version of the tesselation predictor, it turns out that we do marginally better without taking logs. Figure 11 shows the second half of the data set and one-step predictions. Correlations are shown in Figure 12, with somewhat dubious evidence of determinism in the fall off in predictability with number of steps. There are significant differences between embeddings with different lags.

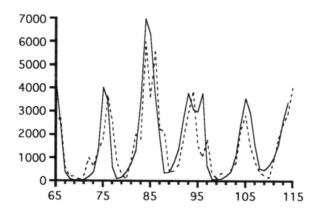

FIGURE 11 Lynx pelt numbers (solid line) and tesselation predictions (dashed).

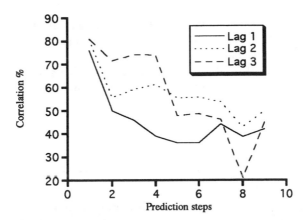

FIGURE 12 Correlation coefficient for the data of Figure 11 as a function of prediction time, for multi-step tesselation predictions. In each case the embedding was in three dimensions; results for lags 1, 2, and 3 are shown.

6. CONCLUSIONS

Triangulations and tesselations have been used give information about both the geometry and the dynamics of attractors and manifolds from experimental data, with explicit reference to detection of folds and branches, and to prediction and interpolation. There are many more possible applications for this and other non-linear modeling algorithms, including a quite remarkable degree of claimed noise reduction.[10] The limitation of using tesselations for modeling is the exponential growth of the time required as the embedding dimension k increases, but the advantages for low-dimensional models are considerable. It is not yet clear whether any of the proposed methods are useful in high dimensions even when the calculations they require are feasible.

The tesselation method has the ability to produce an almost-everywhere-smooth map, though there is no guarantee that it will be a diffeomorphism even when the data comes from a differential equation. In practice, regions of noninvertibility are normally associated with lack of data, or reflect true noninvertibility as in the origami map discussed in section 2.3.

Current research includes attempts to develop better methods for extraction of manifolds than the rather quick and dirty one used in this paper; and implementation of noise-reduction strategies. The latter is particularly important when there is sufficient data that simplex size is of the same order as, or smaller than, noise level; a direct application of the approximations given here will then tend to model the noise rather than the data.

Initial comparisons of prediction quality between this and other methods were given in the examples section, and more detailed comparisons are currently under way.

ACKNOWLEDGMENTS

Some parts of this paper are adapted from a paper to appear in *The International Journal of Bifurcation and Chaos*, December 1991. I thank NATO and the Santa Fe Institute for support for attending this conference, and the Australian Research Council for a grant to continue pursuit of this research.

REFERENCES

1. Aihara, K. "Chaotic Neural Networks." In *Bifurcation Phenomena in Nonlinear Systems and Theory of Dynamical Systems,* edited by H. Kawakami. Singapore: World Scientific, 1990.
2. Albano, A. M., J. Muench, C. Schwartz, A. I. Mees, and P. E. Rapp. "Singular Value Decomposition and the Grassberger-Procaccia Algorithm." *Phys. Rev. A* **38A** (1988): 3017–3026.
3. Barnhill, R. E., and F. F. Little. "Three- and Four-Dimensional Surfaces." *Rocky Mountain J. Math.* **14(1)** (1984): 77–102.
4. Bowyer, A. "Computing Dirichlet Tesselations." *The Computer Journal* **24(2)** (1981): 162–166.
5. Casdagli, M. "Nonlinear Prediction of Chaotic Time Series." *Physica D* **35(3)** (1989): 335–356.
6. Casdagli, M., and S. Eubank, eds. This volume.
7. Crutchfield, J. P., and B. S. McNamara. "Equations of Motion from a Data Series." *Complex Systems* **1** (1987): 417–452.
8. Eckmann, J.-P., and D. Ruelle. "Ergodic Theory of Chaos and Strange Attractors." *Rev. Mod. Phys.* **57(3)** (1985): 617–656.
9. Farmer, J. D., and J. J. Sidorowich. "Predicting Chaotic Time Series." *Phys. Rev. Lett.* **59(8)** (1987): 845–848.
10. Farmer, J. D., and J. J. Sidorowich. "Optimal Shadowing and Noise Reduction." *Physica D* **47** (1990): 373–392.
11. Green, P. J., and R. Sibson. "Computing Dirichlet Tessellations in the Plane." *The Computer Journal* **21(2)** (1978): 168–173.
12. Guckenheimer, J., and P. Holmes. *Nonlinear Oscillations, Dynamical Systems, and Bifurcations of Vector Fields.* New York: Springer-Verlag, 1983.
13. Holden, A. V., and M. A. Muhamad. "Chaotic Activity in Neural Systems." *Cybernetics & Systems Research* **2** (1984): 245–250.
14. Judd, K., A. I. Mees, K. Aihara, and M. Toyoda. "Grid Imaging for a Two-Dimensional Map." *Internat'l. J. Bifurcation & Chaos in Appl. Sci. & Eng.* **1(1)** (1990): 197–210.
15. Mees, A. I. "Modelling Complex Systems." In *Dynamics of Complex Interconnected Biological Systems,* edited by T. Vincent, L. S. Jennings, and A. I. Mees, 104–124. Boston: Birkhauser, 1990.
16. Mees, A. I. "Modelling Dynamical Systems from Real-World Data." In *Measures of Complexity and Chaos,* edited by N. B. Abraham, A. M. Albano, A. Passamente, and P. E. Rapp, 345–349. New York: Plenum, 1990.
17. Mees, A. I. "Dynamical Systems and Tesselations: Detecting Determinism in Data." *Internat'l. J. Bifurcation & Chaos in Appl. Sci. & Eng.* **1(4)** (1991): in press.
18. Packard, N. H., J. P. Crutchfield, J. D. Farmer, and R. S. Shaw. "Geometry from a Time Series." *Phys. Rev. Lett.* **45(9)** (1980): 712–716.

19. Preparata, F. P., and M. I. Shamos. *Computational Geometry: An Introduction.* New York: Springer, 1985.

20. Rapp, P. E., A. M. Albano, and A. I. Mees. "Calculation of Correlation Dimensions from Experimental Data: Progress and Problems." In *Dynamic Patterns in Complex Systems,* edited by J. A. S. Kelso, A. J. Mandell, and M. F. Schlesinger, 191–205. Singapore: World Scientific, 1988.

21. Rapp, P. E., T. R. Bashore, J. M. Martinerie, A. M. Albano, I. D. Zimmerman, and A. I. Mees. "Dynamics of Brain Electrical Activity." *Brain Topography* **2(1/2)** (1989): 99–118.

22. Rapp, P. E., I. D. Zimmerman, A. M. Albano, G. C. Deguzman, and N. N. Greenbaun. "Dynamics of Spontaneous Neural Activity in the Simian Motor Cortex: The Dimension of Chaotic Neurons." *Phys. Lett. A* **110(6)** (1985).

23. Renka, R. J. "Interpolation of Data on the Surface of a Sphere." *ACM Trans Math Software,* **10(4)** (1984): 417–436.

24. Rockafellar, R. T. *Convex Analysis.* Princeton, NJ: Princeton University Press, 1969.

25. Sage, A. P., and J. L. Melsa. *Estimation Theory, with Applications to Communications and Control.* New York: McGraw-Hill, 1971.

26. Sibson, R. "Nonparametric Spatial Regression." Report, Mathematics Department, University of Bath, UK, 1985.

27. Sparrow, C. T. *The Lorenz Equations: Bifurcations, Chaos and Strange Attractors.* New York: Springer, 1982.

28. Sugihara, G., and R. M. May. "Nonlinear Forecasting as a Way of Distinguishing Chaos from Measurement Error in Time Series." *Nature* **344** (1990): 734–741.

29. Takens, F. "Detecting Strange Attractors in Turbulence." In *Dynamical Systems and Turbulence,* edited by D. A. Rand and L. S. Young, 365–281. Berlin: Springer, 1981.

30. Tong, H. *Threshold Models in Non-Linear Time Series Analysis.* New York: Springer-Verlag, 1983.

31. Weigend, A. S., B. A. Huberman, and D. E. Rumelhart. "Predicting the Future: A Connectionist Approach." Report: SSL-90-20, System Sciences Laboratory, Xerox Palo Alto Research Center, 1990.

Sara A. Solla
AT&T Bell Laboratories, Holmdel, New Jersey 07733

Supervised Learning: A Theoretical Framework

INTRODUCTION

Layered neural networks are of interest as a tool to implement input-output maps. This work explores the ability of a highly connected, layered network of simple processing units to perform such task.

The ensemble of all possible network configurations compatible with a fixed architecture is explored to define a probability distribution over the space of input-output maps. Such distribution fully describes the functional capabilities of the chosen architecture. Its entropy measures the intrinsic functional diversity of the network ensemble.

Supervised learning is formulated as an optimization problem resulting in a monotonic decrease of the effective volume of configuration space through the exponential elimination of network configurations incompatible with the examples of the desired map. Such contraction results in increased specificity in the functional capabilities of the ensemble, and a monotonic reduction of the intrinsic entropy.

Nonlinear Modeling and Forecasting, SFI Studies in the Sciences of Complexity,
Proc. Vol. XII, Eds. M. Casdagli & S. Eubank, Addison-Wesley, 1992   **25**

The prior distribution which describes the ensemble of untrained networks is modified multiplicatively through learning, and results in a Gibbs distribution to describe the ensemble of trained networks. A full thermodynamic description thus becomes available.

A monotonic increase of the relative entropy of the ensemble with increasing size of the training set measures the information extracted from the examples of the desired map. Such information gain is achieved through the minimization of the learning error, defined over the training examples.

The emergence of generalization ability is monitored through the distribution of intrinsic generalization abilities of the networks in the ensemble. Learning curves, displaying the dependence of the mean generalization ability on the number of training examples, are shown to depend solely on the distribution of generalization abilities of the untrained networks. It is thus possible to predict learning curves, and to identify features of the prior ensemble which control the emergence of generalization ability.

These topics are explored beginning with a brief description of single-neuron processing. The layered neural network architecture and its associated configuration space are then described, including definitions of functional capabilities and intrinsic entropy. Supervised learning is presented as an optimization problem leading to the contraction of the network ensemble, as described by the Gibbs distribution. The thermodynamic description explores the relation between learning error and relative entropy, and focuses on the emergence of generalization ability.

LAYERED NEURAL NETWORKS

The "neurons" used to construct the networks whose properties are being investigated here are simple information-processing devices, and capture only in crude simplification the complex behavior of biological neurons. The state of a model neuron is specified through a single scalar variable V, with $0 \le V \le 1$. The state $V = 0$ corresponds to an inactive neuron, while $V > 0$ represents the level of activity in reference to a maximum firing rate, $V = 1$. The state of a system of N neurons is described by the N-component vector $\vec{V} = (V_1, V_2, \ldots, V_N)$.

The state V_i of the ith neuron is determined as follows: consider the set of neurons $\{V_j\}$ whose state is communicated to neuron i via synaptic couplings of strength W_{ij}. The total input to neuron i is a linear combination of the states $\{V_j\}$:

$$U_i = \sum_j W_{ij} V_j + W_i \, . \tag{1}$$

The couplings $\{W_{ij}\}$ are the coefficients of the linear combination, and the bias W_i can be interpreted as an additional input due to a neuron always in the $V = 1$ state.

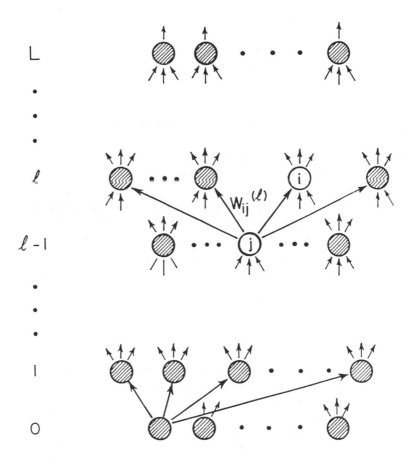

FIGURE 1   A layered feed-forward network with L levels of processing.

The input U_i determines the state V_i via a nonlinear, monotonically increasing, and bounded transfer function $V = y(U)$. The standard choice $y(x) = (1 + \exp(-\lambda x))^{-1}$ confines all state variables to the $[0,1]$ interval.

Consider now a specific network architecture: a layered feed-forward neural network with deterministic parallel dynamics, as shown in Figure 1. The network consists of $L+1$ layers providing L levels of processing.[7] The first layer at $\ell = 0$ is the input field, and contains N_0 units. Subsequent layers are labeled $1 \leq \ell \leq L$; the ℓth layer contains N_ℓ units. The ℓth level of processing corresponds to determining the state of the units in layer ℓ according to the following deterministic and parallel rule:

$$U_i^{(\ell)} = \sum_{j=1}^{N_{\ell-1}} W_{ij}^{(\ell)} V_j^{(\ell-1)} + W_i^{(\ell)},$$

$$V_i^{(\ell)} = g(U_i^{(\ell)}).$$

(2)

The state $V_i^{(\ell)}$ of unit i in layer ℓ is thus determined by the states $\{V_j^{(\ell-1)}\}$, $1 \leq j \leq N_{\ell-1}$, of units in the preceding layer.

The first layer receives input from the external world: a pattern is presented to the network by fixing the values of the state variables $\{V_i^{(0)}\}$, $1 \leq i \leq N_0$. The states of subsequent layers are determined consecutively according to Eq. (2). The state of the last layer at $\ell = L$ is the output. Given $\vec{x} = \vec{V}^{(0)}$, the network produces an output $\vec{y} = \vec{V}^{(L)}$, thus implementing an input-output map $\vec{y} = f(\vec{x})$.

A layered architecture is specified by the number $\{N_\ell\}, 0 \leq \ell \leq L$ of units per layer. The parameters of the network are the couplings $\{W_{ij}^{(\ell)}\}$, $1 \leq i \leq N_\ell$, $1 \leq j \leq N_{\ell-1}$, $1 \leq \ell \leq L$, and the biases $\{W_i^{(\ell)}\}$, $1 \leq i \leq N_\ell$, $1 \leq \ell \leq L$, corresponding to a point $\vec{W}$ in a configuration space of dimension

$$D_{\vec{W}} = \sum_{\ell=1}^{L} N_\ell(1 + N_{\ell-1}).$$

(3)

The dimensionality $D_{\vec{W}}$ thus counts the number of independent parameters needed to specify the network. It follows from Eq. (3) that $D_{\vec{W}} \sim L\langle N_\ell^2 \rangle$, where L is the depth of the network and $\langle N_\ell \rangle$ is a characteristic layer width. Thus $D_{\vec{W}}$ is typically a large number, and it scales as $D_{\vec{W}} \sim N^2/L$, where $N = \sum_{\ell=1}^{L} N_\ell$ is the total number of processing units.

The configuration space $\{\vec{W}\}$ describes the ensemble of all possible networks that can be constructed within the constraint of the specified architecture. Every point $\vec{W}$ in configuration space represents a specific network design, and corresponds to the realization of a specific input-output map $f_{\vec{W}}$. A question that arises is that of characterizing the class of maps $\vec{y} = f(\vec{x})$ that can be implemented by a given network architecture. Such characterization requires a full exploration of configuration space.

Among the various functions that are implementable by the chosen architecture, there is a desired map $\tilde{f}$ to be realized. The process of learning refers to a guided exploration of configuration space so as to determine values of the couplings $\{W_{ij}^{(\ell)}\}$ and biases $\{W_i^{(\ell)}\}$ for which $f_{\vec{W}} = \tilde{f}$. Such search for configurations $\vec{W}$ which implement the map $\tilde{f}$ is guided by whatever information is available on the desired map. Supervised learning requires the availability of examples: input-output pairs $\vec{\xi} = (\vec{x}, \vec{y})$ for which $\vec{y} = \tilde{f}(\vec{x})$.

NETWORK ENSEMBLE

Every point $\vec{W}$ in configuration space selects a specific network design, which implements the map $f_{\vec{W}}$. It is convenient to partition configuration space according to the functional capabilities of the network ensemble.

A prior density $\rho_0(\vec{W})$ constrains the effective volume of configuration space to

$$Z_0 = \int d\vec{W} \rho_0(\vec{W}), \tag{4}$$

usually normalized to $Z_0 = 1$. Regions corresponding to the implementation of a specific map f are selected by a masking function

$$\Theta_f(\vec{W}) = \begin{cases} 1 & \text{if } f_{\vec{W}} = f; \\ 0 & \text{if } f_{\vec{W}} \neq f. \end{cases}$$

The fractional volume occupied by configurations which implement f is thus given by

$$P_0(f) = \int d\vec{W} \rho_0(\vec{W}) \Theta_f(\vec{W}). \tag{5}$$

The functional capabilities of the network ensemble described by the density $\rho_0(\vec{W})$ are quantitatively specified by the probability distribution $P_0(f)$ on the space of functions.[7] The class of functions implementable by the chosen architecture is

$$\mathcal{F} = \{f | P_0(f) \neq 0\}. \tag{6}$$

The realizability of the desired map $\tilde{f}$ corresponds to the requirement $P_0(\tilde{f}) \neq 0$.

It is useful to consider the entropy[2]

$$\tilde{S}_0 = -\sum_{\{f\}} P_0(f) \ln P_0(f) \tag{7}$$

of the prior distribution. Only functions $f \in \mathcal{F}$ contribute to $\tilde{S}_0$. The optimal case of an ensemble devoted to the unique implementation of the desired map $\tilde{f}$ corresponds to $P_0(\tilde{f}) = 1$ and $P_0(f) = 0$ for all $f \neq \tilde{f}$, and results in $\tilde{S}_0 = 0$. But in a typical case, the set $\mathcal{F}$ of realizable functions contains many maps f besides $\tilde{f}$, and $\tilde{S}_0 > 0$.

The prior entropy $\tilde{S}_0$ is an intrinsic property of the chosen network architecture, and measures its functional diversity. The ensemble of possible network configurations described by the density $\rho_0(\vec{W})$ fully determines the distribution $P_0(f)$ and its associated entropy.

Learning refers to a systematic modification of the network ensemble towards specificity in the implementation of the desired map $\tilde{f}$. The starting point is an ensemble characterized by the prior entropy $\tilde{S}_0$; learning results in a systematic entropy reduction toward the goal $\tilde{S} = 0$.

SUPERVISED LEARNING

Supervised learning requires examples of the desired map $\tilde{f}$. The training set contains m input-output pairs $\vec{\xi}^\alpha = (\vec{x}^\alpha, \vec{y}^\alpha)$, $1 \leq \alpha \leq m$, for which $\vec{y}^\alpha = \tilde{f}(\vec{x}^\alpha)$.

Learning the training set results in a monotonic reduction of the effective volume of configuration space. The effect of learning the αth example is described by the masking function

$$\Theta(\vec{W}, \vec{\xi}^\alpha) = \begin{cases} 1 & \text{if } f_{\vec{W}}(\vec{x}^\alpha) = \vec{y}^\alpha; \\ 0 & \text{if } f_{\vec{W}}(\vec{x}^\alpha) \neq \vec{y}^\alpha. \end{cases}$$

The αth example is thus learned by removing from the ensemble all the networks that misclassify it. The prior density $\rho_0(\vec{W})$ is multiplicatively modified into $\rho_0(\vec{W})\Theta(\vec{W}, \vec{\xi}^\alpha)$ by the αth example, and into $\rho_0(\vec{W}) \prod_{\alpha=1}^{m} \Theta(\vec{W}, \vec{\xi}^\alpha)$ by the full training set. The effective volume of configuration space is reduced to

$$Z_m = \int d\vec{W} \rho_0(\vec{W}) \prod_{\alpha=1}^{m} \Theta(\vec{W}, \vec{\xi}^\alpha), \tag{8}$$

with $Z_m \leq Z_0$.

A more realistic description of the learning process requires a measure of the error made by network $\vec{W}$ on the αth example, as given by the distance

$$\varepsilon(\vec{W}, \vec{\xi}^\alpha) = d(\vec{y}^\alpha, f_{\vec{W}}(\vec{x}^\alpha)) \tag{9}$$

between the target $\vec{y}^\alpha$ and the actual output $f_{\vec{W}}(\vec{x}^\alpha)$. To be precise, $d(\vec{y}_1, \vec{y}_2)$, defined for pairs of points in output space $\{\vec{y}\}$, needs not be symmetric or satisfy the triangular inequality. The only requirements are $d(\vec{y}_1, \vec{y}_2) \geq 0$, and $d(\vec{y}_1, \vec{y}_2) = 0$ if and only if $\vec{y}_1 = \vec{y}_2$.

The masking function $\Theta(\vec{W}, \vec{\xi}^\alpha)$ can thus be replaced by a softer survival factor

$$\varphi(\vec{W}, \vec{\xi}^\alpha) = \exp(-\beta\varepsilon(\vec{W}, \vec{\xi}^\alpha)).$$

If network $\vec{W}$ is such that $f_{\vec{W}}(\vec{x}^\alpha) = \vec{y}^\alpha$, then $\varepsilon(\vec{W}, \vec{\xi}^\alpha) = 0$, and both $\Theta(\vec{W}, \vec{\xi}^\alpha) = \varphi(\vec{W}, \vec{\xi}^\alpha) = 1$. The difference between the masking functions Θ and φ arises for networks $\vec{W}$ such that $f_{\vec{W}}(\vec{x}^\alpha) \neq \vec{y}^\alpha$. The use of $\Theta(\vec{W}, \vec{\xi}^\alpha) = 0$ eliminates such networks from the ensemble, while the use of $\varphi(\vec{W}, \vec{\xi}^\alpha)$ keeps them, although exponentially attenuated by a factor controlled by the error $\varepsilon(\vec{W}, \vec{\xi}^\alpha)$. The survival factor $\varphi(\vec{W}, \vec{\xi}^\alpha)$ remains close to one for very small error, and decreases monotonically to zero as the error increases.

The parameter β controls the error tolerance: for a given error $\varepsilon(\vec{W}, \vec{\xi}^\alpha)$, the survival factor $\varphi(\vec{W}, \vec{\xi}^\alpha)$ is increased towards one by using a small β. In the $\beta \to \infty$ limit, the binary masking function Θ is recovered: $\varphi(\vec{W}, \vec{\xi}^\alpha) = 1$ if $\varepsilon(\vec{W}, \vec{\xi}^\alpha) = 0$, but $\varphi(\vec{W}, \vec{\xi}^\alpha) = 0$ if $\varepsilon(\vec{W}, \vec{\xi}^\alpha) > 0$, regardless of the magnitude of the error.

For finite β, the multiplicative effect of learning the αth example is to modify the prior density $\rho_0(\vec{W})$ into $\rho_0(\vec{W})\varphi(\vec{W},\vec{\xi}^\alpha)$, and learning the full training set leads to $\rho_0(\vec{W}) \prod_{\alpha=1}^m \varphi(\vec{W},\vec{\xi}^\alpha) = \rho_0(\vec{W}) \exp(-\beta E_m(\vec{W}))$, with

$$E_m(\vec{W}) = \sum_{\alpha=1}^m \varepsilon(\vec{W},\vec{\xi}^\alpha), \tag{10}$$

the additive error made by network $\vec{W}$ on the full training set. Learning the training set reduces the effective volume of configuration space to

$$Z_m = \int d\vec{W} \rho_0(\vec{W}) e^{-\beta E_m(\vec{W})}. \tag{11}$$

Since $\varepsilon(\vec{W},\vec{\xi}^\alpha) \geq 0$, adding a new example to the training set results in $E_{m+1}(\vec{W}) \geq E_m(\vec{W})$ for all networks $\vec{W}$, and $Z_{m+1} \leq Z_m$. The effective volume of the network ensemble decreases monotonically as the size of the training set is increased.

LEARNING ERROR AND ENTROPY

The ensemble of untrained networks is described by the prior density $\rho_0(\vec{W})$, normalized to $Z_0 = \int d\vec{W} \rho_0(\vec{W}) = 1$. The ensemble of networks trained with examples $\{\vec{\xi}^\alpha\}$, $1 \leq \alpha \leq m$, of the desired map is described by the appropriately normalized posterior density

$$\rho_m(\vec{W}) = \frac{1}{Z_m}\rho_0(\vec{W}) e^{-\beta E_m(\vec{W})}. \tag{12}$$

This density describes a Gibbs canonical ensemble. The normalization denominator Z_m as defined in Eq. (11) is the *partition function* of the ensemble.

A full thermodynamic description is now available. The *free energy*

$$F_m = -\ln Z_m \tag{13}$$

increases monotonically as the size of the training set is increased: $Z_{m+1} \leq Z_m$ implies $F_{m+1} \geq F_m$.

Since networks which do not perfectly fit the training set are retained according to the density $\rho_m(\vec{W})$ of Eq. (12), trained networks are not forced to satisfy $E_m(\vec{W}) = 0$. The ensemble of trained networks exhibits a distribution of possible values for $E_m(\vec{W})$, with mean *learning error*

$$\mathcal{L}_m = \int d\vec{W} E_m(\vec{W})\rho_m(\vec{W}), \tag{14}$$

which follows via a simple derivative from the free energy of Eq. (13),

$$\mathcal{L}_m = \frac{\partial F_m}{\partial \beta}.$$ (15)

It is straightforward to extend the analysis of functional capabilities of Eq. (5) to the ensemble of trained networks described by the density $\rho_m(\vec{W})$. The fractional volume of the trained ensemble occupied by networks which implement f is given by

$$P_m(f) = \int d\vec{W} \rho_m(\vec{W}) \Theta_f(\vec{W}).$$ (16)

The entropy

$$\tilde{S}_m = - \sum_{\{f\}} P_m(f) \ln P_m(f)$$ (17)

monitors the emergence of specificity towards the implementation of the desired map $\tilde{f}$: an entropy reduction with increasing size m of the training set[8] reflects a narrowing of the probability distribution $P_m(f)$ to focus on the implementation of the target function $\tilde{f}$ and a class of functions increasingly similar to $\tilde{f}$.

It is however the *relative entropy*

$$S_m = \sum_{\{f\}} P_m(f) \ln \left[\frac{P_m(f)}{P_0(f)} \right]$$ (18)

which plays a role in the thermodynamic description.[9] It is identical to the entropy of the posterior distribution $\rho_m(\vec{W})$ relative to the prior $\rho_0(\vec{W})$,

$$S_m = \int d\vec{W} \rho_m(\vec{W}) \ln \left[\frac{\rho_m(\vec{W})}{\rho_0(\vec{W})} \right].$$ (19)

The equivalence between Eqs. (18) and (19) is easily established: since all networks $\vec{W}$ for which $\Theta_f(\vec{W}) = 1$ share a common value of $E_m(\vec{W})$, it follows from Eqs. (12) and (16) that

$$\frac{P_m(f)}{P_0(f)} = \frac{\rho_m(\vec{W})}{\rho_0(\vec{W})}$$ (20)

for all $\vec{W}$ such that $\Theta_f(\vec{W}) = 1$. The equality of Eq. (20), which leads to the equivalence between Eqs. (18) and (19) when combined with the identity $\sum_{\{f\}} \Theta_f(\vec{W}) = 1$, establishes that the partition of configuration space $\{\vec{W}\}$ according to its functional capabilities is *sufficient* in the statistical sense[3]: no information is lost in the partition.

The relative entropy of Eq. (19) obeys the usual thermodynamic relations. It is related to the free energy of Eq. (13) and the learning error of Eq. (15) by

$$S_m = F_m - \beta \mathcal{L}_m.$$ (21)

It starts at $S_0 = 0$, and satisfies $S_{m+1} \geq S_m$. The monotonic increase of S_m measures the information gained by the ensemble as new examples are added to the training set. As follows from Eq. (21), such information gain is achieved through the monotonic contraction of the effective volume of configuration space, and through the minimization of the learning error. The thermodynamic relation

$$\frac{\partial S_m}{\partial \mathcal{L}_m} = -\beta \tag{22}$$

confirms that the entropy increases through the minimization of the learning error, and indicates that the information gain is most efficient at high β.

It follows from Eq. (21) that the error-control parameter β is the Lagrange multiplier for the constrained learning error $\mathcal{L}_m$ during minimization of the relative entropy of Eq. (19). The Gibbs distribution (12) is therefore the minimal relative-entropy distribution subject to the mean learning error as a constraint, and the inverse of the error-control parameter β plays the role of the ensemble temperature in statistical mechanics.[3]

GENERALIZATION ABILITY

A description of the generalization ability achieved by supervised learning is based on the generalization ability $g(\vec{W})$ of the individual networks $\vec{W}$, defined as the probability that $f_{\vec{W}}$ will correctly classify a randomly chosen example of the desired map $\tilde{f}$. As an illustration of the intrinsic ability of $f_{\vec{W}}$ to reproduce $\tilde{f}$, consider the simple case of a Boolean function from $N_0 = N$ inputs onto $N_L = 1$ output. The function $\tilde{f}$ is specified by 2^N bits, indicating the output for every possible input. In this case

$$g(\vec{W}) = \frac{2^N - d_H(f_{\vec{W}}, \tilde{f})}{2^N}, \tag{23}$$

where $d_H(f_{\vec{W}}, \tilde{f})$ is the Hamming distance between $f_{\vec{W}}$ and $\tilde{f}$, i.e., the number of bits by which their truth tables differ.

In order to define $g(\vec{W})$ formally, within the framework of a thermodynamic description, it is necessary to consider the desired map $\tilde{f}$ as a probability density $\tilde{P}(\vec{\xi})$ in the space $\vec{\xi} = (\vec{x}, \vec{y})$ of input-output pairs. Examples $\vec{\xi}$ of the desired map are drawn from

$$\tilde{P}(\vec{\xi}) = \tilde{P}(\vec{y}|\vec{x})\tilde{P}(\vec{x}), \tag{24}$$

where $\tilde{P}(\vec{x})$ describes the region of interest in input space, while the functional dependence is described through the conditional probability $\tilde{P}(\vec{y}|\vec{x})$. This description includes but is not restricted to a deterministic functional dependence $\tilde{P}(\vec{y}|\vec{x}) = \delta(\vec{y} - \tilde{f}(\vec{x}))$.

The target distribution $\tilde{P}(\vec{\xi})$ is to be compared to the distribution $P(\vec{\xi}|\vec{W})$ induced on $\{\vec{\xi}\}$ by the hypothesis represented by the network $\vec{W}$. The Gibbs density of Eq. (12) can be inverted using the Bayes formula to obtain the conditional likelihood of $\vec{\xi}$,

$$P(\vec{\xi}|\vec{W}) = \frac{1}{z(\beta)} e^{-\beta\varepsilon(\vec{W},\vec{\xi})}, \tag{25}$$

with

$$z(\beta) = \int d\vec{\xi} e^{-\beta\varepsilon(\vec{W},\vec{\xi})}, \tag{26}$$

a normalization constant independent of $\vec{W}$. The likelihood of Eq. (25) is averaged over the full target distribution to obtain

$$g(\vec{W}) = \int d\vec{\xi}\tilde{P}(\vec{\xi})P(\vec{\xi}|\vec{W}). \tag{27}$$

The ensemble of untrained networks, described by the prior density $\rho_0(\vec{W})$, can be categorized according to the generalization ability of its constituents:

$$\rho_0(g) = \int d\vec{W}\rho_0(\vec{W})\delta\left(g(\vec{W}) - g\right) \tag{28}$$

follows from a full exploration of configuration space consistent with $\rho_0(\vec{W})$, so as to classify networks according to their generalization ability $g(\vec{W})$. The prior density $\rho_0(g)$ thus contains full information about the chosen architecture and available configurations, and about the desired map through the definition of $g(\vec{W})$.

The question that arises is that of monitoring the evolution of such distribution of generalization abilities as the ensemble is modified through learning. The Gibbs density of Eq. (12) describes an ensemble trained with a *specific* set of examples $\{\vec{\xi}^\alpha\}$, $1 \leq \alpha \leq m$. An additional average over all the possible ways in which a training set of size m could be constructed by drawing each example independently from the distribution $\tilde{P}(\vec{\xi})$ results in the density

$$\ll \rho_m(\vec{W}) \gg \equiv \int d\vec{\xi}^1 \ldots d\vec{\xi}^m \tilde{P}(\vec{\xi}^1) \ldots \tilde{P}(\vec{\xi}^m)\rho_m(\vec{W}), \tag{29}$$

which describes the effect of learning a *typical* training set of size m. It is the distribution of generalization abilities of such *typical* trained ensemble, given by

$$\rho_m(g) = \int d\vec{W} \ll \rho_m(\vec{W}) \gg \delta(g(\vec{W}) - g), \tag{30}$$

that is to be monitored as a function of m.

An unexpectedly simple result follows from computing the average density of Eq. (29) in the *annealed approximation*.[4,5] The partition function Z_m, the normalization denominator of $\rho_m(\vec{W})$, is averaged separately

$$\ll Z_m(\vec{W}) \gg \equiv \int d\vec{\xi}^1 \ldots d\vec{\xi}^m \tilde{P}(\vec{\xi}^1) \ldots \tilde{P}(\vec{\xi}^m)Z_m, \tag{31}$$

to obtain

$$\ll Z_m \gg = z^m(\beta) \int_0^1 g^m \rho_0(g) dg, \tag{32}$$

which leads to

$$\rho_m(g) = \frac{g^m \rho_0(g)}{\int_0^1 g^m \rho_0(g) dg}. \tag{33}$$

The result is both simple and powerful: within the annealed approximation, the distribution of generalization abilities $\rho_m(g)$ for the ensemble of trained networks is fully determined by the prior distribution $\rho_0(g)$. Each example of the desired map contributes with a factor of g; the prior density is multiplied by g^m, and appropriately normalized to guarantee $\int_0^1 \rho_m(g) dg = 1$ for all m. Adding a new example to the training set introduces an additional factor of g, thus shifting monotonically the distribution of generalization abilities towards $g = 1$. The monotonic contraction of the effective volume of configuration space with increasing m is not arbitrary: it emphasizes regions of configuration space with intrinsically high generalization ability.

PREDICTION OF LEARNING CURVES

Consider the mean generalization ability

$$G_m = \int_0^1 g \rho_m(g) dg \tag{34}$$

of the ensemble of trained networks. Since the distribution $\rho_m(g)$ is fully determined by the prior $\rho_0(g)$, it follows from Eq. (33) that

$$G_m = \frac{\int_0^1 g^{m+1} \rho_0(g) dg}{\int_0^1 g^m \rho_0(g) dg}. \tag{35}$$

The mean generalization ability after training with m examples of the desired map is simply given by a ratio of two consecutive moments of the prior distribution.

Learning curves display the m-dependence of the error $\mathcal{E}_m = 1 - G_m$. It follows from Eq. (35) that such curves can be computed without performing any learning experiment if $\rho_0(g)$ is given or estimated. An example shown in Figure 2 illustrates the application of Eq. (35) to a problem which allows for an exact computation of $\rho_0(g)$, through the exhaustive exploration of a discrete and finite configuration space.[6]

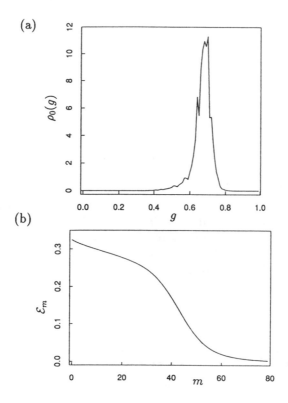

FIGURE 2 (a) Prior distribution of generalization abilities $\rho_0(g)$ for an $L = 2$ layered network to solve the contiguity problem, from Ref. 6. (b) The learning curve $\mathcal{E}_m = 1 - G_m$, resulting from applying Eq. (35) to the $\rho_0(g)$ in (a).

Beyond the use of Eq. (35) as a tool for the numerical prediction of learning curves, it provides a theoretical framework for analyzing the possible outcomes of learning experiments. It is of interest to identify the relevant features of $\rho_0(g)$ which control the shape of learning curves.

The asymptotic form of the moments ratio of Eq. (35) is determined solely by the functional form of $\rho_0(g)$ in the vicinity of $g = 1$. Consider the case

$$\rho_0(g) = h(g)\theta(\hat{g} - g) + 2p\delta(g - 1),\tag{36}$$

for $\hat{g} < 1$, where $\theta(x) = 1$ for $x \geq 0$ and $\theta(x) = 0$ otherwise, and $\int_0^{\hat{g}} h(g)dg = 1 - p$. The prior density of Eq. (36) consists of a continuous part $h(g)$ in the range $0 \leq g \leq \hat{g}$, which integrates to $1 - p$, and a singular contribution of weight p at $g = 1$. There is a gap: $\rho_0(g) = 0$ for $\hat{g} < g < 1$. The average generalization ability of Eq. (35) is easily computed for a density of the form Eq. (36). For large m

$$\mathcal{E}_m = 1 - G_m \sim \left(\frac{1-p}{p}\right)(1 - \hat{g})\,\hat{g}^m,\tag{37}$$

indicating an exponential decay of the form

$$\mathcal{E}_m \sim e^{-m/m_0}, \tag{38}$$

with $m_0^{-1} = -\ln \hat{g}$.

The parameter m_0 controlling the rate of exponential decay is inversely proportional to the gap $\hat{\epsilon} = 1 - \hat{g}$ between $g = 1$ and $g = \hat{g}$. For $\hat{\epsilon} \ll 1$, $m_0 \sim 1/\hat{\epsilon}$. As $\hat{\epsilon} \to 0$, $m_0 \to \infty$, and the exponential decay is replaced by a power law of the form

$$\mathcal{E}_m \sim \frac{\nu}{m + \nu}. \tag{39}$$

Such asymptotic form follows from the moment ratio Eq. (35) for G_m whenever $\rho_0(g) \sim (1 - g)^\nu$ as $g \to 1$, and suggests an interesting possible relation between the value of the exponent ν and the VC-dimension[1] of the layered neural network.

SUMMARY

The Gibbs formulation has been shown to provide a powerful tool for the typical case analysis of supervised learning in layered neural networks. The picture that emerges is that of learning as a monotonic decrease of the effective volume of configuration space, accompanied by an entropy increase which monitors the amount of information provided by the training examples.

The contraction is not arbitrary: it emphasizes regions of configuration space with intrinsically high generalization ability. The iterated convolution with g to obtain $\rho_m(g)$ from $\rho_0(g)$ results in an increasing bias towards $g = 1$, and a monotonic increase of the mean generalization ability with increasing m.

The fundamental result of this paper is to demonstrate that knowledge of the initial distribution $\rho_0(g)$ suffices to predict network performance, Eq. (35). The specific details of the chosen network architecture and the desired map $\vec{y} = \vec{f}(\vec{x})$ matter only to the extent that they influence and determine $\rho_0(g)$. The asymptotic form of the learning curves $\mathcal{E}_m$ vs. m is controlled by the properties of $\rho_0(g)$ close to $g = 1$: the existence of a gap results in exponential decay, while the continuous case leads to power-law decay.

ACKNOWLEDGMENT

This work was done in collaboration with Esther Levin and Naftali Tishby, of AT&T Bell Laboratories, and Daniel B. Schwartz, presently of GTE Laboratories. I thank them all for a very enjoyable interaction.

REFERENCES

1. Blumer, A. , A. Ehrenfeucht, D. Haussler, and M. K. Warmuth. "Learnability and the Vapnik-Chervonenkis Dimension." *JACM* **36** (1989): 929–965.
2. Denker, J. S. , D. B. Schwartz, B. S. Wittner, S. A. Solla, R. E. Howard, L. D. Jackel, and J. J. Hopfield. "Automatic Learning, Rule Extraction, and Generalization." *Complex Systems* **1** (1987): 877–922.
3. Levin, Esther, Naftali Tishby, and Sara A. Solla. "A Statistical Approach to Learning and Generalization in Layered Neural Networks." In *Proceedings of the Second Annual Workshop on Computational Learning Theory*, edited by R. Rivest, D. Haussler, and M. K. Warmuth, 245–260. California: Morgan Kaufmann, 1989.
4. Levin, Esther, Naftali Tishby, and Sara A. Solla. "A Statistical Approach to Learning and Generalization in Layered Neural Networks." *Proc. of the IEEE* **78** (1990): 1568–1574.
5. Levin, Esther, and Sara A. Solla. "Validity of the Annealed Approximation in the Theory of Learning." Unpublished paper, 1991.
6. Schwartz, D. B. , V. K. Samalam, Sara A. Solla, and J. S. Denker. "Exhaustive Learning." *Neural Computation* **2** (1990): 374–385.
7. Solla, Sara A. "Learning and Generalization in Layered Neural Networks: The Contiguity Problem." In *Neural Networks from Models to Applications*, edited by L. Personnaz and G. Dreyfus, 168–177. Paris: IDSET, 1989.
8. Solla, Sara A. "Supervised Learning and Generalization." *Neural Networks: Biological Computers or Electronic Brains*, 21–28. Paris: Springer-Verlag, 1990.
9. Tishby, Naftali, Esther Levin, and Sara A. Solla. "Consistent Inference of Probabilities in Layered Networks: Predictions and Generalization." In *Proceedings of the International Joint Conference in Neural Networks*, II 403–410. New York: IEEE, 1989.

W. C. Mead,†‡ **R. D. Jones,**†‡ **Y. C. Lee,**‡* **C. W. Barnes,**†‡ **G. W. Flake,**†‡**
L. A. Lee,†‡ **and M. K. O'Rourke**†‡

†Applied Theoretical Physics Division, and ‡Center for Nonlinear Studies, Los Alamos National Laboratory, Los Alamos, NM 87545; *Department of Physics and Astronomy, and **Department of Computer Science, University of Maryland, College Park, MD 20740

Prediction of Chaotic Time Series using CNLS-Net–Example: The Mackey-Glass Equation

We use the Connectionist Normalized Local Spline network (CNLS-net) to learn the dynamics of the Mackey-Glass time-delay differential equation, for the case $\tau = 30$. We show the optimum network operating mode, discuss parameter sensitivities, and determine the accuracy and robustness of predictions. We obtain predictions of varying accuracy using some 2–120 minutes of execution time on a *Sun Sparc-1* workstation. CNLS-net is capable of very good performance in predicting the Mackey-Glass time series. We discuss strong sensitivities associated with the embedding structure and basis function centers chosen.

INTRODUCTION

The Mackey-Glass (M-G) equation[13] is a time-delay ordinary differential equation that displays well-understood chaotic behavior with dimensionality dependent upon the chosen value of the delay parameter.[2] The time series generated by the M-G equation has been used as a test bed for a number of new adaptive computing techniques: a local linear (or quadratic) approximation method,[3] a back-propagation neural network,[11] and at least two radial basis function approaches.[1,14]

The Connectionist Normalized Local Spline network[8] (CNLS-net) combines a number of appealing features to yield a capable, versatile adaptive-computing network. The net features normalized radial basis functions, a linear gradient term, and a simple, rapid training algorithm, plus a variety of optional capabilities including Kalman Noise Filtering.[9] The network has been implemented within the code *CNLSTOOL* in the C Language on both *Sun* workstations and the *Cray YMP* supercomputer, although for this work we have used only the workstation version. *CNLSTOOL* provides a flexible, interactive user interface, and a framework that is adaptable to various network approaches and applications requirements. The CNLS network has been successfully applied to a number of fitting, prediction, and control test examples, including a preliminary test of the net's ability to predict the Mackey-Glass equation.[9]

In this work, we have performed extensive studies on the use of CNLS-net to model and predict the Mackey-Glass time series with delay parameter $\tau = 30$, a fully-developed chaotic regime. We find that CNLS-net can provide accurate predictions, comparable with those of previously applied methods,[1,3,11,14] while requiring relatively little training and prediction computational time. Data requirements are more modest than previous work using local linear (or quadratic) predictors[3] or unnormalized radial basis functions,[1,14] and are comparable with those of a back-propagation network[11] or a network using increasing radial basis functions.[1] In the following sections, we discuss our methods and findings.

THE CNLS-NET ARCHITECTURE

Earlier neural-type nets have used feed-forward networks with back propagation of errors to provide supervised training of the network weights.[17] Generally, the back-propagation (BP) networks use sigmoids as basis functions and have nodes organized into several layers. These networks have become very popular, and have provided new versatility and accuracy in predicting the dynamics of nonlinear functions that exhibit low-dimensional chaotic behavior.[11] BP networks, however, generally require long training times, and exhibit several numerical deficiencies that limit robustness and can hinder their application.[19]

Moody and Darken[14] have derived a radial basis function (RBF) network architecture that addresses the learning speed problem. By replacing multiple hidden layers of sigmoid functions with a single layer of localized functions, the learning algorithm can be linear in the coefficients, and learning is much faster. However, the RBF network requires many nodes when the input space is large, and requires much more data for training than does the BP net.

The CNLS-net was developed as an extension of previous adaptive network experience. A natural evolution is to modify radial basis function (RBF) nets in a manner that improves interpolation and reduces the amount of training necessary for accurate learning.[6,8,12]

The CNLS-net architecture begins with the identity,

$$g(\vec{x}) = \frac{\Sigma_{j=1}^{N} g(\vec{x}) \rho_j(\vec{x})}{\Sigma_k \rho_k(\vec{x})}. \tag{1}$$

Here, as in the RBF network, $\rho_j(\vec{x})$ is a localized function of $\vec{x}$ about some center $\vec{x}_j$. Typically, we use gaussian basis functions, with

$$\rho_j(\vec{x}) = \beta_j exp[-\beta_j |\vec{x} - \vec{x}_j|^2]. \tag{2}$$

Here β_j is a width parameter with dimensions of $1/x^2$. Hence, $g(\vec{x})$ on the right of Eq. 1 can be approximated by its Taylor expansion about $\vec{x}_j$. We have then,

$$\phi(\vec{x}) = \Sigma_{j=1}^{N} [f_j + (\vec{x} - \vec{x}_j) \cdot \vec{d}_j] \frac{\rho_j(\vec{x})}{\Sigma_k \rho_k(\vec{x})} \tag{3}$$

for an approximation to $g(\vec{x})$. Here, f_j and d_j are the weights for the jth basis function and linear term. These parameters could be evaluated from the data directly, but we choose to view them as adjustable parameters. In a moment we shall consider a method of evaluating them. The architecture of Eq. 3 can also be derived from information-theoretic principles.[7] The net so defined differs from the RBF net in two ways: (1) the use of normalization and (2) the addition of a linear term, $(\vec{x} - \vec{x}_j) \cdot \vec{d}_j$. The use of a normalization term was suggested but not pursued by Moody and Darken.[14] The addition of these two terms is responsible for reducing the amount of training data needed. As in the case with radial basis functions, the training of f_j and $\vec{d}_j$ is linear and hence very fast.

We will not attempt a rigorous justification of our learning algorithm here. Rather, we will try to motivate the algorithm heuristically and by analogy. Learning algorithms can be either on-line or off-line. Off-line algorithms attempt to calculate weights without any reference to time ordering of the training data. Thus, all the training data must be collected before training can start. On-line algorithms, on the other hand, attempt to modify weights as information in the form of training data flows in. On-line algorithms are able to handle varying amounts of training data and are able to modify the system in the presence of drift in the conditions. This is very difficult with an off-line algorithm. Additionally, on-line algorithms are able to handle amounts of input data that would severely tax memory storage capacities if an off-line method were used. Most neural nets are trained, therefore, with on-line methods.

On-line training methods themselves come in two extremes. The method can remember all the data that it has been shown up to the present or it can only be aware of the training set it is being shown at the present. Conjugate gradient learning and multidimensional Newton's method fall into the former category; we will use a method that falls into the latter. Since we will use less information, we will pay a price in accuracy, but this is compensated by speed and simplicity.

MOTIVATION OF THE LEARNING ALGORITHMS

To train the network, we must invert Eq. (3) to solve for the weights f_j and $\vec{d}_j$. Since the equation is linear in the weights, the solution is basically a matrix inversion. However, the solution technique must account for (1) the dual set of weights, (2) the fact that the number of training data sets typically exceeds the number of weights, and (3) the desirability of an on-line solution. Therefore, let us first consider the following simpler model problem: the on-line inversion of

$$y_i(\vec{x}_p) = \Sigma_{j=1}^N W_{ij} u_j(\vec{x}_p) \tag{4}$$

to find the matrix, W_{ij}, given the set of training vectors, y and $\vec{x}$. Here, u is similar to the ρ vectors in Eq. (3) in that it is a localized vector function of the training input, $\vec{x}_p$. We require the u vectors to be normalized:

$$\Sigma_{j=1}^N u_j = 1. \tag{5}$$

The W_{ij} might correspond to either the f_j or $\vec{d}_j$ quantities in Eq. (3). The vector, $y(\vec{x}_p)$, corresponds to the target value, $f(\vec{x}_p)$. We would like Eq. (4) to be true for any training pair, $[\vec{x}_p, y(\vec{x}_p)]$. This is, in general, not possible when the number of training points exceeds N, the number of dimensions of u. We will show, however, a very good approximation is possible. Equation 4 can be rewritten

$$\Sigma_j W_{ij} u_j(\vec{x}_p) = \Sigma_j W_{ij}^0 u_j(\vec{x}_p) + [y_i(\vec{x}_p) - \Sigma_j W_{ij}^0 u_j(\vec{x}_p)]\frac{\Sigma_k u_k^2(\vec{x}_p)}{\Sigma_k u_k^2(\vec{x}_p)}, \tag{6}$$

where W_{ij}^0 is some arbitrary guess for W_{ij}. If we do not consider information from previous training points, then the best information available is that the change in W lies in the direction of the current vector $u(\vec{x}_p)$ where $\vec{x}_p$ is the current training point. We have, then,

$$W_{ij} = W_{ij}^0 + [y_i(\vec{x}_p) - \Sigma_l W_{il}^0 u_l(\vec{x}_p)]\frac{u_j(\vec{x}_p)}{\Sigma_k u_k^2(\vec{x}_p)}. \tag{7}$$

If one replaces $u(\vec{x}_p)$ with $\vec{x}_p$ in Eq. (7), then the $\alpha - LMS$ algorithm of Widrow[15] is recovered.

All the training algorithms we use are based on Eq. (7), which is equivalent to a one-dimensional Newton's method for finding roots. To see this, define

$$g(W) = \Sigma_j W_{ij} u_j - y. \tag{8}$$

Then Eq. (7) can be written

$$W_{ij} = W_{ij}^0 - g(W^0)\frac{\nabla g(W^0)}{\nabla g(W^0) \cdot \nabla g(W^0)} \tag{9}$$

where the gradient operation is with respect to W^0. This is Newton's method for finding the root of g along a line parallel to the gradient of g.[16] Equation (7) is also equivalent to gradient descent learning[16] with a variable learning rate, $1/\Sigma_k u_k^2$.

The solution obtained with Eq. (7) can be compared with the solution obtained from a least mean squares minimization. We define the cost function

$$I_i = (1/2)\langle[y_i - \Sigma_{j=1}^N W_{ij} u_j]^2\rangle, \tag{10}$$

where the averaging is defined

$$\langle h \rangle = (1/M)\Sigma_{p=1}^M h(\vec{x}_p). \tag{11}$$

Here, M is the number of times a training point is shown to the net. M is greater than or equal to the number of training points. Minimizing Eq. (10) with respect to W_{ij} yields

$$\langle y_i u_j \rangle = \Sigma_k W_{ik}^* \langle u_k u_j \rangle. \tag{12}$$

Here, W^* is the value of W which minimizes Eq. (10). In analogy with Eq. (7), Eq. (12) can be written

$$W_{ij}^* = W_{ik}^0 + \Sigma_k [\langle y_i u_k \rangle - W_{ij}^0 \langle u_j u_k \rangle] \langle u_k u_j \rangle^{-1} \tag{13}$$

where once again W_{ij}^0 is some arbitrary guess for W_{ij}. Since Eq. (7) is shown all the training points, the solution of Eq. (7) can be obtained by averaging:

$$W_{ij} = W_{ij}^0 + \left[\left\langle \frac{y_i u_j}{\Sigma_k u_k^2} \right\rangle - \Sigma_l W_{il}^0 \left\langle \frac{u_l u_j}{\Sigma_k u_k^2} \right\rangle\right]. \tag{14}$$

Comparison of Eqs. (13) and (14) yields the requirements for the convergence of Eq. (7) to the least mean squares solution,

$$\Sigma_k \left\langle \frac{y_i u_k}{\Sigma_k u_k^2} \right\rangle \langle u_k u_j \rangle = \langle y_i u_j \rangle \tag{15}$$

and

$$\left\langle \frac{u_i u_j}{\Sigma_k u_k^2} \right\rangle = \delta_{ij}. \tag{16}$$

A sufficient condition for these requirements to be satisfied is that the components of u should be very localized in $\vec{x}_p$. The normalization of u is very important for this condition to work.

In practice, Eq. (7) is iterated as data is presented to the net. This can be represented as

$$W_{ij}^{p+1} = W_{ij}^p + [y_i(\vec{x}_p) - \Sigma_l W_{il}^p u_l(\vec{x}_p)]\frac{u_j(\vec{x}_p)}{\Sigma_k u_k^2(\vec{x}_p)}. \tag{17}$$

If a very good approximation to Eq. (4) exists for all $\vec{x}_p$ and if the vectors, $u(\vec{x}_p)$ span the space, then it can be shown that Eq. (7) converges to the solution of Eq. (15). To see this, subtract Eq. (17) from W_{ij} and use Eq. (4) to obtain

$$W_{ij}^{p+1} - W_{ij} = \Sigma_l \left[\delta_{jl} - \frac{u_j u_l}{\Sigma_k u_k^2}\right][W_{il}^p - W_{il}]. \tag{18}$$

We see that when a good solution of Eq. (4) exists, Eq. (17) reduces to a projection operator. Thus, if the u vectors span the space, W_{ij}^p converges to W_{ij}.

THE "OLD" LEARNING ALGORITHM

Most of the calculations reported here were performed with the learning algorithm discussed in this section, and referred to as the "old" learning algorithm. In the course of this work, the old algorithm was found to be unstable under some conditions, and therefore we have made a first attempt at an improved learning algorithm, described briefly in the next section. Although the old algorithm may be superseded at some point, it is serviceable (and in some respects superior to the "new" algorithm); thus, we include the following description to document the approach used in most of the Mackey-Glass predictions presented here.

The network of interest, Eq. (3), differs from the model net, Eq. (4), in two important ways: (1) there are two sets of linear weights to train instead of one, and (2) there are two sets of qualitatively different basis vectors instead of one. We can reduce the approximation, Eq. (3), to two problems of the form of Eq. (4). If we assume that we know the optimum gradient, $\vec{d}$, in Eq. (3) then the quantities f_j play the role of W_{ij} and the quantities $\rho_j/\Sigma_k\rho_k$ play the role of u_j. In analogy with Eq. (1)7, we then have

$$f_j^{p+1} = f_j^p + [g(\vec{x}_p) - \Sigma_l(f_l^p + (\vec{x}_p - \vec{x}_l) \cdot \vec{d}_l)\rho_l(\vec{x}_p)]\frac{\rho_j(\vec{x}_p)\Sigma_k\rho_k(\vec{x}_p)}{\Sigma_k\rho_k^2(\vec{x}_p)} \qquad (19)$$

where $\vec{d}_l$ is the optimum value. If we assume that we know f_l but not $\vec{d}_l$, then once again the problem is of the form of Eq. (4) and we have, corresponding with Eq. (17),

$$\vec{d}_j^{p+1} = \vec{d}_j^p + [g(\vec{x}_p) - \Sigma_l(f_l + (\vec{x}_p - \vec{x}_l) \cdot \vec{d}_l^p)\rho_l(\vec{x}_p)]\frac{(\vec{x}_p - \vec{x}_j)\rho_j(\vec{x}_p)\Sigma_k\rho_k(\vec{x}_p)}{\Sigma_k(\vec{x} - \vec{x}_k)^2\rho_k^2(\vec{x}_p)}. \qquad (20)$$

Equations (19) and (20) can be iterated for f_j^{p+1} and $\vec{d}_j^{p+1}$ if we have some approximation for $\vec{d}_l$ in Eq. (19) and f_l in Eq. (20). If we approximate these quantities by $\vec{d}_l^p$ and f_l^p, respectively, then we have an explicit iteration scheme:

$$f_j^{p+1} = f_j^p + [g(\vec{x}_p) - \phi(\vec{x}_p)]\frac{\rho_j(\vec{x}_p)\Sigma_k\rho_k(\vec{x}_p)}{\Sigma_k\rho_k^2(\vec{x}_p)} \qquad (21)$$

and

$$\vec{d}_j^{p+1} = \vec{d}_j^p + [g(\vec{x}_p) - \phi(\vec{x}_p)]\frac{(\vec{x}_p - \vec{x}_j)\rho_j(\vec{x}_p)\Sigma_k\rho_k(\vec{x}_p)}{\Sigma_k(\vec{x} - \vec{x}_k)^2\rho_k^2(\vec{x}_p)}. \qquad (22)$$

Unfortunately, explicit iteration schemes tend to be unstable. Tests indicate that this scheme, in particular, is unstable.

This problem can be corrected by implicitly approximating $\vec{d}_l$ and f_l with $\vec{d}_l^{p+1}$ and f_l^{p+1}. Doing this and solving for the advanced quantities yields

$$f_j^{p+1} = f_j^p + \alpha[g(\vec{x}_p) - \phi(\vec{x}_p)]\frac{\rho_j(\vec{x}_p)\Sigma_k\rho_k(\vec{x}_p)}{\Sigma_k\rho_k^2(\vec{x}_p)} \qquad (23)$$

and

$$\vec{d}_j^{p+1} = \vec{d}_j^p + \alpha[g(\vec{x}_p) - \phi(\vec{x}_p)]\frac{(\vec{x}_p - \vec{x}_j)\rho_j(\vec{x}_p)\Sigma_k\rho_k(\vec{x}_p)}{\Sigma_k(\vec{x} - \vec{x}_k)^2\rho_k^2(\vec{x}_p)}. \tag{24}$$

The difference between the explicit and implicit schemes is that the error term in the implicit scheme is reduced by the factor α. We generally use a *Learning Rate*, $\alpha \leq 1/3$ and find that there is usually a value of α that provides both adequate stability and rapid learning.

The architecture, Eq. (3), and the learning algorithm, Eqs. (23) and (24), form the backbone of the network. Most of the work reported here is based on these three equations or slight variations of them.

The widths β_j of the basis functions in CNLS-net can also, optionally, be trained. However, in the present work, we have used fixed basis function widths, and therefore do not discuss this aspect further. In the remainder of this report, we assume that all basis function widths are identical, and take $\beta_j = 4 \times BetaMultiplier$ for all j.

THE "NEW" LEARNING ALGORITHM

As will be seen in the discussion of the results below, the "old" learning algorithm can be unstable under some circumstances, depending upon exact details of the training set. In this section, we briefly present a "new" algorithm that shows enhanced stability properties.

The training for the linear quantities, f_j and $\vec{d}_j$ is given by

$$f_j^{p+1} = f_j^p + \alpha[g(\vec{x}_p) - \phi(\vec{x}_p)]\frac{\rho_j(\vec{x}_p)\sum_k\rho_k(\vec{x}_p)}{\sum_k[\rho_k^2(\vec{x}_p) + \beta_k(\vec{x} - \vec{x}_k)^2\rho_k^2(\vec{x}_p)]} \tag{25}$$

and

$$\vec{d}_j^{p+1} = \vec{d}_j^p + \alpha[g(\vec{x}_p) - \phi(\vec{x}_p)]\frac{\beta_j(\vec{x}_p - \vec{d}_j)\rho_j(\vec{x}_p)\sum_k\rho_k(\vec{x}_p)}{\sum_k[\rho_k^2(\vec{x}_p) + \beta_k(\vec{x} - \vec{x}_k)^2\rho_k^2(\vec{x}_p)]}. \tag{26}$$

These expressions replace those of Eqs. (23) and (24) to form the "new" learning algorithm. The main difference between the two algorithms is in the denominator of the training function, where the new algorithm has additional terms to reduce the liklihood of a very small value causing the learning to diverge. A few initial tests of this algorithm are discussed in the results section below.

THE MACKEY-GLASS EQUATION

The Mackey-Glass (M-G) equation was first advanced as a model of white blood cell production.[1] It is a time-delay differential equation, namely

$$\frac{dx}{dt} = \frac{ax(t-\tau)}{[1+x^c(t-\tau)]} - bx(t), \tag{27}$$

where the constants are often (and in this work) taken to be $a = 0.2, b = 0.1$, and $c = 10$. The delay parameter τ, determines the nature of the chaotic behavior displayed by the time series.

The behavior of the M-G equation as a function of τ has been studied extensively and is reported by J. D. Farmer in Ref. 2. For $\tau < 4.53$, there is a stable fixed point attractor. For $4.53 < \tau < 13.3$, the equation shows a stable limit-cycle attractor. At $\tau = 13.3$, the period of the limit cycle doubles, and this period doubling continues until $\tau = 16.8$. For $\tau > 16.8$, the equation has a chaotic (strange) attractor with characteristics dependent upon τ. At $\tau = 30$, the value used for all of the studies reported here, the M-G equation's attractor has the following properties:

Non-negative Lyapunov exponents:	3
Fractal dimension:	> 2.94
Information dimension:	3.6
Correlation dimension:	3
Predicted embedding dimension:	$3.6 \leq d_e \leq 8.2$
Empirical embedding dimension:	6

Farmer[2] calculated the Lyapunov exponents and fractal dimension numerically. He then estimated the information dimension from the Kaplan-Yorke conjecture,[4,10] which was in agreement with his numerical simulations for lower-dimensional examples. Lapedes and Farber[11] predicted the embedding dimension from the Takens embedding theorem[18] and found the embedding dimension of 6 to be adequate. Casdagli[1] determined an embedding dimension empirically from radial basis function fits to the time series, and calculated the correlation dimension from the Grassberger-Procaccia algorithm.[5]

Thus, the attractor of the M-G equation with $\tau = 30$ is a good example of low-dimensional chaos. Figure 1 shows a plot of the M-G equation with $\tau = 30$, slightly renormalized to limit its range approximately to the interval $[0, 1]$. The standard deviation of the function so normalized is 0.24. In the results presented below, we use as a performance indicator, the "Error Index" or "Error," defined as the root-mean-squared training or prediction error (RMSE) divided by the standard deviation. With this definition of the Error Index, a constant fit through the mean value of the function leads to a value of 1.0.

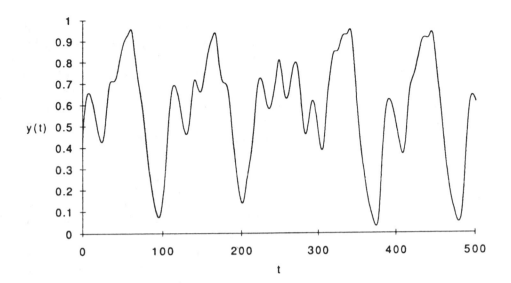

FIGURE 1 A portion of the Mackey-Glass time series, for $\tau = 30$. The plot shows 500 points with a spacing of $\Delta t_{data} = 1$.

INITIAL CHOICES FOR EMBEDDING, ARCHITECTURE, AND OPTIONS

We chose initial running conditions in order to compare CNLS-net's performance with that of the back-propagation net used by Lapedes and Farber.[11] This section discusses our initial choices for the embedding parameters, network architecture, and network operating options.

We initially chose the embedding used in previous work[3,11,14]: the training and test patterns were composed of six inputs, spaced at time intervals of 6 time units each, plus a test output, the point 6 time units after the last entry of the input sequence. Later, we found that embedding is a very sensitive matter, a subject we return to in a subsequent section.

The training and test files generally consisted of 1000–5000 points at fixed time spacing ($\Delta t = 1$, except where specifically noted below). Generally, the M-G equation was initialized as a constant, and 1000 initial points were discarded to allow transients to die out. The training and test files were non-overlapping time sequences, with the test file usually continuing the series begun in the training file. Except where noted, we used 500 training patterns, and we always used 500 test patterns, as did Lapedes and Farber.[11] Generally, the 500 patterns were selected at random from the training file and sequentially from the test file. Also, the selected training patterns were held fixed for the entire training period, but were usually "tumbled," i.e., presented in random, varying sequences for successive training epochs.

The CNLS-net architecture chosen initially used 6 input nodes, 28 hidden nodes (having 7 adjustable weights each), and one output node. This yields about 200 weights, fewer than Lapedes' reported back-propagation net calculation: he used two hidden layers of 14 nodes each, giving about 540 weights, total.[4] Our initial architecture yielded a network that could be trained in about 2 − 6 minutes and tested in about 1 minute (at 21 iterations into the future), which made multiparameter optimization feasible.

Some of the CNLS-net parameters control optional procedures and techniques, and we summarize the usual values used in this work in Table 1, together with brief descriptions of their functions. The value of *Seed* is used by **CNLSTOOL** to initialize the random number generator. A pseudo-random series of numbers is generated to choose arbitrary values in the interval $[0, 1]$ for coordinates of the basis function centers. This aspect of the CNLS-net initialization also proved to be somewhat sensitive and can lead to some difficulties, as discussed below.

TABLE 1 Key parameters of the CNLS-Net as used in this work.[1]

Parameter	Typ.Value	Function
Hidden	28	No. of nodes in hidden layer
FeedInputs	6	No. of inputs per training/test "pair"
PointDistance	6	Spacing of sequential data points from data files
RandomPatterns	On	Use random selection of training patterns
Tumble	On	Present training patterns in new order each epoch
MaxTrainPatterns	500	No. of training patterns
MaxTestPatterns	500	No. of test patterns
InitializationMode	0	Choose random basis function centers
Seed	484	Random number seed
RhoNormalization	On	Use normalized basis function
KalmanFiltering	Off	Optional noise filter
TrainD	On	Train linear gradient term
LearningRate	0.1	Learning rate coefficient
BetaMultiplier	1.6	Gaussian width for basis function

[1] Tabulated values correspond with the model referred to as "Q" in the text. Other models discussed in the text are described by noting their differences from this reference case.

OPTIMIZING CNLS-NET'S CONTINUOUS PARAMETERS

This section discusses the initial optimization and sensitivity tests of the CNLS-net continuous parameters, for fixed embedding and network architecture. When operated in the basic mode used for most of this work, CNLS-net has only two adjustable parameters: *LearningRate* and *BetaMultiplier*. We discuss these next.

The *LearningRate* parameter sets the rate of adjustment of the weights during training. Training occurs in a number of epochs, each epoch consisting of a complete presentation of the selected training patterns. Usually, for fixed *LearningRate*, the fit error decreases rapidly in early epochs, then gradually transitions into a phase of jittering or very slowly improving error. Finally, if the net is over-trained, the fit error increases again. This behavior is shown for net configuration with 15 hidden nodes in Figure 2. Note that the minimum prediction error does not occur where the minimum training error is found.

The value chosen for *LearningRate* determines (together with the training set size and values of some of the other network parameters) the slope of the initial

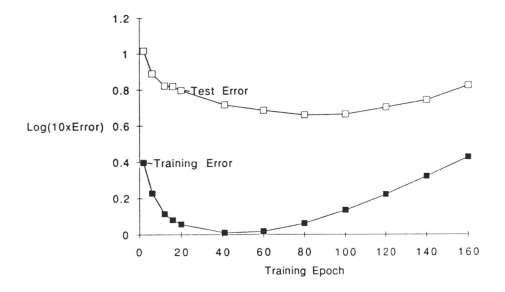

FIGURE 2 Training and test (prediction) errors as a function of the number of training epochs. This network had 15 hidden nodes, and predictions were made to a constant $\Delta t_{pred} = 126$. Training error reaches a definite minimum, and prediction error exhibits a minimum, albeit at a later stage in training.

learning curve, and the extent of the optimum region. Generally, lowering the learning rate makes the initial learning slower, and lengthens the interval when near-optimum training exists. Also, reducing the learning rate can sometimes reduce the fit error in the "fully trained" state of the net, though this is usually a small effect. If the learning rate is too large, the training can rapidly become unstable, possibly even failing to reduce the fit error in early epochs. For the running modes used here, learning rates in the range of 0.01–0.3 were adequate, with 0.03–0.1 the most frequently used values. We return to the subject of learning behavior in a later section.

BetaMultiplier sets the width of the gaussian basis functions. Each basis function ρ is a gaussian with the exponent $[-4 \times BetaMultiplier \times |\vec{x}_p - \vec{x}_j|^2]$. Thus, *BetaMultiplier* sets a characteristic feature size for all the basis functions. The value of *BetaMultiplier* exhibits (for this particular test case) a broad overall optimum, combined with a shallow, sharply localized dip containing the best value, as shown in Figure 3. The best value for *BetaMultiplier* is 1.6 for the initial choices of embedding and net options, with 28 hidden nodes.

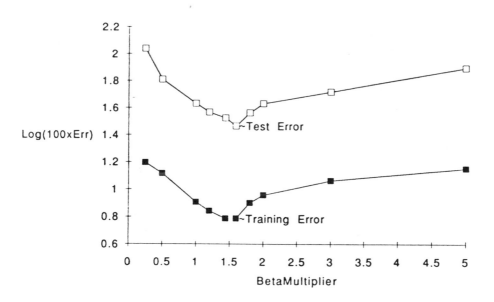

FIGURE 3   Optimization of *BetaMultiplier*, the basis function width, for *Hidden=* 28.

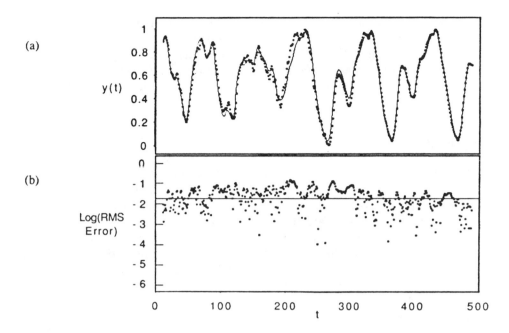

FIGURE 4   (a) Comparison of prediction (square points) with Mackey-Glass function (solid line) for 500 test predictions at $\Delta t_{pred} = 126$ using the parameters of model "Q," shown in Table 1, except that this network had $Hidden = 112$. (b) Log plot of the absolute RMS prediction error (not normalized) for the predictions of (a).

Predictions made using a 112-node network, with $BetaMultiplier = 1.6$ are shown in Figure 4, together with a plot of the RMS prediction error (in this case, not normalized by the standard deviation of the time series). The 500 consecutive predictions shown are at a time $\Delta t_{pred} = 126$ beyond the last data points used as input to the net.

OPTIMIZATION AND BEHAVIOR OF DISCRETE CNLS-NET PARAMETERS

Adaptive networks such as CNLS-net have a number of discrete parameters that must be chosen to suit the problem at hand. We mentioned above the initial choices made for network size, embedding structure, and various other discrete parameters. When we began to explore the optimization of these discrete choices, we encountered Pandora's Box. A single phrase can summarize our findings: the network's learning and predictive performance is not very robust. In this respect, CNLS-net shares some, though not all, of the foibles of back-propagation nets.[19]

In the following subsections, we present the results of these parameter explorations, together with the (limited) understanding we have gained of their behavior. We discuss in turn the effects of different embedding structures, varying network size and initialization, training set size, and other details of the training procedures and learning behavior. At the end of this section, we present two specific tests of versatility and robustness.

EMBEDDING

We first discuss the behavior of a network with 28 hidden nodes under variations of the discrete embedding parameters. The parameter *FeedInputs* controls the number of inputs to the network. For example, a value of *FeedInputs*= 3 would cause the training and test patterns to be fed into the network in groups (vectors) of 3 points. The parameter *PointDistance* controls the spacing of successive data points chosen for input to the net. For example, a value of *PointDistance*= 4 would cause successive input data points extracted from the data files to be separated by 4 time steps. As our initial embedding choice, we used *FeedInputs*= 6 and *PointDistance*= 6, the embedding used in previous work.[11,13]

Because CNLS-net, as used here, is set up to predict a point at a time into the future equal to the time difference between successive input points, *PointDistance* also changes the prediction time interval per predictive iteration. Thus, the input time interval being sampled by the network for a given training or test set is

$$\Delta t_{samp} = (PointDistance \times FeedInputs) \times \Delta t_{data},$$

where Δt_{dat} is the time interval per point in the data files ($\Delta t_{dat}= 1$ for this work). The time of prediction into the future (from the last input point) is set by the iteration control parameter, *FutureTimeStep*, so that

$$\Delta t_{pred} = [PointDistance \times (FutureTimeStep + 1)] \times \Delta t_{data}.$$

In this embedding study, we have varied *PointDistance* and *FeedInputs* over an interesting sample of parameter space. As we varied *PointDistance*, we also changed *FutureTimeStep* to maintain a fixed prediction time of $\Delta t_{pred} \simeq 126$.

The two-dimensional results of the embedding study are shown in Table 2(a). There is a lot of "structure." Note that at many places in the table, adjacent entries differ by a factor of $2\times$ or more. Table 2(b) shows the same parameter space, but we have arbitrarily defined an *ErrorIndex* ≤ 0.5 as a successful fit, and eliminated all unsuccessful entries.

TABLE 2 Results of the embedding study.[1]

(a)

Feed-Inputs:	PointDistance									
	2	3	4	5	6	7	8	10	13	20
2										1.09
3								1.09		
4			1.12		0.98	1.48	0.72	0.69		0.91
5					0.79	0.59	0.63			
6		2.69	1.5	0.59	0.42	0.34	0.45	0.67	0.81	0.94
7		1.09	0.86	0.38	0.45	0.5	0.44	0.75	1.18	
8			3.47	0.42	0.57	0.63	0.62	0.56	0.89	1.01
10		0.5			0.7	0.47	0.86			
12		0.43			0.43	0.76	0.79			
13					0.43	0.5	0.77			
15	0.85									
20					0.67	0.79	1.02			

(b)

Feed-Inputs:	PointDistance									
	2	3	4	5	6	7	8	10	13	20
2										
3										
4										
5										
6					0.42	0.34	0.45			
7				0.38	0.45	0.5	0.44			
8				0.42						
10		0.5				0.47				
12		0.43			0.43					
13					0.43	0.5				
15										
20										

[1] The error index for predictions at $\Delta t_{pred} = 126$ is tabulated as a function of the two embedding parameters, *PointDistance* and *FeedInputs*. Part (a) shows all calculations performed in this study. Note the large variations between neighboring entries in several cases. Part (b) includes only "successful" calculations, i.e., those with error index of 0.5 or less. Most of the successful calculations have $\Delta t_{samp} = 30\text{--}50$.

Figures 5 and 6 show one-dimensional cuts across this parameter space, to help visualize the variations. Clearly, the initial embedding, ($FeedInputs = 6$, $PointDistance = 6$) is near optimum. However, there appear to be some (slightly) better choices, e.g., $(6, 7)$, and there are many embeddings that work almost as well, at least for this case. The variation with $PointDistance$ is fairly benign, but the dependence upon $FeedInputs$ is complicated.

The details of which embeddings work and which fail are not understood. We suspect that, as discussed below in connection with the size of the hidden layer, there are random effects associated with initialization of the basis function centers. We did take a case that failed, $(4, 8)$, and, with a few trials, obtained a run that worked well. In this instance, Kalman Filtering made the formerly unstable training become stable, and the prediction from the trained net was good. We could not seem to do as well as this using reduced learning rate. Note, however, that the use of Kalman Filtering is far from a panacea, and in fact often degrades the performance of the net. The fact that a "failure" can be changed into a "success" with very modest changes in procedure suggests that the learning algorithm is near marginal stability.

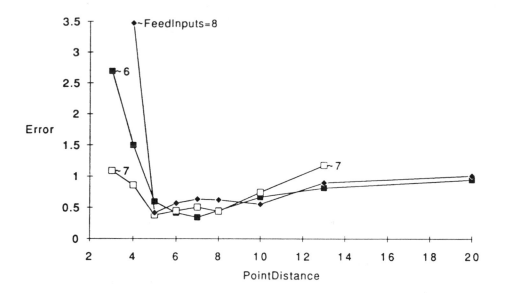

FIGURE 5 Embedding-I: prediction error index (normalized by the standard deviation of the time series) as a function of $PointDistance$, the separation between successive data points. Three one-dimensional cuts across the embedding data of Table 2, labelled by their respective values of $FeedInputs$, the number of successive data points used as input to the network.

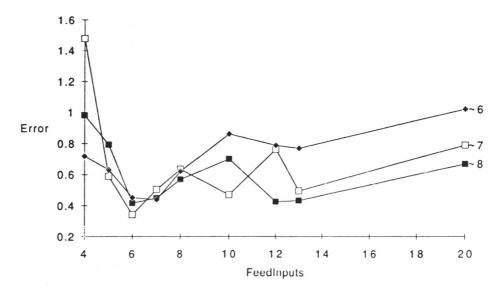

FIGURE 6 Embedding-II: prediction error index (test error) as a function of *FeedInputs*, for three values of *PointDistance*.

We can summarize these results by noting that most of the "successful" embeddings have Δt_{samp} in the range of 30–50. We suspect this is related to the choice of $\tau = 30$ as the Mackey-Glass delay parameter. The parameter τ sets a "coherence timescale," and embeddings for which Δt_{samp} differs greatly from this time interval are either supplying the network with too little or too much information.

The problem of supplying the proper amount of data to the network can be handled by pre-filtering the data using the well-known Sampling Theorem. Starting from an input file that has ample resolution, one can perform a Fast Fourier Transform (FFT). Examining the power spectrum, a reasonable frequency cut-off can be chosen. The truncated power spectrum is reconverted to the input function space using the inverse FFT with the desired cut-off. This reduces the embedding ambiguity by eliminating unnecessary information, while having only slight impact on the function to be fit. For example, in the Mackey-Glass case with $\tau = 30$, a clean cut-off can be made that preserves the function values to within 1% or so. The resulting filtered function is easier to fit and predict. We do not have wide enough experience with this technique to know how much and how generally it can help.

HIDDEN LAYER SIZE AND INITIALIZATION OF BASIS FUNCTION CENTERS

Next we look at the effects of variations in the size of the hidden layer, controlled by the parameter *Hidden*. For this study, we held the embedding and training procedures fixed. However, as a result of the method used here for basis function center initialization, changing the hidden layer size unavoidably moves the basis function centers in an unpredictable fashion. The embedding was that of our initial choice, as described above. For this series, the network was trained until the training error either became nearly constant or increased for two consecutive epochs. This resulted in large variation in the number of training epochs, as the apparent learning rate and stability of the network training algorithm varied.

Two different parameter sets were used to obtain the results shown in Figure 7. The parameter set "F" was taken from early runs, and had $BetaMultiplier=1.0$. The parameter set "Q" was later in the optimization series, and used $BetaMulti$-$plier = 1.6$. For both parameter sets, we used $LearningRate= 0.1$. We find that, for both sets of parameters, the variation of prediction error with the size of the hidden layer fluctuates widely. Also, note that the curves of the two parameter sets cross, with set Q yielding better predictions when $Hidden \geq 20$ and set F performing better when $Hidden < 20$. This is understandable, qualitatively, since the broader set of basis functions should perform better in the case of smaller networks.

Since the basis function center locations are initialized at random for each run, it appeared possible that the fluctuations in performance as a function of hidden layer size were due to the statistics of initialization of the net. If this hypothesis were correct, one would expect different random seeds to yield differing net performance,

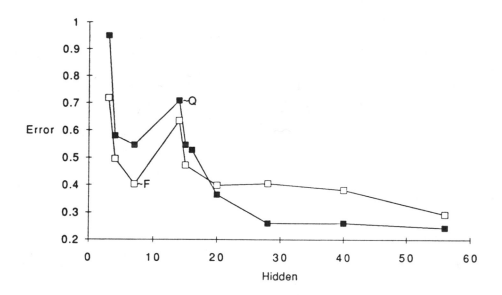

FIGURE 7 Test error as a function of $Hidden$, the number of nodes in the hidden layer. Model "Q" is described in Table 1. Model "F" is identical, except that $BetaMultiplier = 1.0$. Some very small networks are surprisingly successful, but variation with network size fluctuates.

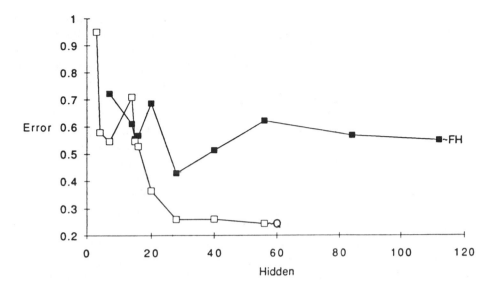

FIGURE 8 Test error as a function of $Hidden$, comparing two identical sets of network architectures, but with different random number seeds. Evidently, the random number sequence used to initialize basis function centers and choose training set data has lasting impact on the trainability of the net. The fact that the model "FH" does not approach model Q's performance, even in the limit of fairly large networks indicates that the effect is not entirely due to statistical changes in the basis function centers.

and that the statistical variations in network performance would be largest for small numbers of basis functions.

We tested this hypothesis by changing the random number seed. Coincidentally, we also used a new set of training and test data files, prepared using a slightly different technique, and containing 5000 data points each. The results, shown in Figure 8, suggest that the statistical effects associated with choosing basis function centers, although probably present, are accompanied by other effects, at least as important. Perhaps the selection of 500 training vectors, also done randomly, leads to varying overall training set quality. This behavior remains to be more fully understood.

TRAINING SET SIZE

Some applications generate data prolifically, while for other applications, data acquisition is severely limited by cost, time, or complexity. Thus it is important to know, and in many cases, to minimize the training data set requirements of a network.

We prepared a training set for $\tau = 30$ that contained 5000 points, and varied the number of data points used in training (by setting $MaxTrainPatt$). The results of predictions at the constant time of $\Delta t_{pred}= 126$ are shown for three different network sizes in Figure 9. Increasing the size of the training set generally decreases the prediction error. However, a given network reaches a minimum error at a certain training set size, and further increases do not consistently improve predictions. The larger networks are capable of more accurate predictions than smaller ones, given a sufficiently large training set.

Of course, the computer time used to train a net increases with the network size and also with the training set size, if the number of training epochs is fixed. The training time per epoch scales slightly faster than linearly with the size of the hidden layer (about the 1.5 power for these calculations), and linearly with training set size. Additionally, to realize the accuracy improvement shown with increasing training set size, it is necessary to increase the number of training epochs. For this study, we scaled the number of training epochs approximately linearly with the network size, and generally held it fixed as the training set size varied. Thus the cost in training time increases rapidly as improved prediction accuracy is sought. The computer time required for a prediction (test) scales linearly with network size and with number of iterations into the future. Test cpu time is independent of

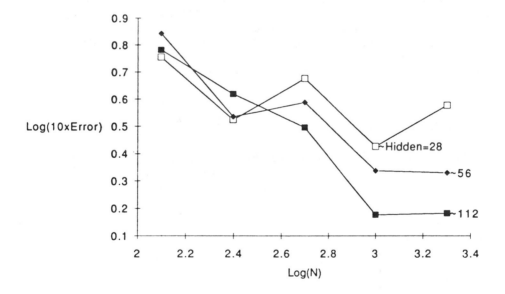

FIGURE 9 Variation of prediction error as a function of N ($MaxTrainPatt$), the number of training sets for three different network sizes. Increasing N generally leads to better prediction performance, up to a point. Larger networks achieve better accuracy.

the training set size, and thus increases much more slowly as increased accuracy is sought.

OTHER TRAINING PROCEDURES AND LEARNING BEHAVIOR

We have already discussed some basic aspects of the training procedures used for this work in previous sections. Here we collect a few additional results and describe the network's learning behavior in more detail.

We showed in Figure 2 that the minimum predictive error does not necessarily occur when the training error is a minimum. The plot of Figure 10 reinforces that point. Here, calculations made as a part of the embedding study show a broad, fluctuating variation in training error as a function of *PointDistance*. However, looking at the prediction error at Δt_{pred}= 126 shows a single, broad minimum. The apparent improvement in training around *PointDistance*= 18 is not reflected in improved predictions. The training that occurs here is "memorization" rather than "learning." Since training error is not a completely reliable indicator of predictive error, we adopted the procedure during all optimizations of training until

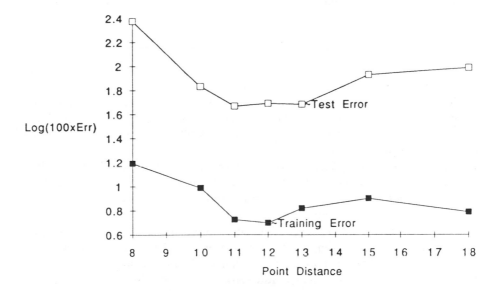

FIGURE 10   Training and test error as a function of *PointDistance*. Often, the training error can be reduced without a corresponding improvement in test error, indicating that the features being learned are not always helpful for prediction.

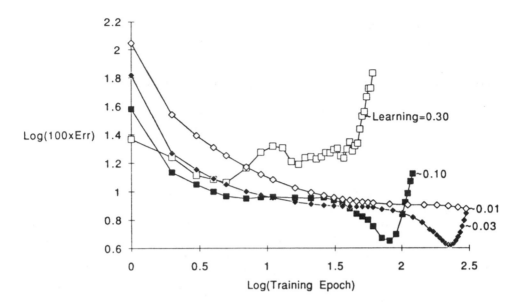

FIGURE 11   Training error curves for four values of *LearningRate*, the learning rate coefficient.

a reasonable approach to a minimum or plateau, then testing the network's prediction at a standard Δt_{pred}. Even this procedure is subject to error, since it is possible to train past a good predictive case. Recently, Weigend et al. reported a verification procedure similar to this that has been automated and tested.[20] Also, they reported a weight-elimination technique that offers another way to deal with network overfitting.

In general, optimization of the net is complicated by changes in effective learning rate. Changes in the number of training patterns per epoch, changes in embedding, and changes in other parameters and options of the net can lead to apparent shifts in learning rate. Although the *LearningRate* parameter is not too sensitive in determining the net's behavior, other effects seem to produce large changes in the effective learning rate, requiring iterative readjustment of *LearningRate*.

The behavior of the network under changes in *LearningRate* can be fairly complicated. Figure 11 shows the training error *vs.* time for a set of calculations with varying *LearningRate*. *LearningRate* = 0.3 is useable, but really too large, leading to early instability of the training. *LearningRate* = 0.01 to 0.03 is satisfactory, but training is somewhat slower. *LearningRate* = 0.1 shows good behavior, plus a slight improvement in trained fit just before the training becomes obviously unstable. These results led us to devise and test the "new" learning algorithm, and we revisit these calculations below in discussing the new algorithm's performance.

Finally, we note that changes in training set size and even in training file preparation can affect the network's performance, and not necessarily in a simple way. Figure 12 shows the prediction curves for runs with training set sizes ($MaxTrainPatt$) of $250, 500$, and 1000. Each calculation was trained for 180 epochs, so each increase in $MaxTrainPatt$ leads to an overall increase in training time. It is possible that the calculations are in different portions of their learning curve. The general trend is toward better training results with larger training sets. However, the two calculations at $MaxTrainPatt= 500$ show markedly different results, even though they differ only in details of preparation of the training and test sets. The three calculations labelled EL, EW, and EK were performed using training and test files prepared by discarding 1000 points, using the next 5000 points in the training file, and using the next 5000 points in the test file. The run labelled EJ is exactly like EW except that its training and test files were prepared by discarding 1000, then using the next 1000 for the training file, and the next 1000 for the test file. This may be another instance of chance, where, in this case, different sets of training and test points yield different trained weights and test results.

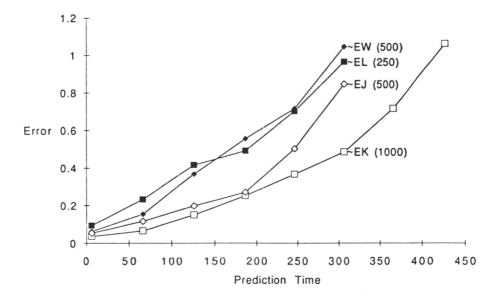

FIGURE 12 Prediction error curves for three different training set sizes, N. Two curves for $N = 500$ differ only in minor details of the preparation of the training and test data sets, indicating that details of the input data can significantly change the success of predictions.

VERSATILITY AND ROBUSTNESS

General concepts of versatility and robustness for numerical algorithms exist. For the purposes of this section, we qualitatively define "versatility" as the extent of the domain over which the network achieves "near-optimum" accuracy. We define "robustness" as the level of performance fluctuations under small variations of input data or code parameters, i.e., data- or parameter-sensitive fluctuations in prediction accuracy. In principle, one could make these definitions quantitative. However, as will be seen shortly, the complexity of code behavior and the fluctuations produced by a lack of robustness can make quantitative determination of versatility difficult. Perhaps in future work, we shall attempt to define and evaluate these qualities quantitatively, but for now we must be satisfied with qualitative accounts.

We devised two additional, specific ways of testing the versatility and robustness of the network: (1) changing the sampling time interval in the training and test files and (2) changing the time delay parameter in the Mackey-Glass (M-G) equation. These tests can be made more or less sensitive by adjusting the continuous parameter excursions to match the versatility of the network being tested.

The first test is closely related to changing the embedding: we varied the time spacing of points in the training and test files. This gives us a continuous, gentle handle on the matching of effective feature size or "wavelength" of the function to

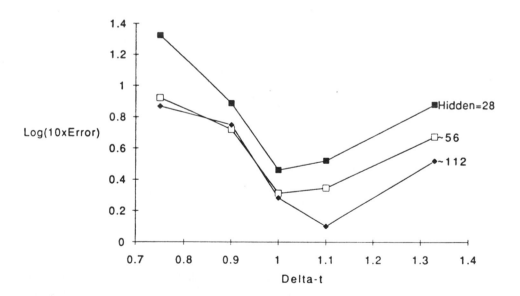

FIGURE 13 Prediction error as a function of Δt_{data}, the time-spacing of data points in the training and test files, for three network sizes. Width of the test-error minimum indicates versatility of the network, which gradually increases as network size increases.

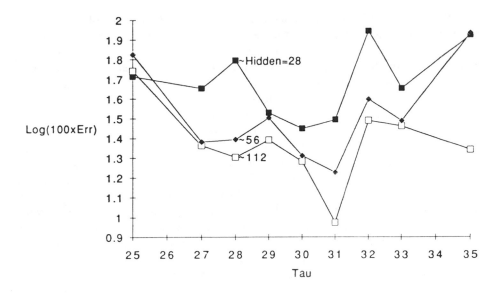

FIGURE 14 Second test of versatility, varying the delay parameter τ of the Mackey-Glass equation. Again, larger networks show somewhat greater versatility.

be fit and the network basis functions. Figure 13 shows the results for three different network sizes. For a fixed set of network parameters, but retraining the net for each Δt_{dat}, we find that variations of order 10% in point spacing significantly affect the ability of the network to obtain predictive fits. Increasing the network size gradually improves the net's versatility.

The second test, changing the M-G time delay parameter τ, is slightly more complex, in the sense that changing τ could change the properties of the M-G equation's attractor. For moderate τ-variations, between $\tau = 25$ and 35, we do not expect these changes to be too important. The results of this test are shown in Figure 14. The results show strong fluctuations in performance, particularly for the smallest network.

Taking the results of these two tests together with the results of simulations reported in other sections, we infer that the larger network shows smoother, more reliable behavior (greater robustness) and maintains this performance over a relatively wider range of input data and network parameters (greater versatility).

BEST RESULTS

In this section, we collect our best results, and compare them with Lapedes' and Farber's successful prediction. Figure 15 shows three sets of predictions, made with three different hidden layer sizes. Each curve represents the best fit obtained at its respective network size. Table 3 summarizes the calculations' network size and performance parameters. All three cases used the initial embedding, the same network parameters (parameter set "Q" with $BetaMultiplier= 1.6$). It is possible that still better results could be obtained by reoptimizing the parameters and embedding for the larger networks, but this was not done.

The Mackey-Glass, $\tau = 30$ calculation of Lapedes and Farber[11] was trained for about 60 minutes on a *Cray XMP* with vectorized coding. Our calculations were performed using *CNLSTOOL*, written in the C language, and executed on a *Sun SPARC-1* workstation. We estimate, without detailed, specific code measurements, that the speed advantage of the *Cray XMP* over the workstation is about 5–10 : 1 for scalar coding, and perhaps 50–100 : 1 if the *Cray XMP* routines were highly vectorized. Lapedes and Farber state that they used a highly vectorized network code, so a speed conversion factor of 40× is probably about right. Thus, our longest run, with predictive accuracy exceeding that of Lapedes' back propagation network, and with its training time of two hours, represents about a factor of 20× improvement in computing resource requirement. Our faster runs, which of course are considerably less accurate in longer-time predictions, showed an additional factor of 20–60× speedup, thus requiring fairly modest computing resources.

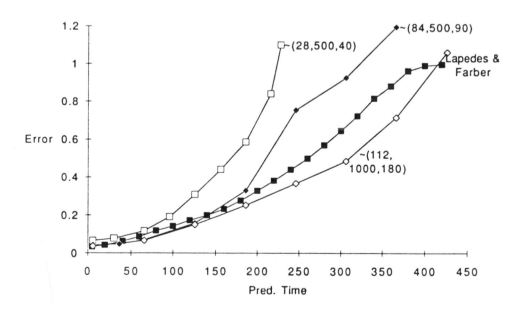

FIGURE 15 Prediction error curves for our best results. Each CNLS-net curve is labelled by the values of $Hidden$, $MaxTrainPatt$, and the number of training epochs used. For comparison, we show the results obtained in by Lapedes and Farber.[11] We obtain greater prediction accuracy with about 20× less training computer time.

TABLE 3 Parameters and performance of CNLS-net for three of our best results.[1]

Run	Hidden	MaxTrPatt	TrEpochs	TrTime	TstErr (126)	TstErr (246)	TstTime
Q	28	500	41	6.4	0.308	1.1+	1.1
EH	84	500	90	24	0.157	0.753	2.1
EK	112	1000	180	120	0.15	0.366	2.7

[1] Prediction accuracy improves as the network is made larger and more training data sets are provided. Predictions obtained using these network and training configurations are shown in Figure 15.

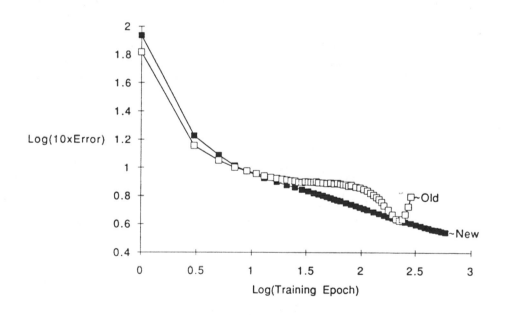

FIGURE 16 Training curves for "old" and "new" learning algorithms, for $LearningRate = 0.03$. In this case, new algorithm is more stable.

INITIAL TESTS OF "NEW" LEARNING ALGORITHM

In response to the tendency of the learning algorithm to be unstable, a "new" algorithm, detailed above, was devised. Here, we compare the new algorithm with the old for a few initial test calculations.

The new learning algorithm has been found to be stable under some conditions that caused the old to fail. Figure 16 shows the training curves for both the old and new algorithms, for the same case shown in Figure 11, with *Learning Rate*= 0.03. The new algorithm leads to steady, stable improvement in the training fit in this case. However, this improvement is not universal, as shown by additional calculations using higher learning values. For *Learning Rate*= 0.3, the new algorithm was immediately unstable. For *Learning Rate*= 0.1, the algorithm appears marginally stable, with one dangerous near-explosion at the second epoch, followed by rapid recovery, as shown in Figure 17.

As an additional test of the new learning algorithm, we compared the new and old algorithms' prediction capabilities for two examples. One such comparison is shown in Figure 18 for the case of a 28-node network with *Learning Rate*= 0.1 and other parameters from set "Q." Although the training error is steadily and stably reduced, the network never reaches a prediction error as low as that obtained using the old algorithm. This behavior is also seen in our best case with a 112-node network: here, too, the new algorithm's prediction is about 25% less accurate than that of the old.

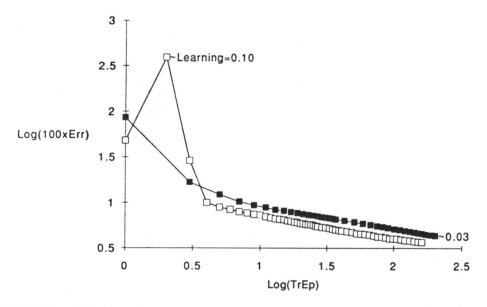

FIGURE 17 New learning algorithm shows near-disaster at the second epoch with *Learning Rate*= 0.1, and is unstable for *Learning Rate*= 0.3, not shown.

A fourth test case, using a 15-node network, could not be made to work satisfactorily with the new learning algorithm. The learning rate that succeeded for the old algorithm was unstable. Further, the learning rate had to be reduced to the point that the network trained very slowly for the new algorithm to be stable, and at the low learning rate, we could not obtain adequate predictions, even with long training times.

After some consideration of the differences between the old and new algorithms, we have made a small adjustment to the new algorithm that improves its performance significantly. We observed that the new algorithm increased the learning rate of the linear term weights (d_j) relative to the weights for the normalized gaussian terms (f_j). Thus, we added an additional learning coefficient that reduced the linear term's learning rate by a relative factor of 10×. We refer to this slightly adjusted new algorithm as the new′ version.

It is instructive to compare results of our third learning algorithm with those of the previous two. Figure 19 should be compared with Figures 17 and 11. The stability of the new′ algorithm has been greatly improved by the reduction of the

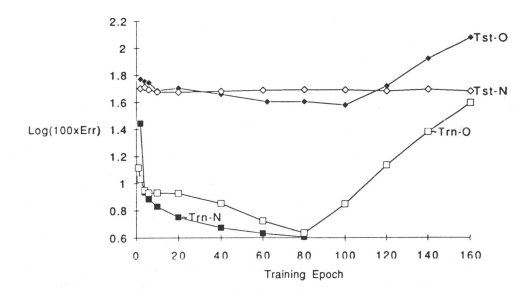

FIGURE 18 Comparison of training and test errors for old and new learning algorithms using the parameters of model Q. New algorithm is not able to find weights that predict as well as the best set obtainable with the old algorithm, but does show better long-training-time stability.

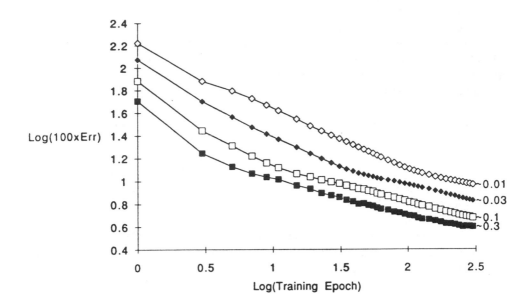

FIGURE 19 Reducing the relative learning rate of the linear gradient term by a factor of ten leads to much-improved stability. Compare these training curves for the new' algorithm with those of Figures's 17 and 11.

linear term's learning rate. Figure 20 should be compared with Figure 18, particularly with respect to prediction accuracy. After 80 training epochs, the new algorithm's prediction error is 1.22× that of the old, while the new' algorithm's prediction is slightly better at 1.15× the old. For the 112 node network, and with the same number of training epochs (160), the prediction accuracy of the new' algorithm yields a factor of 1.6× higher error at $\Delta t_{pred}= 126$. Somewhat surprisingly, this prediction error ratio does not diverge with further iteration, remaining at about 1.4–1.6 over the entire prediction time range of $\Delta t_{pred}= 0 - 400$.

Tests to date suggest that the old algorithm, though sometimes unstable, is still apparently somewhat more successful than either of the new learning algorithm variations. We need to answer the question of whether the optimum network parameters derived using the old learning algorithm are suboptimal for the new algorithm. More work is needed to understand and improve the stability and performance of the CNLS-net learning algorithm. It would be interesting, for example, to see what effect the greater stability of the new' algorithm has on some of the wildly varying results, such as the embedding fluctuations (Figures 5-8), and on the robustness and versatility tests (Figures 13-14). Perhaps most importantly, we need a better way to deal with the differences between trainability and prediction capability (memorization vs. learning), and to find an algorithm with the stability of the new' and the prediction capabilities of the old.

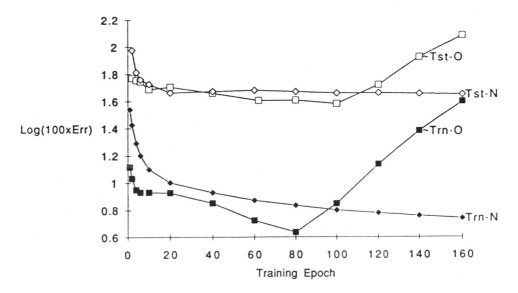

FIGURE 20 Prediction errors of the new' learning algorithm are lower than the new, but still significantly worse than the best results obtainable with the old algorithm.

CONCLUSIONS

The Mackey-Glass time-delay differential equation is a good example of low-dimensional chaotic behavior. With delay parameter $\tau = 30$, the equation provides an irregularly oscillating function having a strange attractor with dimensionality about 3 to 4. The Mackey-Glass equation serves well as a testbed for adaptive computing algorithms that aspire to learning, predicting, and controlling the dynamics of chaotic systems.

CNLS-net has proven able to accurately predict the behavior of the Mackey-Glass equation. We have obtained predictions that match the accuracy of previous work,[3,11] while requiring about 20× lower computational effort.[11] Data requirements are comparable with those of a back-propagation network[11] and much less than those of unnormalized radial basis function nets.[14] Overall, CNLS-net's use of normalized, localized basis functions with linearized correction terms appears to be a successful approach.

Qualitatively, in this low-dimensional space, CNLS-net behaves in ways that are roughly intermediate between typical back propagation (BP) networks and radial basis function (RBF) nets. CNLS-net learns much faster than a BP net, but becomes more readily confused in cases of high dimensionality or with excess data to analyze.

CNLS-net requires less training data and can operate in moderate-dimensional cases with substantially fewer nodes than RBF nets. Accuracy generally improves with larger networks, larger training sets, and greater training times. Versatility and robustness of the net improve somewhat with network size.

CNLS-net also shares some of the foibles of other adaptive networks, and these difficulties indicate some areas where future improvements are possible. The learning behavior is occasionally erratic, and sometimes shows signs of instability. Embedding is a delicate matter, far more prone to performance degradations than is indicated in most of the literature we have encountered, and certainly worthy of additional study. Similar performance fluctuations are associated with initialization of the basis function centers, the size of the network, and other details of training/test file preparation. The success of a given training session depends upon a fairly careful tuning of the network parameters to the problem at hand. Initial trials of a new learning algorithm show promise, but have not yet provided a fully acceptable solution to the learning behavior difficulties. We, and other connectionists, need to continue to work toward greater generality, versatility, and robustness in the future.

ACKNOWLEDGMENTS

The authors appreciate the support of the Center for Nonlinear Studies, the Inertial Fusion and Plasma Theory Group (X-1), the Applied Theoretical Physics Division (X), and Los Alamos National Laboratory. This work was supported in part by the U.S.D.O.E.

REFERENCES

1. Casdagli, M. "Nonlinear Prediction of Chaotic Time Series." *Physica D* **35** (1989): 335.
2. Farmer, J. D. "Chaotic Attractors of an Infinite-Dimensional Dynamical System." *Physica* **4D** (1982): 366.
3. Farmer, J. D., and J. J. Sidorowich. "Predicting Chaotic Time Series." *Phys. Rev. Lett.* **59** (1987): 845.
4. Frederickson, P., J. L. Kaplan, E. D. Yorke, and J. A. Yorke. "The Lyapunov Dimension of Strange Attractors." *J. Diff. Equ.* **49** (1983): 185.
5. Grassberger, P., and I. Procaccia. "Dimensions and Entropies of Strange Attractors from a Fluctuating Dynamics Approach." *Phys. Rev. Lett.* **50** (1983): 346.
6. Howell, J. A., C. W. Barnes, S. K. Brown, G. W. Flake, R. D. Jones, Y. C. Lee, S. Qian, and R. M. Wright. "Control of a Negative-Ion Accelerator Source

Using Neural Networks." In *Proceedings of the International Conference on Accelerator and Large Experimental Physics Control Systems*, to be published, Vancouver, B.C., Canada, October 30-November 3, 1989 and Los Alamos National Laboratory Rpt. No. LA-UR-89-3597, 1989.

7. Jones, R. D., C. W. Barnes, Y. C. Lee, and W. C. Mead. "Information Theoretic Derivation of Network Architecture and Learning Algorithms." Los Alamos National Laboratory Rpt. No. LA-UR-91-325, 1991; and to be published in *IJCNN-'91*.

8. Jones, R. D., Y. C. Lee, C. W. Barnes, G. W. Flake, K. Lee, P. S. Lewis, and S. Qian. "Function Approximation and Time Series Prediction with Neural Networks." Los Alamos National Laboratory Rpt. No. LA-UR-90-21, 1990; and *Proc. of IJCNN, IEEE Cat. No. 90CH2879-5*, I-649. New York: IEEE, NY, 1990.

9. Jones, R. D., Y. C. Lee, S. Qian, C. W. Barnes, et al. "Nonlinear Adaptive Networks: A Little Theory, A Few Applications." Los Alamos National Laboratory Report LA-UR-91-273, 1990.

10. Kaplan, J., and J. Yorke. *Functional Differential Equations and Approximation of Fixed Points*, edited by H. O. Peitgen and H. O. Walther, 228. Berlin: Springer, 1979.

11. Lapedes, A., and R. Farber. "Nonlinear Signal Processing Using Neural Networks: Prediction and System Modeling." Los Alamos National Laboratory Report LA-UR-87-2662, 1987.

12. Lee, Y. C. "Neural Networks With Memory for Intelligent Computations." In *Proceedings of the 13th Conference on the Numerical Simulation of Plasmas*, Santa Fe, New Mexico, September 17-20, 1989.

13. Mackey, M. C., and L. Glass. "Oscillation and Chaos in Physiological Control Systems." *Science* **197** (1977): 287.

14. Moody, J., and C. J. Darken. "Fast Learning in Networks of Locally-Tuned Processing Units." *Neural Comp.* **1** (1989): 281.

15. Nguyen, D., and B. Widrow. "The Truck Backer-Upper: An Example of Self-Learning in Neural Networks." In *Proc. of IJCNN*, Washington, June, 1989, pp. II-357-363.

16. Press, W. H., B. P. Flannery, S. A. Teukolsky, and W. T. Vetterling. *Numerical Recipes in C*. New York: Cambridge Univniversity Press, 1988.

17. Rumelhart, D. E., G. E. Hinton, and R. J. Williams. "Learning Internal Representations by Error Propagation." In *Parallel Distributed Processing*, Vol. 1, edited by D. E. Rumelhart and J. L. McClelland, 318-362. Cambridge: MIT, 1986.

18. Takens, F. "Detecting Strange Attractor in Turbulence." In *Lecture Notes in Mathematics*, edited by D. Rand and L. Young, 366. Berlin: Springer, 1981.

19. Wasserman, P. D. *Neural Computing: Theory and Practice*. New York: Van Nostrand Reinhold, 1989, 56-58.

20. Weigend, A. S., B. A. Huberman, and D. E. Rumelhart. "Predicting the Future: A Connectionist Approach." *Int. J. Neural Sys.* **1** (1990): 193.

K. Stokbro† and D. K. Umberger‡
†Niels Bohr Institute, Blegdamsvej 17, 2100, Copenhagen, Denmark and ‡Nordic Institute for Theoretical Physics, Blegdamsvej 17, 2100, Copenhagen, Denmark

Forecasting With Weighted Maps

We discuss a method for forecasting time series in which a global predictor function is expressed as a weighted average of local maps. These maps are associated with fixed positions in a reconstructed phase space and are distributed in the same manner as the data. Averaging the maps produces smooth and accurate predictors with moderate amounts of data, whose forecasts are stable under iteration. By using highly localized weighting functions, such as Gaussians, a fast computer implementation of the method is possible.

INTRODUCTION

Much of the recent interest in the problem of forecasting chaotic time series has been stimulated by the proposal of two successful, but very different, approaches to the problem. In the first approach, Farmer and Sidorowich[4] demonstrated the utility of local approximation schemes which are methods for approximating the evolution of a system's state from the evolutions of its neighboring states. For low to moderate embedding dimensions and simple local representations, such as linear or quadratic functions, this approach can result in fast prediction algorithms; one needs only to find a set of nearest neighbors of a given state and solve a small (in the sense of the size of the matrix that must be inverted) least-squares problem. The major weakness of the approach is the data requirement; the reconstructed attractor must be covered well enough so that in the vicinity of any state whose evolution is to be predicted there are enough nearby neighbors to construct an accurate local map for the evolution.

In the second approach, Lapedes and Farber[8] demonstrated the utility of a standard, feed-forward neural network for solving the problem. This method is global in the sense that a single forecasting function is built from the known evolutions of states distributed over the entire reconstructed attractor. Because global evolutionary information is used to construct the predictor, accurate predictions can be obtained with much less data than is required by local approximation schemes. However, the method used for fixing the free parameters of the network, that is back propagation, is inefficient and results in extremely slow computer implementations, even when the amount of training data is relatively small. When the amount of data needed to make accurate forecasts is large, the computing time required can be prohibitive.

Each of the above methods works well under circumstances where the other does not; local schemes are fast but need relatively large amounts of data, while the neural network is extremely slow, though requiring smaller amounts of data. Thus, there is a need for fast and accurate forecasting methods that require small to moderate amounts of data. In this respect, techniques based on radial basis functions, first discussed in the context of forecasting by Casdagli,[3] seem promising. For example, Moody and Darken[9] have proposed an algorithm, interpreted as a neural network, in which a global forecasting function expressed as a linear combination of localized radial basis functions is constructed. The algorithm is much faster than the neural network of Lapedes[8] but does not produce very accurate forecasts.

In this paper we discuss a method, motivated by the work of Moody and Darken[9] and detailed in Stokbro et al.,[12] in which a global forecasting function is expressed as a weighted superposition of local maps. The weighting functions used are strongly localized, and this localization can be exploited to speed up the training (i.e., construction) phase considerably. By using linear maps in the superposition, the algorithm produces forecasts that are orders of magnitude more accurate than those produced by the algorithm of Moody and Darken.[9] For moderate amounts of data, the algorithm also produces better forecasts than the local linear predictor of Farmer and Sidorowich.[4]

The method is described in the following section, after which the results of several numerical experiments are presented. These experiments are intended to illustrate how well the method works. We will conclude the paper with a short discussion of our results.

THE APPROACH

Consider the problem of forecasting a univariate time series $\{v_i \equiv v(i\delta t)\}$ obtained from a sequence of measurements of an observable v made at uniformly separated times $t_i = t_{i-1} + \delta t$. Suppose that a phase space has been reconstructed by using a delay embedding with delay time τ and embedding dimension d_E so that the state of the system at time t is given by the d_E-tuple $x_i = (v_i, v_{i-\tau}, \cdots, v_{i-(d_E-1)\tau})$ (the time is now measured in units of δt). Then, the forecasting problem we address is that of finding a real-valued function whose domain is the reconstructed phase space that will yield a good estimate of $y_i \equiv v_{i+T}$ from x_i, where T is the prediction time.

Now, suppose we choose to construct such a function using the following local approximation scheme. First, we choose M fixed positions x^α ($\alpha = 1, \cdots, M$) in the reconstructed phase space. We call each x^α a center, and we associate with it a real-valued map $m^\alpha(x)$. Each map will depend on some set of free parameters; for example, if the maps are chosen to be linear, they will depend on $d_E + 1$ free parameters. To fix the free parameters of $m^\alpha(x)$, we find some number of reconstructed states nearest to x^α and adjust the parameters to minimize the error function

$$E_\alpha^2 = \sum_p (y_{i_p} - m^\alpha(x_{i_p}))^2 \,, \tag{1}$$

where the sum is over the set of nearest neighbors. Then, to forecast the evolution of an arbitrary state x, we find the center closest to it and evaluate that center's map on the state.

Such a winner-take-all scheme is bound to encounter difficulties. Consider, for example, the case where a state x lies in a region mid-way between two centers x^α and x^β. Neither $m^\alpha(x)$ nor $m^\beta(x)$ may yield a good forecast. However, since the dynamics is assumed to be continuous in the phase space, an average of the two maps should yield a better result. Of course, this average should be a weighted average with the weights depending on the distances between x and the two centers. This suggests the following functional form for a predictor:

$$f(x) = \sum_{\alpha=1}^{M} m^\alpha(x) P^\alpha(x) \,, \tag{2}$$

where the $P^\alpha(x)$ are weighting functions that satisfy

$$\sum_{\alpha=1}^{M} P^\alpha(x) = 1 \,. \tag{3}$$

The approach just described suffers from the defect that the free parameters of Eq. (2) are fixed by independently minimizing the M functions of Eq. (1). What should be minimized is the global error function

$$E_f^2 = \sum_i (y_i - f(x_i))^2 , \qquad (4)$$

where the sum is over all members of the training set, i.e., all the reconstructed states used to learn the dynamics. However, the minimization of Eq. (1) is useful as a starting point when an iterative technique is used to minimize Eq. (4).

In this paper, we examine the performances of predictors of the form Eq. (2) where the free parameters of $f(x)$ are found by minimizing the global error function of Eq. (4). We restrict ourselves to cases where the $m^\alpha(x)$ are constant and linear maps. When the constant maps are used, we will refer to the resulting predictor as the Weighted Constant map Predictor (WCP). When linear maps are used, the resulting predictor will be called the Weighted Linear map Predictor (WLP). (The representation of Eq. (2) with linear maps has also been considered in Jones et al.[7])

The weighting functions $P^\alpha(x)$ have not yet been specified. Following Moody and Darken,[9] we choose

$$P^\alpha(x) = \frac{R\left(\frac{\|x - x^\alpha\|}{\sigma^\alpha}\right)}{\sum_{\beta=1}^M R\left(\frac{\|x - x^\beta\|}{\sigma^\beta}\right)} , \qquad (5)$$

where $R(\|x - x^\alpha\|/\sigma^\alpha)$ is a function that has a peak at $x = x^\alpha$ and falls off quickly as $\|x - x^\alpha\|/\sigma^\alpha$ gets larger than 1. We pick $R(x)$ to be a Gaussian $\exp(-x^2)$ because it has short tails, i.e., good localization, though it has the disadvantage that function evaluations are computationally expensive compared to more simple functions like Lorentzians. The effect on the quality of forecasts of various forms for $R(x)$ has been investigated in Gorodkin.[6]

We will construct the function f in two stages. After fixing the parameter M, we will fix the parameters x^α and σ^α according to some criteria discussed below. Next the parameters of the maps $m^\alpha(x)$ will be found by minimizing Eq. (4) using the algorithm described in Appendix A.

Selecting M is equivalent to fixing the number of free parameters. It is important to have fewer free parameters than data points; otherwise, there is a danger of fitting the noise in the system instead of smoothing it out. We will always choose M so that the ratio between free parameters and data points is 1:4. This is not guaranteed to be the optimal choice, but is done merely to have some standard by which different predictors can be compared. An optimal ratio will, among other things, be dependent on the noise level of the time series.

After fixing the number of centers, we must decide how the centers should be distributed. To get an idea of what the optimal distribution is, we try to minimize Eq. (4) with respect to x^α for the WCP using the Euclidean norm, $\|x\|^2 = x^2$. For simplicity, we set $\sigma^\beta = 1$ in the calculations, which does not affect the end result. First,

$$\frac{\partial E_f^2}{\partial x^\alpha} = -2 \sum_{i=1}^N (y_i - f(x_i)) \frac{\partial f(x_i)}{\partial x^\alpha}. \tag{6}$$

Now, a WCP has the form $f(x_i) = Z_i^{-1} \sum_{\beta=1}^M \lambda^\beta e^{-(x_i - x^\beta)^2}$, where $Z_i = \sum_{\beta=1}^M e^{-(x_i - x^\beta)^2}$ is the normalization and the λ^β are constants. Then

$$\frac{\partial f(x_i)}{\partial x^\alpha} = \frac{\partial}{\partial x^\alpha} Z_i^{-1} \sum_{\beta=1}^M \lambda^\beta e^{-(x_i - x^\beta)^2}$$

$$= Z_i^{-1} \lambda^\alpha e^{-(x_i - x^\alpha)^2} 2(x_i - x^\alpha) - Z_i^{-1} e^{-(x_i - x^\alpha)^2} 2(x_i - x^\alpha) f(x_i)$$

$$= 2\Delta_i^\alpha P^\alpha(x_i)(x_i - x^\alpha),$$

with $\Delta_i^\alpha = \lambda^\alpha - f(x_i)$ and $P^\alpha(x_i) = Z_i^{-1} e^{-(x_i - x^\alpha)^2}$. Putting this back into Eq. (6), one obtains

$$\frac{\partial E_f^2}{\partial x^\alpha} = -4 \sum_{i=1}^N \Delta_i \Delta_i^\alpha P^\alpha(x_i)(x_i - x^\alpha), \tag{7}$$

where $\Delta_i = y_i - f(x_i)$ is the error at each data point. For x^α to minimize Eq. (4), we must have $\partial E_f^2 / \partial x^\alpha = 0$. This implies that

$$x^\alpha = \frac{\sum_i W_i^\alpha x_i}{\sum_i W_i^\alpha}, \quad \alpha = 1, \ldots, M \tag{8}$$

with $W_i^\alpha = \Delta_i \Delta_i^\alpha P^\alpha(x_i)$.

Because P^α is localized, W_i^α will be negligible for data points far away from x^α. Let us, therefore, only look at the data points x_{i_p} nearest to x^α. If $\Delta_{i_p} > 0$, the weights in f are too small. The weight which influences $f(x_{i_p})$ the most is λ^α, because x^α is the nearest center to x_{i_p}. Therefore, one should expect Δ_{i_p} and $\Delta_{i_p}^\alpha$ to have the same sign; thus, their product will be positive. This implies that the W_i^α will be large in regions of the reconstructed phase space where there are many data points and high errors, so most of the centers should be placed in these regions. It is difficult to estimate the error distribution $\Delta_i \Delta_i^\alpha$, so, for simplicity, we will assume that the sizes of the errors are independent of the position on the attractor. In this case the centers should be distributed in the same way as are the data points. This can be accomplished, approximately, by splitting the data in a k-d tree structure which sorts the data into a binary tree in such a way that the tree leaves contain clusters of data points.[1,2] The splitting criteria we use sorts the data into clusters having minimal spreads (see Stokbro et al.[12] for details).

After the centers are calculated, the widths must be fixed. The widths are analogous to the band widths in Kernel Density Estimation.[11] Both models can be very sensitive to the actual choice of widths, and there are no simple rules for determining optimal values in either case.[1] However, we will choose the widths from some general forms similar to ones used in Kernel Density Estimation: We will call the widths local if they are chosen from the form

$$\sigma^\alpha(r,p) = \frac{r}{p} \sum_{\beta=1}^{p} \| x^\alpha - x^\beta \|, \tag{9}$$

where the x^β are the p nearest centers of x^α and r is a continuous variable that can be used to vary the widths smoothly. We will call the widths global when an average width

$$\sigma^\alpha = \sigma(r) = \frac{r}{M} \sum_{\alpha=1}^{M} \| x^\alpha - x^\beta \|, \tag{10}$$

is used for all the centers where x^β is the closest center to x^α. Because the centers are set up in a k-d tree, the search for nearest centers can be done very efficiently so that the calculation of the widths is not computationally very expensive.

We are currently investigating the effects of various choices of the widths on the performance of weighted map predictors. We have found that variations in r can affect the WCP considerably, while the WLP is less sensitive (see Stokbro et al.[13]). To have some standard for comparison in the numerical experiments that follow we will use a local width with $p = 1$ and $r = 1$, though this is not necessarily the optimal choice.

NUMERICAL EXPERIMENTS

We will now compare the weighted map approach with other prediction schemes. For this purpose we consider time series generated by the Mackey-Glass delay-differential equation:

$$\frac{dx}{dt}(t) = -.1x(t) + .2\frac{x(t - \Delta t)}{1 + x(t - \Delta t)^{10}} \tag{11}$$

where Δt is a parameter. We will investigate the cases of $\Delta t = 17$ (denoted by MG_{17}) and $\Delta t = 30$ (denoted by MG_{30}) where the system has attractors with fractal dimensions of 2.1 and 3.6, respectively. For each case we will generate a time series from which $N + N_{test}$ state vectors x_i can be extracted. We will also use delay embeddings with delay time $\tau = 6$ and embedding dimensions of $d_E = 4$ (MG_{17}) and $d_E = 6$ (MG_{30}). The first N states will be used to learn, i.e., construct various predictor functions. This set of states will be called the training set. The next

---

[1] A minimization of Eq. (4) with respect to σ^α does not lead to a simple expression like Eq. (8) for the centers.

TABLE 1 Estimated values of $\log_{10} \overline{E}$ are tabulated for eight different predictors, predicting time series, produced by the Mackey-Glass system, $T = 6$ into the future. Training sets of 500 data points sampled with $\delta t = 6$ were used to construct the predictors. The prediction errors for the first six rows were calculated by Casdagli,[3] and we have calculated the rest. The degrees used for the best polynomial and rational predictors (the nominator and denominator had the same degree) are shown in small figures next to the values of the predictor errors.

| | System | |
	MG_{17}	MG_{30}
Casdagli		
Poly.	-1.95^7	-1.40^4
Rational	-1.14^2	-1.33^2
Radial	-1.97	-1.60
N.Net	-2.00	-1.50
LQP	-1.89	-1.42
LLP	-1.48	-1.24
Stokbro/Umberger		
LLP	-1.75	-1.16
WCP	-1.22	-1.00
WLP	-1.89	-1.36

N_{test} states will be used to test the resulting predictors. This set of states will be called the test set. We will also use the normalized root mean square error, or the prediction error, to measure the prediction accuracies. This is defined as

$$\overline{E}^2 = \frac{\langle (y_i - f(x_i))^2 \rangle}{\langle (x_i - \langle x_i \rangle)^2 \rangle}, \tag{12}$$

with the average taken over the test set. To get good statistics we will always use $N_{test} = 1000$.

In Table 1 we list the prediction errors ($\log_{10} \overline{E}$) attained by the WCP, WLP, and some other commonly used predictors. The Poly., Rational, Radial, and N.Net. predictors are global approximation schemes that use polynomial, rational function, radial basis function, and neural network representations, respectively. The LQP (Local Quadratic map Predictor) and the LLP (Local Linear map Predictor) are the local approximation schemes of Farmer and Sidorowich.[4] The first six rows of Table 1 are taken from Table 1 of Casdagli,[3] while we have calculated the entries of the last three rows.

We have tried to make training sets and test sets similar to the ones used in Casdagli,[3] but there are bound to be some differences. These arise from the use of different integrators, initial points for integration, and numbers of iterations used to kill transients. We have repeated the calculations with the LLP in order to get some feeling for the differences between our data sets and those used in Casdagli.[3] As can be seen from the table, the listed values of the prediction errors are almost equal, so it makes sense to compare our numerical experiments with those of Casdagli.[3]

From the table it appears that the neural network and the radial basis function approaches do better than both weighted map schemes. However, the times required for training the former two predictors are much greater than those required for the weighted maps. Also, we do not expect radial basis functions to perform well on data with noise since the method uses the same number of free parameters as training vectors.

The WLP works much better than the WCP in the numerical calculations of Table 1. Further comparisons between the two methods are made in Figure 1, which

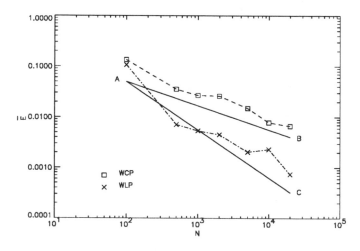

FIGURE 1 The prediction errors for the WCP and WLP when the size of the training set (N) is varied. The time series were produced by MG_{17}. The slopes AB and AC are the scaling law expected from Eq. (13).

shows some runs on MG_{17} using different values of N.[2] The WLP performs better than the WCP for all values of N investigated, and the difference increases with N. For both methods the same number of free parameters were used so that the WCP had more centers.

Farmer and Sidorowich[5] found a scaling law for the prediction error when local polynomials of degree k are used:

$$\overline{E} \sim N^{-(k+1)/d},\tag{13}$$

where d is the dimension of the attractor. In Figure 1 the scaling laws expected for local maps are drawn as the lines AB and AC for $k = 0$ and $k = 1$, respectively. Apparently, the WCP and WLP obey the same scaling laws as their local map counterparts.

How slow is the WLP compared to the WCP? Both algorithms require time to set up and train, and this time increases with the size of the training set. In Figure 2, we plot $\overline{E}$ versus the CPU time needed to set up and train the two predictors. In each case the training was done by going through the training set 200 times. This was enough to get within .001 of the minimum of E_f, defined in Eq. (4). The training could have been stopped several steps earlier without affecting the prediction errors noticeably, so the CPU times in the figure represent upper bounds.

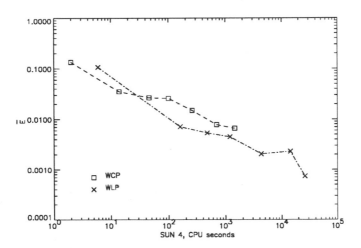

FIGURE 2 The prediction error as a function of the CPU time needed to set up and train the WCP and the WLP. The calculations are for the same cases as those of Figure 1.

[2] In this and the following numerical experiments, the sampling time is always fixed at $\delta t = 1.0$.

Figure 2 shows that when high-accuracy predictions are desired, using the smallest amount of computing time, one should use the WLP rather than the WCP. With this in mind, we now abandon the WCP in favor of the WLP for the rest of this paper.

Let us now compare the WLP with the LLP. In the numerical calculations of Table 1, the WLP worked slightly better than the LLP, when the time series were predicted one time step into the future. We will now try to get a more detailed picture by comparing the quality of the two approximations to the underlying dynamical system. The first test will be to see how well iterated forecasting works.

In Figure 3 we have used a time series produced by MG_{17}, and constructed maps which predict $T = 6$ into the future with $N = 100$ and $N = 500$. The figure shows the prediction error as a function of the number of iterations. When $N = 100$ the LLP is not capable of producing a valid map; the prediction error blows up almost from the beginning. The WLP does not have this problem; the prediction error grows smoothly as the map is iterated. When the size of the training set is increased to $N = 500$, the LLP produces a valid map. Now the errors are comparable to those of the WLP.

To study a higher-dimensional case, we now consider MG_{30}. Because the dimension of the attractor is larger than that of MG_{17}, we use 1000 and 5000 training points to set up predictors which predict $T = 6$ into the future. The results of iterating the predictors are shown in Figure 4. As can be seen from the figure, even 5000 training points are not enough for the LLP to make good predictions. The

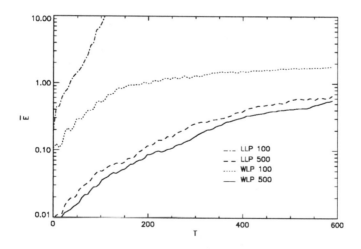

FIGURE 3 The prediction error as a function of the prediction time T for the WLP and the LLP using time series produced by MG_{17}. The predictions were made by iterating maps built with 100 and 500 training points that predict $T = 6$ into the future.

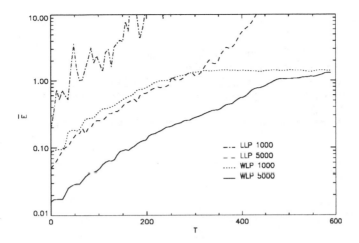

FIGURE 4 Similar to Figure 3, except that the time series were produced by MG_{30}. The number of training points was increased to 1000 and 5000, respectively.

prediction error increases dramatically if one tries to forecast beyond a certain time. For both the $N = 1000$ and $N = 5000$ cases, the WLP produces a map with prediction errors that are well behaved when the map is iterated. Furthermore, the error in the first iteration is one order of magnitude less than the one obtained with the LLP.

Recall that local predictions can be done on the fly. This means that to make a single prediction, the local map must be built and evaluated. Thus, making a single prediction with the LLP requires slightly more time than a global method which uses a function that has already been constructed. However, the only time needed to set up a local predictor is the time it takes to construct a k-d tree or some other appropriate data structure that aids in searches. If one is concerned with computation speed, the choice of method should depend on how many predictions will be made. To get an idea of how much computation time is required, we have listed in Table 2 the CPU seconds it took to perform the numerical calculations for Figures 3 and 4 on a SUN-4 computer.

In Appendix A it is argued that the training time t_t for the WLP should scale as

$$t_t \sim k_t (N \log N)^{\frac{3}{2}} . \tag{14}$$

TABLE 2 CPU time in SUN-4 seconds for making the numerical experiments in Figures 3 and 4, each involving 100,000 predictions. From these CPU times we have calculated the scaling constants k_t and k_p of Eqs. (14) and (15).

	System			
	MG_{17}	MG_{17}	MG_{30}	MG_{30}
LLP				
N	100	500	1000	5000
Train	0	0	0	3
Pred.	1700	1811	4166	4815
Total	1700	1811	4166	4815
WLP				
Train	6	160	429	9329
Pred.	394	766	1290	2781
Total	400	926	1719	12110
LLP				
k_t	0.0000	0.0000	0.0000	0.0000
k_p	0.0010	0.0008	0.0010	0.0010
WLP				
k_t	0.0006	0.0009	0.0007	0.0011
k_p	0.0002	0.0003	0.0003	0.0005

We expect that the computation time for making one prediction, denoted by t_p, for the LLP and the WLP to scale as

$$t_p \sim k_p d_E \log N , \qquad (15)$$

where the constant k_p are different for the two methods. The reason for this is that, for the LLP, $\log N$ is the number of steps needed to find the state vectors used in the construction of a local map when the training vectors are set up in a k-d tree. At each step in the search routine, distances between d_E-dimensional vectors are calculated, hence the factors $\log N$ and d_E in Eq. (15). For the weighted maps the scaling law comes from the fact that the number of maps contributing to a prediction, on average, is $\log N$ (see Appendix A). These have to be evaluated, and this involves calculating d_E-dimensional distances, hence the factors $\log N$ and d_E in Eq. (15) for the WLP.

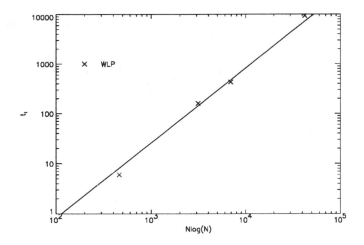

FIGURE 5 The prediction times for the WLP tabulated in Table 2. The solid line shows the scaling behavior expected from Eq. (14) using the average value of k_t in Table 2.

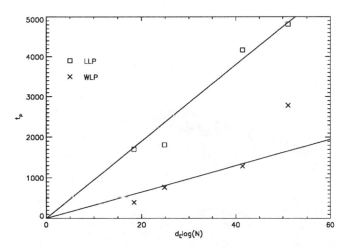

FIGURE 6 The prediction times for the LLP and WLP tabulated in Table 2. The solid lines show the scaling behavior expected from Eq. (15) using the average values of k_p in Table 2.

For each numerical experiment in Table 2, we have estimated the scaling constants k_t and k_p. The average values are $k_t = 0.000825$ and $k_p = 0.000325$ for the WLP, and $k_p = 0.00095$ for the LLP. In Figures 5 and 6 we have plotted the scaling laws of Eqs. (14) and (15), respectively, using the above average values of the scaling constants together with the CPU times of Table 2. Except for the scaling of the prediction time t_p for the WLP, the scaling laws of Eqs. (14) and (15) seem to hold.

The numerical results on iterated forecasting, discussed above, show that the WLP performs better than the LLP when data are limited. To understand why this is the case, we will look at how the two different representations approximate the dynamics in a region of the phase space instead of just evaluating the prediction errors on some points lying on the attractor. For this purpose we will use a time series from the Hńon map

$$x_{n+1} = 1 - ax_n^2 + y_n$$
$$y_{n+1} = bx_n$$

with $a = 1.4$ and $b = 0.3$. We have used a two-dimensional embedding of x to reconstruct a phase space where the dynamics is given by the equivalent second-order map

$$x_{n+1} = 1 - 1.4x_n^2 + 0.3x_{n-1} \,. \tag{16}$$

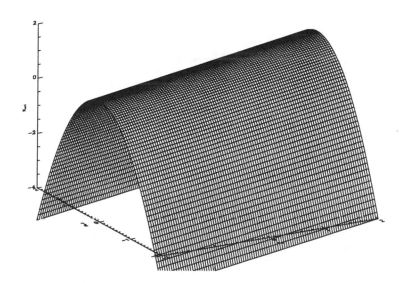

FIGURE 7   Surface plot of the function defined in Eq. (16).

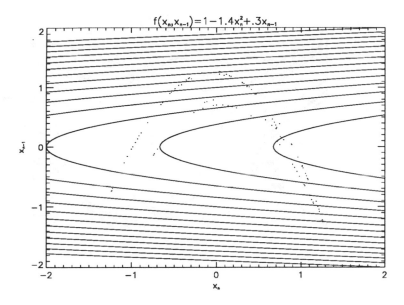

FIGURE 8 Contour plot of the surface in Figure 7. The dots show the positions of the training data.

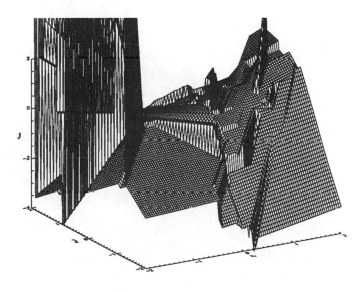

FIGURE 9 The approximation of the surface in Figure 7 obtained with an LLP.

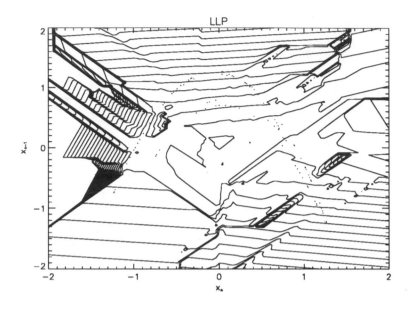

FIGURE 10 Contour plot of the surface in Figure 9.

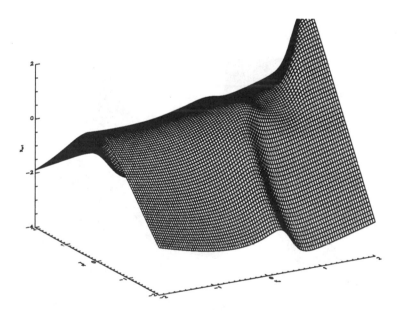

FIGURE 11 The approximation of the surface in Figure 7 obtained with a WLP.

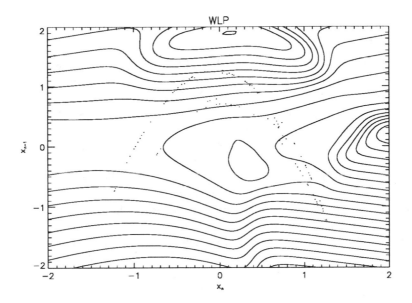

FIGURE 12 Contour plot of the surface in Figure 11.

Figure 7 shows a surface plot of the dynamics in the region $([-2,2],[-2,2])$ of the phase space, and Figure 8 shows the corresponding contour plot. A training set of 200 iterates, shown as dots in Figure 8, was used to construct an LLP and a WLP predictor. Figures 9 and 10 show the resulting approximation obtained with the LLP. The approximation is highly discontinuous and blows up in regions where there are no training data. Comparing Figure 10 with Figure 8, we see that the approximation is quite good in the vicinity of the data points but poor in regions short distances away from them. This should be compared with Figure 11 and Figure 12, which are the corresponding plots for the WLP. The approximation is continuous and quite similar to the real dynamics. If we compare contour plots, Figures 8 and 12, we see that the WLP approximation holds over a larger region than the LLP; while the two approximations might be comparable in accuracy near data points, the WLP obtains a much better overall approximation.

This influences the quality of iterated forecasts. In the LLP case, when an iterate falls in a region where there are few nearby data points, the next approximation can be very poor; this often makes the succeeding iterates go off to infinity. This is not a problem with the WLP since it produces a smooth function; iterates that fall off the attractor are brought back. This implies that the WLP should perform well on noisy data. We are currently investigating this possibility and preliminary results indicate that this is the case.[13]

CONCLUSIONS

In this paper, we have investigated a global forecasting method that utilizes a superposition of maps weighted by strongly localized weighting functions. We have used 0- and 1-degree polynomials for the local maps and found that the 1-degree polynomials are superior. Perhaps some other representation for the local maps, such as rational functions, would work even better.

The numerical experiments we presented indicate that weighted map schemes work better than their local map counterparts for moderate amounts of data. This is not surprising since the global schemes use all the data points for adjusting the free parameters, while local approaches only use a fraction of them. We believe that any representation that will work in a local scheme can be improved on by incorporating it into a weighted map scheme.

The problem with the global approach is computation time. If the representation is linear the training time will be $O(d_\lambda^3)$ where d_λ is the number of free parameters. However, by exploiting the localization properties of the basis functions, the training time for the weighted maps has been reduced to $O(d_\lambda^{3/2} \log^{3/2} d_\lambda)$. The computer resources limit how many free parameters a global model can contain. If the amount of training data is large, it might only be possible to use a crude global model because training of a refined model would not be possible. Taking

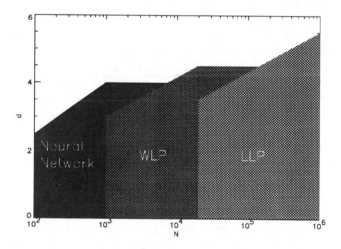

FIGURE 13 The figure shows which predictor we expect will give the best forecasts for a time series of length N produced by a chaotic dynamical system with an attractor of dimension d. The figure is based on experience from our numerical experiments and those published in Farmer and Sidorowich, and Lapedes and Farber.[4,5,8]

computation time into account we have made a guess, illustrated by Figure 13, about whether a neural network, a WLP, or an LLP will give the best predictions on a time series of length N produced by a chaotic system with an attractor of dimension d. Further comparisons have to be made to establish accurate guidelines on which representation to use. In 1991 there will be a prediction contest that will, hopefully, clear up a lot of these questions.

ACKNOWLEDGMENTS

We would like to thank M. C. Casdagli, J. D. Farmer, J. Hertz, A. Krogh, J. J. Sidorowich, and S. Solla for valuable discussions. One of the authors, K. Stokbro, would like to thank J. D. Farmer for making it possible to visit his group in Los Alamos National Laboratory. We would also like to thank the city of Prague for making this work possible.

APPENDIX

A. TRAINING LINEAR REPRESENTATIONS

A representation is called linear, when it can be rewritten as a linear combinations of some set of basis functions Φ^α

$$f(x) = \sum_{\alpha=1}^{d_\lambda} \lambda^\alpha \Phi^\alpha(x) \tag{17}$$

where d_λ is the number of free parameters λ^α in the approximation scheme. The weighted map representation can be cast into this form when the local map is a polynomial. We will now find the free parameters λ^α by minimizing the cost function (Eq. (4))

$$E_f^2 = \sum_{i=1}^{N}(y_i - \sum_{\alpha=1}^{d_\lambda} \lambda^\alpha \Phi^\alpha(x_i))^2 \, .$$

By multiplying it out we see that it is a quadratic form in the free parameters λ^α

$$E_f^2 = \left[\sum_i y_i^2\right] - 2\sum_\alpha \lambda^\alpha \left[\sum_i y_i \Phi^\alpha(x_i)\right] + \sum_{\alpha,\beta} \lambda^\alpha \lambda^\beta \left[\sum_i \Phi^\alpha(x_i)\Phi^\beta(x_i)\right] \, . \tag{18}$$

A quadratic form has only one minimum, so there exists only one set of parameters λ^α which minimizes Eq. (4). This assures us that the problem is well defined. There are several algorithms for finding the solution. Most of them find the solution by inverting the $N \times d_\lambda$-matrix $M_{i\alpha} = \Phi^\alpha(x_i)$, called the design matrix. When the data set is large, storage can be a problem; therefore, we will work directly with the quadratic form of Eq. (18) because, then, the required storage can be reduced from $O(Nd_\lambda)$, which is the size of the design matrix, to $O(d_\lambda^2)$, the size of the matrix part in Eq. (18).

The minimum of Eq. (18) can be found very effectively by using conjugate gradient descent. The starting point for the algorithm is some initial guess for λ^α, which is improved iteratively by minimizing Eq. (18) in the steepest direction conjugate to all the directions previously minimized. A direction is conjugate to the previous directions when it does not destroy the minimizations done in these other directions. This assures us that the minimum will be found in less than d_λ steps. We have implemented a slightly modified version of the conjugate gradient descent algorithm found in Press, Flannery, Teukolsky, and Vetterling.[10] In our revision the line minimization procedure is changed so that it makes use of the quadratic form of the error function. This makes the algorithm 5–10 times faster.

If the basis functions Φ^α are localized, many entries in the matrix part of Eq. (18) will essentially be zero, so it is not necessary to store and calculate these entries when conjugate gradient descent is applied. Therefore, localization properties of the basis functions can easily be exploited with this method and will highly reduce the training time. The gain will depend on how much the basis functions are localized. For the weighted maps we have found that in most cases the storage can be reduced from $O(d_\lambda^2)$ to $O(d_\lambda \log d_\lambda)$. We initialize the free parameters in the weighted maps by fitting the maps locally to the nearest data points, this together with localization reduces the number of iteration steps to $O(d_\lambda^{1/2} \log^{1/2} d_\lambda)$, so the overall training time is reduced from $O(d_\lambda^3)$ (for nonlocalized basis functions) to $O(d_\lambda^{3/2} \log^{3/2} d_\lambda)$. A numerical verification is shown in Table 2.

REFERENCES

1. Bentley, J. H. "Multidimensional Binary Search Trees in Database Applications." *IEEE Transactions on Software Engineering* **SE-5(4)** (1979): 333.
2. Bentley, J. H. "Multidimensional Binary Search Trees used for Associative Searching." *Communications of the ACM* **18** (1975): 509.
3. Casdagli, M. C. "Nonlinear Prediction of Chaotic Time Series." *Physica D* **35** (1989): 335.
4. Farmer, J. D., and J. J. Sidorowich. "Predicting Chaotic Time Series." *Phys. Rev. Lett.* **59** (1987): 845.
5. Farmer, J. D., and J. J. Sidorowich. "Exploiting Chaos to Predict the Future and Reduce Noise." In *Evolution, Learning and Cognition*, edited by Y.C. Lee. Singapore: World Scientific, 1988.
6. Gorodkin, J. "Predicting One-Dimensional Chaotic Time Series." Internal report, Niels Bohr Institute, 1991.
7. Jones, R. D., Y. C. Lee, C. W. Barnes, G. W. Flake, K. Lee, P. S. Lewis, and S. Qian. "Function Approximation and Time Series Prediction with Neural Networks." Technical Report LA-UR 90-21, Los Alamos National Laboratory, Los Alamos, New Mexico, 1990.
8. Lapedes, A., and R. Farber. "Nonlinear Signal Processing Using Neural Networks: Prediction and System Modeling." Technical Report LA-UR 87-2662, Los Alamos National Laboratory, Los Alamos, New Mexico, 1987.
9. Moody, J., and C. Darken. "Fast Learning in Networks of Locally-Tuned Processing Units." *Neural Comp.* **1** (1989): 281.
10. Press, W. H., B. P. Flannery, S. A. Teukolsky, and W. T. Vetterling. *Numerical Recipes.* Cambridge: Cambridge University Press, 1988.
11. Silverman, B. W. "Density Estimation for Statistics and Data Analysis." *Monographs on Statistics and Applied Probability* **26**. [city]: Chapman and Hall, 1986.
12. Stokbro, K., D. K. Umberger and J. A. Hertz. "Exploiting Neurons with Localized Receptive Fields to Learn Chaos." Nordita Preprint # 28, 1990. To appear in the *J. of Complex Systems* (1991).
13. Stokbro, K. "Predicting Chaos with Weighted Maps." Nordita preprint #10, Thesis for the Danish Master degree in Physics, The Niels Bohr Institute, 1991.

Grace Wahba
Department of Statistics, University of Wisconsin, Madison, WI 53706
e-mail: wahba@stat.wisc.edu

Multivariate Function and Operator Estimation, Based on Smoothing Splines and Reproducing Kernels

We review a general approach to multivariate function estimation based on optimal approximation and smoothing in reproducing kernel spaces. Learning with radial basis functions (including thin plate splines) is an important special case. In this context we describe a test for linearity of a functional iteration, and a general class of model-building methods based on sums and products of reproducing kernels. We describe the use of (radial) basis functions in the context of regularization, for use with very large data sets. We generalize some of the results from the estimation of real-valued functions to the estimation of vector-valued functions. Finally, we generalize from the estimation of vector-valued functions to the estimation of function-valued functions on arbitrary index sets, thereby proposing a theory of regularized estimates of nonlinear operators.

1. INTRODUCTION

We are interested in the problem of multivariate function estimation, particularly when the data are noisy, and (primarily) when the function to be estimated is believed to be "smooth" in some sense. Abstractly, consider some index set $\mathcal{T}$, which may be the integers $\mathcal{T} = \ldots - 1, 0, 1, \ldots$, the unit interval, $\mathcal{T} = [0, 1]$, Euclidean d space, $\mathcal{T} = E^d$, for example. We also consider more general $\mathcal{T}$. For the first part of this article we consider functions $f : \mathbf{t} \in \mathcal{T} \rightarrow f(\mathbf{t})$, where $f(\mathbf{t})$ is a real number; that is, f is a real-valued function on $\mathcal{T}$. Later we will generalize to $f(\mathbf{t}) = \mathbf{f}(\mathbf{t}) \in E^p$, for each $\mathbf{t}$; that is, $\mathbf{f}$ is a vector-valued function on $\mathcal{T}$, with p components. Finally, we will suppose that $\mathbf{f}$ is a Hilbert-space valued function of $\mathbf{t}$; that is, for each $\mathbf{t} \in \mathcal{T}$, $\mathbf{f}(\mathbf{t})$ is an element of a Hilbert space (which may depend on $\mathbf{t}$). By this generalization we initiate what we believe is a new approach to a theory of the estimation of nonlinear operators by regularization methods.

We begin with real-valued functions f. The mathematical problem is to estimate f, given pairs $(y(i), \mathbf{t}(i), i = 1, \ldots, n)$, where

$$y(i) = f(\mathbf{t}(i)) + \epsilon(i), \quad i = 1, \ldots, n, \tag{1}$$

and the $\epsilon(i)$ are independent disturbances with some properties. In the special case

$$\mathbf{t}(i) = (y(i-1), \ldots, y(i-d)), \tag{2}$$

we will have a nonlinear dynamical system. In that case, the class of methods for estimating f that will be under discussion here will include those proposed by Casdagli,[5] under the rubric "inverse methods." See also Nychka et al.[28] In this case (and in general) the set of $\mathbf{t}$'s available may have some special properties, and we have to be able to recognize that we can only estimate f when we are "sufficiently close" to the data points.

We can think of the estimation of f as a "learning" process, which is also the point of view of neural net methods. Quoting Cybenko,[10] "loosely speaking, an artificial neural network is formed from compositions and superpositions of a single, simple nonlinear activation or response function." To be specific, for the moment let $\mathbf{t} = (t^{(1)}, \ldots, t^{(d)})'$, that is, $\mathcal{T} = E^d$, and let $\mathbf{u}(j) = (u(j)^{(1)}, \ldots, u(j)^{(d)})' \in E^d$, and α_j and θ_j be fixed real numbers. Let $\sigma(\cdot)$ be a univariate sigmoidal function; that is $\sigma(\tau) \rightarrow 1$ as $\tau \rightarrow \infty$, and $\sigma(\tau) \rightarrow 0$ as $\tau \rightarrow -\infty$. Then a neural net approximation to f will be a function formed from compositions and superpositions of functions of the form

$$G(\mathbf{t}) = \sum_{j=1}^{N} \alpha_j \sigma(\mathbf{t}'\mathbf{u}(j) + \theta_j) \tag{3}$$

where the $N, \alpha_j, \mathbf{u}(j)$ and θ_j are fixed, but will be chosen from the data $\{y(i), \mathbf{t}(i)\}$. In most applications in the neural network literature, emphasis is placed on recursive updating as new data arrives, so that very large data sets can be handled. Cybenko[10] showed that, under mild conditions on σ, linear combinations of functions of the form Eq. (3) can uniformly approximate any continuous function with support in the unit hypercube of E^d. (That is, compositions, or multilayer networks, are not necessary for the approximation of continuous functions on E^d.) As Cybenko[9] notes, at some level of abstraction "multilayer feedforward continuous networks are biologically meaningful models of intelligent behavior and their study sheds light on the neurophysiological foundation of intelligence." The lay reader will find Hubel's[21] discussion of the visual system of the brain, as well as Penrose's[30] discussion of the brain as computer, fascinating in this connection. However, our purpose here is not to discuss neural nets, but to note that much work in neural networks has been motivated either by an attempt to use the rudimentary but unfolding knowledge of biological neural networks to obtain better computer learning algorithms, or, conversely, to use an increasing understanding of computer learning algorithms to attempt to understand the function of brain architecture.

The purpose of this paper is to describe just one of a group of approaches to the estimation of f, whose motivating sources are in approximation theory and statistics, and which have developed, until recently, more or less independently of the neural net literature or any reference to brain function. The point of view that we will be expounding goes back at least to Golomb and Weinberger[15]; see also Kimeldorf and Wahba.[23] The methods we will be talking about here are initially, at least, based on a global rather than a local approach, and for large data sets, they require relatively large computer resources that are only recently becoming available. Our discussion here will be considering methods that were unthinkable a few years ago, are doable at various levels on present equipment, and will probably be a snap on equipment available in a few years. See Corcoran.[7]

Statisticians have seriously addressed themselves only relatively recently to what is called in the statistics literature multivariate nonparametric function estimation, but the field is very active. We only mention representative papers from a few approaches: tree structured algorithms (CART, Brieman et al.[4]), projection pursuit (PP, Friedman et al.[13]), multivariate adaptive regression splines (MARS, Friedman[12]). (See also the survey Barron.[2]) These approaches as well as the one we will describe have certain common features—they postulate "noisy" data, and they all have some method for making the tradeoff between goodness of fit to the data and amount of structure in the solution. In particular, CART, MARS, and the smoothing splines and their generalizations (GSS) to be described here all use some form of cross validation to make this trade. It is also possible to use one of the various forms of cross validation in PP. The resulting algorithms are not generally intended as interpolation algorithms, that is, algorithms which produce functions that go through the data exactly. The GSS algorithms below, in fact, have very nice optimum theoretical properties when used as interpolants, but are likely to be numerically unstable for moderate or larger data sets. The smoothing process improves numerical stability. For neural nets the tradeoff between goodness of fit to the data and structure in the solution of Eq. (3) is controlled by the various parameters, particularly N, and some method for choosing them must be provided, either explicitly or implicitly.

Poggio and Girosi[31] have discussed Tihonov regularizations methods, which are the quintessential generalized smoothing splines, in the context of multilayer neural nets. Bertero, Poggio, and Torre[3] have given a broad overview of the use of regularization in ill-posed problems in early vision. Girosi and Poggio[14] have argued that the radial basis functions have a "best approximation property" that is not possessed by classical neural network schemes. The approach we will be describing coincides in broad outlook with the regularization methods of these very interesting papers, which are bridging the regularization and brain function literature.

An overview of GSS, Tihonov regularization methods, and a form of cross validation known as GCV (generalized cross validation) may be found in Wahba,[36] along with 30 pages of references. Most of the results quoted here without attribution, may be found in Wahba[36] and the references cited there.

The boundaries of these various approaches to multivariate function estimation are becoming blurred, as workers associated with the various approaches make comparisons and "borrow" methods from one approach that can improve another approach. It will be exciting to see how it all turns out.

The rest of the paper is organized as follows: In section 2 we give a few well-known facts concerning reproducing kernel Hilbert spaces (rkhs) and reproducing kernels (rk's). We also describe a test for whether or not the function f is linear based on Cox et al.[8] In section 3 we describe isotropic rk's and conditionally positive definite functions, and generalize slightly the results of Cox et al. In section 4 we describe a class of model-building methods based on tensor sums and products of reproducing kernels proposed by Gu and Wahba.[18,19,20] In section 5 we note the role of radial (and other) basis functions in computing smoothing spline and other regularized estimates of f. Lastly, in section 6 we describe generalizations from f real-valued, to $f = \mathbf{f}$ vector-valued to $\mathbf{f}$ function-valued. In this latter case $\mathbf{f}$ may be an *operator* from one reproducing kernel space to another. Approximation of this *operator* via the generalized smoothing spline paradigm can be described.

2. POSITIVE DEFINITE FUNCTIONS, RK SPACES, REGULARIZATION, AND SMOOTHING SPLINES

Let $\mathcal{T}$ be an index set, and let $Q(\cdot, \cdot)$ be a (real-valued) symmetric positive definite function on $\mathcal{T} \otimes \mathcal{T}$; that is, for every $\mathbf{t}(1), \ldots, \mathbf{t}(N) \in \mathcal{T}$ and $a_1, \ldots, a_N$, $N = 1, 2, \ldots,$

$$\sum_{i,j=1}^{N} a_i a_j Q(\mathbf{t}(i), \mathbf{t}(j)) \geq 0.$$

With each such positive definite function, there is associated a unique Hilbert space, call it $\mathcal{H}_Q$, with the property that, for each $\mathbf{t} \in T$,

$$Q(\cdot, \mathbf{t}) \in \mathcal{H}_Q$$

$$\langle f(\cdot), Q(\cdot, \mathbf{t}) \rangle_Q = f(\mathbf{t})$$

where $\langle \cdot, \cdot \rangle_Q$ is the inner product in $\mathcal{H}_Q$. The subscript Q on the inner product will be dropped when it is clear which inner product is meant. $Q(\cdot, \cdot)$ is called the reproducing kernel for $\mathcal{H}_Q$ because $\langle Q(\cdot, \mathbf{s}), Q(\cdot, \mathbf{t}) \rangle_Q = Q(\mathbf{s}, \mathbf{t})$. $\mathcal{H}_Q$ can be constructed as all the finite linear combinations of functions of the form

$$\sum_j a_j Q(\cdot, \mathbf{t}(j))$$

and their limits under the scalar product $\langle Q(\cdot, \mathbf{s}), Q(\cdot, \mathbf{t}) \rangle = Q(\mathbf{s}, \mathbf{t})$. Conversely, let $\mathcal{H}$ be a reproducing kernel Hilbert space (rkhs), that is, a Hilbert space in which all the evaluation functionals are bounded. This means that, for each $\mathbf{t} \in T$, there exists $Q_{\mathbf{t}} \in \mathcal{H}$ such that $\langle Q_{\mathbf{t}}, f \rangle = f(\mathbf{t})$. Let $Q(\mathbf{s}, \mathbf{t}) = \langle Q_{\mathbf{s}}, Q_{\mathbf{t}} \rangle$. Then $\mathcal{H}$ is the rkhs with reproducing kernel (rk) Q. Note that $Q_{\mathbf{t}}(\cdot) = Q(\cdot, \mathbf{t})$. See Aronszajn[1] and Weinert.[38]

We always have an interpolation formula in $\mathcal{H}_Q$ as follows: given $f \in \mathcal{H}_Q$ we can always find $g \in \mathcal{H}_Q$ to minimize $\| g \|_Q^2$ subject to $g(\mathbf{t}(i)) = f(\mathbf{t}(i)), i = 1, \ldots, n$. It is

$$g(\cdot) = (Q(\cdot, \mathbf{t}(1)), \ldots, Q(\cdot, \mathbf{t}(n))) Q_n^\dagger (f(\mathbf{t}(1)), \ldots, f(\mathbf{t}(n))' \qquad (4)$$

where Q_n is the $n \times n$ matrix with ijth entry $Q(\mathbf{t}(i), \mathbf{t}(j))$ and $\dagger$ is the inverse or generalized inverse. We remark that if $f \in \mathcal{H}_Q$, then $(f(\mathbf{t}(1)), \ldots, f(\mathbf{t}(n)))'$ is in the range of Q_n. g is the orthogonal projection $P_n f$ onto the subspace of $\mathcal{H}_Q$ spanned by $Q_{\mathbf{t}(i)}, i = 1, \ldots, n$. Therefore we have the interpolation error bounds

$$|f(\mathbf{t}) - g(\mathbf{t})| = |\langle (I - P_n)f, Q_{\mathbf{t}} \rangle| \leq \| f \| \, \| (I - P_n)Q_{\mathbf{t}} \| \, .$$

As the $\mathbf{t}(i)$ become dense in T, $\| (I - P_n)Q_{\mathbf{t}} \| \to 0$. The rate at which $\| (I - P_n)Q_{\mathbf{t}} \|$ goes to 0 is known in many cases; for results directly related to the properties of the rk Q, see Wahba,[32,33] and Micchelli and Wahba.[26] However, the calculation of Eq. (4) will in general be unstable for large n.

Let $\mathcal{H}$ be an rkhs which has an orthogonal decomposition

$$\mathcal{H} = \mathcal{H}_0 \oplus \mathcal{H}_1$$

where $\mathcal{H}_0$ is M dimensional, spanned by $\phi_1, \ldots, \phi_M$, and $\mathcal{H}_1$ has rk $R(\mathbf{s}, \mathbf{t})$. Then the smoothing spline paradigm is: given data

$$y(i) = f(\mathbf{t}(i)) + \epsilon(i), \quad i = 1, \ldots, n, \qquad (5)$$

where $\epsilon(i)$ are independently normally distributed with common unknown variance σ^2, find $f_\lambda \in \mathcal{H}$ to minimize

$$\frac{1}{n} \sum_{i=1}^{n} \left(y(i) - f(\mathbf{t}(i)) \right)^2 + \lambda \parallel P_1 f \parallel^2, \tag{6}$$

where P_1 is the orthogonal projection onto $\mathcal{H}_1$. The solution is well known to be unique provided the $n \times M$ matrix T with $i\nu$th entry $\phi_\nu(\mathbf{t}(i))$ is of rank M, and is given by

$$f_\lambda(\cdot) = \sum_{\nu=1}^{M} d_\nu \phi_\nu(\cdot) + \sum_{i=1}^{n} c_i R(\cdot, \mathbf{t}(i)) \tag{7}$$

where the coefficient vectors $\mathbf{d} = (d_1, \dots, d_M)'$ and $\mathbf{c} = (c_1, \dots, c_n)'$ are given by

$$(R_n + n\lambda I)\mathbf{c} + T\mathbf{d} = \mathbf{y},$$
$$T'\mathbf{c} = 0;$$

here R_n is the $n \times n$ matrix with ijth entry $R(\mathbf{t}(i), \mathbf{t}(j))$.

The most celebrated example is the famous polynomial smoothing spline of degree $2m - 1$. For this case, one takes $M = m$, $\mathcal{T} = [0, 1]$, the ϕ_ν span the polynomials of degree less than m, and $\parallel P_1 f \parallel^2 = \int_0^1 (f^{(m)}(t))^2 dt$. $\mathcal{H}$ is the Sobolev space W_2^m of functions with square integrable mth derivative on $[0, 1]$. See Wahba[36] for a reproducing kernel. In this particular case it is not particularly convenient to use a reproducing kernel to compute f_λ because of special properties possessed by f_λ.

Returning to the general case, Cox, Koh, Wahba and Yandell[8] proposed two tests (locally most powerful and GCV tests) for the hypothesis that f is in the span of the ϕ_ν (the null space of the penalty functional $\parallel P_1 f \parallel^2$) versus the alternative $f \in \mathcal{H}$. A third test (a generalized likelihood ratio test) is proposed in Wahba.[36] We will outline the results here for the GCV test. Existing examples refer to univariate polynomial splines and bivariate thin plate splines. In the next section we will fill in a few details to show how the GCV test can be applied to test for *linearity* of f in several d-dimensional examples. The reader familiar with Cox et al. will have no trouble extending the results of the next section to the locally most powerful and likelihood ratio test also. Which test the user might prefer would depend on whether they are concerned with local alternatives, alternatives in $\mathcal{H}$, or random alternatives with covariance $R(\cdot, \cdot)$

First, the smoothing parameter λ is estimated by GCV, which consists of finding $\hat{\lambda}$ to minimize

$$V(\lambda) = \frac{\frac{1}{n} \parallel (I - A(\lambda))\mathbf{y} \parallel^2}{[\frac{1}{n} tr(I - A(\lambda)]^2}. \tag{8}$$

where $A(\lambda)$ is the influence matrix, which satisfies

$$\begin{pmatrix} f_\lambda(\mathbf{t}(1)) \\ \vdots \\ f_\lambda(\mathbf{t}(n)) \end{pmatrix} = A(\lambda)\mathbf{y}. \tag{9}$$

The test statistic is

$$t_{GCV} = \frac{V(\hat{\lambda})}{V(\infty)}.$$ (10)

This test statistic has a representation

$$t_{GCV} = inf_{\gamma} \frac{\sum_{\nu=1}^{n-M} \frac{z_{\nu}^2}{(1+\gamma\lambda_{\nu})^2}}{\left(\sum_{\nu=1}^{n-M} \frac{1}{(1+\gamma\lambda_{\nu})}\right)^2} \bigg/ \frac{\sum_{\nu=1}^{n-M} z_{\nu}^2}{(n-M)^2}$$ (11)

where $\mathbf{z} = (z_1, \ldots, z_{n-M})'$, γ plays the role of $1/n\lambda$ and the λ_{ν} are defined as follows: Let T and R_n be as before and let F be any $n \times (n-M)$ matrix satisfying $F'F = I$ and $T'F = 0$. F may be found with the Q-R decomposition. Let UDU' be the eigenvalue-eigenvector decomposition of $F'R_nF$. Then $\mathbf{z} = U'F'\mathbf{y}$ and the λ_{ν} are the diagonal entries of D. $\sum_{\nu=1}^{n-M} z_{\nu}^2$ is the residual sum of squares after least-squares regression onto the ϕ_{ν}. Under the hypothesis that f is in the span of the ϕ_{ν} and the $\epsilon(i)$ are independent zero-mean Gaussian random variables with a common variance, then the z_{ν} are also independent zero mean Gaussian random variables with common variance, and the distribution of t_{GCV} is independent of their variance. In this case the distribution can easily be found by a Monte Carlo simulation by generating the z_{ν} as independent normal $(0,1)$ pseudo-random variables. $V(\lambda)$ will have a (possibly local) minimum at $\lambda = \infty$ (that is, at $\gamma = 0$) if $\sum_{\nu=1}^{n-M} \lambda_{\nu}z_{\nu}^2 \leq 1/n - M\lambda_{\nu} \sum_{\nu=1}^{n-M} z_{\nu}^2$. Thus, the distribution of t_{GCV} will have a mass point at 1, which will become small for large n for distant alternatives. Some Monte Carlo results for univariate polynomial and two-dimensional thin plate splines can be found in Wahba.[36]

---

3. ISOTROPIC RK'S ON E^d

$Q(\mathbf{s},\mathbf{t}), \mathbf{s},\mathbf{t} \in E^d$ is said to be an isotropic rk if $Q(\mathbf{s},\mathbf{t}) = E(\|\mathbf{s} - \mathbf{t}\|)$ where $\|\mathbf{s} - \mathbf{t}\|$ is the Euclidean distance between $\mathbf{s}$ and $\mathbf{t}$, and E is positive definite; that is, $\sum_{i,j} E(\|\mathbf{t}(i) - \mathbf{t}(j)\|) \geq 0$ for any $\mathbf{t}(1), \ldots, \mathbf{t}(n)$. Micchelli[25] gives the following examples:

$$E(\tau) = \exp^{-\alpha|\tau|^{\beta}},$$
$$\alpha \geq 0, 0 < \beta \leq 2$$
$$E(\tau) = (\alpha + |\tau|^{\beta})^{-\gamma},$$
$$\alpha, \gamma \geq 0, 0 < \beta \geq 2,$$
$$E(\tau) = (1 - |\tau|^{\delta})_+,$$
$$\delta \geq [d/2 + 1].$$

Here $(x)_+ = x$ for $x \geq 0$ and 0 otherwise. See Micchelli for a characterization of all such E's. The associated reproducing kernel spaces consist of f's in $\mathcal{L}_2(E^d)$ whose Fourier transforms satisfy

$$\int \cdots \int \|\tilde{E}(\omega)\|^{-1} \|\tilde{f}(\omega)\|^2 d\omega < \infty \tag{12}$$

where $\tilde{f}, \tilde{E}$ are the Fourier transforms of f and E respectively. Minimal norm interpolation and smoothing in these spaces involve the radial basis functions

$$Q_{\mathbf{t}(i)}(\mathbf{t}) = E(\|\mathbf{t}(i) - \mathbf{t}\|), \quad i = 1, \ldots, n.$$

Now, let $\phi_1, \ldots, \phi_M$ be M functions on E^d, with the property that the $n \times M$ matrix T with $i\nu$th entry $\phi_\nu(\mathbf{t}(i))$ is of rank M. Suppose we are interested in considering f of the form

$$f = \sum_{\nu=1}^{M} d_\nu \phi_\nu + f_1 \tag{13}$$

where $f_1 \in \mathcal{H}_R$, and we want to decide whether f_1 is 0 or not. In particular, if $M = d + 1$ and the ϕ_ν span the linear functions in d variables, then we are trying to decide whether f is linear or whether it is linear plus some function f_1 satisfying Eq. (12). The partial spline paradigm[36] then says to estimate f of Eq. (13) as

$$min_{d; f_1 \in \mathcal{H}_R} \sum_{i=1}^{n} (y(i)) - \sum_{\nu=1}^{M} d_\nu \phi_\nu(\mathbf{t}(i)) - f_1(\mathbf{t}(i))^2 + \lambda \|f_1\|_R^2. \tag{14}$$

It is known that the minimizer f_λ is (also) given by Eq. (7) (with R replaced by Q), and it will follow that t_{GCV} is again given by Eq. (10).

We now proceed to the popular thin plate splines. Let $\phi_1, \ldots, \phi_M$ be the $M = \binom{d+m-1}{d}$ polynomials of total degree less than m in d variables, and let the matrix T be as before. A function $E(\tau)$ defined on $[0, \infty)$ is said to be conditionally (strictly)positive definite of order m on E^d if, for any distinct points $\mathbf{t}(i), \ldots, \mathbf{t}(n)$ and scalars $c_1, \ldots, c_n$, satisfying

$$\sum_{\nu=1}^{n} c_i \phi_\nu(\mathbf{t}(i)) = 0, \quad \nu = 1, \ldots, M,$$

then $\sum_{i,j} c_i c_j E(\|\mathbf{t}(i) - \mathbf{t}(j)\|)$ is (positive) non-negative. Let $2m - d > 0$. The functions

$$E(\tau) = |\tau|^{2m-d}$$

$$2m - d \text{ not an even integer}$$

$$E(\tau) = |\tau|^{2m-d} \log |\tau| \tag{15}$$

$$2m - d \text{ an even integer}$$

are conditionally positive definite of order m. They appear in the construction of the thin plate splines.

Letting $\mathbf{t} = (t_1, \ldots, t_d)$, let

$$J_m^d(f) = \sum_{\alpha_1 + \cdots + \alpha_d = m} \frac{m!}{\alpha_1! \cdots \alpha_d!} \int_{-\infty}^{\infty} \cdots \int_{-\infty}^{\infty} \left(\frac{\partial^m f}{\partial t_1^{\alpha_1} \cdots \partial t_d^{\alpha_d}} \right)^2 dt_1 \cdots dt_d \quad (16)$$

wherever it exists. The thin plate splines are obtained as minimizers, in an appropriate function space which we will call X (see Duchon[11]), of

$$\frac{1}{n} \sum_{i=1}^{n} \left(y(i) - f(\mathbf{t}(i)) \right)^2 + \lambda J_m^d(f). \quad (17)$$

Loosely speaking, this function space consists of span $\{\phi_\nu\}$ and functions in $\mathcal{L}_2$ whose Fourier transform satisfies

$$\int \cdots \int ||\omega||^{2m} ||\tilde{f}(\omega)||^2 d\omega < \infty. \quad (18)$$

The solution is well known to have a representation

$$f_\lambda(\mathbf{t}) = \sum_{\nu=1}^{M} d_\nu \phi_\nu(\mathbf{t}) + \sum_{i=1}^{n} c_i E(||\mathbf{t} - \mathbf{t}(i)||)$$

where E is as in Eq. (15),

$$(K_n + n\lambda I)\mathbf{c} + T\mathbf{d} = \mathbf{y},$$
$$T'\mathbf{c} = 0;$$

here K_n is the $n \times n$ matrix with ijth entry $E(||\mathbf{t}(i) - \mathbf{t}(j)||)$.

Letting F be as before, it follows that although K_n is not positive definite, $F'K_n F$ is. To test

$$f = \sum_{\nu=1}^{M} d_\nu \phi_\nu + f_1, \quad J_m^d(f_1) \leq \infty \quad (19)$$

versus

$$f = \sum_{\nu=1}^{M} d_\nu \phi_\nu, \quad (20)$$

then t_{GCV} is as in Eq. (11) with R_n replaced by K_n.

If we are particularly interested in the case f linear, then in this setup we would take $m = 2$. This can be done for $d = 2, 3$, but not for $d \geq 3$ since we need $2m - d \geq 0$ for X to be a reproducing kernel space. Suppose $d \geq 3$. One possibility is to use a two-stage procedure; that is, the first stage tests whether f_1 is 0 or not as before. If it is decided that f_1 is 0, then, letting $\phi_1, \ldots, \phi_{d+1}$

span the linear functions in d variables, the second stage is the usual F test for $f \in span\{\phi_\nu\}_{\nu=1}^M$ vs. $f \in span\{\phi_\nu\}_{\nu=1}^{d+1}$.

Another possibility is to move $\phi_{d+2}, \ldots, \phi_M$ out of $\mathcal{H}_0$ and into $\mathcal{H}_1$. Then our model becomes

$$f = \sum_{\nu=1}^{d+1} d_\nu \phi_\nu + f_2 \tag{21}$$

where

$$f_2 = \sum_{\nu=d+2}^{M} d_\nu \phi_\nu + f_1 \tag{22}$$

and where we are now trying to decide whether or not $f_2 = 0$. We can do this as before by constructing a reasonable reproducing kernel for functions of the form Eq. (22).

There are various ways this can be done. We describe one of the constructions given in Gu and Wahba.[18,19] First, define $(f, g)_n = 1/n \sum_{i=1}^n f(\mathbf{t}(i)) g(\mathbf{t}(i))$ and choose the basis $\phi_1, \ldots, \phi_M$ to be orthonormal with respect to $(\cdot, \cdot)_n$ with $\phi_1, \ldots, \phi_{d+1}$ spanning the linear functions. This can always be done numerically using the Q-R decomposition. Then we can take as the squared norm on the Hilbert space $X \setminus \{\phi_1, \ldots, \phi_{d+1}\}$ (functions of the form Eq. (22)),

$$\|f\|_R^2 = \frac{1}{\theta} \sum_{\nu=1}^{M} (f, \phi_\nu)_n + J_m^d(f), \tag{23}$$

where θ is a positive constant to be discussed shortly. Using the construction in Gu and Wahba, the rk for this space can be found and we give it here. To do this, define, for any $g \in X$,

$$P_0 g = \sum_{\nu=1}^{M} (g, \phi_\nu)_n \phi_\nu. \tag{24}$$

Then the reproducing kernel for functions of the form Eq. (22) with square norm Eq. (23) can be shown to be

$$R(\mathbf{s}, \mathbf{t}) = \theta \sum_{\nu=d+1}^{M} \phi_\nu(\mathbf{s}) \phi_\nu(\mathbf{t}) + (I - P_{0(\mathbf{s})})(I - P_{0(\mathbf{t})}) E(\|\mathbf{s} - \mathbf{t}\|) \tag{25}$$

where $P_0(\mathbf{s})$ means P_0 is applied to what follows considered as a function of $\mathbf{s}$.

If θ is given, then t_{GCV} (which will depend on θ) can be computed and the null distribution generated by Monte Carlo methods as before. More generally both λ and θ can be obtained by GCV, using the code in Gu,[17] which is available through netlib@ornl.gov. See also Gu and Wahba.[20] In principle the distribution of t_{GCV} under the null hypothesis can still be obtained by Monte Carlo methods by setting $\lambda_\nu = \lambda_\nu(\theta)$ and minimizing over θ as well as γ in Eq. (10), since t_{GCV} does not depend on $U = U(\theta)$. Each new θ requires a new matrix decomposition; however, Gu[17] has provided some shortcuts, and the iterative search method in RKPACK converges rapidly in θ in the examples we have tried. It remains to be studied how a one-stage test like this would compare to the two-stage test described before.

4. ADDITIVE AND INTERACTION SPLINES

Let $\alpha = 1, 2, \ldots, \Gamma < \infty$ and for each α, let $\mathcal{H}^{(\alpha)} = \mathcal{H}_0^{(\alpha)} \oplus \mathcal{H}_1^{(\alpha)}$ be an rkhs of real-valued functions of $\mathbf{t}^{(\alpha)} \in \mathcal{T}^{(\alpha)}$, with $\mathcal{H}_0^{(\alpha)}$ of finite dimension $M(\alpha)$, and let the reproducing kernels (in an obvious notation) be $R_0^{(\alpha)}(\mathbf{s}^{(\alpha)}, \mathbf{t}^{(\alpha)})$ and $R_1^{(\alpha)}(\mathbf{s}^{(\alpha)}, \mathbf{t}^{(\alpha)})$. Then $\mathcal{H} = \otimes^\alpha \mathcal{H}^{(\alpha)} = \otimes^\alpha (\mathcal{H}_0^{(\alpha)} \oplus \mathcal{H}_1^{(\alpha)})$ is an rkhs of functions of $\mathbf{t} = (\mathbf{t}^{(1)}, \ldots, \mathbf{t}^{(\Gamma)}) \in \otimes^\alpha \mathcal{T}^{(\alpha)}$, with rk

$$Q(\mathbf{s}, \mathbf{t}) = \prod_\alpha \left(R_0^{(\alpha)}(\mathbf{s}^{(\alpha)}, \mathbf{t}^{(\alpha)}) + R_1^{(\alpha)}(\mathbf{s}^{(\alpha)}, \mathbf{t}^{(\alpha)}) \right). \tag{26}$$

This follows since the reproducing kernels of tensor products and (orthogonal) sums of reproducing kernel spaces may be obtained from the corresponding products and sums of the individual reproducing kernels. If each $\mathcal{H}_0^{(\alpha)}$ contains the constant functions, then smoothing in various subspaces of $\mathcal{H}$ consisting of functions of the form

$$f(\mathbf{t}) = \sum_\alpha f_\alpha(\mathbf{t}^{(\alpha)}) \tag{27}$$

(additive functions), or

$$f(\mathbf{t}) = \sum_\alpha f_\alpha(\mathbf{t}^{(\alpha)}) + \sum_{\alpha \leq \beta} f_{\alpha\beta}(\mathbf{t}^{(\alpha)}, \mathbf{t}^{(\beta)}) \tag{28}$$

(two-factor interactions), and so forth, can be carried out by extracting the rk's corresponding to subspaces of functions of the given form. Usually only low-order interactions would be included if γ is large. See Gu and Wahba[18,19] and references cited there for model-building strategies with additive and interaction splines.

5. MORE ON RADIAL BASIS FUNCTIONS

Consider the minimization of Eq. (17) and the GCV estimate of λ when n is very large. Wahba[34] suggested that, instead of minimizing Eq. (17) in X, it be minimized in an approximating subspace, generated as follows. Choose $\mathbf{s}(1), \ldots, \mathbf{s}(N)$ spaced "nicely" in the region of interest of E^d, and such that the $N \times M$ matrix S with $k\nu$th entry $\phi_\nu(\mathbf{s}(k))$ is of rank N. N basis functions $B_1, \ldots, B_N$ are then selected as follows: $B_k = \phi_k, k = 1, \ldots, M$, $B_{M+l}(\mathbf{t}) = \sum_{k=1}^N u_{lk} E(\|\mathbf{t} - \mathbf{s}(k)\|)$, $l = 1, \ldots, N - M$ where the $N \times (N - M)$ matrix U with lkth entry u_{lk} satisfies $U'S = 0_{N \times M}$, and E is as in Eq. (15). Thus, one seeks $f \in span \{B_k\}$ to minimize

$$\frac{1}{n} \sum_{i=1}^n \left(y(i) - f(\mathbf{t}(i)) \right)^2 + \lambda J_m^d(f). \tag{29}$$

Letting K_N be the $N \times N$ matrix with klth entry $E(\|\mathbf{s}(k) - \mathbf{s}(l)\|), U'K_N U$ is positive definite and $J_m^d(\sum_{k=1}^{N-M} c_k B_k) = \mathbf{c}'U'K_N U\mathbf{c}$. The remaining calculations are straightforward. The general idea is that if n is very large and N is reasonably large, the minimizer of Eq. (29) in *span* $\{B_k\}$ with λ chosen appropriately is an excellent approximation to the minimizer in X(with λ chosen appropriately), and it will be cheaper to compute. A number of authors have pursued this or similar approaches; see for example Hutchinson and Bischof,[22] O'Sullivan,[29] and Constable and Parker.[6]

In the case where one has $Q(\mathbf{s}, \mathbf{t}) = E(\|\mathbf{s} - \mathbf{t}\|)$ positive definite and one wishes to minimize

$$\frac{1}{n}\sum_{i=1}^{n}\left(y(i) - f(\mathbf{t}(i))\right)^2 + \lambda\|f\|_Q^2, \tag{30}$$

the natural basis functions to take are simply

$$B_k(\mathbf{t}) = E(\|\mathbf{t} - \mathbf{s}(k)\|); \tag{31}$$

here $\|\sum_{k=1}^{N} c_k B_k\|_Q^2 = \mathbf{c}'K_N\mathbf{c}$.

Hutchinson and Bischof[22] proposed a scheme for selecting the $\mathbf{s}(k)$ as approximately a representative subsample of $1/6$ of the data points $\mathbf{t}(i)$ modified by adding or deleting points where an especially poor fit was obtained or the data did not change much. Poggio and Girosi[31] and other authors discuss methods of moveable centers $\mathbf{s}(k)$ in the context of nonlinear least squares on a cost functional. Poggio and Girosi also mention the k-means clustering procedure of MacQueen,[24] and some other approaches. Choosing the $\mathbf{s}(k)$ to optimize a cost functional can be a very difficult (ill-posed) numerical problem. Provided one is going to let λ do most of the smoothing (rather than a small N), heuristic procedures for obtaining the $\mathbf{s}(k)$ may work well. There is no "cost" in terms of the accuracy of the approximation for N too large (provided λ is chosen well), only cost in computing time. For example, if the $\mathbf{t}(i)$ arrive sequentially, create a new basis function at $\mathbf{t}(i)$ if $\mathbf{t}(i)$ does not fall within a ball of radius r surrounding the existing basis functions; otherwise do not. Note that we can measure distance by $\|\mathbf{s}(k) - \mathbf{s}(l)\|$ as well as by $\|B_k - B_l\|$.

6. RKHS OF VECTOR- AND FUNCTION-VALUED FUNCTIONS

We wish to consider next $\mathcal{H}$ a Hilbert space of vector-valued functions of $\mathbf{t} \in T$; that is, for each $\mathbf{t}$, $f(\mathbf{t}) = \mathbf{f}(\mathbf{t}) \in E^p$, for some given p. Let k index the kth component of $\mathbf{f}(\mathbf{t}), k = 1, \ldots, p$. We can define an rkhs of E^p-valued functions of $\mathbf{t}$ once we are given an rk of the form

$$K(k, \mathbf{s}; l, \mathbf{t}) \quad k, l \in \{1, \ldots, p\}, \quad \mathbf{s}, \mathbf{t} \in T. \tag{32}$$

Letting $T^p = \{1, \ldots, p\}$, K must be positive definite on $\{T^p \otimes T\} \otimes \{T^p \otimes T\}$, i.e.,

$$\sum_{k,i;\ell,j} a_{ki} a_{\ell j} K(k, \mathbf{t}(i); \ell, \mathbf{t}(j)) \geq 0 \tag{33}$$

for every $\{k, \mathbf{t}(i)\}$ and $\{a_{ki}\}$.

For each fixed $(\ell_*, \mathbf{t}_*) \in \{T^p \otimes T\}$, Eq. (34) defines an E^p-valued function of $\mathbf{t} \in T$ by the formula

$$\mathbf{K}_{\ell_*, \mathbf{t}_*}(\cdot, \mathbf{t}) = \begin{pmatrix} K(1, \mathbf{t}; \ell_*, \mathbf{t}_*) \\ \vdots \\ K(p, \mathbf{t}; \ell_*, \mathbf{t}_*) \end{pmatrix}. \tag{34}$$

The desired rkhs $\mathcal{H}_K$ will consist of the linear manifold of all finite linear combinations of functions of the form Eq. (34) as $(\ell_*, \mathbf{t}_*)$ varies in $T^p \otimes T$, and its closure with respect to the scalar product

$$\langle \mathbf{K}_{k, \mathbf{s}}, \mathbf{K}_{\ell, \mathbf{t}} \rangle_K = K(k, \mathbf{s}; \ell, \mathbf{t}). \tag{35}$$

This approach to rkhs of vector-valued functions of $\mathbf{t}$ is similar to but not exactly the same as that of Golosov and Tempelman.[16]

Denote by $f(k, \mathbf{t})$ the kth component of $\mathbf{f}(\mathbf{t}) \in \mathcal{H}_K$, and let $\mathcal{O}(\mathbf{t})$ be a subset of one or more elements of T^p, possibly depending on $\mathbf{t}$. We observe

$$y(k, i) = f(k, \mathbf{t}(i)) + \epsilon(k, i), \quad i = 1, \ldots, \tilde{n}, \quad k \in \mathcal{O}(\mathbf{t}(i)). \tag{36}$$

We now have a total of $n = \sum_{i=1}^{\tilde{n}} \mathcal{O}(\mathbf{t}(i))$ data points. We can obtain the regularized, i.e., generalized smoothing spline estimate of $\mathbf{f} \in \mathcal{H}_K$ as the minimizer $\mathbf{f}_\lambda$ of

$$\sum_{i=1}^{\tilde{n}} \sum_{k \in \mathcal{O}(\mathbf{t}(i))} \left(y(k, i) - f(k, \mathbf{t}(i)) \right)^2 + \lambda \|f\|_K^2. \tag{37}$$

The minimizer will be an E^p-valued function of $\mathbf{t}$ of the form

$$\mathbf{f}_\lambda(\cdot, \mathbf{t}) = \sum_{i=1}^{\tilde{n}} \sum_{k \in \mathcal{O}(\mathbf{t}(i)} c_{ki} \mathbf{K}_{k, \mathbf{t}(i)}(\cdot, \mathbf{t}). \tag{38}$$

The coefficient vector $\mathbf{c}$ satisfies

$$(K_n + n\lambda I)\mathbf{c} = \mathbf{y} \tag{39}$$

where the indices (k, i) have been arranged in some linear order, and K_n is the $n \times n$ matrix with $ki, \ell j$th entry $K(k, \mathbf{t}(i); \ell, \mathbf{t}(j))$. If one desires not to penalize some M-dimensional subspace of vector functions spanned by $\{\phi_1, \ldots, \phi_M\}$, then one minimizes

$$\sum_{i=1}^{\tilde{n}} \sum_{k \in \mathcal{O}(\mathbf{t}(i))} \left(y(k,i) - \sum_{\nu=1}^{M} d_\nu \phi_\nu(k, \mathbf{t}(i)) - f(k, \mathbf{t}(i)) \right)^2 + \lambda \|f\|_K^2. \tag{40}$$

and it can be shown that the minimizer has the representation

$$\mathbf{f}_\lambda(\cdot, \mathbf{t}) = \sum_{\nu=1}^{M} d_\nu \phi_\nu(\cdot, \mathbf{t}) + \sum_{i=1}^{\tilde{n}} \sum_{k \in \mathcal{O}(\mathbf{t}(i)} c_{ki} \mathbf{K}_{k, \mathbf{t}(i)}(\cdot, \mathbf{t}) \tag{41}$$

where

$$(K_n + n\lambda I)\mathbf{c} + T\mathbf{d} = \mathbf{y}, \tag{42}$$
$$T'\mathbf{c} = 0; \tag{43}$$

here, T is the matrix with ki, νth entry $\phi_\nu(k, \mathbf{t}(i))$.

We can further generalize by replacing T^p by an arbitrary index set $\mathcal{V}$. (This section is based on joint work with C. Gu.) To do this we need an rk of the form

$$K(\mathbf{u}, \mathbf{s}; \mathbf{v}, \mathbf{t}), \mathbf{u}, \mathbf{v} \in \mathcal{V}, \mathbf{s}, \mathbf{t} \in \mathcal{T}, \tag{44}$$

that is, a symmetric function on $\{\mathcal{V} \otimes \mathcal{T}\} \otimes \{\mathcal{V} \otimes \mathcal{T}\}$ such that

$$\sum_{k,i;\ell,j} a_{ki} a_{\ell j} K(\mathbf{u}(k), \mathbf{t}(i); \mathbf{u}(\ell), \mathbf{t}(j)) \geq 0 \tag{45}$$

for every $\{\mathbf{u}(k), \mathbf{t}(i)\}$ and $\{a_{ki}\}$. For fixed $\mathbf{t}$, let $k_\mathbf{t}(\mathbf{u}, \mathbf{v}) = K(\mathbf{u}, \mathbf{t}; \mathbf{v}, \mathbf{t})$. Note that $k_\mathbf{t}$ is a positive definite function on $\mathcal{V} \otimes \mathcal{V}$. We claim that for each fixed $\mathbf{v}_*, \mathbf{t}_* \in \mathcal{V} \otimes \mathcal{T}$, Eq. (44) defines an $\mathcal{H}_{k_\mathbf{t}}$-valued function of $\mathbf{t} \in \mathcal{T}$, where $\mathcal{H}_{k_\mathbf{t}}$ is the rkhs with rk $k_\mathbf{t}$. That is, letting $K_{\mathbf{v}_*, \mathbf{t}_*}(\mathbf{u}, \mathbf{t}) = K(\mathbf{u}, \mathbf{t}; \mathbf{v}_*, \mathbf{t}_*)$, $\mathbf{K}_{\mathbf{v}_*, \mathbf{t}_*}(\cdot, \mathbf{t})$ is an element of $\mathcal{H}_{k_\mathbf{t}}$ for each $\mathbf{t} \in \mathcal{T}$. A proof will be given momentarily.

$\mathcal{H}_K$ is the linear span of all such $\mathcal{H}_{k_\mathbf{t}}$-valued functions of $\mathbf{t}$,

$$\mathbf{K}_{\mathbf{v}_*, \mathbf{t}_*}(\cdot, \mathbf{t}) = K(\cdot, \mathbf{t}; \mathbf{v}_*, \mathbf{t}_*) \tag{46}$$

as $(\mathbf{v}_*, \mathbf{t}_*)$ varies in $\mathcal{V} \otimes \mathcal{T}$, closed with respect to the scalar product

$$\langle \mathbf{K}_{\mathbf{u}, \mathbf{s}}, \mathbf{K}_{\mathbf{v}, \mathbf{t}} \rangle_K = K(\mathbf{u}, \mathbf{s}; \mathbf{v}, \mathbf{t}). \tag{47}$$

We now sketch the argument that, for each fixed $\mathbf{t}$, $\mathbf{v}_*, \mathbf{t}_*$; $\mathbf{K}_{\mathbf{v}_*, \mathbf{t}_*}(\cdot, \mathbf{t}) \in \mathcal{H}_{k_\mathbf{t}}$. Let $f_{*, \mathbf{t}}(\cdot) = \mathbf{K}_{\mathbf{v}_*, \mathbf{t}_*}(\cdot, \mathbf{t})$. We want to show that $f_{*, \mathbf{t}}(\cdot) \in \mathcal{H}_{k_\mathbf{t}}$. Given $\mathbf{u}(1), \ldots, \mathbf{u}(N) \in \mathcal{V}$, let $\mathcal{H}_{N, \mathbf{t}}$ be the span of the representers of evaluation at $\mathbf{u}(1), \ldots, \mathbf{u}(N) \in \mathcal{H}_{k_\mathbf{t}}$; that is, span$\{k_\mathbf{t}(\cdot, \mathbf{u}(1)), \ldots, k_\mathbf{t}(\cdot, \mathbf{u}(N))\}$ and let $P_{N, \mathbf{t}} f_{*, \mathbf{t}}$ be the minimal norm interpolant to $f_{*, \mathbf{t}}(\mathbf{u}(i)), i = 1, \ldots, N$ in $\mathcal{H}_{k_\mathbf{t}}$. By the theory of reproducing kernels

$f_{*,t}$ will be in $\mathcal{H}_{k_t}$ if and only if $\|P_{N,t}f_{*,t}\|_{k_t}^2$ remains bounded as $\mathbf{u}(1),\ldots,\mathbf{u}(N)$ become dense in $\mathcal{V}$.

Letting $K_{N,t}$ be the $n \times n$ matrix with ijth entry $k_t(\mathbf{u}(i),\mathbf{u}(j)) = K(\mathbf{u}(i),\mathbf{t};\mathbf{u}(j),\mathbf{t})$, we can easily compute that

$$\|P_{N,t}f_{*,t}\|_{k_t}^2 = \mathbf{c}'K_{N,t}^{-1}\mathbf{c} \tag{48}$$

where

$$\mathbf{c} = (K(\mathbf{u}(1),\mathbf{t};\mathbf{v}_*,\mathbf{t}_*),\ldots,K(\mathbf{u}(N),\mathbf{t};\mathbf{v}_*,\mathbf{t}_*))' . \tag{49}$$

Now consider $\mathcal{H}_N$, the N-dimensional subspace of $\mathcal{H}_K$ spanned by $\mathbf{K}_{\mathbf{u}(i),\mathbf{t}}$, $i = 1,\ldots,N$, and let $f_* = \mathbf{K}_{\mathbf{v}_*,\mathbf{t}_*} \in \mathcal{H}_K$. It is not hard to show that $\|P_{N,t}f_{*,t}\|_{k_t}^2$, given by Eq. (48) is equal to $\|P_N f_*\|_K^2$; however, $\|P_N f_*\|_K^2$ is always bounded by $\|f_*\|_K^2 = K(\mathbf{v}_*,\mathbf{t}_*;\mathbf{v}_*,\mathbf{t}_*) \le \infty$, completing the argument. For the reader concerned with symmetry questions, the same argument can be used to show that $f_{*,t}$ is also in $\mathcal{H}_{k_{t_*}}$.

In the special case that $K(\mathbf{u},\mathbf{t};\mathbf{v},\mathbf{t})$ does not depend on $\mathbf{t}$ (for example, if $K(\mathbf{u},\mathbf{s};\mathbf{v},\mathbf{t}) = R(\mathbf{u},\mathbf{v})E(\|\mathbf{s}-\mathbf{t}\|))$, then $\mathcal{H}_{k_t}$ will not depend on $\mathbf{t}$.

Now consider an arbitrary element $\mathbf{f}$ in $\mathcal{H}_K$, recall that for each $\mathbf{t}$, $\mathbf{f}(\mathbf{t})$ is an $\mathcal{H}_{k_t}$-valued function of $\mathbf{u} \in \mathcal{V}$. Denote by $f(\mathbf{u},\mathbf{t})$, $\mathbf{f}(\mathbf{t})$, evaluated at $\mathbf{u}$. As before, let $\mathcal{O}(\mathbf{t})$ be a finite subset of $\mathcal{V}$, generally depending on $\mathbf{t}$. Given observations

$$y(\mathbf{v},i) = f(\mathbf{v},\mathbf{t}(i)) + \epsilon(\mathbf{v},i), i = 1,\ldots,\tilde{n}, \mathbf{v} \in \mathcal{O}(\mathbf{t}(i)), \tag{50}$$

we can obtain an estimate of $\mathbf{f}$ in $\mathcal{H}_K$ as the minimizer $\mathbf{f}_\lambda$ of

$$\sum_{i=1}^{\tilde{n}} \sum_{\mathbf{v}\in\mathcal{O}(\mathbf{t}(i))} (y(\mathbf{v},i) - f(\mathbf{v},\mathbf{t}(i)))^2 + \lambda\|\mathbf{f}\|_K^2 \tag{51}$$

and $\mathbf{f}_\lambda$ will have a representation

$$\mathbf{f}_\lambda(\cdot,\mathbf{t}) = \sum_{i=1}^{\tilde{n}} \sum_{\mathbf{v}\in\mathcal{O}(\mathbf{t}(i))} c_{\mathbf{v},i}\mathbf{K}_{\mathbf{v},\mathbf{t}(i)}(\cdot,\mathbf{t}), \tag{52}$$

where, as before, the coefficient vector $\mathbf{c}$ satisfies

$$(K_n + n\lambda I)\mathbf{c} = \mathbf{y} . \tag{53}$$

Here the indices $(\mathbf{v},i)$ have been lined up in some order, and the $(\mathbf{v},i;\mathbf{v}',j)$th entry of K_n is $K(\mathbf{v},\mathbf{t}(i);\mathbf{v}',\mathbf{t}(j))$. If one desires not to penalize some M-dimensional subspace of vector functions spanned by $\{\phi_1,\ldots,\phi_M\}$, then one minimizes Eq. (40) as before, except now the integers $k \in \mathcal{O}(\mathbf{t}(i))$ are replaced by the $\mathbf{v} \in \mathcal{O}(\mathbf{t})$, the formulas for $\mathbf{f}_\lambda(\cdot,\mathbf{t})$ will then be the same as Eqs. (41), (42), and (43) with this substitution.

We remark that the "Bayesian confidence intervals" discussed in Nychka[27] and Wahba[35,37] can be extended to the generalizations considered here. The results described there may in principle be used to make accuracy statements about $f_\lambda(\mathbf{u}, \mathbf{t})$; equivalently, these results may be used to identify regions in $(\mathbf{u}, \mathbf{t})$ where the estimate is likely to be trustworthy or untrustworthy.

So far we have assumed that $\mathbf{t}(i)$ is known "exactly." For $\mathbf{t} \in E^d$ this is a reasonable assumption. Suppose $\mathcal{T}$ is itself a Hilbert space $\mathcal{H}$, say. Then $\mathbf{f}(\cdot) \in \mathcal{H}_K$ is a map (operator) from $\mathbf{t} \in \mathcal{H} \to \mathbf{f}(\mathbf{t}) \in \mathcal{H}_{k_t}$. In this case we would generally not know $\mathbf{t}(i)$ "exactly," since that might require an infinite set of components. We may just know that $\mathbf{t}(i)$ is in some small ball centered at a point which only requires a finite set of numbers to describe. Here we have a twist on the "errors in variables" problem, which opens up further interesting questions. Similar remarks apply to $\mathbf{v} \in \mathcal{O}(\mathbf{t}(i))$.

ACKNOWLEDGMENTS

Research supported in part by NSF under Grant DMS-9002566, and AFOSR under Grant AFOSR-90-0103.

REFERENCES

1. Aronszajn, N. "Theory of Reproducing Kernels." *Trans. Am. Math. Soc.* **68** (1950): 337–404.
2. Barron., A. "Statistical Learning Networks: A Unifying View." *Computing Science and Statistics: Proceedings of the 20th Symposium on the Interface*, edited by E. Wegman, D. Gantz, and J. Miller, 192–203. American Statistical Assoc., 1988.
3. Bertero, M., T. Poggio, and V. Torre. "Ill-Posed Problems in Early Vision." *Proc. of the IEEE* **76** (1988): 869–889.
4. Brieman, L., J. Friedman, R. Olshen, and C. Stone. *Classification and Regression Trees.* Wadsworth, 1984.
5. Casdagli, M. "Nonlinear Prediction of Chaotic Time Series." *Physica D* **35** (1989): 335–356.
6. Constable, C., and R. Parker. "Smoothing, Splines and Smoothing Splines; Their Application in Geomagnetism." *J. Comp. Phys.* **78** (1988): 493–508.
7. Corcoran, E. "Calculating Reality." *Sci. Amer.* **264** (1991): 100–109.
8. Cox, D., E. Koh, G. Wahba, and B. Yandell. "Testing the (Parametric) Null Model Hypothesis in (Semiparametric) Partial and Generalized Spline Models." *Ann. Statist.* **16** (1988): 113–119.

9. Cybenko, G. "Continuous Valued Neural Networks: Approximation Theoretic Results." *Computing Science and Statistics: Proceedings of the 20th Symposium on the Interface*, edited by E. Wegman, D. Gantz, and J. Miller, 174–183. American Statistical Assoc., 1988.

10. Cybenko, G. "Approximations by Superpositions of a Sigmoidal Function." Technical Report 856, University of Illinois Center for Supercomputing Research and Development, Urbana, IL, 1989.

11. Duchon, J. "Splines Minimizing Rotation-Invariant Semi-Norms in Sobolev Spaces." In *Constructive Theory of Functions of Several Variables*, 85–100. Berlin: Springer-Verlag, 1977.

12. Friedman, J. "Multivariate Adaptive Regression Splines." *Ann. Statist* **19** (1991): 1–141.

13. Friedman, J. H., E. Grosse, and W. Stuetzle. "Multidimensional Additive Spline Approximation." *SIAM J. Sci. Stat. Comput.* **4** (1983): 291–301.

14. Girosi, F., and T. Poggio. "Networks and the Best Approximation Property." Technical Report 1164, MIT Artificial Intelligence Laboratory, Boston, MA, 1989.

15. Golomb, M., and H. F. Weinberger. "Optimal Approximation and Error Bounds." *Proc. Symp. on Numerical Approximation*, edited by R. E. Langer, 117–190. University of Wisconsin Press, 1959.

16. Golosov, J., and A. Tempelman. "On Equivalence of Measures Corresponding to Gaussian Vector-Valued Functions." *Soviet Math. Dokl.* **10** (1969): 228–231.

17. Gu, C. "RKPACK and its Applications: Fitting Smoothing Spline Models." In *Proceedings of the Statistical Computing Section*, 42–51. American Statistical Association, 1989.

18. Gu, C., and G. Wahba. "Semiparametric ANOVA with Tensor Product Thin Plate Splines." Technical Report 90-61, Department of Statistics, Purdue University, Lafayette, IN, 1990.

19. Gu, C., and G. Wahba. "Comments to Multivariate Adaptive Regression Splines, by J. Friedman." *Ann. Statist* **19** (1991): 115–123.

20. Gu, C., and G. Wahba. "Minimizing GCV/GML Scores with Multiple Smoothing Parameters via the Newton Method." *SIAM J. Sci. Statist. Comput.* **12** (1991): 383–398.

21. Hubel, D. *Eye, Brain, Vision*. Scientific American Library, 1988.

22. Hutchinson, M., and R. Bischof. "A New Method for Estimating the Spatial Distribution of Mean Seasonal and Annual Rainfall Applied to the Hunter Valley, New South Wales." *Aust. Met. Mag.* **31** (1983): 179–184

23. Kimeldorf, G., and G. Wahba. "Some Results on Tchebycheffian Spline Functions." *J. Math. Anal. Applic.* **33** (1971): 82–95.

24. MacQueen, J. "Some Methods for Classification and Analysis of Multivariate Observations." *Proceedings of the Fifth Berkeley Symposium on Mathematical Statistics and Probability*, edited by J. Neyman, 281–297. University of California Press, 1966.

25. Micchelli, C. "Interpolation of Scattered Data: Distance Matrices and Conditionally Positive Definite Functions." *Constructive Approximation* **2** (1986): 11–22.

26. Micchelli, C., and G. Wahba. "Design Problems for Optimal Surface Interpolation." *Approximation Theory and Applications*, edited by Z. Ziegler, 329–348. New York: Academic Press, 1981.

27. Nychka, D. "Confidence Intervals for Smoothing Splines." *J.A.S.A.* **83** (1988): 1134–1143.

28. Nychka, D., D. McCaffrey, S. Ellner, and A. Gallant. "Estimating Lyapunov Exponents with Nonparametric Regression." Technical Report 1977, North Carolina State University Institute of Statistics, Raleigh, NC, 1990.

29. O'Sullivan, F. "An Iterative Aproach to Two-Dimensional Laplacian Smoothing with Application to Image Restoration." *J. Amer. Statist. Assoc.* **85** (1990): 213–219.

30. Penrose, R. *The Emperor's New Mind*. Oxford University Press, 1989.

31. Poggio, T., and F. Girosi. "Regularization Algorithms for Learning that are Equivalent to Multilayer Networks." *Science* **247** (1990): 978–982.

32. Wahba, G. "Convergence Rates of Certain Approximate Solutions to Fredholm Integral Equations of the First Kind." *J. Approx. Theory* **7** (1973): 167–185.

33. Wahba, G. "Practical Approximate Solutions to Linear Operator Equations when the Data are Noisy." *SIAM J. Numer. Anal.* **14** (1977): 651–667.

34. Wahba, G. "Spline Bases, Regularization, and Generalized Cross Validation for Solving Approximation Problems with Large Quantities of Noisy Data." *Approximation Theory III*, edited by W. Cheney, 905–912. New York: Academic Press, 1980.

35. Wahba, G. "Bayesian 'Confidence Intervals' for the Cross-Validated Smoothing Spline." *J. Roy. Stat. Soc. Ser. B* **45** (1983): 133–150.

36. Wahba, G. *Spline Models for Observational Data*. CBMS-NSF Regional Conference Series in Applied Mathematics, vol. 59. SIAM, 1990.

37. Wahba, G. "Multivariate Model Building with Additive, Interaction and Tensor Product Thin Plate Splines." In *Curves and Surfaces*, edited by P.-J. Laurent, A. LeMéhauté, and L. L. Schumaker, 491–504. New York: Academic Press, 1991.

38. Weinert, H., ed. *Reproducing Kernel Hilbert Spaces: Application in Signal Processing*. Stroudsburg, PA: Hutchinson Ross, 1982.

Section II: Statistics

Richard L. Smith
Department of Statistics, University of North Carolina, Chapel Hill, N.C. 27599-3260, U.S.A.

Optimal Estimation of Fractal Dimension

This paper is devoted to variants of the Grassberger-Procaccia procedure for estimating correlation dimension based on inter-point distances. An alternative estimator suggested by Takens is examined, and shown to perform well in cases where the data are from a smooth underlying distribution, though even in this case the required sample size grows very rapidly with true dimension. In more complex cases where the data are concentrated on a genuine fractal, the distribution typically has a "lacunarity" property which makes Takens' procedure sub-optimal and even inconsistent. For this situation an alternative estimator is proposed which is equivalent to Takens' estimator in those cases where Takens' estimator works well, but which also copes with lacunarity. The paper is illustrated by examples taken from several well-known chaotic systems.

INTRODUCTION

The context of this paper is that we observe a discrete-time, scalar time series $\{X_n, \ n = 1, 2, \ldots\}$ and we are interested in determining whether it is a realization of a chaotic, deterministic system as opposed to a process containing a significant random component. According to Takens'[23] embedding theorem, if $\{X_n\}$ is a discrete-time scalar cross-section across a continuous-time multidimensional system with an attractor contained in a manifold of dimension I, there will in general exist an embedding dimension $J \leq 2I + 1$ such that the J-vectors $X_{n-1}, \ldots, X_{n-J}$ fill out a set of the same dimension as the underlying attractor. The problem of detecting "chaos" in a time series is then reduced to one of deciding whether a sample of vector observations lies within a subset of possibly fractional dimension.

Most of the literature on dimension estimation has developed from the procedure proposed by Grassberger and Procaccia.[10,11] They focussed on a particular definition of dimension, the *correlation dimension*, and estimated it by a procedure based on *nearest-neighbor distances*. Numerous other dimension concepts are known,[9] and they do not all give the same numerical values,[14] but in this paper we focus on correlation dimension because of its ease of estimation. An alternative estimation procedure was proposed by Takens,[3,24] which attempted to exploit the optimality of the maximum likelihood method of statistical estimation. However, many fractal sets have a "lacunarity" property which acts contrary to the optimality claimed by Takens,[26] and it has also been argued[18,20] that there are quite severe inherent limitations on the Grassberger-Procaccia method. Other recent studies include numerical and approximate assessments of the bias and variance in dimension estimators,[16,17,27] modifications to reduce boundary effects,[8] and more detailed statistical studies.[6,7,25]

The purpose of the present paper is to explore in more detail the relation between the Grassberger-Procaccia and Takens' procedures, to establish conditions under which the latter really does improve on the former, and to propose a third procedure which copes with the lacunarity pointed out by Theiler. In each case, asymptotic bias and variance calculations are given, and theoretical calculations made for the optimal performance of the procedures.

CHARACTERIZING DIMENSION VIA NEAREST-NEIGHBOR DISTANCES

For much of the paper, we ignore the time-series context described in the introduction, and assume we are dealing with independent random vectors $\{\mathbf{Y}_n, \ n = 1, 2, ..\}$ in d-dimensional space. Assume that the distribution of the $\mathbf{Y}$'s is concentrated on a subset of dimension p, where $p \leq d$ and, in general, p is fractional. We wish to estimate p. In the time-series context, we will have $d = J + 1$ where J is the embedding dimension, and verification that $p < d$ will be considered confirmation that the series is deterministic chaos rather than random.

In this section we depart from what has been the traditional approach in the literature on this topic, which has been to assume that the true distribution is concentrated on a fractal, and suppose that p is an integer and the data are from a continuous distribution in p-dimensional space. Although at first sight this assumption has nothing to do with dynamical systems, it is in fact an important case because, when $p = d$, it represents the "devil's advocate" position of a random, non-chaotic system. Alternatively, if we want to apply the Grassberger-Procaccia procedure and its relatives to distinguish chaos from randomness, we must first understand how they behave when randomness is the true state of affairs.

In this context a theoretical result for nearest-neighbor distributions was derived by Silverman and Brown,[19] and our calculations are a generalization of theirs.

Suppose the observations $\mathbf{Y}_n$, $1 \leq n \leq N$ are independent, identically distributed p-vectors with twice continuously differentiable density f. Let $D_{mn} = \|\mathbf{Y}_m - \mathbf{Y}_n\|$ denote the distance between $\mathbf{Y}_m$ and $\mathbf{Y}_n$. Here $\|\mathbf{y}\|$ is the L_2 norm, i.e., if $\mathbf{y} = (y_1, \ldots, y_p)$, then $\|\mathbf{y}\| = (\sum y_i^2)^{1/2}$, though similar calculations could be made for other norms. Then

$$\Pr\{D_{mn} < r\} = \int \int f(\mathbf{x}) f(\mathbf{x} + \mathbf{y}) I(\|\mathbf{y}\| < r) d\mathbf{x} d\mathbf{y} \tag{1}$$

where I denotes the indicator function. Let us expand the integrand in Eq. (1) about $\mathbf{y} = 0$, letting $f_i = \partial f / \partial y_i$, $f_{ij} = \partial^2 f / \partial y_i \partial y_j$, etc. We find that

$$\Pr\{D_{mn} < r\} = \int f(\mathbf{x}) \int \{f(\mathbf{x} + \sum y_i f_i(\mathbf{x}) + \frac{1}{2} \sum \sum y_i y_j f_{ij}(\mathbf{x}) \\ + o(r^2)\} I(\|\mathbf{y}\| < r) d\mathbf{y} d\mathbf{x}. \tag{2}$$

where $o(r^2)$ denotes a term which is negligible in comparison to r^2 as $r \to 0$. However, it can be shown that

$$I_1 = \int I(\|\mathbf{y}\| < r) d\mathbf{y} = \frac{\pi^{p/2} r^p}{\Gamma(1 + p/2)},$$

$$\int y_i I(\|\mathbf{y}\| < r) d\mathbf{y} = 0,$$

$$I_2 = \int y_i^2 I(\|\mathbf{y}\| < r) d\mathbf{y} = \frac{I_1 r^2}{2 + p},$$

$$\int y_i y_j I(\|\mathbf{y}\| < r) d\mathbf{y} = 0 \text{ whenever } i \neq j.$$

Here, of course, Γ denotes the gamma function, $\Gamma(x) = \int_0^\infty t^{x-1} e^{-t} dt$.

Hence Eq. (2) becomes

$$
\Pr\{D_{mn} < r\}
$$
$$
= \frac{\pi^{p/2} r^p}{\Gamma(1 + p/2)} \int f^2(\mathbf{x}) d\mathbf{x} \left\{ 1 + \frac{r^2}{2(2+p)} \frac{\int f(\mathbf{x}) \sum f_{ii}(\mathbf{x}) d\mathbf{x}}{\int f^2(\mathbf{x}) d\mathbf{x}} + o(r^2) \right\}. \tag{3}
$$

As an example, suppose the distribution is multivariate normal with mean zero and covariance matrix Σ, i.e.,

$$
f(\mathbf{x}) = (2\pi)^{-p/2} |\Sigma|^{-1/2} \exp\left(-\frac{1}{2} \mathbf{x}^T \Sigma^{-1} \mathbf{x}\right). \tag{4}
$$

In this case it can be checked that

$$
\int f^2(\mathbf{x}) d\mathbf{x} = 2^{-p} \pi^{-p/2} |\Sigma|^{-1/2},
$$

$$
\frac{\int f(\mathbf{x}) \sum f_{ii}(\mathbf{x}) d\mathbf{x}}{\int f^2(\mathbf{x}) d\mathbf{x}} = -\frac{1}{2} \mathrm{tr}(\Sigma^{-1}) = -\frac{1}{2} \sum \lambda_i^{-1},
$$

where $\lambda_1, \ldots, \lambda_p$ are the eigenvalues of Σ. Thus

$$
\Pr\{D_{mn} < r\} = \frac{r^p |\Sigma|^{-1/2}}{2^p \Gamma(1 + p/2)} \left\{ 1 - \frac{r^2}{4(2+p)} \sum \lambda_i^{-1} + o(r^2) \right\}. \tag{5}
$$

Incidentally, Eq. (5) could be derived more quickly from the representation

$$
D_{mn}^2 = 2 \sum_{i=1}^{p} \lambda_i W_i
$$

where $W_1, \ldots, W_p$ are independent χ_1^2 random variables, but this method is specific to the multivariate normal distribution whereas Eq. (3) is general.

Now let us consider the consequences of Eq. (3) for the estimation of dimension. We may rewrite Eq. (3) in the form

$$
C(r) = \Pr\{D_{mn} < r\} = ar^p \{1 + br^2 + o(r^2)\} \tag{6}
$$

for constants a and b. Here $C(r)$ is the *correlation integral*. Equation (6) shows that this is asymptotic to ar^p as $r \to 0$ and this is a special case of the more general relation

$$
\lim_{r \to 0} \log C(r) / \log r = p, \tag{7}
$$

which, for an arbitrary distribution concentrated on a possibly fractal set, serves to define p as the *correlation dimension* of the set.

For the moment, however, we stick with Eq. (6). There are N data points, hence $N(N-1)/2$ interpoint distances D_{mn}. These will not all be independent, for example, because D_{lm} and D_{mn}, for indices l, m and n, are correlated. Let us ignore this feature for the moment and assume we have $N(N-1)/2$ independent D's. Let us further approximate Eq. (6) and assume we can write

$$C(r) = ar^p, \quad r < \epsilon \tag{8}$$

for some small (assumed known) ϵ. Suppose we order the D's as $D_1 \le D_2 \le \ldots D_{N(N-1)/2}$ and suppose the first M of these are less than ϵ. In that case the estimation of p under Eq. (8) based on $D_1, \ldots, D_M$ becomes a simple exercise in maximum likelihood estimation, with solution

$$\widehat{p} = M / \sum_{j=1}^{M} \log(\epsilon/D_j). \tag{9}$$

This estimate was obtained by Takens[24] as an alternative to the direct Grassberger-Procaccia method of plotting $\log C(r)$ against $\log r$, but an equivalent estimator has appeared earlier in the statistical literature for estimating distributions with power-law tails.[13,15,22,28] These papers motivate the following development.

Suppose Eq. (6) holds and define $Z_{mn} = \log(\epsilon/D_{mn})$ which is, of course, positive if and only if $D_{mn} < \epsilon$. Calculation from Eq. (6) shows that

$$\begin{aligned}
\Pr\{Z_{mn} > z | Z_{mn} > 0\} &= \Pr\{Z_{mn} > z\} / \Pr\{Z_{mn} > 0\} \\
&= C(\epsilon e^{-z})/C(\epsilon) \tag{10} \\
&= e^{-pz}\{1 + b(e^{-2z} - 1)\epsilon^2 + o(\epsilon^2)\}
\end{aligned}$$

and by direct integration of Eq. (10)

$$E\{Z_{mn} | Z_{mn} > 0\} = \frac{1}{p}\left\{1 - \frac{2b\epsilon^2}{p+2} + o(\epsilon^2)\right\}. \tag{11}$$

Equation (10) confirms that, to the first order of approximation, the distribution of Z_{mn} given $Z_{mn} > 0$ is exponential with mean $1/p$ and hence variance $1/p^2$. It is then automatic that, still considering only the first order of approximation, Eq. (9) is the maximum likelihood estimator of p and this has approximate variance p^2/M. Equation (11) computes the mean to second order and leads to the conclusion that the bias of $\widehat{p}$ is approximately $2pb\epsilon^2/(p+2)$. Since we also have $M \sim \frac{1}{2}N^2 a\epsilon^p$, the mean squared error (MSE), or the sum of squared bias and variance, is given approximately by

$$\text{MSE} \approx \left(\frac{2pb}{p+2}\right)^2 \epsilon^4 + \frac{2p^2}{N^2 a\epsilon^p}. \tag{12}$$

So far we have assumed the D_{mn}'s independent. In fact it can be shown that, for small ϵ, the number of overlapping pairs D_{lm}, D_{mn} for which both $D_{lm} < \epsilon$

and $D_{mn} < \epsilon$, is small compared with M and so this feature of the problem is insignificant. There is, however, another way of looking at it. The formula

$$\begin{aligned}
\Pr\{D_{lm} < r, D_{mn} < s\} \\
&= \int \int \int f(\mathbf{x})f(\mathbf{x}+\mathbf{y})f(\mathbf{x}+\mathbf{z})I(\|\mathbf{y}\| < r)I(\|\mathbf{z}\| < s)d\mathbf{x}d\mathbf{y}d\mathbf{z} \\
&\sim \int \int \int f^3(\mathbf{x})I(\|\mathbf{y}\| < r)I(\|\mathbf{z}\| < s)d\mathbf{x}d\mathbf{y}d\mathbf{z} \\
&\sim cr^p s^p,
\end{aligned}$$

where $c = \pi^p \int f^3(\mathbf{x})d\mathbf{x}/\Gamma^2(1+p/2)$, leads to the asymptotic formula

$$\begin{aligned}
\Pr\{Z_{lm} > y, Z_{mn} > z | Z_{lm} > 0, Z_{mn} > 0\} &\sim \frac{c(\epsilon e^{-y})^p(\epsilon e^{-z})^p}{c\epsilon^p \epsilon^p} \\
&= e^{-py} e^{-pz} \quad (y > 0, \ z > 0)
\end{aligned}$$

which shows that Z_{lm} and Z_{mn} are conditionally independent to the first order of calculation, provided ϵ is small. Thus, the bias and variance calculations we have presented are, to the order of accuracy given, not affected at all by dependence among the Z's.

Now let us return to Eq. (12). Suppose we decide that we want the MSE to be no less than η^2, for some predetermined η. We can achieve this by minimizing MSE with respect to ϵ, setting the result equal to η^2, and solving for N. This will tell us what sample size we need to achieve a root mean squared error no more than η. The result is

$$N = \left(\frac{4}{p}\right)^{p/8} \left(1 + \frac{p}{4}\right)^{(p+4)/8} \left(\frac{2p|b|}{p+2}\right)^{p/4} \left(\frac{2p^2}{a}\right)^{1/2} \eta^{-(p+4)/4}. \tag{13}$$

For a specific example, suppose we specify $\eta = 0.1$, and that the true distribution is the p-dimensional multivariate normal distribution with covariance matrix $\Sigma = I_p$. This might seem to be already a considerable restriction, but in fact it is only assuming that all p components of the random noise are independent with common variance, which is a natural enough assumption when talking about random noise; the further assumption that the common variance is 1 is irrelevant, since for estimating dimension the whole problem is scale invariant. In this case we see from Eq. (5) that $a = 1/\{2^p \Gamma(1+p/2)\}$, $b = -p/(8+4p)$, and substitution into Eq. (13) gives the results in Table 1. The third and fourth columns in this table will be explained later in the paper.

TABLE 1 Optimal N, B, and Ratio for
Various Values of p

p	N	B	Ratio
1	23	9.008	1.17
2	102	3.625	1.15
3	391	2.551	1.14
4	1423	2.101	1.13
5	5092	1.853	1.13
6	18190	1.697	1.13
7	65262	1.589	1.12
8	2.36×10^5	1.510	1.12
10	3.17×10^6	1.403	1.12
20	2.57×10^{12}	1.196	1.11
30	4.92×10^{18}	1.130	1.11
40	1.82×10^{25}	1.097	1.11

It is interesting to compare these results with the calculations of L. A. Smith.[20] Smith's conclusion was that, for the estimation of fractal dimension to be accurate to within 5%, when p is the true dimension, a sample size of at least 42^p is needed! Although the preceding calculations are based on quite different principles to those of Smith, the broad conclusion—that the sample size needed grows exponentially fast as the dimension increases—is the same. However, the sensitivity of this to η should be pointed out—for example, the same formula applied to $\eta = 1$ implies a necessary sample size of only 1000 for $p = 10$, and about 45000 for $p = 15$. This is clearly much less stringent. Thus, while my calculations confirm Smith's claims about the exponential growth in sample size with p, from a practical point of view I would regard his conclusions as too pessimistic.

LACUNARITY; GENERALIZED BAKER'S TRANSFORMATION

The preceding calculations are valid whenever an expansion of the form of Eq. (6) holds. This is not necessarily the case, however. Indeed, for many fractal sets $C(r)$ is of the form $r^p H(r)$ where $H(r)$ is a bounded function but does not tend to any constant a as $r \to 0$. In this case, Eq. (7) still holds (so, in particular, p is the correlation dimension), but Eq. (6) does not. The oscillatory or *lacunary* [21,26] behavior of $H(r)$ destroys the optimality properties of the estimator (9) and indeed may make it inconsistent. Theiler[26] illustrated these possibilities using as an example a random variable concentrated on the middle-third Cantor set. For a slightly more general but basically very similar example, I consider the generalized baker's transformation[9] which is complicated enough to illustrate many typical features of chaotic dynamical systems, but at the same time simple enough to allow many interesting quantities, including numerous dimension concepts, to be calculated analytically.

The generalized baker's transformation (henceforth GBT) is a deterministic transformation of the unit square $\{(x, y) : 0 \leq x \leq 1, 0 \leq y \leq 1\}$ given by

$$x_{n+1} = \begin{cases} \lambda_a x_n, & \text{if } y_n < \alpha, \\ \frac{1}{2} + \lambda_b x_n, & \text{if } y_n > \alpha, \end{cases} \tag{14}$$

$$y_{n+1} = \begin{cases} y_n/\alpha, & \text{if } y_n < \alpha, \\ (y_n - \alpha)/(1 - \alpha), & \text{if } y_n > \alpha, \end{cases} \tag{15}$$

where $\alpha, \lambda_a, \lambda_b$ are all $\leq 1/2$ and $\lambda_a \leq \lambda_b$. (There is an alternative version in which λ_a and λ_b are $> 1/2$; this is called the *fat baker's transformation*,[1] and its dynamics are quite a bit more complicated.)

The GBT is a deterministic map, so that given (x_0, y_0) we can determine the whole sequence, though it has the property of sensitive dependence on initial conditions which is often taken to be the definition of chaos. From our point of view, however, it is more convenient to treat (x_0, y_0) as random so that the whole sequence becomes a realization of a stochastic process. Specifically, suppose (x_0, y_0) are uniformly distributed on the unit square. Then, for each $n \geq 1$, y_n is also uniformly distributed on (0,1). The y_n's are not, of course, independent, but the indicators $I(y_n < \alpha)$ are. Hence the x_n process viewed on its own may be represented in the form

$$x_{n+1} = \begin{cases} \lambda_a x_n & \text{with probability } \alpha, \\ \frac{1}{2} + \lambda_b x_n & \text{with probability } 1 - \alpha, \end{cases} \tag{16}$$

the choices being independent for each n.

For the moment we consider only this process. Suppose x_n and x'_n are two independent realizations, then

$$x_{n+1} - x'_{n+1} = \begin{cases} \lambda_a(x_n - x'_n) & \text{with probability } \alpha^2, \\ \frac{1}{2} + \lambda_b x_n - \lambda_a x'_n & \text{with probability } \alpha(1 - \alpha), \\ -\frac{1}{2} + \lambda_a x_n - \lambda_b x'_n & \text{with probability } \alpha(1 - \alpha), \\ \lambda_b(x_n - x'_n) & \text{with probability } (1 - \alpha)^2. \end{cases} \tag{17}$$

The problem is further simplified if we further restrict ourselves to $\lambda_a = \lambda_b = \lambda$. Then, writing $z_n = 2(x_n - x'_n)$, we have

$$z_{n+1} = \lambda z_n + v_{n+1} \tag{18}$$

where v_n, $n \geq 1$ are independent and equal to 0, +1, or -1 with probabilities δ, $(1 - \delta)/2$, and $(1 - \delta)/2$. Here

$$\delta = \alpha^2 + (1 - \alpha)^2.$$

Now suppose F_n is the distribution function of $|z_n|$. From Eq. (18) it can be shown that

$$F_{n+1}(x) = \begin{cases} (1+\delta)/2 + (1-\delta)F_n((x-1)/\lambda)/2, & x > 1, \\ \delta F_n(x/\lambda) + (1-\delta)\{1 - F_n((1-x)/\lambda)\}/2, & x < 1, \end{cases} \quad (19)$$

from which it is possible to calculate F_n recursively. Moreover, as $n \to \infty$ so $F_n \to F$ say, and the limit of Eq. (19) then gives a functional equation satisfied by F.

Suppose $\lambda < 1/2$. (The argument when $\lambda = 1/2$ is only slightly more complicated.) The range of $|z_n|$ is $(0, 1/(1-\lambda))$ so $F_n((1-x)/\lambda) = 1$ whenever $x < (1-2\lambda)/(1-\lambda) = x_0$ say, so for $x < x_0$ we have

$$F(x) = \delta F(x/\lambda).$$

Define $H(x) = x^{-p}F(x)$ where p will be specified shortly; then for $x < x_0$

$$H(x) = \delta\lambda^{-p}H(x/\lambda).$$

This simplifies to $H(x) = H(x/\lambda)$ on $x \le x_0$, provided we define

$$p = \log\delta/\log\lambda. \quad (20)$$

In this case, then, p is the correlation dimension and $H(x)$ is a periodic function of $\log x$, but it is easily seen that $H(x)$ is not a constant and so does not tend to a constant; i.e., it is a lacunary function.

The more general case in which $\lambda_a \ne \lambda_b$ is harder to analyze exactly, but Eq. (17) shows that for sufficiently small x we have

$$F(x) = \alpha^2 F(x/\lambda_a) + (1-\alpha)^2 F(x/\lambda_b). \quad (21)$$

Define p by the relation

$$\alpha^2\lambda_a^{-p} + (1-\alpha)^2\lambda_b^{-p} = 1 \quad (22)$$

and let $H(x) = x^{-p}F(x)$. Equation (21) shows that for sufficiently small x we have

$$H(x) - \alpha^2\lambda_a^{-p}H(x/\lambda_a) + (1-\alpha)^2\lambda_b^{-p}H(x/\lambda_b)$$

which expresses $H(x)$ as a convex combination of $H(x/\lambda_a)$ and $H(x/\lambda_b)$. This establishes that $H(x)$ is bounded away from both 0 and ∞ as $x \to 0$, but again, it does not tend to a constant, so it is again a lacunary function.

Figure 1 shows a plot of $H(x)$ against $\log x$ for $\lambda_a = \lambda_b = 0.4$, $\alpha = 0.3$. The plot is similar to Figure 1(b) of Theiler,[26] who made the computation for the middle-thirds Cantor set.

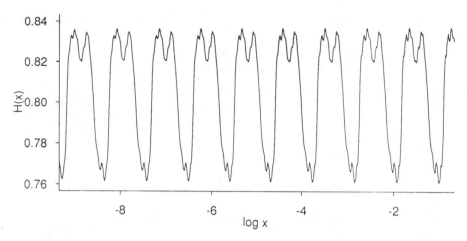

FIGURE 1 H-function for generalized baker's transformation.

Returning now to the original GBT in which both the x and y coordinates are considered, since the attractor is the product of a Cantor set for the x variable and the interval $(0,1)$ for the y variable, it is natural to expect that the correlation dimension of the two-dimensional attractor will be $1 + p$, where p is given by Eq. (22).[14] The Lyapunov dimension, information dimension, and pointwise dimension of the attractor are all[9] equal to

$$1 + \frac{\alpha \log \alpha + (1 - \alpha) \log(1 - \alpha)}{\alpha \log \lambda_a + (1 - \alpha) \log \lambda_b}$$

which is larger than the correlation dimension $1+p$, but there is no contradiction.[14]

Theiler[26] discussed the consequences of lacunarity on the Takens method of estimating fractal dimension. By easy adaptation of a formula in Theiler's paper, the bias in $1/\hat{p}$ as an estimator of $1/p$, where $\hat{p}$ given by Eq. (9), is

$$\int_0^\infty e^{-pz} \left\{ \frac{H(\epsilon e^{-z})}{H(\epsilon)} - 1 \right\} dz.$$

In periodic cases, this tends to 0 if and only if H is constant. Despite this short-coming, Theiler suggested that Takens' estimator could still be superior to the Grassberger-Procaccia procedure in practice. However this seems a rather odd suggestion: the direct comparison made by Theiler was based only on a very inefficient version of the Grassberger-Procaccia procedure, and it is difficult to defend the use of an inconsistent estimator when alternatives without this shortcoming are available.

THREE ALTERNATIVE ESTIMATORS

In this section three alternative estimators are discussed. The first is equivalent to Takens' procedure, and suffers from the same disadvantage when the true distribution is lacunar. The second and third versions represent attempts to preserve the best features of Takens' estimator while also handling the lacunar case.

All three procedures take as their starting point the obvious sample estimate of $C(r)$,

$$C_N(r) = \frac{\text{Number of interpoint distances } D_{mn} < r}{N(N-1)/2}$$

Suppose $C_N(r)$ is evaluated at $r = r_0, r_1, \ldots, r_K$ where $r_0 = \epsilon$ and $r_k = r_0 \phi^k$ for $k \geq 1$, where ϕ is between 0 and 1. Specifically, suppose there are N_k interpoint distances less than r_k. We write $\epsilon' = r_K = \epsilon/B$ where $B = \phi^{-K}$. As before, we treat the interpoint distances as if they were $N(N-1)/2$ independent random variables with distribution function C, N being the total sample size.

Within this set up, the conditional distribution of N_k given N_{k-1} is binomial with parameters N_{k-1} and θ_k, where $\theta_k = C(r_k)/C(r_{k-1})$. Let us first suppose that $C(r)$ satisfies the relation $C(r) \sim ar^p$ as $r \to 0$ for constants a and p. Then, provided ϵ is small enough, for all k we will have $\theta_k \approx \theta = \phi^p$. This motivates what we call the *binomial model*, in which we assume that, given N_{k-1}, N_k has a binomial distribution with parameters N_{k-1} and $\theta = \phi^p$.

Estimation of θ and hence p may proceed by the method of maximum likelihood. Under the assumptions we have made, the joint distribution of $N_1, \ldots, N_K$ given N_0 is a product of the conditional distributions of N_1 given N_0, N_2 given N_1 and so on up to N_K given N_{K-1}. The likelihood for θ is therefore proportional to

$$\prod_{k=1}^{K} \theta^{N_k}(1-\theta)^{N_{k-1}-N_k}. \tag{23}$$

The maximum likelihood estimator $\hat{\theta}$ is found by maximizing Eq. (23), to give

$$\hat{\theta} = \frac{\sum_{k=1}^{K} N_k}{\sum_{k=1}^{K} N_{k-1}}, \tag{24}$$

and we then have $\hat{p} = \log \hat{\theta}/\log \phi$.

It is possible to obtain approximate expressions for the bias and variance of this estimator. It turns out that, for ϕ close to 1 and K large, these are equivalent to the expressions obtained for Takens' estimator but, again, the estimator is inconsistent in the lacunar case.

Now let us consider a *nonhomogeneous binomial* model, in which we do not assume equality of θ_k for $k = 1, \ldots, K$, but allow these to vary arbitrarily. It will be seen that this avoids the difficulty in Takens' or the binomial estimator, but at the cost of inefficiencies elsewhere.

Since the conditional distribution of N_k given N_{k-1} is binomial with parameters N_{k-1} and θ_k, the natural estimator is $\widehat{\theta}_k = N_k/N_{k-1}$ with estimated variance $\widehat{\theta}_k(1 - \widehat{\theta}_k)/N_{k-1}$. We then estimate $\log \theta_k$ by $\log \widehat{\theta}_k$ with approximate variance $(1 - \widehat{\theta}_k)/(\widehat{\theta}_k N_{k-1})$. Consider $\widehat{\theta}$ defined by

$$\log \widehat{\theta} = \frac{1}{K} \sum_{k=1}^{K} \log \widehat{\theta}_k. \tag{25}$$

Let us consider the bias and variance of this. We have $\theta_k = C(r_k)/C(r_{k-1}) = \phi^p H(r_k)/H(r_{k-1})$ so that

$$\frac{1}{K} \sum \log \theta_k = p \log \phi + \frac{1}{K} \{\log H(r_K) - \log H(r_0)\}. \tag{26}$$

Hence the asymptotic bias in $\widehat{p} = \log \widehat{\theta}/\log \phi$ is

$$\frac{1}{\log B} \{\log H(\epsilon) - \log H(\epsilon')\}. \tag{27}$$

If H is lacunar but bounded, then Eq. (27) is guaranteed to tend to 0 as $B \to \infty$.
The variance of $\log \widehat{\theta}$ is given approximately by

$$\frac{1}{K^2} \sum \frac{1 - \theta_k}{\theta_k N_{k-1}}.$$

In the case $C(r) \sim ar^p$, we have $N_{k-1} \sim \frac{1}{2} N^2 a \epsilon^p \phi^{(k-1)p}$ so the variance as $\phi \to 1$ is asymptotically

$$\frac{1}{K^2} \frac{1 - \theta}{\theta} \frac{2}{N^2 a \epsilon^p} \sum_{k=1}^{K} \phi^{-(k-1)p} \sim \frac{2(1 - \phi^p)}{a N^2 \epsilon^p K^2} \frac{\phi^{-Kp} - 1}{\phi^{-p} - 1}$$

$$\sim \frac{2}{N^2 a \epsilon^p} \frac{\log^2 \phi}{\log^2 B} (B^p - 1)$$

so that the variance of $\widehat{p}$ is asymptotically

$$\frac{2}{N^2 a \epsilon^p} \frac{B^p - 1}{\log^2 B}. \tag{28}$$

If C satisfies Eq. (6), then we find $H(\epsilon)/H(\epsilon/B) = 1 + b\epsilon^2(1 - 1/B^2) + o(\epsilon^2)$ and the bias expression (27) reduces to

$$\frac{b\epsilon^2}{\log B} \left(1 - \frac{1}{B^2}\right). \tag{29}$$

We can now combine Eqs. (28) and (29) to form an expression for the mean squared error similar to Eq. (12), and in similar fashion, may minimize with respect to ϵ to obtain the sample size N needed to achieve root mean squared error η. In this case, however, there is the additional minimization with respect to B. The third and fourth columns of Table 1 show, respectively, the optimal value of B associated with a particular p, and the ratio of resulting optimal values of N, i.e., the optimal N under this procedure divided by the optimal N under the Takens procedure.

It can be seen that, for large p, this ratio is in the range 1.1–1.13. In other words, in the smooth case when Eq. (6) holds, the nonhomogeneous binomial procedure is between 10% and 13% less efficient than the Takens (or homogeneous binomial) procedure, and it has the additional advantage of giving a good estimate in the lacunar case. The loss of efficiency is relatively small and might be considered a good "insurance" against the lacunar case.

However, things are not quite as simple as that. It is still necessary to choose a suitable value of B in practice, and it can be shown that the mean squared error increases very rapidly as B moves away from its optimal value. The optimal value of B in Table 1 is close to 1 for large p, but from Eq. (27) it can be seen that, for the lacunar case, the bias of $\hat{p}$ is proportional to $1/\log B$. In other words, to obtain reasonable efficiency in the smooth case, we require small B, whereas to obtain small bias in the lacunar case, we require large B. These are clearly incompatible requirements.

Now we introduce our third estimator. A common device in analyzing binomial data, where the "success probability" parameter θ_k varies from one sample to another, is to assume the θ_k's are random with a common distribution, which (since $0 \le \theta_k \le 1$) may conveniently be taken to be a beta distribution. The resulting model is called a *beta-binomial* model.[4,5,12]

If the beta parameters are α and β, then the density of θ_k is given by

$$f(\theta_k) = \frac{\Gamma(\alpha+\beta)}{\Gamma(\alpha)\Gamma(\beta)}\theta_k^{\alpha-1}(1-\theta_k)^{\beta-1}$$

with mean $\alpha/(\alpha+\beta)$ and variance $\alpha\beta/\{(\alpha+\beta)^2(\alpha+\beta+1)\}$. The probability mass function of N_k given θ_k and N_{k-1} is proportional to

$$\theta_k^{N_k}(1-\theta_k)^{N_{k-1}-N_k}$$

which, after integrating out the density of θ_k, becomes

$$\frac{\Gamma(\alpha+\beta)}{\Gamma(\alpha)\Gamma(\beta)}\frac{\Gamma(\alpha+N_k)\Gamma(\beta+N_{k-1}-N_k)}{\Gamma(\alpha+\beta+N_{k-1})}$$
$$= \frac{\alpha(\alpha+1)\dots(\alpha+N_k-1)\beta(\beta+1)\dots(\beta+N_{k-1}-N_k-1)}{(\alpha+\beta)(\alpha+\beta+1)\dots(\alpha+\beta+N_{k-1}-1)}. \tag{30}$$

The likelihood function is proportional to the product of Eq. (30) over $k = 1,\dots,K$, and maximum likelihood estimates of α and β can be obtained by numerical maximization of that.

To interpret the parameters α and β, define $\theta = \alpha/(\alpha+\beta)$ and $\tau = 1/(\alpha+\beta+1)$. Then θ_k has mean θ and variance $\theta(1-\theta)\tau$; i.e., the parameter τ represents the additional variability in θ_k under the beta-binomial model, the case $\tau = 0$ being the binomial case. As in earlier models, once we have estimated θ by a maximum likelihood estimator $\widehat{\theta} = \widehat{\alpha}/(\widehat{\alpha} + \widehat{\beta})$, the estimate of p is given by

$$\widehat{p} = \log \widehat{\theta}/\log \phi.$$

Detailed asymptotic properties of this estimator have not been worked out, but preliminary results indicate the following:

a. If $C(r) \sim ar^p$, then the model estimates τ close to 0; i.e., the results approximate the binomial model.

b. In the lacunar case, the bias remains small.

c. The properties do not depend critically on the correct choice of B. (Intuitively, the reason for this is that in both the binomial and beta-binomial models, the estimate automatically downweights the large values of k, i.e., those where the data are sparse, whereas the nonhomogeneous binomial model treats all k as of equal importance so that when using that estimator it is important not to take K and hence B too large.)

EXAMPLES

The binomial and beta-binomial estimates are now applied to some artifically generated data sets in two-dimensional space.

Figure 2 gives the four data sets. Figure 2(a) is a cloud of independent points from a bivariate normal distribution. Figure 2(b) is generated from the generalized baker's transformation with $\lambda_a = \lambda_b = 0.3333$, $\alpha = 0.4$. In this case the information dimension is 1.6125, and the correlation dimension 1.5952. Figure 2(c) is a more complicated but still easily understood example of a fractal—the Sierpinski gasket with unequal weights (cf. Figure 9.1.3, pages 338 and 339 of Barnsley[2]). Finally, Figure 2(d) illustrates the well-known Hénon attractor generated by iterating the system $x_{n+1} = y_n + 1 - 1.4x_n^2$, $y_{n+1} = 0.3x_n$. This is an example of a set which is believed to be fractal but without any known self-similarity structure.

The binomial and beta-binomial estimates of $1/p$ for these four data sets, with associated confidence bands, are illustrated in Figure 3. Figure 3(a) and 3(b) illustrate the binomial and beta-binomial estimates of $1/p$ for the normal data; here, of course, $p = 2$. The abscissa (here and on all subsequent plots) is -log ϵ. The value of B is not given because the estimates are insensitive to this provided B is reasonably large.

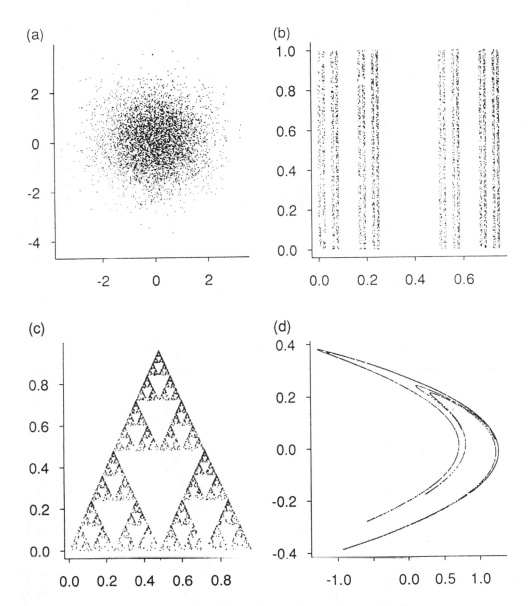

FIGURE 2 (a) Normal data; (b) GBT; (c) Sierpinski gasket; and (d) Hénon map.

Figure 3(a) illustrates typical behavior that might be expected with smooth data. For very small ϵ (right-hand end of plot), the variability of the estimate is large as indicated by the 95% confidence limits (dotted curves). The left-hand side

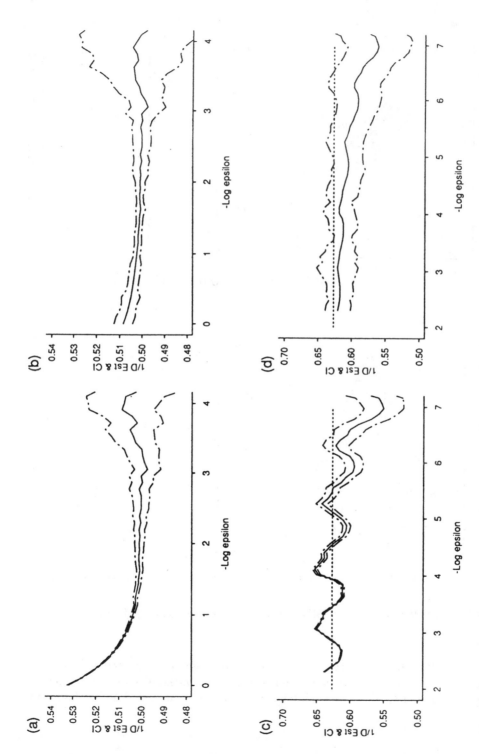

FIGURE 3   (a) Normal data, binomial estimate; (b) normal data, beta-binomial estimate; (c) GBT, binomial estimate;
(d) GBT, beta-binomial estimate. (continued)

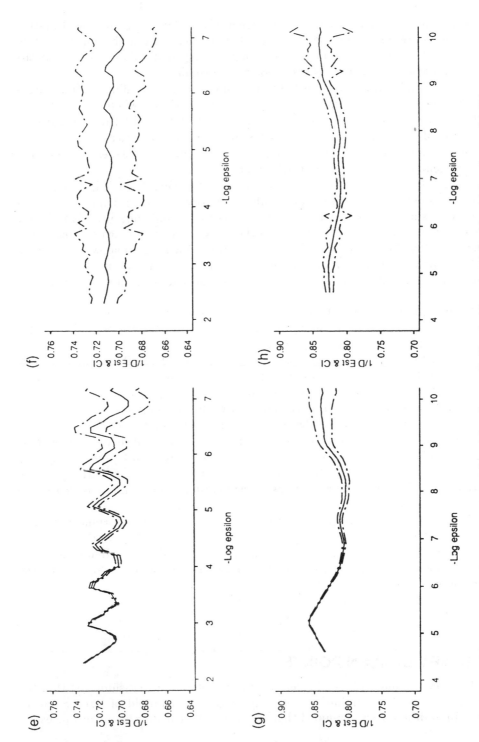

FIGURE 3 (continued) (e) Sierpinski gasket, binomial estimate; (f) Sierpinski gasket, beta-binomial estimate; (g) Hénon map, binomial estimate; (h) Henon map, beta-binomial estimate.

illustrates the systematic bias that results when ϵ is too large—in fact, the deviation that occurs here almost exactly matches that predicted from the Eq. (6). From inspection of the plot, one might guess that the optimal value of ϵ is near the middle and this is confirmed by the theoretical formulae.

In contrast, Figure 3(b) shows both the estimate and the confidence limits much less sensitive to the choice of ϵ, except at the right-hand end where the confidence limits become very wide. In the middle, i.e., where the binomial estimate is optimal, the binomial and beta-binomial estimates give virtually identical estimates and confidence limits. It is believed that this behavior holds good generally when the data are generated by a smooth distribution.

Figure 3(c) shows the binomial estimate for the generalized baker's transformation data, which as can be seen is highly oscillatory with variation in ϵ. The confidence limits here are very narrow except at the right-hand end, but this is clearly completely misleading since the estimates themselves vary to a much greater extent than the width of the confidence limits. In contrast, Figure 3(d) shows much more stable estimates, admittedly with much wider confidence limits, but this time with some confidence in the reality of those limits. The dotted horizontal line is the true value.

Figure 3(e) and (f) are the corresponding plots for the Sierpinski map—something with a structure which is, again, very similar to a Cantor set. The general conclusions are the same as for the Bakers map.

Finally, Figure 3(g) and (h) illustrate the behavior of the Hénon map. In this case the true behavior is unknown, but it is believed to be lacunar though without the periodicity associated with self-similar fractals. Figure 3(g) again shows variability much greater than allowed by the nominal confidence limits, while Figure 3(h) shows the estimate and its confidence limits greatly smoothed.

These four examples illustrate what I believe to be a general theme; namely that in the case of smooth distributions (the normal data), the beta-binomial estimator performs just as well as the binomial estimator but with less sensitivity in the choice of ϵ, while, in the other three cases, the beta-binomial estimate is clearly the superior of the two.

Finally, it should be mentioned that there are numerous important statistical issues that this paper has not touched upon at all. Among these are the effect of noise in the data (for example, measurement error added to the Hénon set), the effect of time-series correlations which have been ignored in this analysis, and the choice of embedding dimension.

SUMMARY OF MAIN POINTS

The formulae leading to Eqs. (12) and (13) were based on the assumption that the data are independent random vectors contained in a manifold of integer dimension p. In addition it was assumed that the M smallest integer point distances are independent random variables. The second assumption can be justified by asymptotic arguments for large sample sizes and relatively small M. The first is strictly speaking incompatible with chaotic dynamics, but its purpose is to form a reference standard against which dimension estimates can be judged. Under these assumptions we have argued that there are limits to the precision it is possible to achieve, but these are not nearly as bad as argued in earlier work of L. A. Smith.[20]

These results assume Takens'[24] estimator in place of the original Grassberger-Procaccia procedure. Amplifying an argument due to Theiler,[26] it has been pointed out that a "lacunarity" property of many self-similar sets invalidates this estimator. This naturally clouds the question of whether our bias-variance results are applicable to chaotic systems.

Elaborating this theme, three new estimators have been proposed. The first, Eq. (24), is effectively equivalent to the Takens procedure (and suffers from the same difficulty regarding lacunarity) but is based on the correlation integral evaluated at only a finite number of values of r, which is more convenient for computation. The second estimator (25) goes to the opposite extreme by assuming that nothing is known about the function $H(x)$ besides the fact that it is bounded. The theoretical bias-variance results for this are not very much different from those of Takens' estimator, but it is argued that optimal performance would be difficult to achieve in practice because of the need to achieve a good lower bound, as well as a good upper bound, for the r values considered. This leads to our third estimator, the beta-binomial estimator derived by maximizing Eq. (30), which does not assume $H(x)$ to be constant but does implicitly assume some regularity, as represented by the beta density for the θ_k's. Confidence limits may be obtained by using standard statistical results for maximum likelihood estimators. Results with various artificial data sets support the efficacy of the beta-binomial estimator.

ACKNOWLEDGMENTS

The original stimulus for this paper came from the SERC-funded research workshop in Non-Linear Time Series, held at Edinburgh in July 1989, organized by Professor H. Tong. I would like to thank Doyne Farmer and Martin Casdagli for conversations during that workshop which helped to stimulate my interest. The bulk of the paper was written at Surrey University and supported by the Wolfson Foundation in the form of a personal research fellowship. The work employed computational facilities provided by the SERC through its Complex Stochastic Systems initiative.

REFERENCES

1. Alexander, J. C., and J. A. Yorke. "Fat Baker's Transformations." *Ergodic Theory and Dynamical Systems* **4** (1984): 1–23.

2. Barnsley, Michael. *Fractals Everywhere.* Boston: Academic Press, 1988.

3. Broomhead, D. S., and R. Jones. "Time-Series Analysis." *Proc. R. Soc. Lond. A* **423** (1989): 103–121.

4. Crowder, M. J. "Beta-Binomial Anova for Proportions." *Applied Statistics* **27** (1978): 34–37.

5. Crowder, M. J. "Inference about the Intraclass Correlation Coefficient in the Beta-Binomial ANOVA for Proportions." *J.R. Statist. Soc. B* **41** (1979): 230–234.

6. Cutler, D., and D. A. Dawson. "Estimation of Dimension for Spatially Distributed Data and Related Limit Theorems." *J. Multivariate Analysis* **28** (1989): 115–148.

7. Cutler, D., and D. A. Dawson. "Nearest-Neighbor Analysis for a Family of Fractal Distributions." *Annals of Probability* **18** (1990): 256–271.

8. Dvorak, I., and J. Klaschka. "Modification of the Grassberger-Procaccia Algorithm for Estimating the Correlation Exponent of Chaotic Systems with High Embedding Dimension." *Phys. Lett. A* **145** (1990): 225–231.

9. Farmer, J. D., E. Ott, and J. A. Yorke. "The Dimension of Chaotic Attractors." *Physica D* **7** (1983): 153–180.

10. Grassberger, P., and I. Procaccia. "Characterization of Strange Attractors." *Phys. Rev. Lett.* **50** (1983): 346–349.

11. Grassberger, P., and I. Procaccia. "Measuring the Strangeness of Strange Attractors." *Physica D* **9** (1983): 189–208.

12. Griffiths, D. A. "Maximum Likelihood Estimation for the Beta-Binomial Distribution and an Application to the Household Distribution of the Total Number of Cases of a Disease." *Biometrics* **29** (1973): 637–648.

13. Hall, P. "On Some Simple Estimates of an Exponent of Regular Variation." *J.R. Statist. Soc. B* **44** (1982): 37–42.

14. Hentschel, H. G. E., and I. Procaccia. "The Infinite Number of Generalized Dimensions of Fractals and Strange Attractors." *Physica D* **8** (1983): 435–444.

15. Hill, B. M. "A Simple General Approach to Inference about the Tail of a Distribution." *Annals of Statistics* **3** (1975): 1163–1174.

16. Ramsey, J. B., and H.-J. Yuan. "Bias and Error Bars in Dimension Calculations and Their Evaluation in Some Simple Models." *Phys. Lett. A* **134** (1989): 287–297.

17. Ramsey, J. B., and H.-J. Yuan. "The Statistical Properties of Dimension Calculations using Small Data Sets." *Nonlinearity* **3** (1990): 155–176.

18. Ruelle, D. "Deterministic Chaos: The Science and the Fiction (The Claude Bernard Lecture 1989)." *Proc. R. Soc. Lond. A* **427** (1990): 241–248.

19. Silverman, B. W., and T. C. Brown. "Short Distances, Flat Triangles and Poisson Limits." *J. Appl. Probability* **15** (1978): 815–825.

20. Smith, L. A. "Intrinsic Limits on Dimension Calculations." *Phys. Lett. A* **133** (1988): 283–288.

21. Smith, L. A., J.-D. Fournier, and E. A. Speigel. "Lacunarity and Intermittency in Fluid Turbulence." *Phys. Lett. A* **114** (1986): 465–468.

22. Smith, R. L. "Estimating Tails of Probability Distributions." *Annals of Statistics* **15** (1987): 1174–1207.

23. Takens, F. "On the Numerical Determination of the Dimension of an Attractor." In *Lecture Notes in Mathematics*, edited by D.A. Rand and L.S. Young, vol. 808, 366–381. New York: Springer Verlag, 1981.

24. Takens, F. "On the Numerical Determination of the Dimension of an Attractor." In *Lecture Notes in Mathematics*, vol. 1125, 99–106. New York: Springer Verlag, 1984.

25. Taylor, C. C., and S. J. Taylor. "Estimating the Dimension of a Fractal." *J. R. Statist. Soc. B* **53(1)** (1991): to appear.

26. Theiler, J. "Lacunarity in a Best Estimator of Fractal Dimension." *Phys. Lett. A* **133** (1988): 195–200.

27. Theiler, J. "Statistical Precision of Dimension Estimators." *Phys. Rev. A* **41** (1990): 3038–3051.

28. Weissman, I. "Estimation of Parameters and Large Quantiles based on the *k* Largest Observations." *J. Amer. Statist. Assoc.* **73** (1978): 812–815.

W. A. Brock† and S. M. Potter‡

†Department of Economics, University of Wisconsin-Madison, 1180 Observatory Drive, Madison, WI 53706 and ‡Department of Economics, University of California, Los Angeles, 405 Hilgard Avenue, Los Angeles, CA 90024

Diagnostic Testing for Nonlinearity, Chaos, and General Dependence in Time-Series Data

"The plain fact is that an investigator into oscillations in time series nowadays is very much in the position of the detective in the modern crime novel. By the time he arrives on the scene to inspect the corpse, so many feet have trampled all round it that he can easily find a footprint to fit any suspect he likes to choose. The main difference is that under the rules of criminal fiction the detective must not be the culprit. In the theory of time-series he frequently is."

—Kendall (1945) as quoted by Matthew Richardson in "Temporary Components of Stock Prices: A Skeptic's View." November, 1988.

1. INTRODUCTION

A major theme of this conference is the development of methods for prediction of time-series observations. The methods are "trained" on a subset of the available data and evaluated by predictive performance on the rest of the data. Many of the methods discussed during the conference were designed to perform well when the data is assumed to be generated by a deterministically chaotic recurrence which may be contaminated with error in the observer function (e.g., Eubank and Farmer,[16] and references to work on prediction of chaotic systems by Abarbanel, Broomhead, Casdagli, Eubank, Farmer, King, Lapedes, Packard, Sidorowich, et al.).

It is helpful to state some notation at the outset. Denote random variables by capitals and their sample values by lower case. To be precise let A, X, M, N, and e be vector-valued random variables. To avoid confusion with the expectation operator we use only lower case e.

DEFINITION 1.1

We shall say the observed data process $\{A(t)\}$ is generated by a noisy deterministically chaotic explanation if

$$A_t = h(X_t, M_t),\qquad(1.1.a)$$

where $\{M_t\}$ is a mean zero, finite variance, Independent and Identically Distributed (IID)[1] process. Hence, $\{M_t\}$ represents measurement error and $h(x, m)$ is a noisy observer function of the state X_t. Let

$$X_t = G(X_{t-1}, N_t).\qquad(1.1.b)$$
$$N_t = f(\mathcal{F}^X_{t-1})e_t,\quad e_t \text{ IID}, E[e_t] = 0,\quad e_t \text{ independent of } \mathcal{F}^X_{t-1}.\qquad(1.1.c)$$

Here N_t is a strictly stationary stochastic process satisfying a martingale-difference-sequence-type property relative to the sigma fields generated by X_t (we use the notation $f(\mathcal{F}^X_{t-1})$ to stand for $f(\mathbf{Y_{t-1}})$, where $\mathbf{Y_{t-1}}$ is measurable with respect to the sigma algebra generated by $(X_{t-1}, X_{t-2}, \ldots)$ and $f(\cdot)$ is a measurable function, cf. Hall and Heyde[17]). That is, the expectation of N_t conditioned on the values of $(X_{t-1}, X_{t-2}, \ldots)$ is zero. However, unlike an independent sequence, other aspects of N_t (e.g., $N_t^2 = f^2(\mathcal{F}^X_{t-1})e_t^2$) might be more predictable than implied by the unconditional distribution. We label $\{N_t\}$ propagated noise in order to clearly distinguish it from measurement error.

When $\{N_t\}$ is fixed at its unconditional mean, the resulting dynamics $x_t = G(x_{t-1}, 0)$ are deterministically chaotic, i.e., the largest Lyapunov exponent exists, is constant almost surely with respect to the assumed unique natural invariant measure of $G(\cdot, 0)$, and is positive.[2]

We take the definition of chaos used here to be positive largest Lyapunov exponent of the underlying deterministic map. The reconstruction techniques discussed

[1] We do not wish to take a stand on the meaning of randomness; hence, the reader should feel free to interpret IID as a deterministic system with a sufficiently high dimension such that prediction is prohibitively expensive.

[2] The case of more than one lag appearing in G can be treated by the usual device of redefining variables and enlarging the state space to obtain a first-order system. Similar devices apply to the case where the noise processes are finite-order Markov processes, although we shall concentrate on the martingale difference sequence case.

by Eubank and Farmer[16] are designed for best use when the propagated noise processes are fixed at their means. In this paper we concentrate on the alternative case where the measurement error is assumed to be fixed at its mean of zero, and h is the identity function. Thus, we have:

$$A_t = h(X_t, M_t) = h(X_t, 0) = X_t \qquad (1.1.a')$$

However, we provide some indications of how our methods can help in separating signal from noise and distinguishing measurement error from propagated noise.

We develop a diagnostic testing methodology to test for the presence of dynamic, possibly stochastic nonlinear structure which is potentially exploitable for prediction purposes. Hence, we concentrate on the properties of $\{n_t\}$ sequences generated by the whole class (i.e., linear through chaotic) of difference equations $H(\cdot)$ applied to the observed data $\{X_t\}$. To focus the discussion we assume that the propagated noise enters additively:

$$X_t = H(X_{t-1}) + N_t. \qquad (1.1.b')$$

We concentrate on two types of problem.

Firstly, in view of the simplicity of linear methods, i.e., Autoregressive Moving Average (ARMA), it seems worthwhile to develop a simple diagnostic test for nonlinearity. This is especially so in a science like economics or biology where the signal may be weak relative to noise, so that linear methods may pick up most of the structure that is usable for out-of-sample prediction.[3] The suggested diagnostics are easy to use and could be thought of as a first hurdle that a time-series sample must pass before expending the extra effort to attempt to find and exploit nonlinear structure or as a warning sign that a linear model may perform comparably to a nonlinear one in a prediction exercise.

Secondly, suppose we find a linear model inadequate; then, in order to construct optimal forecasts in the mean squared error sense, it will be necessary to model exactly any higher-order dependencies in the propagated noise term.[4] For linear models this is not the case because linearity allows the interchange of expectation and function operators, hence always zeroing out the noise. The testing framework below has the advantage that it will be sensitive to higher-order dependence in the noise left over after estimating nonlinear models. Furthermore, by generating a sequence of predicted values ignoring the problems of interchanging expectation and function operators, we are able to suggest techniques for distinguishing between measurement error and propagated noise. The methods we develop to provide a statistical measure for IID are based on the theory of V-statistics and their close relatives U-statistics. A theme we develop here is that many objects studied in

---

[3] However, there maybe other aspects of the time series for which a nonlinear model provides a superior explanation.[22]

[4] Ignoring the propagated noise or assuming linearity allows for much quicker computation of forecasts, however, with modern computing power numerical integration or Monte Carlo integration is feasible and not much slower.

nonlinear science such as correlation integrals and certain lower bounds to Kolmogorov entropy are V-statistics. Hence, V-statistic theory allows one to develop some theory of statistical inference for nonlinear science and chaos theory. Much of the theory we use builds on Denker and Keller.[14,15] A central idea we shall exposit is a diagnostic test for temporal dependence or "left out structure" which may be applied to forecast errors of fitted models. A very important property of the U- and V-statistics-based diagnostic test for temporal dependence that we develop is invariance of the first-order asymptotic distribution to estimation error. In particular our main diagnostic test can be applied to estimated residuals or forecast errors of a prediction algorithm to test for the presence of potentially usable "left out" structure without the need to correct for the dependencies introduced by estimation.

The paper is organized as follows. In section 2, we outline some notions of stochastic linearity and nonlinearity. Section 3 develops the diagnostics for testing for extra predictive content over a null model. Section 4 outlines the argument behind the invariance of the asymptotic sampling distribution of the diagnostic to replacing unobserved time series with estimated residuals. Section 5 discusses some potential extensions of the statistical techniques presented. Section 6 briefly describes some applications to financial time series. Section 7 concludes the paper.

2. SOME NOTIONS OF STOCHASTIC LINEARITY AND NONLINEARITY

Let $\{x_t\}$ be a time-series sample from a strictly stationary stochastic process $\{X_t\}$. In applications to economics and finance, it is assumed that the series has been reduced to stationarity by estimation of trend and seasonal components and applying the theory to the residuals. There are many important issues in the estimation of the stationary-inducing transform which we assume away here. Without loss of generality we set the mean to zero. In order to test the null of stochastic linearity, one first needs to decide what exactly it is. It is not obvious what the best working definition of linearity is for stochastic processes. The first definition we use is taken from Priestley[24] and without loss of generality is for a univariate series.

DEFINITION 1

$\{X_t\}$ is a stochastic linear process if X_t can be written

$$X_t = \sum_{j=1}^{\infty} a_j N_{t-j} = a(B)N_t \, , \quad a_0 = 1 \text{ and } B^j N_t = N_{t-j} \, , \tag{2.1}$$

where $\{N_t\}$ is an independent, identically distributed (IID) continuously distributed stochastic process with distribution function $F(\cdot)$ and zero mean. We assume either

$$\{N_t\} \text{ has a finite variance, and } \sum_{j=0}^{\infty} |\ a_j\ | < \infty; \qquad (2.2.a)$$

or

$$x^{\alpha}(1 - F(x)) = x^{\alpha} P(N_t > x) \rightarrow pC, \text{ as } x \rightarrow \infty$$
$$x^{\alpha} F(-x) = x^{\alpha} P(N_t < -x) \rightarrow qC, \text{ as } x \rightarrow \infty$$
$$\text{and } \sum_{j=1}^{\infty} j\ |\ a_j\ |^k < \infty, \text{ for some } 0 < k < \alpha, k \leq 1, \qquad (2.2.b)$$

where $1 < \alpha < 2$, $0 \leq p = 1$ and C is a finite constant.

We further assume that $a(B)$ is a rational polynomial in B, i.e.,

$$a(B) = \frac{\theta(B)}{\phi(B)} \qquad (2.3)$$

where both $\theta(\cdot)$ and $\phi(\cdot)$ are finite polynomial operators. We restrict $\phi(z) \neq 0$ for $|\ z\ | \leq 1$ and, in order to obtain invertibility, we shall assume ϕ and θ to have no common zeros and $\theta(z)$ to have no roots inside the unit circle in the complex plane.[11] Thus, the representation (2.1) makes sense under Eq. (2.2.b) (see Brockwell and Davis[11] Propositions 12.5.1 and 12.5.2). Equation (2.3) is also an additional assumption that the time series only exhibits short-range dependence. We discuss the case of long-range dependence in section 5. We have an ARMA representation of Eq. (2.1) as

$$\phi(B)X_t = \theta(B)N_t . \qquad (2.4)$$

If $\{N_t\}$ has a finite variance, then X_t will be a conventional covariance stationary time series. The case of infinite variance is considered because financial time series have bursting behavior that might be symptomatic of lack of finite moments from a stochastic point of view. Figure 1 illustrates the point showing a sample path from an AR(3) with $\alpha = 1.9$. Notice how there are similarities to the intermittency found in complex deterministic dynamic systems.

The generalization to higher-order systems will allow contemporaneous dependence between the N_t components but no dependence across time. Definition 1 has the desirable property that, using $\{N_t\}$ as "basis," if we add two processes that are linear, then we obtain another linear process.

Notice that on the surface this definition of "linear" looks empirically empty. This is so because every stationary time series under modest regularity conditions[23] possesses a "Wold" representation of the form (2.1) where the process $\{N_t\}$ is uncorrelated. The word "uncorrelated" is key.

What gives the definition empirical content is the assumption of IID on the "innovation" process $\{N_t\}$. Remember independence implies lack of correlation but lack of correlation does not imply independence. In order to introduce a weaker definition of linear stochastic process, let us say a few words about motivation of Definition 1 from a formal Taylor series perspective called the Volterra expansion.[24,27]

Since space is at a premium, we proceed heuristically in the context of a simple pedagogical example. Consider the process

$$X_t = H(X_{t-1}, N_t),\tag{2.5}$$

$\{N_t\}$ IID, with all moments finite. If $H(x, n)$ were linear, e.g.,

$$H(x, n) = \phi x + n,\tag{2.6}$$

then we may recursively solve Eq. (2.5) backwards to obtain, in the limit, provided that $|\phi| < 1$, an expression of a form similar to Eq. (2.1):

$$X_t = \sum_{j=0}^{\infty} a_j N_{t-j} = a(B) N_t, \; a_j = \phi^j.\tag{2.7}$$

The idea of the Volterra expansion is similar. Recursively "solve" Eq. (2.5) backwards as we did for the linear autoregression case to obtain some function

$$X_t = G(N_t, N_{t-1}, \ldots).\tag{2.8}$$

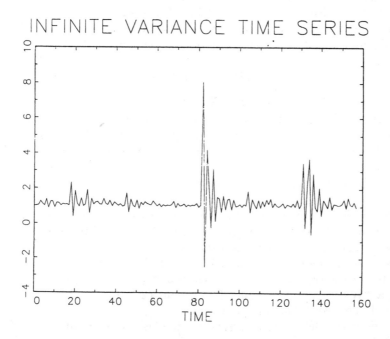

FIGURE 1 Infinite variance time series.

Now formally expand Eq. (2.8) in a "Taylor" series around a "point," e.g., the infinite vector of means $\mathbf{N} \equiv (N, N, \ldots)$ or the infinite vector of zeros $\mathbf{0} \equiv (0, 0, \ldots)$. For specificity expand around 0. We obtain

$$X_t = G(0) + DG(0) \cdot \mathbf{N}_t + \frac{1}{2}\mathbf{N}_t' \cdot D^2 G(0) \cdot \mathbf{N}_t + \ldots ; \mathbf{N}_t \equiv (N_t, N_{t-1}, \ldots), \quad (2.9)$$

where "$\cdot$" denotes formal matrix-vector product, and D, D^2 denote formal first and second derivatives. The above "expansion" which is purely formal and not justified is intuitive. Making it precise is rather subtle. See, for example, Rosenblatt.[27] Nevertheless we may use Eq. (2.9) as a pedagogical vehicle. The "$DG(0) \cdot \mathbf{N}_t$" term satisfies Definition 1 for linear process and we see how the Volterra expansion motivates Definition 1 in the same way linearity is defined for mappings from $\mathbf{R}^k$ to $\mathbf{R}^k, k < \infty$; that is to say:

The best linear approximation to the function at a point is the same as the function itself.

One can rewrite Eq. (2.9) in a more intuitive manner by appropriately defining coefficients in the next expression:

$$
\begin{aligned}
Y_t = \sum_{u=0}^{\infty} g_u N_{t-u} &+ \sum_{u=0}^{\infty}\sum_{v=0}^{\infty} g_{uv} N_{t-u} N_{t-v} \\
&+ \sum_{u=0}^{\infty}\sum_{v=0}^{\infty}\sum_{w=0}^{\infty} g_{uvw} N_{t-u} N_{t-v} N_{t-w} + \ldots
\end{aligned}
\quad (2.10)
$$

Suppose our objective was to form optimal mean squared error predictions of Y_{T+1} given observations on $\{n_s\}_{s=-\infty}^{T}$. It is possible that the first sum would be sufficient. If the sums other than the "Wold," one only involve terms that satisfy (2.11):

$$g_{uv} = g_{uvw} = g_{uvw\ldots} = 0, \text{ if } u > 0. \quad (2.11)$$

Then, using the symmetry implicit in the g coefficients,

$$E[Y_{T+1} \mid \mathcal{F}_T^N] = \sum_{u=1}^{\infty} g_u n_{t-u+1} . \quad (2.12)$$

Where, $\mathcal{F}_t^N$ is the sigma algebra generated by $(N_t, N_{t-1}, \ldots)$. Indeed at all forecast horizons under the restrictions in Eq. (2.11), we have:

$$E[Y_{T+j} \mid \mathcal{F}_T^N] = \sum_{u=j}^{\infty} g_u N_{t-u+j} . \quad (2.13)$$

Thus, a time series might fail to satisfy Definition 1 but not have any extra forecastible structure in the mean squared error sense. Potter[22] calls such processes linear in mean because their impulse response functions are the same as a linear process with the same moving average infinite polynomial. It is possible to view the impulse response function defined as the difference between two conditional expectations of the process (with the most recent conditioning event or starting value varying) as a stochastic analogue of the comparison between two trajectories of a deterministic map from different starting values. A deterministic map is linear if the difference between the two trajectories can always be expressed as a linear function of the difference between the starting values. Similarly one can think of a stochastic time series as having linear dynamics if its impulse response function can be expressed as a linear function of the difference between the starting values. Thus, we present an alternative definition of stochastic linearity.

DEFINITION 2

$\{X_t\}$ is a "linear in mean" stochastic process if

$$\phi(B)X_t = \theta(B)N_t, \; N_t = f(\mathcal{F}_{t-1}^X)e_t, \; e_t \text{ IID}, \; E[N_t^4] < \infty. \qquad (2.14)$$

Definition 2 informs us that if $\{X_t\}$ is linear in mean, then optimal mean squared error one-step-ahead forecasts can be produced by taking linear combinations of previous realizations. Further, under this restriction we can use the "Law of Iterated Expectations" to form j-step-ahead forecasts as linear combinations of observed $\{X_t\}$ without having to take account of any forecastible structure in the higher moments of the noise term.[5] If Definition 2 does not hold, then to form optimal MSE j-step ahead forecasts ($j > 1$), the interaction of the propagated noise process and the structure of the nonlinear difference equation $G(\cdot)$ must be taken into account.

---

3. TESTING FOR PREDICTIVE STRUCTURE BEYOND STOCHASTIC LINEARITY

Returning to the quote at the start of our paper, suppose that the detective has to interrogate the obvious first suspect produced by circumstantial evidence. For many trained in the techniques of Box and Jenkins time-series analysis (cf. Brockwell and Davis[11]), the obvious first suspect for a stationary time series is an ARMA model and an ARIMA model with possible seasonalities for a nonstationary time series. For such a detective the equivalent of checking footprints would be, for example, to

---

[5] In many applications MSE forecasts of $\{X_t\}$ will not be sufficient as, for example, in economics we might be interested in the expected utility (i.e., general concave functions) effects of future changes in $\{X_t\}$ In general, there is no reason for goal functions which are used to evaluate the "quality" of forecasts to be the mean squared error function.

perform a Box-Ljung test for serial correlation on the residuals from an estimated ARMA model. See Brockwell and Davis,[11] Chapters 8 and 9.

In this section, we consider a more powerful form of footprint testing on the residuals based on the properties of the correlation integral. One should think of our procedure as a nonlinear analogue of the Box-Ljung "portmanteau" test on estimated residuals of fitted models as exposited by Brockwell and Davis,[11] section 9.4. We concentrate on the case where $\{n_t\}$ is either directly observable or can be obtained from the observed series $\{x_t\}$ by a known transformation. In the next section we examine the behavior of the diagnostic when the transformation must be estimated. The procedure will be to use U- and V-statistics theory (cf. Denker and Keller[14,15]) to create a statistical test of the IID null hypothesis that has good power against nonlinear departures from IID such as deterministic chaos or noisy chaos. This statistical test due to Brock, Dechert, and Scheinkman,[13] hereafter "BDS," can also be used on estimated residuals of fitted models as a general nonlinear goodness of fit test in analogy with the Box-Ljung test for linear models mentioned above.

The BDS class of tests for IID are based upon functions of measures of spatial correlation. These functions are zero under IID but are nonzero for many non-IID processes. In order to explain the BDS class of tests, we must first explain U- and V-statistics theory.

3.1 V-STATISTICS

V-statistics and their close relatives, U-statistics, are generalizations of single sum averages to multiple sum "averages" in such a way that classical ergodic theory, central limit theory, laws of the iterated logarithm, etc., are preserved. Let $\{Y_t\}$ be a stationary, $\mathbf{R}^k$ valued stochastic process. Consider the statistics (cf. Serfling,[29] Chapters 5 and 6),

$$V(T) = \frac{1}{T^2} \sum_{1 \leq s} \sum_{t \leq T} h(T_t, Y_s),$$

$$U(T) = \frac{2}{T(T-1)} \sum \sum_{1 \leq s < t \leq T} h(Y_t, Y_s), \qquad (3.1.1)$$

where $h(y, z) = h(z, y)$ is a symmetric function, called a "kernel."

Serfling,[29] p. 176, shows that $U(T)$ is a minimum variance unbiased estimator of $E[h(y, z)]$ over the class of all unbiased estimators of $E[h(y, z)]$. Since we are doing asymptotic theory in this article, the difference between $V(T)$ and $U(T)$ will vanish in the limit. For ease of exposition we concentrate on V-statistics for the article.

By unbiasedness, the expectation of both $V(T)$ and $U(T)$ is the same as the expectation of h. Under conditions that require a measure of the dependence of Y on Y_{t+i} to tend to zero as $i \to \infty$ and some bounds on moments of the kernel function, Denker and Keller[14] prove the following alternative representation for the V statistic in Eq. (3.1.1)

$$V(T) = \left(\frac{2}{T}\right) \sum_{s=1}^{T} h_1(Y_s) + R(T), \qquad (3.1.2)$$

where $h_1(Y_s) = E[h(Y_t, Y_s) \mid Y_s]$, and $T^{1/2} R(T) \to 0$ in distribution. We shall refer to Eq. (3.1.2) as the "projection method" because one is reducing a double sum statistic to a single sum statistic by "projection" on one of the variables by use of the conditional expectation, h_1.

Note that the mean of h_1 is the mean of h. Central limit theory and ergodic theory for single sum averages can now be applied. Examples of statistics of the form (3.1) include the Grassberger-Procaccia-Takens correlation integral and estimators of correlation dimension (cf. Brock and Baek,[7] Denker and Keller[15]) and estimates of lower bounds to Kolmogorov entropy and measures of decay of h-step-ahead predictors (Brock and Baek,[7]). More conventional examples of U- and V-statistics are in Serfling.[29]

We examine a class of functions of correlation integrals due to Brock, Dechert and Scheinkman[11] which produces functions of V-statistics which are zero for IID processes. This is the basis of the class of tests for IID contained in BDS. Let us explain. For $\varepsilon > 0, m = 1, 2, \ldots \mathbf{M}$,

$$C_m(\varepsilon, T) = \frac{1}{T^2} \sum_{1 \le s, t \le T} \sum h(\mathbf{y}_t^m, \mathbf{y}_s^m), \qquad (3.1.3)$$

$$D_m(\varepsilon, T) = C_m - C_1^m \qquad (3.1.4)$$

where $\mathbf{y}_t^m = (n_{t-1}, \ldots, n_{t-m}) \in \mathbf{R}^m$ for C_m. Here $h(y, z)$ is a symmetric indicator function equal to one if the two m histories are within ε of each other in the sup norm and equal to zero otherwise.[6] Hence, C_m, C_1 are V-statistics, D is a smooth function of two V-statistics, and so the theory of Denker and Keller[14] applies.

3.2 A DIRECT TEST FOR IID

First, consider the case where a time series $\{n_t\}$ is directly observed. Then, under the null hypothesis of IID with no restrictions on moments of $\{N_t\}$ (the moment conditions needed by Denker and Keller[14] are trivially satisfied for indicator kernels) and provided that the variance V_m is positive, we have

$$T^{1/2} D_m \to N(0, V_m), \quad \text{as } T \to \infty \qquad (3.2.1)$$

[6] Throughout the paper we assume that the relevant asymptotic behavior of the test statistic using the indicator function and the sup norm can be well approximated by use of a symmetric "kernel" function twice continuously differentiable defined on $\mathbf{R}^m \times \mathbf{R}^m$ with values in the nonnegative reals (see the Appendix to Brock, Hsieh and LeBaron[9] for the formal argument and for simulation evidence).

where the convergence is in distribution, and $N(0, V)$ denotes a normal distribution with mean zero and variance V. As shown by Brock, Dechert and Scheinkman[11] the variance V_m can be consistently estimated by functions of V-statistics.

This is a good place to give an heuristic argument so that the reader may see how the BDS test statistic is derived: allowing the reader to program up the BDS test for themselves or modify it in directions they find desirable.

Recall that

$$D_m(\varepsilon) \equiv C_m(\varepsilon) - C_1(\varepsilon)^m \equiv C_m - C_1^m . \qquad (3.2.2)$$

Now use the estimator (3.1.3) to estimate C_m and C_1 for m and 1, respectively. We parenthetically remark that if one wants to achieve the minimum-variance unbiased estimation of each component of D_m, then one should use the U-statistic estimator rather than the V-statistic estimator (3.1.3). Of course, it is not true that this would give minimum variance unbiased estimation of D_m itself under the null hypothesis that $\{N_t\}$ is IID.

We now show how to derive the limit distribution of $T^{1/2} D_m$ from use of the projection method on each of its components and by expansion in a second-order Taylor series around the means C_m, C_1. Put $H(x, y) \equiv x - y^m$. Then

$$
\begin{aligned}
T^{1/2} D_m \equiv & T^{1/2} H(C_m(\varepsilon, T), C_1(\varepsilon, T) \\
= & T^{1/2} [H + H_1(C_m(\varepsilon, T) - C_m(\varepsilon)) \\
& + H_2(C_1(\varepsilon, T)) - C_1(\varepsilon) + H.O.T.],
\end{aligned}
\qquad (3.2.3)
$$

where H, H_1, H_2, denote H, the partial derivative of H w.r.t. the first argument; the partial derivative of H w.r.t. the second argument, all evaluated at the point $(C_m(\varepsilon), C_1(\varepsilon))$, respectively. Note that under the null hypothesis, $\{N_t\}$ IID, H is zero. Here $H.O.T.$ denotes higher-order terms in the second-order Taylor expansion. Because $T^{1/2}(C_i(\varepsilon, T) - C_i(\varepsilon)) \to N(0, v_i), i = 1, \ldots, m$, modest sufficient conditions can be found to ensure these go to zero in probability as $T \to \infty$. Bounded second-order partial derivatives of H are more than enough regularity for this. Now use the projection method and compute the partial derivatives $H_i, i = 1, 2$ to rewrite $T^{1/2} D_m$ under the null hypothesis, $\{N_t\}$ IID; thus, up to terms which converge to zero in probability as $T \to \infty$, we have

$$
\begin{aligned}
T^{1/2} D_m = & [\frac{2}{T^{1/2}}]\{\sum [h_m(\mathbf{y_s^m}) - C_m(\varepsilon)] - m C_1(\varepsilon)^{m-1}[h_1(\mathbf{y_s^1}) - C_1(\varepsilon)]\} \\
\equiv & [\frac{2}{T^{1/2}}]\{\sum g_m(\mathbf{y_s^m})\},
\end{aligned}
\qquad (3.2.4)
$$

where the $\sum$ runs from $s = 1, 2, \ldots, T$.

Under the null of $\{N_t\}$ IID this may be further simplified. Recall that $\mathbf{y_t^m} \equiv (n_{t-1}, \ldots, n_{t-m}), m = 1, 2, \ldots$. Hence

$$h_m(\mathbf{y_t^m}) = \Pi_{i=1}^m h_1(n_{t-i}) . \qquad (3.2.5)$$

Furthermore, in all cases

$$h_1(x) = F(x+\varepsilon) - F(x-\varepsilon), \text{ where } F(x) = \text{prob}\{N_t \le x\}. \qquad (3.2.6)$$

It is now easy to work out the formulae for the limiting mean and variance under the null hypothesis, $\{N_t\}$ IID. The mean of g_m is zero. Using stationarity, compute the variance of $\sum g_m$ and take T to infinity to obtain

$$v_m \equiv \lim var(T^{1/2}D_m) = 4E\{g_m(\mathbf{y_1^m})^2 + 2\sum_{i>1} g_m(\mathbf{y_1^m})g_m(\mathbf{y_{1+i}^m})\}. \qquad (3.2.7)$$

Now use the definition of g_m and $\{N_t\}$ IID to work out the formula

$$\frac{1}{4}V_m = K^m + 2\sum_{j=1}^{m-1} K^{m-j}C^{2j} + (m-1)^2 C^{2m} - m^2 K C^{2m-2}, \qquad (3.2.8)$$

where $C \equiv C_1(\varepsilon)$, $K \equiv E\{I_\varepsilon(n_r, n_s)I_\varepsilon(n_s, n_t)\}$, and $I_\varepsilon(a, b)$ is unity if the distance of a from b is less than ε and is zero otherwise.

The BDS test statistic is given by

$$\frac{T^{1/2}D_m}{V_m(T)^{1/2}} \to N(0, 1) \text{ as } T \to \infty, \qquad (3.2.9)$$

where $N(0, 1)$ denotes the normal distribution with mean zero and variance one, $V_m(T)$ denotes a consistent estimator of the variance V_m, and the convergence takes place under the null hypothesis $\{N_t\}$ IID.

It is important to realize that the above argument crucially depends upon the variance V_m being positive. Certain distributions such as the uniform distribution on the unit circle may not satisfy this requirement.[31] Nevertheless, unless there is some special reason to expect this type of "hairline" degeneracy, it seems natural to assume the variance is positive. In any event, if the variance is zero, a more delicate central limit theory can be developed (cf. Serfling,[29] Chapter 5.).

Here is a candidate for a consistent estimator of the variance that is used in the IBM PC software of W. D. Dechert[13] which computes BDS statistics.

$$V_m(T) = V(C_1(\varepsilon, T), K(\varepsilon, T), m), \qquad (3.2.10)$$

where $C_1(\varepsilon, T)$ is given by Eq. (3.1.3) above and

$$K(\varepsilon, T) \equiv \left(\frac{1}{T^3}\right)\sum\sum\sum I_\varepsilon(n_r, n_s)I_\varepsilon(n_s, n_t), \qquad (3.2.11)$$

where the sums run from $1 \le r, s, t \le T$. Here $V(\cdot, \cdot, \cdot)$ is defined in Eq. (3.2.8).

The BDS test statistic has a natural interpretation in terms of the Grassberger-Procaccia-Takens dimension plots which are popular in natural science (e.g., Eubank and Farmer,[16] p. 155.). Here one plots $\log[C_m(\varepsilon, T)]$ against $\log \varepsilon$ and tries to identify a constant slope zone on this plot to "estimate" the correlation dimension. Except for the log transformation and scaling by the variance and $T^{1/2}$, the BDS statistic just compares C_m with C_1 as a measure of departure from IID. One can obtain a picture of this on a Grassberger-Procaccia-Takens (GPT) plot by super-imposing a plot of $\log[C_m(\varepsilon, T, r)]$ (here "r" denotes "random") computed on an approximate IID process with the same mean and variance as in Brock et al.[4] or the same stationary distribution as the empirical distribution estimated from the series under scrutiny by drawing a sample of length T from the empirical distribution as in Scheinkman and LeBaron.[28] For a fixed ε the difference between $\log[C_m(\varepsilon, T)]$ for the original data and $\log[C_m(\varepsilon, T, r)]$ for the shuffled counterpart gives a proxy for $C_m(\varepsilon, T) - C_1(\varepsilon, T)^m$. This is so because $C_m(\varepsilon, T, r)$ is approximately equal to $C_1(\varepsilon, T)^m$. Indeed one can base a test for IID on the statistic:

$$T^{1/2}\{\log[C_m(\varepsilon, T)] - \log[C_1(\varepsilon, T)^m]\}, \qquad (3.2.12)$$

if one wishes a direct correspondence between GPT plots for the data and its shuffled counterpart.

Tests for IID which are based upon the correlation integral are performance evaluated by Monte Carlo work as well as applied to economics and finance in the forthcoming book by Brock, Hsieh, and LeBaron.[9]

It is instructive to consider the relationship between the above test for IID and predictive ability. Throughout, we restrict ourselves to strictly stationary processes and consider population values (i.e., the sample is long enough for the asymptotic results to be exact). Using the indicator function kernel, we can write

$$C_{m+1}(\varepsilon, \infty) = P[|\, n_{t+m} - n_{s+m}\, |< \varepsilon, \text{ given } \|\, \mathbf{y}_t^m - \mathbf{y}_s^m\, \|< \varepsilon]C_m(\varepsilon, \infty). \quad (3.2.13)$$

By induction on m, we see that $C_1^m \neq C_m$ implies that for some $r < m + 1$,

$$P[|\, n_{t+r} - n_{s+r}\, |< \varepsilon, \text{ given } \|\, \mathbf{y}_t^r - \mathbf{y}_s^r\, \|< \varepsilon] \neq P[|\, n_{t+r} - n_{s+r}\, |< \varepsilon]. \quad (3.2.14)$$

Inequality (3.2.14) captures the notion of the r-past helping to predict the future by using nearest neighbors with nearness ε.

3.3 TESTING FOR STOCHASTIC LINEARITY

To test for linearity as in Definition 1, one can use an ARMA model to remove linear dependence in the data. The BDS test is ideally suited to test linearity as given by Definition 1 because, as we shall see in section 4 below, the asymptotic distribution of the BDS test statistic is not changed by estimation error. However, the sampling distribution of tests on estimated residuals which are based upon the autocorrelation function (e.g., Brockwell and Davis,[11] section 9.4) are affected by estimation error.

From Eq. (2.4) we have $\phi(B)X_t = \theta(B)N_t$, which can be expressed as

$$\pi(B)X_t = N_t, \pi(B) = \frac{\phi(B)}{\theta(B)}. \tag{3.3.1}$$

There are two cases:

i. $\pi(B)$ is a finite polynomial in L of order r, then one can apply the BDS test to $\{n_t\}_{t=r+1}^T$.

ii. $\pi(B)$ is an infinite-order polynomial in B. Here we can only use a truncated version of $\pi(B), \pi^*(B)$. If w represents the order of the truncated polynomial, then one applies the BDS test to $\{n_t^*\}_{t=w+1}^T$. Depending on the circumstances this might not be a problem. For example, if the time series has a well-defined starting point, as in an experiment, then we can set all $\{x_t : t < 0\}$ equal to zero and use x_0 as a starting value in the recursion to generate $\{n_t\}_{t=1}^T$. In the case where there is not a fixed starting value, the assumption of short run dependence implicit in the ARMA representation means that a truncation that grows slowly in relation to the sample size T, will provide approximations to the unobserved noise terms that converge to the true noise terms as $T \to \infty$.

In the case where Definition 2 applies, then, we first apply either (i) or (ii) above to obtain an $\{n_t\}$ sequence. Then take transformations of $\{n_t\}, \{f(\mathcal{F}_{t-1}^X)\}$, and $\{e_t\}$ into the strictly positive reals. Finally, take logarithms to separate out the transformation of $\{e_t\}$. That is, $h(n_t) = h(f(\mathcal{F}_{t-1}^X))h(e_t), \ln(h(e_t)) = \ln(h(n_t)) - \ln(h(f(\mathcal{F}_{t-1}^X)))$. The assumption of independence (by definition) applies to all measurable functions of a stochastic process; hence, the test can be applied to this transformation. Such a procedure may appear somewhat cumbersome compared to testing the ratio $n_t/f(\mathcal{F}_{t-1}^X)$, but it allows us to ignore estimation error in $f(\cdot)$.

An example is instructive here. A common form of higher-order nonlinear dependence is known as Autoregressive Conditional Heteroskedasticity (ARCH) which, in the simplest case, is defined by the following,

$$n_t = e_t h_t^{1/2}, \ h_t = \alpha_0 + \alpha_1 n_{t-1}^2. \tag{3.3.2}$$

If n_t is directly observed and α_0, α_1 known, then we can construct h_t and test it for IID by taking squares, and then logarithms. Hence, in the ARCH first-order case $\ln(e_t^2) = \ln(n_t^2) - \ln(\alpha_0 + \alpha_1 n_{t-1}^2)$. Now suppose one has estimated a model of the form in Eq. (3.3.2) and examines the estimated residuals $\hat{n}_t \equiv \hat{e}_t \hat{h}_t^{1/2}$. Under the null hypothesis, Eq. (3.2.2) we have

$$\log(\hat{e}_t^2) = \log(e_t^2) + \log(h_t^2) - \log(\hat{h}_t^2). \tag{3.3.3}$$

Under the null hypothesis (3.3.2), one can apply the BDS test to $\{\log(\hat{e}_t^2)\}$ and the invariance theorem of the next section to prove the asymptotic distribution converges to the normal distribution with mean zero and variance one. Here the change of units by first squaring and then taking logs reduces the model (3.3.2) with multiplicative noise into one with additive noise that can be tested by BDS without correcting for the error produced by estimation of $f(\mathcal{F}_{t-1}^X)$.

3.4 TESTING FOR EXTRA PREDICTIVE STRUCTURE IN NONLINEAR STOCHASTIC MODELS

We now move onto the testing of noise terms from a nonlinear model for the IID property. In order to back out the noise term from the observed series $\{x_t\}$, we assume that:

$$X_t = H(X_{t-1}, X_{t-2}, \ldots, X_{t-p}) + N_t. \qquad (3.4.1)$$

Thus, N_t is equal to the one-step-ahead forecast error.

One would first test $\{n_t\}_{t=p+1}^{T}$ for the IID property and, if that fails, attempt to find a function $f(\mathcal{F}_{t-1}^X)$ to describe the higher-order dependence in the noise. The correctness of $f(\cdot)$ is evaluated by examining the resulting transformation of the underlying $\{e_t\}$ sequence. Note that in principle there is no difference from the application of the test in the linear case.

3.5 DISTINGUISHING MEASUREMENT ERROR FROM PROPAGATED NOISE

All of the above discussion was predicated on the assumption of no measurement error. We briefly consider how the same diagnostic test can be used to aid in distinguishing measurement error $\{M_t\}$ from propagated noise $\{N_t\}$ and as a diagnostic for noise-reduction algorithms in the case where the underlying map is assumed chaotic and is known to the investigator.

Suppose an investigator examines a time series by starting with a *known* chaotic map $G(\cdot)$ and she uses shadowing or noise-reduction techniques (e.g., Eubank and Farmer[16]) to uncover the trajectory of $\{X_t\}$ from $\{A_t \equiv h(X_t, M_t)\}$. Given an observed sample $\{a_t\}$, call the resulting trajectory $\hat{x}_t$ and assume that the measurement error M_t is additive. Consider, first, the one-step-ahead forecast error v_t from applying $G(\cdot)$ to $\hat{x}_t$

$$v_t = m_t + G(x_{t-1}) + n_t - G(\hat{x}_{t-1}). \qquad (3.5.1)$$

Take a first-order exact Taylor series expansion:

$$v_t = m_t + n_t - DG(x_{t-1}^*)(x_{t-1} - \hat{x}_{t-1}). \qquad (3.5.2)$$

Note that if the noise process, $\{N_t\}$ from which the sample $\{n_t\}$ was drawn, in the law of motion is identically zero, then examination of v_t for IID is one method of assessing the accuracy of the trajectory $\{X_t\}$. The sample $\{v_t\}$ should look like it came from the IID process $\{M_t\}$ if the noise-reduction/optimal shadowing procedure were very accurate on average, i.e., $x_s \cong \hat{x}_s, s = 1, 2, \ldots, T$.

Indeed, maintaining the assumption that $\{N_t\}$ is identically zero, the j-step-ahead forecast errors by iterating on the known map j times would provide a very stringent test of the accuracy of the trajectory $\{\hat{x}_t\}$ because of the chaotic nature of the map. Of course, for j sufficiently large, a chaotic map would introduce dependence due to sensitive dependence on initial conditions. A Monte Carlo experiment with no measurement error could be used on the particular chaotic map to assess the role of rounding error, etc.

If the noise process $\{N_t\}$ is not zero and is IID, then under an assumption of small measurement error, the BDS diagnostic for IID would not find large evidence against the IID measurement error null for one-step-ahead forecast errors. However, now consider the two-step-ahead forecast error $\{u_t\}$ produced by applying $G(\cdot)$ twice to $\hat{x}_t$

$$u_t = G(G(x_{t-2} + n_{t-1}) + n_t + m_t - G(G(\hat{x}_{t-2})). \qquad (3.5.3)$$

Taking a mean value expansion, we would have:

$$u_t = n_t + m_t + g_1(x_{t-2}^*, n_{t-1}^*)(x_{t-2} - \hat{x}_{t-2}) + g_2(x_{t-2}^*, n_{t-1}^*)n_{t-1} \qquad (3.5.4)$$

where g_i are the appropriate partial derivatives.

Hence, even if $\hat{x}_{t-2}$ was very close to x_{t-2} but there was propagated noise present, we would find evidence of predictability in $\{u_t\}$. As a method of attempting to distinguish a rejection due to an inaccurate $\{\hat{x}_t\}$ from propagated noise, one could use the estimated measurement errors in a Monte Carlo experiment as follows:

Use the known chaotic map to generate K trajectories of $\{x_t\}$ of length T. Then randomly sample from the estimated measurement errors to produce K shuffled series of $\{m_t\}$ of length T. Form $\{a_t^K\}$ by adding the measurement error to the chaotic trajectories (i.e., generate time series that satisfy the null hypothesis of no propagated noise). Run the K samples through the noise-reduction algorithm and generate the K BDS statistics for both the one- and two-step forecast errors. The K BDS statistics can then be used to generate a sampling distribution under the null of no propagated noise error to assess the significance of the test statistics on the actual sample $\{a_t\}$.

4. ESTIMATION ERROR

Many diagnostic tests based upon estimated residuals require a correction for estimation error to the asymptotic distribution of the test statistic under the null hypothesis. We give an heuristic discussion of this problem here. For a complete discussion with simulation evidence, see the book by Brock, Hsieh, and LeBaron[9] and Brock, Dechert, Scheinkman, and LeBaron.[8]

4.1 EXAMPLE OF AN AUTOREGRESSION OF ORDER ONE

Consider the expression (3.2.4) which we record here for convenience,

$$\hat{W}(T) \equiv \left(\frac{2}{T^{1/2}}\right)\sum g_m(\hat{y}_s^m), \quad \hat{y}_s^m \equiv (\hat{n}_{s-1}, \dots, \hat{n}_{s-m}), \qquad (4.1.1)$$

where $\hat{}$ over a symbol denotes its estimated counterpart.

In order to locate sufficient conditions for the limit distribution of Eq. (4.1) to be independent of whether estimated residuals, $\hat{y}_s^m$, or true residuals, y_s^m are used, let us examine the case of an AR(1) below,

$$X_s = bX_{s-1} + N_s, \quad |b| < 1, \tag{4.1.2}$$

where we assume $\{N_s\}$ is IID with mean zero and finite variance or satisfies Eq. (2.2.b).

Let b_T be any root T consistent estimate of b, i.e., assume either:

$$T^{1/2}(b_T - b) \to N(0, \sigma^2) \tag{4.1.3.a}$$

under Eq. (2.2.a) where convergence is in distribution, or

$$T^{1/2}(b_T - b) \to 0 \text{ a.s.} \tag{4.1.3.b}$$

under (2.2b).

Equation (4.1.3.a) is standard. Equation (4.1.3.b.) is not so well known but is discussed in Hannan and Kanter.[18] Note also that a Law of Large Numbers can be applied to the sample mean of X_t under Eq. (2.2.b) (see, for example, Davis and Resnick[12]). The assumption that $\alpha > 1$ implies that $E[|X_t| < \infty]$.

Suppose we estimate b from a sample of length T of data generated by Eq. (4.1.2). Then Eq. (4.1.2) implies

$$\hat{n}_s = (b - b_T)x_{s-1} + n_s , \tag{4.1.4}$$
$$\hat{y}_s^m = y_s^m + (b - b_T)x_{s-1}^m . \tag{4.1.5}$$

Now insert Eqs. (4.1.4) and (4.1.1), expand in a second-order Taylor series around the true y_s^m, to obtain

$$\hat{W}(T) = W(T) + T^{1/2}(b - b_T) \left[\frac{2}{T} \sum Dg_m(y_s^m) \times X_{s-1}^m \right] \equiv W(T) + D(T) , \tag{4.1.6}$$

where, in the general case, one would have second-order terms.

A general principle is illustrated by Eq. (4.1.6): The distribution of W is the distribution of W corrected for estimation error by adding the extra term $D(T)$. One sees right away from Eq. (4.1.6) in the following Theorem.

THEOREM If time averages converge, and

$$E[Dg_m(t_s^m) \times X_{s-1}] = 0 , \tag{4.1.7}$$

then the limit distribution of $\hat{W}$ is the same as the limit distribution of W.

PROOF The proof is obvious under the hypothesis that time averages converge since $T^{1/2}(b - b_T)$ converges to a random variable with mean zero and finite or zero variance.

Let us examine the moment condition (4.1.7). For the case under scrutiny where the innovations $\{N_s\}$ are IID, the condition

$$E[Dh_1(N)] = 0 \,, \tag{4.1.8}$$

where D denotes derivative, is sufficient for the moment condition (4.1.7). But, surprisingly perhaps, Eq. (4.1.8) is not very restrictive. Let us explain. Recall that in the case where the kernel $h(x, y)$ is the indicator function $I_\varepsilon(x, y)$, we have

$$h_1(N) \equiv E[I_\varepsilon(N, M) \mid N] = F(N + \varepsilon) - F(N - \varepsilon), F(x) \equiv \text{prob}\{N \le x\}. \tag{4.1.9}$$

Computing Eq. (4.1.8), and assuming F has a density f, we obtain

$$E\{Dh_1(N)\} = \int [f(n + \varepsilon) - f(n - \varepsilon)] \, f(n)dn = 0 \,. \tag{4.1.10}$$

The last integral is easily seen to equal zero by noticing that

$$\int f(n + \varepsilon)f(n)dn = \int f(n - \varepsilon)f(n)dn$$

by a change of variable argument.

Parenthetically we remark that the proof of the theorem does not apply to the indicator function I_ε because it is not differentiable. However, under modest regularity conditions, one can find C^∞ symmetric approximations to I_ε such that Eq. (4.1.10) approximately holds and, hence, our theory is valid for each of these approximations.[7] The procedure appears to be practical as well. See Brock, Hsieh, and LeBaron[9] for the details. Much of this theory is in Brock and Baek.[5] Turn now to more general models than AR(1)'s.

4.2 USE OF THE BDS STATISTIC AS A DIAGNOSTIC TEST OF FITTED MODELS

Consider the class of parametric models with additive IID errors:

$$X_s = H(J_s, \mathbf{a}) + N_s, \quad N_s \text{ independent of } J_s \text{ and } E[N_s^4] < \infty, \tag{4.2.1}$$

where J_s is an information set of "regressor" variables which may include past X_s, $\mathbf{a}$ is a vector of parameters which can be estimated root T consistently, i.e.,

---

[7]Perhaps, Randles[25] can be generalized to prove Eq. (4.1.7) directly for the nondifferentiable kernel.

$T^{1/2}(\mathbf{a} - \hat{\mathbf{a}}) \to N(\mathbf{0}, \sum)$, where $\mathbf{0}$ is the zero vector and $\sum$ is the variance covariance matrix, and $N(\mathbf{0}, \sum)$ denotes the normal distribution with mean vector $\mathbf{0}$ and variance-covariance matrix $\sum$. It is easy to use the BDS test as a test of the adequacy of this class of models.

First, estimate the parameter vector $\mathbf{a}$ by a root T consistent method (many estimation methods are root T consistent under the fourth moment condition). Now under modest regularity conditions, one may repeat the same Taylor expansion argument we did above to show that the asymptotic distribution of the BDS statistic is the same on the estimated residuals $\{n_S\}$ as on the true residuals provided the data are generated by a member of the null class (4.2.1). The key step is showing that the "distortion term" $D(T)$ in the Taylor expansion around the true vector $\mathbf{a}$ goes to zero in distribution as $T \to \infty$. This is done by using the root T estimation of $\mathbf{a}$ and independence of N_s and J_s to show, using much the same type of argument as in the AR(1) case, that $D(T) \to 0$ in distribution as $T \to \infty$.

5. EXTENSIONS

5.1 LINEARITY TESTING

It may be possible to extend correlation integral-based methods to test the hypothesis of a linear process even when the process may exhibit long-range dependence or for general ARMA models satisfying Eq. (2.2.b). The key step in the proof of the invariance of the asymptotic distribution of the BDS statistic to estimated residuals was to show that the distortion term $D(T)$ in Eq. (4.1.6) converged to zero in probability.

In the infinite variance case, one can adapt the intuition of the AR(1) case by taking only a first-order Taylor series expansion and, instead of letting the moment condition (4.1.8) bound the terms in Eq. (4.1.7), use the almost sure convergence of the autoregressive estimators to zero in Eq. (4.3.b). A similar argument may be applicable in the case when $\alpha \leq 1$ (i.e., the first moment of the noise term is not defined) by normalizing the sum in (4.1.7) by an extra $T^{1/2-\alpha+\delta}$ for some $\delta > 0$ and still maintaining the almost sure convergence of the parameter estimates to the true values.

Long-term dependence (i.e., the infinite sum of absolute values of the moving average coefficients does not converge or converges at a nongeometric rate even after standard detrending) can be dealt with by estimating fractionally integrated time-series models. Instead of the standard integer differencing operation (i.e., $(1 - B)^d$, d an integer), one considers the fractional difference operator (i.e., $(1 - B)^d$, $-.5 \leq d \leq .5$). In time-series analysis it is standard to assert rather than estimate the order of integer differencing (although there has recently been a heated debate in macroeconomics about the assertion). For fractional differencing an estimator of d is required. Brockwell and Davis[11] (chapter 12, section 4) discuss some candidate estimators but so far there appears to be no general proof of root T consistency of *d.* If one had a root-T-consistent estimator of d, then it might be possible to use a truncation to the infinite binomial expansion of the fractional difference operator to recover good approximations to the underlying noise process.

Another important class of extensions is using correlation integral-based tests to test the IID null hypothesis for vectors of time series as well as to test for the presence of "causal predictive chains" in a vector of time series. That is to say, the correlation integral-based tests exposited here can be used to test the hypothesis that a series $\{X_t\}$ incrementally helps to predict future values of another series $\{Y_t\}$ above and beyond the predictive content for future Y from past values of Y. See Baek[3] for this kind of work.

5.2 SIMULATION METHODS

There are a large class of test statistics which may be built up from the correlation integral which do not directly test for IID. The assumption of IID has a substantive role as the previous discussion has made clear but is also very important in the calculation of the asymptotic sampling distribution.

Asymptotic sampling distribution theory is available for many other cases with the proviso that the mean and variance of the sampling distribution may not be known. Here we discuss examples where simulation of parametric null models can be used to determine the sampling variability of test statistics under the null hypothesis. One might use a correlation integral-based statistic to test for chaos in the style of Sugihara and May.[30] For example, given a scalar stochastic process $\{X_t\}$, look at the probability:

$$P[|\, X_{t+p} - X_{s+p}\,| < \varepsilon \text{ and } |\, X_t - X_s\,| < \varepsilon] \equiv C_\varepsilon(p,0) , \qquad (5.2.1)$$

which, by analogy, may be estimated by the U statistic

$$\hat{C}_\varepsilon(p,0) \equiv \frac{1}{T(T-1)} \sum_{t=1\, t\neq s}^{T} \sum_{s=1}^{T} I_\varepsilon(x_{t+p}, x_{s+p}) I_\varepsilon(x_t, x_s) , \qquad (5.2.2)$$

which is a minimum variance unbiased estimator of $C(p,0)$ over the class of unbiased estimators (cf. Serfling,[29] Chapter 5). If one has a deterministic chaos, for example, the tent map

$$x_t = G(x_{t-1}) ,$$

$$G(x) = \begin{cases} 2x & \text{if } x \le .5, \\ 2 - 2x & \text{if } x \ge .5, \end{cases} \qquad (5.2.3)$$

we would expect, for the tent map, $C(p,0) \cong P[|\, X_t - X_s\,| < \varepsilon/2^p]$ for ε small and $p = 1, 2, 3, \ldots, m$, where m is not large.

It is convenient to normalize $\hat{C}(p,0)$ by using $\hat{C}_1(\varepsilon)^2$:

$$\hat{C}_1(\varepsilon) \equiv \frac{1}{T(T-1)} \sum_{t=1\, t\neq s}^{T} \sum_{s=1}^{T} I_\varepsilon(x_t, x_s) . \qquad (5.2.4)$$

We call the resulting statistic $\hat{D}_\varepsilon(0,p)$. We would expect, using the tent map example for intuition, that deterministic chaotic maps would have a pattern of $\hat{D}_\varepsilon(0,p)$ that for small p and ε would decline away from one at an exponential rate. However, if $\{X_t\}$ is IID, then $\hat{D}_\varepsilon(0,p) \cong 1$ for all p as $T \to \infty$. If $\{X_t\}$ is a linear first-order autoregression (4.2.1), then we can write the p-step ahead difference as:

$$X_{t+p} - X_{s+p} \cong \sum_{j=0}^{p-1} b^j \{N_{t-j+p} - N_{s-j+p}\} + b^p\varepsilon . \qquad (5.2.5)$$

For p greater than a certain number n, short-term dependence and stationarity will mean that the event $|X_{t+p} - X_{s+p}| < \varepsilon$ conditioned on $|X_t - X_s| < \varepsilon$ will occur approximately as many times as the event $|X_t - X_s| < \varepsilon$. The size of n would depend on the persistence of the autoregression and the signal-to-noise ratio. However, it is unlikely that an autoregression would show short-term exponential decay for small p. For example, if the noise is small compared to the signal, then the global contraction property of $|b| < 1$ would produce a large number of cases where the future trajectories were within ε. Alternatively, if the noise is strong compared to the signal, then the immediate link at $p = 1$ might be very weak but even for small p the statistic would return to 1.

For various null models fitted to the observed data, one could use the estimated residuals to "bootstrap" K samples from the null model. Then calculate the $\{\hat{D}(0,p): p = 1, \ldots m\}$ statistic on the K samples. The histograms of the simulated statistics could then be compared with the statistics calculated on the observed data. Brock, Lakonishok, and LeBaron[6] used such a bootstrap approach on statistics calculated from stock market technical trading rule to test various null models of stock market returns. Savit and Green[32] discuss a closely related test statistic. Their goal is to find a nonparametric test for the presence of additional explanatory power at higher embedding dimensions.

The power of $\hat{D}_\varepsilon(0,p)$ test against nonlinear stochastic alternatives could be increased by examining a sequence of ε values. Linear time series driven by IID noise would have a similarity across the ε sequence from their global contraction property not found in many nonlinear cases (see Liu, Granger, and Heller[20] for evidence on the value added from ε sequences).

Another way to give statistical precision to the Sugihara-May testing methodology would be to bootstrap the distribution of Sugihara-May[30] quantities under a parametric null hypothesis in the manner suggested above. They were interested in showing how measures of forecast quality of p-step-ahead forecasts fell off exponentially in p for deterministic chaos but fell off non-exponentially for stochastic processes such as autoregressions with seasonalities. The distribution of cleverly chosen functions of their measures under the null hypothesis of, for example, an autoregression could be bootstrapped to give a precise test of null hypothesis in favor of alternatives that share the exponential fall-off property.

6. SOME APPLICATIONS IN FINANCE

The methods exposited here have been applied in several recent articles in economics and finance (see Brock and Potter[10] for a review). Many of the applications have been to provide a general test of adequacy of a parametric null class of models to describe the data under scrutiny. Here is an example from finance from a recent paper of David Hsieh.[19] Hsieh looked at stock returns data at the weekly frequency from 1962–1989, daily frequency from 1983–1989, and 15-minute frequency for 1988. He also looked at various subsamples. He used the BDS test to test the null hypothesis that stock returns, $\log[P(t+h)] - \log[P(t)]$, were IID where $P(t)$ is the price of the stock or portfolio at time t, and h is the period length. The null hypothesis was strongly rejected for all frequencies and subperiods. Hsieh examined four classes of possibilities for the rejections of the null: (i) deterministic chaos; (ii) nonstationarity; (iii) linear process in the sense of Definition 1; and (iv) nonlinear stochastic process. The last class includes deterministic chaos with measurement noise or noise in the law of motion. The linear null hypothesis was rejected by fitting linear models and testing the estimated residuals using BDS as described above. The pattern of rejections did not change. Nonstationarity may well be present but cannot explain the pattern of rejections because it was similar at all frequencies including the 15-minute frequency. Unless one believes that the structure changes every 15 minutes, nonstationarity is not a plausible explanation for the results.

The hypothesis of deterministic chaos with no noise of either type was tested by conducting a short-term forecasting exercise using a nonparametric regression method called locally weighted regression (LWR) which is a generalization of the nearest-neighbors technique. LWR was tested by Hsieh and shown to do very well on short-term out-of-sample forecasting on a class of chaotic maps.

He then conducted an out-of-sample LWR forecasting exercise on stock returns and showed that it failed to outperform simple random walk forecasts. This conclusion was not changed by shortening the forecast length which one would expect if chaos were present. Hence the evidence is not consistent with a low-dimensional deterministic chaos unless the chaos is so "complex" that it is not short-term forecastible with methods tailor made to do short-term forecasting in such a situation. The conclusion is that returns are described by a type of stochastic nonlinear process where the returns themselves are not predictable out of sample by any known forecasting technique although measures of volatility of returns appear somewhat predictable.

LeBaron[21] has shown that there appear to be periods where returns may be predictable following periods of low volatility. Brock, Lakonishok, and LeBaron[9] have shown that certain technical trading rules have apparent predictive power following buy and sell signals measured relative to received models in finance. These effects are small, however, and probably would not cover transactions costs and adjustments for extra risk bearing. While these efforts to adduce evidence for deterministic chaos in stock markets have led to negative conclusions, they have enhanced our understanding of the conditional probability distribution of stock returns and that is the main goal of financial science.

7. SUMMARY AND CONCLUSIONS

This article has exposited how the theory of U-statistics can be used to create diagnostic tests for nonlinearity and chaos. These diagnostic tests are easy to use and should be thought of as a first hurdle to which one should subject a data set of forecast errors from simple procedures before using more complicated nonlinear forecasting procedures.

ACKNOWLEDGMENTS

W. A. Brock thanks the Wisconsin Graduate School, the Wisconsin Alumni Research Foundation, the Vilas Trust, and the National Science Foundation (Grant# SEC-8420872) for essential financial support of this work. S. M. Potter thanks the Academic Senate of UCLA for financial support. None of the above are responsible for errors.

REFERENCES

1. Baek, E. "Three Essays in Nonlinearity in Economics." Ph.D. thesis, Department of Economics, The University of Wisconsin, Madison, 1987.
2. Brock, W., "Distinguishing Random and Deterministic Systems: Abridged Version." *J. Econ. Theor.* **40** (1986): 168–195.
3. Brock, W., and W. Dechert. "A General Class of Specification Tests: The Scalar Case." In *Proceedings of the American Statistical Society* (1989): 70–79.
4. Brock, W., J. Lakonishok, and B. LeBaron. "Simple Technical Trading Rules and the Stochastic Properties of Stock Returns." Social Systems Research Institute Working Paper # 9022, Department of Economics, The University of Wisconsin, Madison, 1990.
5. Brock, W., and E. Baek. "Some Theory of Statistical Inference for Nonlinear Science." *Rev. Econ. Studies* (1991).
6. Brock, W., W. Dechert, J. Scheinkman, and B. LeBaron. "A Test for Independence Based on the Correlation Dimension." Unpublished, 1991. Revised version of W. Brock, W. Dechert, and J. Scheinkman, working paper, Department of Economics, University of Wisconsin, University of Houston, University of Chicago, 1987.
7. Brock, W., E. Hsieh, and B. LeBaron. *A Test for Nonlinear Dynamics and Chaos.* Cambridge: MIT Press, 1991.
8. Brock, W., and S. Potter. "Nonlinear Time Series and Macroeconometrics." In *Handbook of Statistics Volume 10: Econometrics*, edited by G. Maddala. New York: North-Holland, fourthcoming.
9. Brockwell, P., and R. Davis. *Time Series: Theory and Methods.* New York: Springer-Verlag, 1987.
10. Davis, R,. and S. Resnick. "Limit Theory for Moving Averages of Random Variables with Regularly Varying Tail Probabilities." *Ann. of Prob.* **13** (1985): 179–195.
11. Dechert, W. "A Program to Calculate BDS statistics for the IBM PC." Technical paper, Department of Economics, The University of Houston, 1987.
12. Denker, M., and G. Keller. "On U-Statistics and von Mises Statistics for Weakly Dependent Processes." *Zeitschrift Fur Wahrscheinlichkeitstheorie and Verwandte Gebiete* **64** (1983): 505–522.
13. Denker, M., and G. Keller. "Rigorous Statistical Procedures for Data from Dynamical Systems." *J. Stat. Phys.* **44** (1986): 67–93.
14. Eubank, S., and D. Farmer. "An Introduction to Chaos and Prediction." In *1989 Lectures in Complex Systems*, edited by E. Jen. Santa Fe Institute Studies in the Sciences of Complexity, Lec. Vol. II. Redwood City, CA: Addison-Wesley, 1989.
15. Hall, P., and C. Heyde. *Martingale Limit Theory and Its Application.* New York: Academic Press, 1980.

16. Hannan, E., and M. Kanter. "Autoregressive Processes with Infinite Variance." *J. App. Prob* **14** (1977): 411–415 .

17. Hsieh, D. "Chaos and Nonlinear Dynamics: Application to Financial Markets." Working Paper 90-109, Fuqua School of Business. *J. Finance* (1991).

18. Liu, T., G. Granger, and W. Heller. "Using the Correlation Exponent to Decide if an Economic Series is Chaotic." Working paper, Department of Economics, University of California, San Diego, 1991.

19. LeBaron, B. "Some Relations Between Volatility and Serial Correlations in Stock Market Returns." Social Systems Research Institute Working Paper #9002, Department of Economics, The University of Wisconsin, Madison, 1990.

20. Potter, S. "Nonlinear Time Series and Economic Fluctuations." Ph.D thesis, Department of Economics, The University of Wisconsin, Madison, 1990.

21. Priestley, M. *Spectral Analysis and Time Series*. New York: Academic Press, 1981.

22. Priestley, M. *Non-Linear and Non-Stationary Time Series*. New York: Academic Press, 1988.

23. Randles, R. "On The Asymptotic Normality of Statistics with Estimated Parameters." *Ann. Stat.* **10(2)** (1982): 462–474.

24. Richardson, M. "Temporary Components of Stock Prices: A Skeptic's View." Wharton School Working Paper, University of Pennsylvania.

25. Rosenblatt, M. *Markov Processes Structure and Asymptotic Behavior*. New York: Springer-Verlag, 1977.

26. Savit, R., and M. Green "Time Series and Dependent Variables." *Physica D*, fourthcoming.

27. Scheinkman, J., and B. LeBaron "Nonlinear Dynamics and Stock Market Returns." *J. Bus.* **62** (1989): 311–337

28. Serfling, R. *Approximation Theorems of Mathematical Statistics*. New York: Wiley, 1980.

29. Sugihara, G., and R. May. "Nonlinear Forecasting as a Way of Distinguishing Chaos from Measurement Error in Time Series." *Nature* **344** (1990): 734–741.

30. Theiler, J. "Statistical Precision of Dimension Estimators." *Phys. Rev. A* **41** (1990): 3038 3051.

James Theiler, Bryan Galdrikian, André Longtin, Stephen Eubank, and J. Doyne Farmer
Theoretical Division and Center for Nonlinear Studies, Los Alamos National Laboratory, Los Alamos, NM, 87545 and the Santa Fe Institute, 1660 Old Pecos Trail, Santa Fe, NM 87501

Using Surrogate Data to Detect Nonlinearity in Time Series

We address the issue of reliably discriminating between chaos and noise from a time series. In particular, we are interested in avoiding claims of chaos when simpler models (such as linearly correlated noise) can explain the data. We take a statistical approach, and use a form of bootstrapping to detect nonlinearity by showing that a given linear model is unlikely to have produced the data. Our method requires the careful statement of a null hypothesis which characterizes a candidate linear process, the generation of an ensemble of "surrogate" data sets which are similar to the original time series but consistent with the null hypothesis, and the computation of a discriminating statistic for the original and for each of the surrogate data sets. The idea is to test the original time series *against* the null hypothesis by checking whether the discriminating statistic computed for the original time series differs significantly from the statistics computed for each of the surrogate sets. We present algorithms for generating surrogate data under various null hypotheses, and we show the results of numerical experiments on artificial data using correlation dimension, Lyapunov exponent, and forecasting error as discriminating statistics. Finally, we consider a number of experimental time series—including sunspots, electroencephalogram (EEG) signals, and fluid convection—and evaluate the statistical significance of the evidence for nonlinear structure in each case.

1. INTRODUCTION

With the advent of inexpensive desktop computers, the consequent rise in the science of experimental mathematics, and the popularization of the phenomenon of chaos, it has become widely appreciated that nonlinear systems with only a few degrees of freedom can exhibit stationary aperiodic time series.[10,13] On the other hand, noisy linear systems also exhibit stationary aperiodic time series,[1] so it is of considerable value to determine the nature of the underlying dynamics (is it chaos or is it noise?) in the practical situation where all that is available is a time series of data. Algorithms have been developed for making this distinction, but they are notoriously unreliable, and usually involve considerable human judgement.

Distinguishing nonlinear deterministic from linear stochastic processes is ultimately a statistical problem, for it is always *possible* for any finite length time series to be a particular realization of a noise process, just as it is possible for a quite random-looking time series to come from a low-dimensional deterministic process (witness the pseudorandom number generator). From a finite time series, one can only *infer* whether the underlying dynamics is stochastic or deterministic. Numerous authors have emphasized the importance of the statistical approach in quantifying and characterizing nonlinear phenomena from a time series.[2,3,12,25,26,29,35,37,44,49] A brief review of some of these approaches in a language accessible to a physicist can be found in Brock and Dechert.[4]

The issue is further complicated by the fact that chaotic time series taken from experimental systems are always contaminated by some (usually unknown) level of noise, both observational and dynamical. So the real problem is not simply one of distinguishing chaos and noise, but of disentangling them. While we are motivated by the prospect of ultimately disentangling chaos and noise, our goal for this paper will be more simply to *detect* nonlinear structure in a time series.

The usual approach to detecting low-dimensional behavior in experimental data requires numerically estimating the dimension, and then checking if this value is small. With a finite amount of data, and especially if the data are noisy, the dimension estimated by the algorithm will at best be approximate (and at worst, outright wrong). What is commonly done in this case is to attempt to identify the various sources of error (both systematic and statistical), and then to put error bars on the estimate.[24,35,38,40,42,43,44] But this can be problematic for nonlinear algorithms like dimension estimators not only because of difficulties with analytical tractability, but also because some model of the underlying process must be assumed. Indeed, these algorithms are often quite easily fooled by linear correlations in the data.

The approach we will take here detects nonlinearity by showing that the time series is inconsistent with a linear stochastic process. *Detecting* nonlinearity is easier than describing it; we do not need to exhibit the underlying nonlinear dynamics, but merely to demonstrate the inadequacy of a linear model. The value of testing data against a null hypothesis, or "plausible alternative," was recently emphasized by Ellner.[16]

We determine the distribution of the quantity we are interested in (dimension, say) for an ensemble of "surrogate" data sets which are just different realizations of the particular noise process that is our null hypothesis. Then, instead of putting error bars on the estimated dimension, we put error bars around the value that we wish to distinguish it from (the value that noise gives). This can be done reliably because we know the model of the null hypothesis, and furthermore we bypass the issue of analytic tractability by computing the error bar numerically (from the standard deviation of all the numerically estimated dimensions of all the surrogate data sets). The estimation of the error bar by this method is one of a class of computationally intensive statistics, which includes the related technique of "bootstrapping," and which (again, with the advent of cheap computing) are becoming increasingly popular. See Efron[15] for an inspirational essay on the value of these methods.

This paper describes preliminary work that is part of an ongoing project to more reliably characterize nonlinear structure from a time series. While some data sets very cleanly exhibit low-dimensional chaos, there are many cases where the evidence is sketchy and difficult to evaluate. We hope to provide a framework within which such claims of nonlinearity can be evaluated.

In section 2, we discuss the statistical problem, develop useful null hypotheses, describe how we generate surrogate data, and introduce the "significance." Section 3 demonstrates the usefulness of this technique on several computer-generated examples. In section 4, we apply the technique to several real data sets, including sunspots, electroencephalograms (EEG), and fluid dynamics data.

2. TESTING FOR SIGNIFICANCE

Statistical hypothesis testing involves two components: a null hypothesis against which observations are tested, and a discriminating statistic. The null hypothesis is the too-simple explanation that we seek to show as inadequate for explaining the data. The discriminating statistic, in general, is just a *number* which quantifies some aspect of the time series. If this number is different for the observed data than would be expected under the null hypothesis, then the null hypothesis can be rejected.

The method of surrogate data basically follows this recipe, but it permits us to treat these two components separately, which gives added flexibility. The choice of a null hypothesis does not (in principle or in practice) restrict the choice of discriminating statistic.

Our null hypotheses have the form, "the underlying dynamics is of a particular type," where the particular type might be, for example, white noise. Our aim is to show that the actual data, D, differs significantly from any realization that would be generated under the null hypothesis. For this purpose, we employ a discriminating statistic denoted by Q. This can, in principle, be anything from a moment to an estimated fractal dimension. Basically, we want to show that this statistic is different for the original time series than it is for the realizations consistent with the null hypothesis. Roughly speaking, we say $Q_D \neq Q_H$. Since the value of Q will vary for different realizations under the null hypothesis, we must concern ourselves with the *distribution* of Q on the null hypothesis. If we know this distribution, the idea of a significant difference can be made statistically meaningful.

Surrogate data is an ensemble of data sets similar to the observed data, but consistent with the null hypothesis. For a given null hypothesis, we generate many realizations, $H_1, \ldots, H_k$, of time series which are of the specified type. For instance, if the null hypothesis is that the underlying dynamics is white noise, then the surrogate data will be realizations of white noise. For each null hypothesis, one must design a surrogate data generator, but once this has been done, no more analytical work is needed.

Let Q_D denote the statistic computed for the original time series. Let μ_H and σ_H denote the mean and standard deviation of the distribution of Q under the null hypothesis. In practice, we just use the sample mean and sample standard deviation from a finite sample of realizations of the null hypothesis (these are the surrogate data sets). We also write μ_D as the mean value of Q for the original data.

If many realizations are available of both the surrogate and the original data, then the next logical step is to compare the two distributions. We can compare the full distributions, using for instance the Kolmogorov-Smirnov statistic; or just their means, in which case a Student-t statistic is appropriate. Often, however, only a single realization exists for the original data. Indeed, this is the case that we will primarily consider in this paper. In that case μ_D obviously equals Q_D, but a direct comparison of distributions is not sensible with only one sample from the data distribution. Instead, we define the "significance" by the difference between the original and the mean surrogate value of the statistic, divided by the standard deviation of the surrogate values.

$$S \equiv \frac{|\mu_D - \mu_H|}{\sigma_H}. \tag{1}$$

The significance is a dimensionless quantity, but it is common parlance among physicists to call the units of S "sigmas." Thus, one might speak of a "two-sigma effect" as not especially significant, but ten sigmas as very significant.

2.1 CHOICE OF NULL HYPOTHESIS

Some restrictions on the null hypothesis are desirable. For example, if we generate white Gaussian noise with a different mean and standard deviation from the original data, it will be no surprise if the discriminating statistic is able to distinguish the two. We want our null hypothesis to mimic all of the *trivial* features of the raw data set. Among these features we certainly include the mean, and variance, but might also include the Fourier power spectrum. Other features may be important as well. For instance, if the raw data is discretized to integer values, then it is recommended that the surrogate data be similarly discretized.

Ultimately we envision a hierarchy (perhaps even a hierarchical tree) of null hypotheses against which time series might be compared. Beginning with the simplest hypotheses, and increasing in generality, the following list outlines several possibilities.

2.1.1 TEMPORALLY UNCORRELATED NOISE

The null hypothesis of no temporal correlations is of particular interest in circumstances (e.g., stock market returns, or outcomes on a roulette wheel) where *any correlation at all* can potentially be exploited for profit. The simplest null hypothesis in this case is that the observed data are totally uncorrelated; that is, that they are fully described by independent and identically distributed (IID) random variables. Surrogate data in this case are readily generated from any standard pseudorandom number generator, normalized to the mean and variance of the original data. The null hypothesis in this case further assumes that the distribution of the random variable is the same as the random number generator, usually gaussian. A clever and more general approach was used by Scheinkman and LeBaron[37] in an analysis of stock market returns. To test the null hypothesis of uncorrelated noise (of arbitrary amplitude distribution), they just scrambled the original time series. This more closely mimics the original data, but it destroys any temporal correlations that may have been in the data.

2.1.2 ORNSTEIN-UHLENBECK NOISE

The null hypothesis that there are no temporal correlations at all is unfortunately not very useful for most physical systems. Often it is obvious that there *are* temporal correlations, but the nature of these correlations may not be so clear. The simplest case of non-white noise is that given by the Ornstein-Uhlenbeck process.[50] For a discrete time series, this can be produced by

$$x_t = a_1 x_{t-1} + \sigma e_t \tag{2}$$

where e_t is uncorrelated gaussian noise of unit variance and the coefficient a_1 is a constant less than one. The autocorrelation function is an exponential; that is,

$$A(\tau) \equiv \langle\, x_t x_{t-\tau}\, \rangle / \langle\, x_t^2\, \rangle = e^{-\lambda \tau} \tag{3}$$

where $\langle\ \rangle$ denotes an average over time t, and $\lambda = -\log a_1$. The coefficient a_1 is chosen so that $A(\tau = 1)$ is the same as for the original data, and σ is chosen so that the variance of the surrogates matches the variance of the original data.

Grassberger[18] used a single surrogate data set based on an Ornstein-Uhlenbeck process to reject an earlier claim of low correlation dimension in a time series of global climate.

2.1.3 LINEARLY CORRELATED NOISE

The null hypothesis of Ornstein-Uhlenbeck noise is arguably restrictive, and can readily be generalized to higher order. This can be done by fitting coefficients a_k and σ to a process

$$x_t = \sum_{k=1}^{q} a_k x_{t-k} + \sigma e_t \tag{4}$$

which mimics the original time series in terms of autocorrelation function and variance. The null hypothesis in this case is that all the structure in the time series is given by the autocorrelation function or, equivalently, by the Fourier power spectrum. The use of surrogate data based on this hypothesis has previously been advocated,[7,42,43] and an algorithm for generating surrogate data under this null hypothesis is described as Algorithm I of the appendix. This algorithm does not directly employ Eq. (4), but the generated surrogate data is guaranteed to have the same Fourier spectrum as the original data.

We should point out here that this is the null hypothesis that researchers have in mind when they advocate subtracting out the linear component and leaving residuals. We emphasize that this is a very common technique, and that most traditional tests for nonlinearity are based on it, as well as the more recently advocated BDS test.[2] We argue in subsection 3.3 that it is preferable to work with the raw data directly, using this null hypothesis, instead of working with residuals.

2.1.4 STATIC NONLINEAR FILTER OF LINEARLY CORRELATED NOISE

In this paper, we will also consider a slightly more general null hypothesis, that the time series arises from a linear stochastic process, but the observed time series may be a nonlinear distortion of the underlying linear time series. In particular, we suppose that the underlying time series $\{y_t\}$ derives from an arbitrary linear process, such as the following infinite-order autoregressive process

$$y_t = \sum_{k=1}^{\infty} a_k y_{t-k} + \sigma e_t \tag{5}$$

where a_k are coefficients of the process, and e_t is uncorrelated noise with unit variance. Further, we take the observed time series to be a nonlinearly distorted image of y_t, but the distortion is "static," by which we mean that it depends only on the current value y_t. That is,

$$x_t = h(y_t) \,. \tag{6}$$

This null hypothesis further assumes that h is a monotonic function.

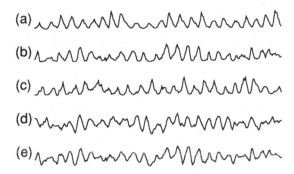

FIGURE 1 The Wolfer sunspot numbers and some surrogates under various null hypotheses. Note that differences are often detectable by eye. The true series is in (a). The series in (b) and (c) are generated by Algorithm II. The series in (d) and (e) are generated by Algorithm I. Surrogate data generated by Algorithm I have a gaussian distribution, which is clearly distinguishable from the nongaussian amplitude distribution of the original data in (a).

An algorithm for generating surrogate data corresponding to this null hypothesis has been developed which shuffles the data but in such a way as to preserve the temporal correlations (see Algorithm II of the appendix). By preserving temporal correlations, strictly we mean that if $\hat{x}_t$ is the surrogate data, then the ("underlying") time series $h^{-1}(\hat{x}_t)$ has the same Fourier spectrum as the time series $y_t = h^{-1}(x_t)$.

Note that time series in this class are strictly speaking nonlinear, but the nonlinearity is not in the dynamics. Thus, while many conventional tests for nonlinearity might indicate that the time series is nonlinear, by using surrogate data that have been tailored to this case of a nonlinear time series with linear dynamics, we are able to make finer distinctions about the underlying dynamics. The method of surrogate data permits us to choose flexibly exactly what null hypothesis we will attempt to reject.

Ultimately, we would like to extend this list to include more general null hypotheses. Foremost in our minds is the noisy limit cycle, which cannot be described by a linear process, even if viewed through a nonlinear static filter. Yet, it is often of great interest, particularly in systems driven by seasonal cycles, to determine the nature of the inter-seasonal variation.

2.2 DISCRIMINATING STATISTICS

The choice of discriminating statistic is, formally speaking, arbitrary. However, some choices will be better than others. Since we are motivated by the possibility that the underlying dynamics may be chaotic, our first choices for discriminating statistics are just the conventional discriminants of nonlinear dynamics: correlation dimension, Lyapunov exponent, and forecasting error. Indeed, one of our eventual interests in this project is to outline the conditions in which one or the other of these methods will be more effective. But the method in principle can be used with any discriminating statistic. For example, it is possible to use the correlation integral itself, that is, the $C(r)$ function from which the correlation dimension is defined, instead of the dimension (this is the basis of the BDS statistic[2]). Also, one may use two-sided forecasting (predicting the "present" x_t from the "past" $x_{t-1}, \ldots$ *and* the "future" $x_{t+1}, \ldots$), instead of the usual forecasting which seeks to predict the future from the past. Other candidates which we have not investigated include the embedding criterion of Liebert and Schuster,[30] and the dimension statistic of Čenys and Pyragas.[8]

Below, we describe how we used the three particular discriminating statistics that we chose for the numerical experiments in this paper. Ideally, dimension counts degrees of freedom, Lyapunov exponent quantifies the sensitivity to initial conditions, and forecasting error tests for determinism. These are three different aspects of low-dimensional chaotic systems. Now, we are not explicitly looking for low-dimensional chaos but just trying to detect nonlinearity, so any nonlinear statistic is a viable candidate. But our choice of statistic is *motivated* by the notion that the underlying process *might* be chaotic, and so (we hope) the statistics which characterize such processes might be most adept at detecting them. In general, we advocate using a battery of statistics, not only to increase the opportunity of rejecting the null hypothesis (since we expect some will be sensitive where others are not), but also to have some qualitative notion of "how" the data set differs from the surrogates.

2.2.1 CORRELATION DIMENSION, ν

Dimension is an exponent which characterizes the scaling of some bulk measure with linear size. To compute a dimension, it is necessary to choose some range of sizes over which the scaling is to be estimated. Algorithms abound[27,43] for estimating the dimension of an underlying strange attractor from a time series; we chose a box-assisted variation[41] (see Grassberger[19] for an elegant alternative) of the Grassberger-Procaccia-Takens algorithm[20,21,39] to compute a correlation integral, and the best estimator of Takens[40] for the dimension itself. The Takens estimator requires an upper cutoff size; we used one-half of the rms variation in the time series for this value. This is rather large if our aim is to make our best guess of the fractal dimension (in that case we would choose the characteristic scale small enough to minimize finite size effects, but still large enough to get good statistics: this tradeoff is discussed in Theiler[44]) but it gives us good values for statistical significance.

2.2.2 LYAPUNOV EXPONENT, λ   Following Sano and Sawada,[36] we compute the Lyapunov exponent by multiplying Jacobian matrices along a trajectory, with the matrices computed by local linear fits. We use the QR decomposition method of Eckmann et al.[14] to maintain orthogonality. For the results in this paper, we consider only the largest exponent.

We have found that numerical estimation of Lyapunov exponents in the presence of noise can be problematic. Indeed, for our surrogate data sets, for which the linear dynamics is contracting, we often obtain positive Lyapunov exponents. It may be possible to remedy this by using more near neighbors, but remember that it is not the best estimate of the Lyapunov exponent itself that we are seeking, but only a statistic which distinguishes the original data from surrogate data. We would prefer to use a discriminating statistic which correctly quantifies some feature of the dynamics, as this provides more qualitative information, but the method of surrogate data does not formally require this.

2.2.3 FORECASTING ERROR, ϵ   A direct test for determinism comes from quantifying the forecasting errors one gets from nonlinear modeling. The method we use entails first splitting the time series into a fitting set of length N_f, and a testing set of length N_t, with $N_f + N_t = N$, the length of the time series; then fitting a local linear model[17] to the fitting set, locality given by the number of neighbors k; and finally, using this model to forecast the values in the testing set, and comparing them with the actual values.

If $e_t = x_t - \hat{x}_t$ is the difference between the actual value of x and the predicted value, $\hat{x}$, then we define our discriminating statistic as the mean log absolute prediction error.

$$\epsilon = \frac{1}{N_t} \sum_{t=N_f+1}^{N_f+N_t} \log |e_t|. \tag{7}$$

Several modeling parameters must be chosen, including the partitioning of the data set into fitting (N_f) and testing (N_t) segments, the number of steps ahead to predict (T), and number of neighbors (k) used in the local linear fit. We arbitrarily chose to divide the fitting and testing sets equally, with $N_f = N_t = N/2$, and to predict one step ahead, so $T = 1$. More important is the choice of k. For the results in this paper, we set k to 1.5 times the minimum number needed for a fit, but we note that this is often not optimal. Indeed, Casdagli[5,6] has advocated sweeping the parameter k in a local linear forecaster as an exploratory method to look for nonlinearity in the first place. For few neighbors, this models noise-free low-dimensional determinism; for many neighbors, the forecasting method is effectively a global linear predictor. When the optimal k is some intermediate value, this indicates nonlinearity with noise.

2.3 COMPARISON TO THE TRADITIONAL STATISTICAL APPROACH

Traditional statistical tests for nonlinearity (see Tong[47] for a review) generally require the use of a discriminating statistic and a null hypothesis that together allow analytic calculation of the distribution of the statistic. Once this analytical calculation has been done, all the experimenter has to do is compute the statistic for the time series under question, and then look up in a standard statistical table the significance of the deviation of the statistic from the null hypothesis.

An example of one of the simpler traditional tests for nonlinearity is attributed by Tong[47] to McLeod and Li. This is not touted as a particularly powerful test, but it illustrates the approach. Let $\{e_t\}$ be the fitted residuals from a linear model, as in Eq. (9). The Wold decomposition theorem assures us that the residuals will be linearly uncorrelated, but if the underlying dynamics really is linear, then the residuals will furthermore be independent and identically distributed (IID). The squares of the residuals will also be IID, and in particular will be linearly uncorrelated. Now let $A_{e^2}(\tau)$ be the sample autocorrelation of the squared residuals $\{e_t^2\}$, and compute the statistic

$$ Q = N(N+2) \sum_{\tau=1}^{m} \frac{A_{e^2}^2(\tau)}{(N-\tau)} . \tag{8} $$

For a linear model, in the limit of large N, the statistic Q will approach a chi-squared distribution with m degrees of freedom.

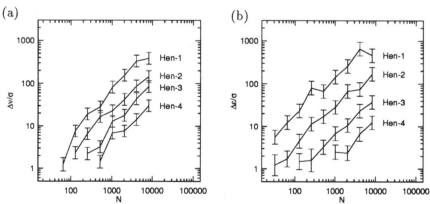

FIGURE 2 Significance as a function of the number N of data points for time series generated by adding n independent trajectories of the Hénon map. The discriminating statistic is (a) correlation dimension and (b) forecasting error. Note that significance increases with number of data points and decreases with the complexity of the system.

The advantage of the traditional approach is that the statistic Q is only computed once for a given time series. If it is outside a certain range, then the result is significant, and the null hypothesis is rejected. To the extent that computational resources are precious, this is an advantage.

The disadvantage of the traditional approach is that the discriminating statistics and null hypotheses are restricted to those for which analytical computation of the distribution is possible. The method of surrogate data is more computationally intensive than the traditional approaches, but the tradeoff is that the test is more flexible: one can design null hypotheses and choose discriminating statistics to suit one's taste. While we hope that this flexibility will permit researchers to design more powerful statistical tests, we feel that the more important applications will be to more general "plausible alternatives" than just linearity.

---

3. NUMERICAL EXPERIMENTS

3.1 VARIATION WITH NUMBER OF DATA POINTS

The significance with which nonlinearity can be detected in a chaotic time series increases with the number of points in the time series, and in general decreases with the complexity of the time series. This is shown in Figure 2 for the attractor of Hénon,[23] using estimations of dimension and forecasting error as the discriminating statistic. Here 'Hen-n' corresponds to the sum of n independent trajectories of the Hénon map; thus, it is a time series whose underlying strange attractor will have dimension $n\nu$ where $\nu \approx 1.25$ is the dimension of a single Hénon trajectory.

3.2 EFFECT OF OBSERVATIONAL AND DYNAMICAL NOISE

The method of surrogate data, far from being *sensitive* to noise, permits us to work with data that are very noisy. We can detect nonlinearity even in cases where the noise dominates the signal.

For chaotic determinism, the effect of noise in general is to degrade the significance of the nonlinearity.

For nonchaotic determinism, the situation is more complicated. Nonchaotic noiseless time series can in general be modeled by linear processes, and the method of surrogate data does not identify these as being nonlinear. Adding observational noise in this case has little effect; the detection of nonlinearity remains insignificant. Dynamical noise however does have an effect. For small dynamical noise, the effect is small, but as the noise amplitude is increased, more of the nonlinearity becomes evident. In some cases it is also possible for dynamical noise to induce a transition to chaotic behavior[9]; in this case a large increase is observed in the significance.

(a)

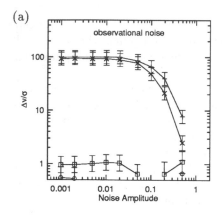

(b)

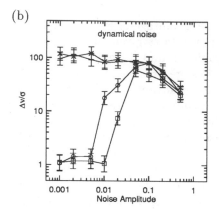

FIGURE 3 Significance of nonlinearity measured for time series of length $N = 1024$ obtained from a noisy logistic map: $x_{t+1} = \lambda x_t(1 - x_t)$ (mod 1). (The mod 1 is needed to keep the data inside the basin of attraction [0,1].) We considered parameter values $\lambda = 3.5$ (□), 3.5699 (○), 3.8 (×), and 4.0 (+), corresponding respectively to a period-four orbit, the period-doubling critical cascade point, an evidently chaotic orbit, and the provably chaotic orbit at $\lambda = 4.0$. For the chaotic attractors, both (a) observational noise and (b) dynamical noise have the effect of degrading significance, although evidence for nonlinearity is still significant with quite large noise amplitude. For the nonchaotic attractors with low noise, the time series did not test as significantly nonlinear. Increasing observational noise did not affect this conclusion, but increasing dynamical noise increased the nonlinear significance. The significance in this figure is based on correlation dimension, ν.

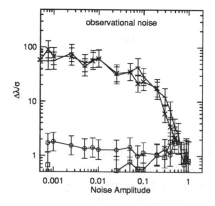

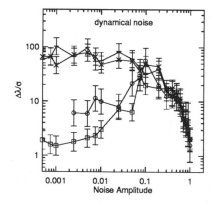

FIGURE 4 Significance of nonlinearity measured for a noisy logistic map. Same as the previous figure, except that significance is based on estimation of the Lyapunov exponent.

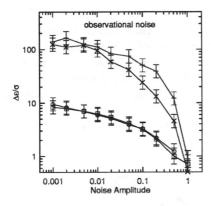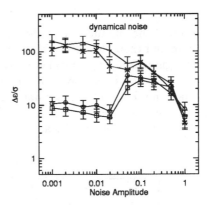

FIGURE 5 Significance of nonlinearity measured for a noisy logistic map. Same as the previous figure, except that the significance in this figure is based on prediction error.

As the noise amplitude becomes very large, however, the effect of the dynamics is again lost, and the significance drops. These effects are observed in Figures 3, 4, and 5 for all three discriminating statistics.

3.3 DON'T BLEACH CHAOTIC DATA

A common approach to testing for nonlinearity involves first "subtracting out" the linear component, and then working with what is left, the "residuals." Given a time series x_t, the residuals are given by

$$e_t = x_t - \left[a_o + \sum_{k=1}^{q} a_k x_{t-k} \right], \qquad (9)$$

where the coefficients a_k are chosen to minimize the variance $\sum_t e_t^2 / N$ of the residuals. Here q is the order of the linear model. This is an attractive approach because the Wold decomposition theorem guarantees that the residuals will have a white spectrum; that is, there will be no linear correlations. Thus, *any* temporal correlations that are detected will be nonlinear correlations.

Because the residuals e_t are spectrally white (equal power at all frequencies), the process of determining residuals is sometimes called "pre-whitening" or "bleaching." However, linear filtering of chaotic data is not without its pitfalls. The linear map from x_t to e_t in Eq. (9) will not formally change the structure of the attractor for finite q; for example, if x_t lies on a low-dimensional attractor, then e_t will lie on an attractor of the same dimension. However, in practice, the distortion induced by the linear map can drastically affect the *appearance* of the attractor and can likewise affect *estimates* of its dimension. The effect of linear filtering on the Hénon attractor is shown in Figure 6. The determinism which is obvious in the unfiltered data ceases to be so obvious in the filtered case.

We computed the significance of the nonlinearity in time series obtained from the Hénon map and then bleached with ever larger values of q. We show in Figure 7 a decrease in significance, as quantified by the method of surrogate data, computed with statistics based on dimension, forecasting, and Lyapunov exponent. However, this result may only hold for chaotic data. Townshend[48] has described a situation with data from human speech in which nonlinear predictions of linearly filtered data were superior to direct nonlinear predictions of the original time series.

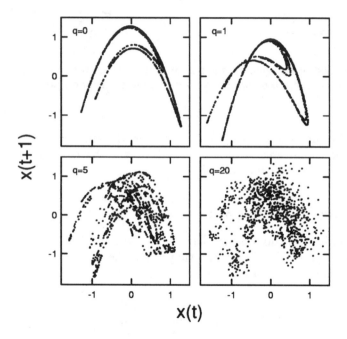

FIGURE 6   Residuals of the Hénon map, as fitted with Eq. (9). As q increases, the deterministic nature of the map becomes less evident.

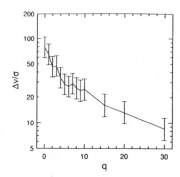

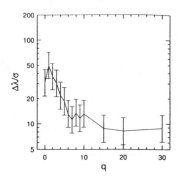

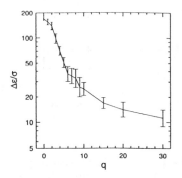

FIGURE 7 Effect of bleaching on significance for a time series of length $N = 1024$ derived from the Hénon map. For $q = 0$, the raw data set is used. For $q > 0$, the qth order residuals (as computed by Eq. (9)) is used. It is apparent that attempting to "subtract out" the linear component only decreases the power of the test. The discriminating statistic is (a) correlation dimension, (b) largest Lyapunov exponent, and (c) forecasting error.

4. REAL DATA

We report preliminary results on some experimental time series from various sources. These results should be taken as anecdotal, and not necessarily representative of the class which they represent (the sunspot time series is an exception, of course). In particular, we have not yet attempted to "normalize" our findings with others that have previously appeared in the literature. We have only included cases for which nonlinearity is positively detected, as these illustrate the variety of ways that a hypothesis can be rejected. To paraphrase Tolstoy, all negative results are alike, but positive results are all positive in their own way. Unless otherwise noted, the surrogate data for the results in this section were generated by Algorithm II of the appendix, which corresponds to a static nonlinear filter of a linear time series.

4.1 SUPERFLUID CONVECTION

Data from a superfluid convection cell[22] provides an example where the evidence for low-dimensional chaos is quite clear. Using discriminating statistics of dimension and forecasting error, we obtain about fifteen sigmas of significance. This data was also analyzed by Farmer and Sidorowich,[17] who found sizable increases in predictability using nonlinear rather than linear predictors.

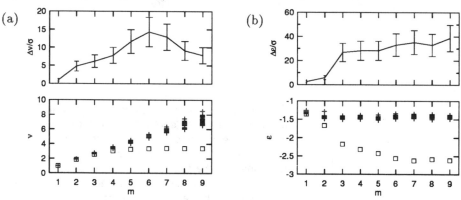

FIGURE 8 Data from a fluid convection experiment exhibits very significant nonlinear structure, using (a) dimension and (b) forecasting error. The top panel in these figures show the significance, measured in "sigmas," and the bottom panel shows the values of the statistics, with squares (□) for the original data and pluses (+) for the surrogates. Both panels plot these statistics against the embedding dimension m. Not only is the evidence for nonlinear structure statistically significant, but the estimated dimension of about $\nu = 3.8$ suggests that the underlying dynamics is in fact low-dimensional chaos.

4.2 ELECTROENCEPHALOGRAM (EEG)

That the brain should exhibit chaos is an idea that some authors have been unable to resist. Our own investigations so far have been mixed; some data sets exhibit nonlinear structure and some do not. A more systematic survey is clearly in order. In the meantime, we present in Figure 9 the "best" result, the one time series we looked at which exhibited the most statistically significant evidence for nonlinear structure. The time series was obtained from the brain of an individual who was staring at a spot on a wall. Here we see almost ten sigmas of significance with the dimension algorithm (around five sigmas with the forecasting algorithm), but it is worth pointing out that we do not see any evidence that the time series is in fact low-dimensional (the correlation dimension ν does not converge with increasing embedding dimension m). We are only able to reject the null hypothesis that the data arise from a linear stochastic process; we cannot assert that low-dimensional chaos is evident.

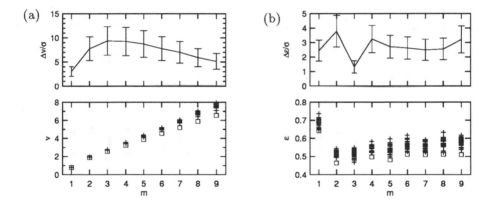

FIGURE 9 Data from an electroencephalogram time series exhibits significant nonlinear structure, using both (a) the dimension statistic and (b) the forecasting error. The evidence for low-dimensional chaos, however, is weak, since the estimated dimension increases almost as rapidly with embedding dimension for the original time series as it does for the surrogates.

4.3 SUNSPOTS

The sunspot cycle has attracted perhaps more attention than any other time series, due to its interesting mixture of regularity and irregularity.[11,28,32,46,51,52]

Using both dimension and forecasting error, we can quite confidently reject the null hypothesis that the time series itself is linear stochastic; this is in agreement with the numerous authors[11,28,32,46,51] who obtained better agreement using nonlinear models instead of linear models. However, when we expand the null hypothesis to include a static nonlinear observation of an underlying linear stochastic process, the evidence (for dynamical nonlinear structure) is less dramatic. Using the dimension statistic, there is virtually no significance (of order one sigma). Using a forecasting algorithm, on the other hand, we do see significantly more predictability in the sunspot data than in surrogates, at about the five sigma level. This illustrates the advantage of having a battery of tests: one statistic may be more sensitive than another to certain kinds of nonlinear structure.

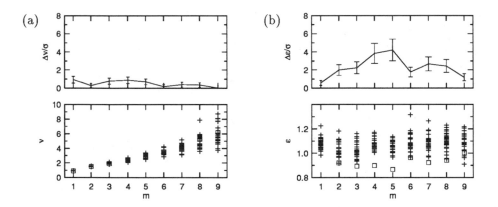

FIGURE 10 Significance of nonlinearity in the Wolfer sunspot series, (a) using correlation dimension and (b) using forecasting error. As with all the experimental time series reported here, Algorithm II was used to generate the surrogate data.

5. DISCUSSION

In this paper, we have provided a framework for evaluating the statistical significance of evidence for nonlinearity in a stationary time series. (We do *not* seek to characterize *non*-stationary time series—see Mayer-Kress,[31] Osborne et al.,[33] Provenzale et al.,[34] and Theiler[45] for a discussion of some of the problems arising in the estimation of nonlinear statistics from nonstationary data.)

By construction, the test will fail to find nonlinear structure in linear stochastic systems. We have found that the test correctly identifies nonlinearity in several well-known examples of low-dimensional chaotic time series, even when contaminated with considerable observational and dynamical noise.

Our experiments with chaotic data found that using linear pre-processing to "bleach" out the linear correlations decreases the power of the test to detect nonlinearity. Consequently, we advocate a direct application of the method of surrogate data to the raw time series, instead of to a time series of residuals.

Finally, we tested several experimental data sets for nonlinear structure.

We have described the method of surrogate data as a formal test for quantifying the statistical significance of the rejection of a particular null hypothesis. It is useful, however, to take a more informal approach, and view surrogate data as a control experiment. Having exhibited that some data gives a certain dimension, say, it is wise to compute the dimension for surrogate data as well, to make sure that the estimator is not being fooled by some feature (such as linear autocorrelation) that is also present in the random surrogate data.

In this case, there is some room for human judgment. For example, if the estimated dimension for the original data and the surrogate data are approximately equal and both small (or worse yet, if the surrogate data exhibit a *lower* dimension than the original data), then the conclusion that the data arises from low-dimensional dynamics is doubtful. It may be that the significance as defined in Eq. (1) is large enough that the particular null hypothesis can be positively rejected, but that does not automatically imply low-dimensional chaos. In general, we advocate using a battery of statistical tests, not only to increase the chances of rejecting the null hypothesis, but to provide some intuitive insight into the nature of the nonlinearity. Reducing the role of human judgment requires the development of more general null hypotheses whose rejection makes a stronger statement.

Any test which fails to reject the null hypothesis is strictly speaking inconclusive. Just because the original and surrogate time series have the same value for the discriminating statistic, that does not imply that they have the same underlying dynamics. On the other hand, if one observes evidence of low fractal dimension, but surrogate data shows the same low dimension, then claims based on the original evidence can be dismissed as not well founded.

Much work remains to make the method of surrogate data a powerful and flexible tool for nonlinear analysis. In particular, we hope to expand the hierarchy of null hypotheses and to broaden the battery of discriminating statistics. The algorithms should also be extended to deal with multivariate time series. Further, by using significance as a measure of "goodness," we have a quantitative way to optimize parameters for a given statistic and also a way to compare different discriminating statistics.

Indeed, we feel that the investigation of the effectiveness of various statistics for different null hypotheses and in different situations will be valuable not only for increasing our ability to reject null hypotheses, but also for the more qualitative task of characterizing the nature of the nonlinearity that might be evidenced by one statistic but not another.

APPENDIX

ALGORITHMS FOR GENERATING SURROGATE DATA

In this section, we describe two algorithms we use for generating surrogate data.

ALGORITHM I

The first algorithm is based on the null hypothesis that the data come from a linear stochastic process; the assumption is that there is no nonlinearity either in the dynamics or in the observation of the data. The surrogate data are designed to have the same Fourier spectra as the raw data.

1. Input the original data into an array $y[t]$, $t=1,\ldots,N$.
2. Compute the Discrete Fourier Transform: $z[t] = DFT y[t]$.
3. Note $z[t]$ has real and imaginary components.
4. Randomize the phases: $z'[t] = z[t]e^{i\phi[t]}$.
5. Here, $\phi[t]$ is uniformly distributed between 0 and 2π.
6. Symmetrize the phases:
7. Re $z''[t]$ = Re($z'[t]$ + $z'[N+1-t]$)/2.
8. Im $z''[t]$ = Im($z'[t]$ - $z'[N+1-t]$)/2.
9. Invert the Discrete Fourier Transform: $y'[t] = DFT^{-1}z''[t]$.
10. Note that because of the symmetry of the phases, the resulting time series $y'[t]$ is real; this is the surrogate data.

ALGORITHM II

The second algorithm creates data that are realizations of the null hypothesis that the observed time series is a nonlinear static transformation of a linear stochastic process. Our approach is first to rescale values of the original time series so that they are gaussian, then to use the first algorithm to create a surrogate time series which has the same Fourier spectrum as the rescaled original. This surrogate is then rescaled to have the same values as the original time series.

1. Input the original data into an array $x[t]$, $t=1,\ldots,N$.
2. Sort the array $Sx[k]$, $k=1,\ldots,N$.
3. Make ranked time series $Rx[t]$, defined to satisfy $Sx[Rx[t]] = x[t]$.
4. Note $Sx[k]$ is a monotonic function with a well-defined inverse; so $Rx[t] = Sx^{-1}[t]$ is a static rescaling of $x[t]$.
5. Create a random gaussian data set $g[t]$, $t=1,\ldots,N$.
6. Sort the gaussian random numbers $Sg[k]$, $k=1,\ldots,N$.
7. Define new time series: $y[t] = Sg[Rx[t]]$.
8. The new time series is a static rescaling of $x[t]$ with the property that the amplitude distribution is gaussian.

9. Use Algorithm I to make a surrogate of this gaussian time series: `y'[t]`.
10. Make a ranked time series for `y'[t]`, call it `Ry'[t]`.
11. The surrogate time series is then given by `x'[t]` = `Sx[ Ry'[t] ]`.

Note that the surrogate time series `x'[t]` is just a shuffling of the original time series `x[t]`, so it obviously has the same amplitude distribution. Further if we define G as the transformation from the amplitude distribution exhibited by $\mathbf{x}$ to a gaussian amplitude distribution, then we have the property that $G(\mathbf{x})$ has the same Fourier power spectrum (and hence, the same autocorrelation) as $G(\mathbf{x}')$. Note $G = h^{-1}$ where h is the observation function of our null hypothesis.

ESTIMATING ERROR BARS ON SIGNIFICANCE

Recall, our measure of significance is given by

$$S = \frac{|\mu_H - \mu_D|}{\sigma_H}. \tag{10}$$

The "error bar" on S is given by ΔS and can be computed by standard propagation of errors methodology. Here

$$\left(\frac{\Delta S}{S}\right)^2 = \left(\frac{\Delta|\mu_H - \mu_D|}{|\mu_H - \mu_D|}\right)^2 + \left(\frac{\Delta\sigma_H}{\sigma_H}\right)^2$$

$$= \frac{(\Delta\mu_H)^2 + (\Delta\mu_D)^2}{(\mu_H - \mu_D)^2} + \left(\frac{\Delta\sigma_H}{\sigma_H}\right)^2. \tag{11}$$

Now the error of the sample mean based on N observations is given by $(\Delta\mu)^2 = \sigma^2/N$, so we can write

$$(\Delta\mu_H)^2 = \frac{\sigma_H^2}{N_H},$$

$$(\Delta\mu_D)^2 = \frac{\sigma_D^2}{N_D}. \tag{12}$$

Further, the error of the sample standard deviation is given by $(\Delta\sigma)^2 = 2\sigma^2/N$, so that

$$(\Delta\sigma_H)^2 = \frac{2\sigma_H^2}{N_H}. \tag{13}$$

Combining all of these gives a relative error of

$$\left(\frac{\Delta S}{S}\right)^2 = \frac{\sigma_H^2/N_H + \sigma_D^2/N_D}{(\mu_H - \mu_D)^2} + \frac{2}{N_H}. \tag{14}$$

The absolute error bar is then given by

$$\Delta S = \sqrt{(1 + 2S^2)/N_H + (\sigma_D/\sigma_H)^2/N_D}. \tag{15}$$

Usually, we will consider the case that only a single realization of the time series is available. Here we take $\sigma_D = 0$ and ignore the second term in the above equation. Note also that when the significance is large, the error bar in S is dominated by the error in the estimation of the standard deviation of the hypotheses' statistics, σ_H.

ACKNOWLEDGEMENTS

We are grateful to Martin Casdagli, Xiangdong He, Peter Grassberger, Mark Berge, and Tony Begg for useful discussions. We also thank Bob Ecke for providing convection data, and Paul Nuñez and Arden Nelson for providing EEG data. This work was partially supported by the National Institute for Mental Health under grant 1-R01-MH47184-01, and performed under the auspices of the Department of Energy. We urge the reader to use these results for peaceful purposes.

REFERENCES

1. Box, G. E. P., and G. M. Jenkins. *Time Series Analysis Forecasting and Control.* San Francisco: Holden-Day, 1970.
2. Brock, W. A. "Distinguishing Random and Deterministic Systems." *J. Econ. Theo.* **40** (1986): 168–195.
3. Brock, W. A., and C. L. Sayers. "Is the Business Cycle Characterized by Deterministic Chaos?" Technical Report 8617, Social Systems Research Institute, University of Wisconsin, Madison, 1986.
4. Brock, W. A., and W. D. Dechert. "Statistical Inference Theory for Measures of Complexity in Chaos Theory and Nonlinear Science." In *Measures of Complexity and Chaos*, edited by N. Abraham et al. New York: Plenum, 1989.
5. Casdagli, M. "Nonlinear Forecasting, Chaos and Statistics." In *Modeling Complex Phenomena, Proceedings of the Third San Jose State University Woodward Conference, April, 1991*, edited by L. Lam and V. Naroditsky. Springer-Verlag, to appear.
6. Casdagli, M. "Chaos and Deterministic Versus Stochastic Nonlinear Modeling." *J. Roy. Stat. Soc.*, to appear.
7. Casdagli, M., D. Des Jardins, S. Eubank, J. D. Farmer, J. Gibson, N. Hunter, and J. Theiler. "Nonlinear Modeling of Chaotic Time Series: Theory and Applications." In *EPRI Workshop on Applications of Chaos*, to appear.
8. Cenys, A., and K. Pyragas. "Estimation of the Number of Degrees of Freedom From Chaotic Time Series." *Phys. Lett. A* **129** (1988): 227.
9. Crutchfield, J. P., J. D. Farmer, and B. A. Huberman. "Fluctuations and Simple Chaotic Dynamics." *Phys. Rep.* **92(2)** (1982).
10. Crutchfield, J. P., J. D. Farmer, N. H. Packard, and R. S. Shaw. "Chaos." *Sci. Am.* **254** (1986): 46–57.
11. Currie, D. Personal Communication.
12. Denker, M., and G. Keller. "Rigorous Statistical Procedures for Data from Dynamical Systems." *J. Stat. Phys.* **44** (1986): 67–93.
13. Eckmann, J.-P., and D. Ruelle. "Ergodic Theory of Chaos and Strange Attractors." *Rev. Mod. Phys.* **57** (1985): 617.
14. Eckmann, J.-P., S. Oliffson Kamphorst, D. Ruelle, and S. Ciliberto. "Liapunov Exponents from a Time Series." *Phys. Rev. A* **34** (1986): 4971–4979.
15. Efron, B. "Computers and the Theory of Statistics: Thinking the Unthinkable." *SIAM Rev.* **21** (1979): 460–480.
16. Ellner, S. "Detecting Low-Dimensional Chaos in Population Dynamics Data: A Critical Review." In *Does Chaos Exist in Ecological Systems*, edited by J. Logan and F. Hain. University of Virginia Press, to appear.
17. Farmer, J. D., and J. J. Sidorowich. "Exploiting Chaos to Predict the Future and Reduce Noise." In *Evolution, Learning and Cognition*, edited by Y. C. Lee, 277–330. Singapore: World Scientific, 1988.
18. Grassberger, P. "Do Climatic Attractors Exist?" *Nature* **323** (1986): 609.

19. Grassberger, P. "An Optimized Box-Assisted Algorithm for Fractal Dimensions." *Phys. Lett. A* **148** (1990): 63–68.
20. Grassberger, P., and I. Procaccia. "Characterization of Strange Attractors." *Phys. Rev. Lett.* **50** (1983): 346.
21. Grassberger, P., and I. Procaccia. "Measuring the Strangeness of Strange Attractors." *Physica D* **9** (1983): 189–208.
22. Haucke, H., and R. Ecke. "Mode Locking and Chaos in Rayleigh-Benard Convection." *Physica D* **25** (1987): 307.
23. Hénon, M. "A Two-Dimensional Mapping with a Strange Attractor." *Comm. Math. Phys.* **50** (1976): 69.
24. Holzfuss, J., and G. Mayer-Kress. "An Approach to Error-Estimation in the Application of Dimension Algorithms." In *Dimensions and Entropies in Chaotic Systems—Quantification of Complex Behavior*, 114–122. Springer Series in Synergetics, Vol. 32. Berlin: Springer-Verlag, 1986.
25. Hsieh, D. A. "Testing for Nonlinear Dependence in Daily Foreign Exchange Rate Changes." *J. Business* **62** (1989): 339–368.
26. Hseih, D. A. "Chaos and Nonlinear Dynamics: Application to Financial Markets." Preprint; *J. Finance*, to appear.
27. Kostelich, E. J., and H. L. Swinney. "Practical Considerations in Estimating Dimension from a Time Series." In *Chaos and Related Natural Phenomena*, edited by I. Procaccia and M. Shapiro, 141–156. New York: Plenum, 1987.
28. Kurths, J., and A. A. Ruzmaikin. "On Forecasting the Sunspot Numbers." *Solar Physics* **126** (1990): 407–410.
29. Lee, T.-H., H. White, and C. W. J. Granger. "Testing for Neglected Nonlinearity in Time Series Models: A Comparison of Neural Network Methods and Alternative Tests." Preprint, Department of Economics, University of California, San Diego, 1989.
30. Liebert, W., and H. G. Schuster. "Proper Choice of the Time Delay for the Analysis of Chaotic Time Series." *Phys. Lett. A* **142** (1988): 107–111.
31. Mayer-Kress, G. "Application of Dimension Algorithms to Experimental Chaos." In *Directions in Chaos,* edited by Hao Bai-Lin, Vol. I. Singapore: World Scientific, 1988.
32. Mundt, M., W. B. Maguire, and R. P. Chase. "Chaos in the Sunspot Cycle: Analysis and Prediction." *J. Geophys. Res.* **96** (1991): 1705 1716.
33. Osborne, A. R., and A. Provenzale. "Finite Correlation Dimension for Stochastic Systems with Power-Law Spectra." *Physica D* **35** (1989): 357–381.
34. Provenzale, A., A. R. Osborne, and R. Soj. "Convergence of the K_2 Entropy for Random Noises with Power Law Spectra." *Physica D* **47** (1991): 361–372.
35. Ramsey, J. B., and J.-J. Yuan. "Bias and Error Bars in Dimension Calculations and Their Evaluation in Some Simple Models." *Phys. Lett. A* **134** (1989): 287.
36. Sano, M., and Y. Sawada. "Measurement of the Lyapunov Spectrum from Chaotic Time Series." *Phys. Rev. Lett.* **55** (1985): 1082.
37. Scheinkman, J. A., and B. LeBaron. "Nonlinear Dynamics and Stock Returns." *J. Business* **62** (1989): 311–338.

38. Smith, R. L. "Optimal Estimation of Fractal Dimension." This volume.

39. Takens, F. "Invariants Related to Dimension and Entropy." In *Atas do 13º*. Rio de Janeiro: Colóqkio Brasiliero de Matemática, 1983.

40. Takens, F. "On the Numerical Determination of the Dimension of an Attractor." In *Dynamical Systems and Bifurcations, Groningen, 1984*. Lecture Notes in Mathematics, Vol. 1125. Berlin: Springer-Verlag, 1985.

41. Theiler, J. "Efficient Algorithm for Estimating the Correlation Dimension from a Set of Discrete Points." *Phys. Rev. A* **36** (1987): 4456–4462.

42. Theiler, J. "Quantifying Chaos: Practical Estimation of the Correlation Dimension." Ph.D. thesis, California Institute of Technology, 1988.

43. Theiler, J. "Estimating Fractal Dimension." *J. Opt. Soc. Am. A* **7** (1990): 1055–1073.

44. Theiler, J. "Statistical Precision of Dimension Estimators." *Phys. Rev. A* **41** (1990): 3038–3051.

45. Theiler, J. "Some Comments on the Correlation Dimension of $1/f^\alpha$ Noise." *Phys. Lett. A* **155** (1991): 480–493.

46. Tong, H., and K. S. Lim. "Threshold Autoregression, Limit Cycles and Cyclical Data." *J. Roy. Stat. Soc. B* **42** (1980): 245–292.

47. Tong, H. *Non-linear Time Series: A Dynamical System Approach*. Oxford: Clarendon Press, 1990.

48. Townshend, B. "Nonlinear Prediction of Speech Signals." This volume.

49. Tsay, R. S. "Nonlinear Time Series Analysis: Diagnostics and Modelling." Preprint, Department Statistics, Carnegie-Mellon University, 1988.

50. Uhlenbeck, G. E., and L. S. Ornstein. "Theory of Brownian Motion." *Phys. Rev.* **36** (1930): 823–841. Reprinted in *Noise and Stochastic Processes*, edited by N. Wax. New York: Dover, 1954.

51. Weigend, A., B. Huberman, and D. Rumelhart. "Predicting the Future: A Connectionist Approach." *Int. J. Neural Systems* **1** (1990): 193–206.

52. Yule, G. U. "On a Method of Investigating Periodicities in Disturbed Series with Special Reference to Wolfer's Sunspot Numbers." *Philos. Trans. Roy. Soc. London A* **226** (1927): 267–298.

Clive W. J. Granger and Timo Teräsvirta
Department of Economics, University of California, La Jolla, CA 92093-0508

Experiments in Modeling Nonlinear Relationships Between Time Series

1. INTRODUCTION

Economists are chiefly interested in relationships between economic variables. They ask, for instance, how a change in money will affect prices or how a change in interest rates will affect the unemployment rate. Many of these relationships are believed to be nonlinear, although the precise specification is often not known. An example is the production function, in which production is taken to be nonlinearly related to the labor input and the capital input. It should be noted that labor is fairly easy to measure (person hours), but that capital is difficult to measure, and another input, level of technology, is extremely difficult to measure. Many economic theorists produce deterministic theory but we think that virtually all econometricians and applied economists believe that the economy is stochastic, both because the shocks to the economy are unforecastable (not merely linearly unforecastable) and because there is always measurement error in the data. It follows from this belief, which has been tested in Liu, Granger, and Heller[5] and elsewhere, and found to be correct, that deterministic (chaotic) models are of little relevance in economics and so we will consider only stochastic models.

The basic model will take the form

$$Y_t = f(\underline{X}_t) + e_t,\qquad(1.1)$$

where Y_t is a single output, X_t is a vector of inputs, usually including lagged Y_t, and e_t is a series of stochastic independent, identically distributed (IID) shocks, including the effects of unobserved variables. If Y_t includes only lagged Y's, one has a univariate model, actually a nonlinear autoregressive model, but if X includes current and lagged values of some other variable, x_t, then one has a bivariate nonlinear model.

Some examples are (excluding e_t terms):

i.   polynomial:

$$Y_t = \underline{\beta}'\underline{X}_t + \underline{X}_t'\underline{\Gamma}\underline{X}_t + \text{ Cubic Terms},$$

where β is a vector and $\underline{\Gamma}$ a square matrix;

ii. bilinear:

$$Y_t = \underline{\beta}'\underline{X}_t + \sum_{j,k}\underline{\alpha}_{jk}e_{t-j}X_{t-k},$$

where β is a vector and each $\underline{\alpha}$ is a square matrix;

iii. flexible Fourier form:

$$Y_t = \sum_j[\beta_{1j}\cos(\underline{\gamma}_j'\underline{X}_t) - \beta_{2j}\sin(\underline{\gamma}_j'\underline{X}_t)],$$

where $\underline{\gamma}_j$ are all vectors (the model may also include linear and quadratic terms from the polynomial model);

iv. neural networks:

$$Y_t = \underline{\beta}'\underline{X}_t + \sum_{j=1}^{p}\alpha_j\phi(\underline{\gamma}_j'\underline{X}_t),$$

where $\underline{\beta}$ and $\underline{\gamma}_j$ are vectors and $\phi(x)$ is the squashing function, here always taken to be the logistic

$$\phi(x) = \frac{1}{1+e^{-x}},\text{ and}$$

v.   the STR model (smooth transition regression):

$$Y_t = \underline{\alpha}'\underline{X}_t + (\beta'\underline{X}_t)[1 + \exp(-\gamma'\underline{X}_t - C]^{-1},$$

with α, β and γ all vectors, and C scaler. Some coefficient restrictions may apply.

The STR model essentially moves between two regimes, each of which is a linear regression, as $\underline{\gamma}'\underline{X}_t$ is large and positive or large and negative. This model is particularly useful as it can be used to approximate other *parametric* models shown above; for applications see Teräsvirta.[8]

An important class of models are the nonparametric models, an example being projection pursuit (see Friedman and Stuetzle,[2] and Huber.[3]).

$$Y_t = \sum_{j=1}^{r} \phi_j(\underline{\gamma}'_j \underline{X}_t) + e_{rt},$$

is a model fitted to Y_t using nonparametric functions, and then the error is modeled as

$$e_{rt} = \phi_{r+1}(\underline{\gamma}'_{r+1} \underline{X}_t) + \eta_{rt},$$

where ϕ is a nonparametric function, using a cubic spline, super-smoother or nearest-neighbor technique, say, and a stopping rule is used to fix the largest value of r. The effective number of parameters involved can be much greater than in the parametric models.

A number of questions naturally arise:

a. how many lags of the explanatory variables to use in the model;

b. how to test for nonlinearity;

c. how to model nonlinearity if appropriate; and

d. how to evaluate these models, particularly how to compare their post-sample forecasting ability.

The above questions are considered in some detail in a forthcoming book *Modeling Dynamic Nonlinear Relationships*, by ourselves, to be published by Oxford University Press, 1991. In economics, much of the data is long-memory series, but in this paper we only consider short-memory or stationary series. In this paper we describe simple experiments to investigate some of these questions. It should be emphasized that these are just experiments, involving a few specifications and a small number of replications.

2. FIRST EXPERIMENT: MAXIMAL SERIAL CORRELATION

In linea analysis of a single series, the autocorrelations $\rho_j = corr(Y_t, Y_{t-j})$ are useful in deciding the length of the memory of the series when considering alternative model specifications. A possible generalization is the maximum correlation between Y_t and Y_{t-j}, for a given lag j, so that;

$$C_j = corr(Y_t, \hat{f}(Y_{t-j})),$$

TABLE 1 corr $(Y_t, \hat{f}(Y_{t-j}))$ Estimated maximum correlation n = Sample Size, 100 Replications

j	white noise (IID)	bilinear	STAR
n	200	200	100
1	0.10	0.275	0.88
2	0.098	0.209	0.62
3	0.11	0.248	0.44
4	0.108	0.128	0.41
5	0.111	0.111	0.58
6	0.106	0.114	0.60
7	0.107	0.096	0.58
8	0.113	0.116	0.55
9	0.112	0.107	0.44
10	0.111	0.109	0.41
:			
15	0.127	0.105	0.50
:			
20	0.116	0.116	0.63

where $\hat{f}(\ )$ is a nonparametric function, here estimated by using a super-smoother (as described by Friedman[1]), chosen to maximize this correlation. The maximum correlation for a pair of random variables X, Y measures the extent to which X can be explained by (some function of) Y. It thus provides an upper bound to the goodness of fit of a regression with X the dependent variable and any function of Y as the explanatory variable. In such an equation, the goodness of fit is usually measured as R^2, the square of the correlation between X and the explanatory part of the equation, here $f(Y)$. The maximum correlation is thus the maximum value that R^2 can take for alternative choices of Y. It does not measure the information in Y that can occur with combinations of other variables Z, when trying to explain X. In this experiment three types of data were generated, using sample sizes 100 or 200 and 100 replications. The three model specifications used are:

i. zero mean Gaussian independent series (called here white noise), denoted e_t, sample size = 200;

ii. a bilinear series:

$$Y_t = 0.8Y_{t-1}e_{t-2} + e_t,$$

where e_t is zero mean white noise, and sample size = 200; and

iii. a STAR (smooth transition autoregressive) series

$$Y_t = 1.8Y_{t-1} - 106Y_{t-2} + (0.02 - 0.9Y_{t-1} + 0.795Y_{t-2})F(Y_{t-1}) + e_t,$$

where $F(Y_{t-1}) = [1 + \exp(-100(Y_{t-1} - 0.02)]^{-1}$, and $e_t \approx nid(0, 0.02^2)$ (here sample size = 100). Teräsvirta[8] used the same model in a specification example.

Table 1 shows estimated C_j values for $j = 1, \ldots, 20$ for the three series. It is seen that the maximum correlation is biased upward, to a value of around 0.11 for the first two series, whereas the true value should be zero for the first series, and for higher lags of the second series. This bias is potentially misleading, but it corresponds to an R^2 of only about 0.01. How this bias changes with sample size has not yet been investigated. The bilinear series suggests a memory of three periods, which is reasonable given the actual model. The third series from the STAR model suggests that series is long memory. This agrees with a simple analysis of the model. If $F(Y) = 0$, the regime is slightly explosive, with a modulus of 1.03, and if $F(Y) = 1$, the regime has modulus 0.52, both regimes having periods of 12.4 units. The model has the stable point $Y_\infty = 0.055$.

3. PROJECTION PURSUIT EXPERIMENT

The model considered takes the form

$$Y_t = \alpha_0 + \underline{\alpha}' \underline{X}_{1t} + \sum_{j=1}^{r} \phi_j(\underline{\gamma}_j' \underline{X}_{2t}) + e_t,$$

where $\underline{X}_{1t}$ is used in the linear component and is $\underline{X}_{1t} = (x_{t-i}, i = 1, \ldots, kx; y_{t-i}, i = 1, \ldots, ky)$ and $\underline{X}_{2t}$ is used in the nonparametric components and is $\underline{X}_{2t} = (x_{t-1}, i = 1, \ldots, l_x; y_{t-i}, i = 1, \ldots l_y)$. Thus, the lags used can differ in the two components. The criteria used to decide on the number of lags used was to minimize Bayesian Information Criterion (BIC) $= \log var_k(e_t) + k \log T/T$ when k parameters are used in a model with a sample of size T. The number of lags is searched over a relevant range to minimize BIC. This criterion is known to be conservative in the sense that it chooses a smaller number of parameters than most other criteria like Akaike's Information Criterion (AIC) or Generalized Cross Validation (GCV). It is dimension consistent; if the model has a finite number of parameters, asymptotically BIC chooses this model with probability one.

The data was generated by the bivariate STR model

$$Y_t = \frac{\theta x_{t-1} - 0.9x_{t-1}}{1 + \exp(-\gamma d_{t-1})} + \sigma \epsilon_t, \quad Y > 0$$

TABLE 2

$\pi = 0$		kx	ky	In Sample lx	ly	R^2	$\hat{\sigma}_I$	Out of Sample $\hat{\sigma}_0$	$\hat{\sigma}_0/\hat{\sigma}_i$
	L	1	2	—	—	0.42	0.66	0.62	0.94
$\sigma = 0.5$	PP	—	—	1	3	0.68	0.49	0.56	1.14
	LP	1	1	3	2	0.60	0.55	0.56	1.02
	L	1	1	—	—	0.12	1.51	1.58	1.05
$\sigma = 1.5$	PP	—	—	2	1	0.31	1.34	1.74	1.30
	LP	1	2	2	1	0.23	1.41	1.62	1.15
	L	1	1	—	—	0.05	2.92	3.08	1.05
$\sigma = 3$	PP	—	—	3	1	0.27	2.56	3.20	1.25
	LP	1	2	1	3	0.20	2.67	3.34	1.25

where $d_{t-1} = y_{t-1} - \theta x_{t-2}$ is the lagged deviation from the linear path $y_t = \theta x_{t-1}$, and x_t is AR(1);

$$x_t = 0.5x_{t-1} + e_t, \sigma_e^2 = 1.$$

A variety of parameter values have been tried, but the only case shown here has $\theta = 0, \gamma = 100$ and three values of $\sigma, 0.5, 1.5, 3$, giving different noise levels.

Three models are fitted

a. linear component only (L),

b. projection pursuit component only (PP),

c. both parts (LP),

The in-sample size was $T = 300$, and this was used to estimate the model. A further 100 out-of-sample terms were used to evaluate the models. The summary statistics used in Table 2 are R^2, σ_I (the estimated in sample standard deviation of the residual), $\hat{\sigma}_0$ (the corresponding out of sample standard deviation of one step forecast errors), and the ratio $\hat{\sigma}_0/\hat{\sigma}_I$. The table shows these statistics and the lags chosen by the technique for a single set of data. For the lowest-noise level, the linear model fits poorly, with $\hat{\sigma}_I > 0.5$, the true value, whereas the models using projection pursuit fit well in sample and also forecast better. However, for larger σ values, when the noise dominates the nonlinear component, the PP models overfit in sample, achieving an estimate of $\hat{\sigma}_I$ lower than the true value, and then forecast less well out of sample. These results are found throughout our experiments; when nonlinearity is strong PP fits and forecasts very well, but when nonlinearity is weak PP data mines, overfits in sample, and forecasts poorly out of sample. These results suggest that one should first test for nonlinearity. If found, then projection pursuit models can be used with some confidence; if not found, then one should use only linear models. Discussion of some alternative test procedures can be found in Luukkonen, Saikkonen, and Teräsvirta[6,7]; Lee, White, and Granger[4]; Teräsvirta[9]; Tong,[10] chapter 5; and in the book mentioned earlier.

4. EXPERIMENT USING PARAMETRIC MODELS

Three of the classes of parametric models listed in section 1 were used in the experiment: polynomials (of orders 1, 2, 3), neural networks (with one and two nonlinear terms), and the Fourier flexible form (with one and two nonlinear terms but including linear and quadratic terms also). The polynomials are estimated by least-squares regressions; in the other models all parameters are estimated by maximum likelihood. Some severe identification problems are found to occur when there are more than one nonlinear terms in the neural network model, but this problem was not investigated. The data used in this illustration were generated by

$$Y_t = -\frac{0.9X_{t-1}}{1 + \exp(-100X_{t-1})} + 0.5e_t ,$$

TABLE 3 Results of single replication[1]

	$\hat{\sigma}_I$	$\hat{\sigma}_0$
Polynomial		
linear	0.55	0.57
quadratic	0.50	0.52←
cubic	0.50	0.52
Neural Network		
p= 1	0.59	0.59
p = 2	0.53	0.52←
Fourier		
flexible form	0.48	0.51

[1] Arrow indicates model with lowest BIC value in sample.

where $e_t \sim N(0, 1)$ and X_t is generated by the same AR(1) model as in the previous section. The in-sample size was 300 and out-of-sample 200. The results are shown for a single data set and $\hat{\sigma}_I, \hat{\sigma}_0$ are defined as before.

The number of lags used in the models were determined by BIC from the linear model, which is efficient but not necessarily optimal. The results in Table 3 find that all of the models chosen by minimizing BIC work very well both in and out of sample. There is little to choose between them.

5. PROVISIONAL CONCLUSIONS

The conclusions reached from these experiments, plus others not shown, are that:

i.   nonparametric techniques are inclined to overfit and then forecast poorly, particularly where the noise level is high;

ii. parametric techniques fit and forecast well and are similar; and

iii. several sound tests of linearity exist and should be used before attempting to model.

Clearly, further experiments are required but the evidence currently available suggests that series of the extent found in economics, with, say, 200 to 300 terms, can indicate nonlinearity of a stochastic form if it is present, and in that case useful nonlinear models can be found.

ACKNOWLEDGMENTS

We would like to thank Chien-Fu Lin, Jin-Lung Lin, Zhuanxin Ding, Jyotsne Jalan, and Chor-Yin Sin for helping perform the experiments reported here. The research of Clive W. J. Granger was supported by NSF Grant SES-902950. Timo Teräsvirta acknowledges support from the Yrjö Jahnsson Foundation.

REFERENCES

1. Friedman, J. H. "A Variable Span Smoother." Technical Report No. 5, Laboratory for Computational Statistics, Department of Statistics, Stanford University, Stanford, CA, 1986.
2. Friedman, J. H., and W. Stuetzle. "Projection Pursuit Regression." *J. Amer. Stat. Assoc.* **76** (1981): 817–823.
3. Huber, P. J. "Projection Pursuit." *Ann. Stat.* **13** (1985): 435–475.
4. Lee, T.-H., H. White, and C. W. J. Granger. "Testing for Neglected Nonlinearity in Time Series Models. A Comparison of Neural Network Methods and Alternative Tests." Economics Discussion Paper No. 8942, University of California, San Diego, CA, 1989.
5. Liu, T., C. W. J. Granger, and W. Heller. "Using the Correlation Exponent to Decide if an Economic Series is Chaotic." Economics Discussion Paper No. 91–21, University of California, San Diego, CA, 1990.
6. Luukkonen, R., P. Saikkonen, and T. Terasvirta. "Testing Linearity Against Smooth Transition Autoregressive Models." *Biometrika* **75** (1988): 491–499.
7. Luukkonen, R., P. Saikkonen, and T. Terasvirta. "Testing Linearity in Univariate Time Series Models." *Scandinavian J. Stat.* **15** (1988): 161–175.
8. Teräsvirta, T. "Specification, Estimation, and Evaluation of Smooth Transition Autoregressive Models." Economics Discussion Paper No. 90–39, University of California, San Diego, CA, 1990.
9. Teräsvirta, T. "Power Properties for Linearity Tests for Time Series." Economics Discussion Paper No. 90–15, University of California, San Diego, 1990.
10. Tong, H. *Non-Linear Time Series. A Dynamical System Approach*. Oxford: Oxford University Press, 1990.

T. Subba Rao
Department of Mathematics, University of Manchester Institute of Science and Technology, Manchester M60 1QD, ENGLAND

Analysis of Nonlinear Time Series (and Chaos) by Bispectral Methods

Second-order spectra have played a very significant role in the analysis of linear (Gaussian) time series. When the series is non-Gaussian (and nonlinear), it is important to study higher-order spectra. Here, we briefly consider the estimation of the bispectrum, and discuss the usefulness of this function in several situations, for example, when estimating the parameters of the signal in the presence of noise, for discriminating "deterministic" nonlinear models and chaotic models, etc. We also consider bilinear models which were introduced recently in time-series literature, and discuss their estimation and study the forecasts obtained in a specific example. Lastly we point out how this model can be generalized to deal with nonlinear long-range dependence.

INTRODUCTION

One of the usual assumptions that is often made when analyzing time series is that the time series is linear, and perhaps even Gaussian. If the process is Gaussian, then the second-order covariances (and second-order spectra) will contain all the useful information. If not, one has to calculate higher-order spectra to study the departures from linearity and Gaussianity. The simplest higher-order spectrum is bispectrum.

Bispectrum was originally proposed by Hasselman, Munk, and Macdonald[12] for investigating nonlinear interaction of ocean waves, and Godfrey[9] used it for analysis of economic time series. Sato, Sasaki, and Nakamura[23] have analyzed acoustic gear noise by bispectra. Lii and Rosenblatt[18] have used it for deconvolution of seismic signals. In a series of papers, Lii, Rosenblatt, and Van Atta,[17] Van Atta,[33] Holland, Lii, and Rosenblatt[14] have described how bispectrum could be used to study nonlinear spectral transfer of energy in turbulence.

Subba Rao and Gabr[29] and Hinich[13] have proposed tests for linearity and Gaussianity based on bispectrum, and possible applications seem to be unlimited. In this paper, we briefly review a method of estimation of bispectrum and study possible applications to non-Gaussian signals, and especially we will study the phenomenon of "chaos." Our investigations in this new field are preliminary, and it seems, on the basis of the results we obtained, the estimated higher-order spectra possibly could be used to distinguish between nonlinear deterministic stable systems and nonlinear deterministic chaotic systems. On the basis of these investigations, it seems that for stable nonlinear deterministic systems, the energy in the estimates of the spectrum and bispectrum seems to be in the low frequency range, whereas for the "chaotic" systems, the energy seems to be distributed over a wide band in the high frequency range and this may be consistent with "bifurcation" phenomenon of chaotic systems. In sections 2 and 3 we concentrate on frequency domain approach for analyzing non-Gaussian time series and time domain approaches are considered in later sections. In section 4 we consider the properties of one nonlinear model called bilinear model and study their properties and aspects of forecasting. In the final section, 5, we point out the possibility of extending the bilinear models to deal with long-range dependence.

2. SECOND-ORDER AND HIGHER-ORDER SPECTRA

Let $x(t)$ be a discrete-parameter, real-valued time series. Let the process $\{x(t)\}$ satisfy the following conditions.

i. $E(x(t)) = \mu$ independent of t.

ii. Var $(x(t)) = \sigma^2$ independent of t;
    Cov $(x(t), x(t+s)) = R(s)$ a function of s only.

iii. Cum $(x(t), x(t+s_1), x(t+s_2))$
    $= E(x(t) - \mu)(x(t+s_1) - \mu)(x(t+s_2) - \mu)$
    $= C(s_1, s_2)$ is a function of s_1 and s_2 only.

The time series $\{x(t)\}$ which satisfies the above three conditions is said to be third-order stationary.

We note that $R(s)$ and $C(s_1, s_2)$ satisfy the following symmetry conditions.

$$R(s) = R(-s);$$
$$C(s_1, s_2) = C(s_2, s_1) = c(-s_1, -s_2 - s_1) = C(s_1 - s_2, -s_2).$$

The second-order spectrum $f(\omega)$ and the third-order spectrum $f(\omega_1, \omega_2)$ are defined as follows:

$$f(\omega) = \frac{1}{2\pi} \sum_{-\infty}^{\infty} R(s)e^{-is\omega}, \quad -\pi \leq \omega \leq \pi;$$

$$f(\omega_1, \omega_2) = \frac{1}{(2\pi)^2} \sum_{-\infty}^{\infty} \sum_{-\infty}^{\infty} C(s_1, s_2)e^{-i(s_1\omega_1 + s_2\omega_2)}.$$

$$(2.1)$$

From Eq. (2.1) we have

$$R(0) = E(X(t) - \mu)^2 = \int_{-\pi}^{\pi} f(\omega)d\omega,$$

$$c(0,0) = E(X(t) - \mu)^3 = \int_{-\pi}^{\pi} \int f(\omega_1, \omega_2)d\omega_1 d\omega_2.$$

In other words, second-order spectral density $f(\omega)$ can be interpreted as frequency decomposition of the variance (total power in the signal) and similarly the bispectra is frequency decomposition of the third-order moments. Similar interpretation can be given for higher-order spectra. Just second-order covariances measure linear dependence, third-order covariances (bispectra) measure quadratic dependence, fourth-order cubic dependence etc.

In view of the above symmetries, we have

$$f(\omega) = f(-\omega) \text{ and }$$
$$f(\omega_1, \omega_2) = f(\omega_2, \omega_1) = f(-\omega_1, -\omega_1 - \omega_2) = f(-\omega_1 - \omega_2, \omega_2) \quad (2.2)$$
$$= f(-\omega_1, -\omega_2)$$

and because of Eq. (2.2), the bispectrum $f(\omega_1, \omega_2)$ is completely specified on any one of the twelve sectors, including the boundaries (see Subba Rao and Gabr,[27] Figure 1.3, page 13). The bispectrum is usually complex valued. It is easy to evaluate the theoretical forms of the spectrum and the bispectrum if we know the model the series $x(t)$ satisfies. At least this is true in the case of linear models. For example, if $x(t)$ satisfies the finite-parameter, autoregressive, moving average model of order (p, q) of the form

$$x(t) + a_1 x(t-1) + \ldots + a_p x(t-p) = e(t) + b_1 e(t-1) + \ldots + b_q e(t-q)$$

where $\{e(t)\}$ is a sequence of independent, identically distributed random variables with $E(e(t)) = 0, var(e(t)) = \sigma^2, \mu_3 = E(e^3(t))$, then we know

$$f(\omega) = \frac{\sigma_e^2}{2\pi} \times |\Gamma(\omega)|^2,$$

$$f(\omega_1, \omega_2) = \frac{\mu_3}{(2\pi)^2}\Gamma(e^{-i\omega_1})\Gamma(e^{-i\omega_2})\Gamma(e^{i(\omega_1+\omega_2)}) \qquad (2.3)$$

where

$$\Gamma(e^{-i\omega}) = \frac{(1 + b_1 e^{-i\omega} + \ldots + b_q e^{-iq\omega})}{1 + a_1 e^{-i\omega} + \ldots + a_p e^{-ip\omega}}.$$

In general, if $x(t)$ satisfies the linear representation

$$x(t) = \sum_{u=0}^{\infty} g_u e(t - u),$$

then $f(\omega)$ and $f(\omega_1, \omega_2)$ are given as above with the transfer function $\Gamma(e^{-i\omega}) = \sum_{u=0}^{\infty} g_u e^{-iu\omega}$.

We observe that if the time series $\{x(t)\}$ is Gaussian, then $f(\omega_1, \omega_2) = 0$ for all ω_1 and ω_2. If $x(t)$ is linear, but not Gaussian, then

$$\frac{|f(\omega_1, \omega_2)|^2}{f(\omega_1)f(\omega_2)f(\omega_1 + \omega_2)} = \text{constant for all } \omega_1 \text{ and } \omega_2. \qquad (2.4)$$

In other words, the bispectrum can be used for testing departure from Gaussianity and linearity, and this is the basis of Subba Rao and Gabr's test[25] and Hinich's test.[13]

If the series $\{X(t)\}$ are mutually uncorrelated, i.e., $R(s) = 0$ for all $s \neq 0$, then $f(\omega) = \sigma^2/2\pi$ for all ω. Because of this, the time series $\{X(t)\}$ is said to be a white noise. The second-order correlations, $R(s)$, measure "linear" dependence between successive observations. It is possible that for some series (usually nonlinear), the second-order covariances can be zero, but not the higher-order covariances. This implies that there is still "structure" in the series which can be exploited by nonlinear models for prediction purposes. If the series are mutually independent, then, of course, the spectra of all orders will be constant for all frequencies. This constancy of all spectra for all frequencies can be taken as an indication of independence of the series, and if they are not constant, it is possible that the underlying structure can be a nonlinear one. Though one has to compute spectra of as many orders as possible, it is sufficient to calculate up to fourth order (trispectra). Here we consider bispectra only. The computation of higher-order spectra, though feasible, may be quite expensive.

We now consider, briefly, the estimation of $f(\omega)$ and $f(\omega_1, \omega_2)$. For details of this estimation, see Priestley[20] and Subba Rao and Gabr.[27]

ESTIMATION OF $f(\omega)$

Let $x(1), x(2), \ldots, x(N)$ be a sample from $\{x(t)\}$. Then the natural estimates of $\mu, R(s)$, and $C(s_1, s_2)$ are given by

$$\bar{x} = \frac{1}{N} \sum x(t), \quad \hat{R}(s) = \frac{1}{N} \sum_{t=1}^{N-s} (x(t) - \bar{x})(s(t+s) - \bar{x}), s \geq 0$$

and $\hat{C}(s_1, s_2) = 1/N \sum_{t=1}^{N-\gamma} (x(t) - \bar{x})(x(t+s_1) - \bar{x})(x(t+s_2) - \bar{x})$, where $\gamma = \max (0, s_1, s_2), s_1 \geq 0, s_2 \geq 0$. A form of the spectral estimate of $f(\omega)$ is

$$\hat{f}(w) = \frac{1}{2\pi} \sum_{-(N-1)}^{N-1} \lambda\left(\frac{\tau}{M}\right) \hat{R}(\tau) \cos \omega\tau,$$

where $M = M(N)$ and $\lambda(\cdot)$ is a lag window generator. If $\lambda(s) = 0$, for $\mid s \mid \geq 1, M$ corresponds to the truncation point. We assume that the function $\lambda(s)$ is a bounded, even, and square integrable such that $\lambda(0) = 1$. The integer M is chosen such that as $M \rightarrow \infty, N \rightarrow \infty$, and $M/N \rightarrow 0$. It is well known that

$$var(\hat{f}(w)) \simeq \frac{M}{N} f^2(w) \int \lambda^2(s) ds, \quad w \neq 0, \pi.$$

In Table 1, we give some standard lag window generators.

ESTIMATION OF $f(\omega_1, \omega_2)$

Let $K_0(\theta_1, \theta_2)$ be a bounded and a non-negative function satisfying

i. $\int \int K_0(\theta_1, \theta_2) d\theta_1 d\theta_2 = 1$.

ii. $\int \int K_0^2(\theta_1, \theta_2) d\theta_1 d\theta_2 < \infty, \int \int \theta_i^2 K_0(\theta_1, \theta_2) d\theta_1 d\theta_2 < \infty, (i = 1, 2)$.

iii. $K_0(\theta_1, \theta_2) = K_0(\theta_2, \theta_1) = K_0(\theta_1, -\theta_1, -\theta_2) = K_0(-\theta_1 - \theta_2, \theta_2)$.

Then the bispectral estimate $\hat{f}(\omega_1, \omega_2)$ is given by

$$\hat{f}(\omega_1, \omega_2) = \frac{1}{(2\pi)^2} \sum \sum \lambda(\frac{\tau_1}{m}, \frac{\tau_2}{M}) \hat{C}(\tau_1, \tau_2) e^{-i\tau_1\omega_1 - i\tau_2\omega_2}$$

TABLE 1 Lag Window Generators

Deniell Window	$\lambda_D(s) = \frac{\sin s\pi}{s\pi}$
Tukey-Hanning Window	$\lambda_T(s) = \begin{cases} 0.54 + 0.46 \cos s\pi & \lvert s \rvert \le 1 \\ 0 & \text{otherwise} \end{cases}$
Parzen Window	$\lambda_p(s) = \begin{cases} 1 - 6s^2 + 6\lvert s \rvert^3 & \lvert s \rvert \le 1/2 \\ 2(1 - \lvert s \rvert)^3 & 1/2 \le s \le 1 \\ 0 & \text{otherwise} \end{cases}$
Bartlett-Priestley Window	$\lambda_{Bp}(S) = \frac{3}{(\pi s)^2} \left\{ \frac{\sin \pi s}{\pi s} - \cos \pi s \right\}$

where M, a function of N, is a window parameter chosen such that $M^2/N \to 0$ as $M \to \infty, N \to \infty$. Then (see Brillinger and Rosenblatt[2] and Rosenblatt and Van Ness):[21]

$$var(\hat{f}(\omega_1, \omega_2)) = \frac{M^2}{N} \times \frac{V_2}{2\pi} f(\omega_1) f(\omega_2) f(\omega_1 + \omega_2), (0 < \omega_2 < \omega_1)$$

where

$$V_2 = \int \int \lambda^2(u_1, u_2) du_1 du_2 = (2\pi)^2 \int \int K_0^2(\theta_1, \theta_2) d\theta_1 d\theta_2.$$

One choice of the two-dimensional window $\lambda(u_1, u_2)$ is to consider a product of three one-dimensional windows of the form

$$\lambda(s_1, s_2) = \lambda(s_1)\lambda(s_2)\lambda(s_1 - s_2).$$

An optimum window has been obtained by Subba Rao and Gabr,[27] which produces the smallest mean-square error.

One can also estimate the bispectrum using Fast Fourier transforms.[16]

3. APPLICATIONS OF BISPECTRA

As pointed out earlier, second-order spectral analysis is not always sufficient to draw all the information contained in the series, if the series happens to be non-Gaussian. In fact, if one stops at second-order analysis, it is possible to draw wrong conclusions as the following example shows.

Consider the nonlinear process $\{x(t)\}$ satisfying the model

$$x(t) = e(t) + \beta e(t-1)e(t-2),$$

where $\{e(t)\}$ are i.i.d. random variables. We can easily show that $cov(x(t),$ $x(t+s)) = 0, s \neq 0$, and this implies that the time series $\{x(t)\}$ is white noise when, in fact, $\{x(t)\}$ is highly correlated. If we compute the third-order cumulants, $cum(x(t), x(t+s_1), x(t+s_2))$, we find they are not zero for some values of s_1 and s_2 confirming that the underlying structure is nonlinear. Also, it may be important to mention here that some series which may look nonstationary may, in fact, be nonlinear.

We now consider a simple nonlinear model recently introduced called bilinear model which has been studied extensively.[10,26,27] Let $x(t)$ satisfy the equation

$$x(t) = ax(t-1) + bx(t-1)e(t-1) + e(t).$$

It is well known that the second-order properties of this model are similar to the ARMA (1,1) model or, in otherwords, the second-order covariances cannot distinguish between a linear model and a bilinear model.

TEST FOR LINEARITY AND GAUSSIANITY

If $x(t)$ satisfies the linear relation, then we know that $f(\omega_1, \omega_2)$ is given by Eq. (2.3), and this function is zero if the series is Gaussian. If it is linear and non-Gaussian, then the ratio given by Eq. (2.4) is constant for all ω_1 and ω_2. Using these properties, Subba Rao and Gabr[25] and later Hinich [13] and Brockett et al.,[5] have constructed statistical tests for testing linearity and Gaussianity.

To illustrate these, we consider two well-known data sets.

CANADIAN LYNX DATA Consider the annual number of Canadian lynx trapped in the Mackenzie river district of North West Canada for the years 1821–1834.

The bispectral density function of the logarithmically transformed data is computed using the optimum lag window (with M = 16) and the modulus is plotted in Figure 1 for the frequencies $\omega_1, \omega_2 = 0.0, 0.01\pi, 0.02\pi, \ldots, 0.26\pi$. There is a clear peak at $\omega_1 = \omega_2 = 0.20\pi$ which corresponds to the periodicity of ten years. Since the bispectral values are non-zero, it is clear that the process is non-Gaussian. It may be interesting to observe that the graph is smooth with a single dominant peak at the low frequency, and this is in contrast to what we will observe for chaotic models. It is reasonable to conclude that Canadian lynx data is nonlinear, but not chaotic.

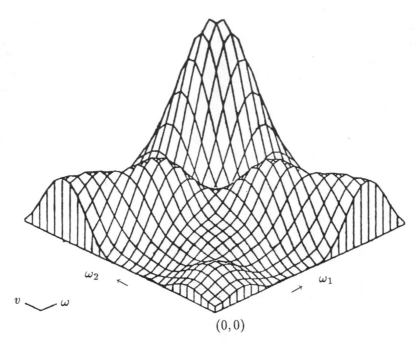

FIGURE 1   Bispectrum of log Canadian lynx data.

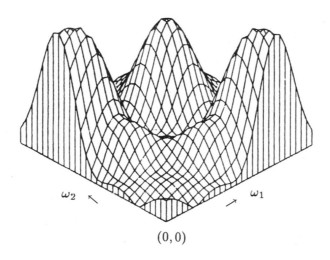

FIGURE 2   Bispectrum of sunspot numbers.

SUNSPOT NUMBERS For our second illustration, we consider the annual Wolfer sunspot numbers for the years 1700–1955, giving 256 observations. The modulus of the bispectral density function is plotted for frequencies $\omega_1, \omega_2 = 0.0, 0.01\pi, 0.02\pi,$ $\ldots, 0.27\pi$ in Figure 2. There is a clear peak at the frequency at $\omega_1 = \omega_2 = 0.18\pi$ corresponding to approximately 11 years. Once again, we can confirm that the data is nonlinear, but not chaotic.

HIGHER-ORDER CUMULANTS AND NOISE In many practical situations, the signal $x(t)$, which we will assume to be non-Gaussian and perhaps nonlinear, is corrupted by noise $N(t)$. Actually, for each t, we observe $Z(t)$,

$$Z(t) = x(t) + N(t),\qquad(3.1)$$

where the noise $N(t)$ is assumed to be independent of $x(t)$ and is stationary up to the rth order.

In view of the relation Eq. (3.1), we have the rth order cumulant of $Z(t)$.

$$cum(Z(t), Z(t+s_1), \ldots, Z(t+s_{r-1})) = cum(x(t), x(t+s_1), \ldots, x(t+s_{r-1}))$$
$$+ cum(N(t), N(t+s_1), \ldots, N(t+s_{r-1})).$$

If we take Fourier transforms of both sides, we obtain the relation between higher-order cumulant spectra

$$f_z(\omega_1, \omega_2, \ldots, \omega_{r-1}) = f_x(\omega_1, \omega_2, \ldots, \omega_{r-1}) + f_N(\omega_1, \omega_2, \ldots, \omega_{r-1}).\qquad(3.2)$$

Our main interest is when $r = 2$ and 3. When $r = 2$, we have the relation between the second-order spectra

$$f_z(\omega) = f_x(\omega) + f_N(\omega)$$

and when $r = 3$, we have the relation between bispectra

$$f_z(\omega_1, \omega_2) = f_x(\omega_1, \omega_2) + f_N(\omega_1, \omega_2).\qquad(3.3)$$

In general, if $N(t)$ is Gaussian (and $x(t)$ is non-Gaussian), we have the important relationship between the high-order spectra

$$f_z(\omega_1, \omega_2, \ldots, \omega_{r-1}) = f_x(\omega_1, \omega_2, \ldots, \omega_{r-1}), \quad r > 2.$$

This extremely useful relation suggests that if one wishes to study the properties of the signal (which is non-Gaussian and may satisfy a nonlinear relation of the form $x_{t+1} = f(x_t)$ as is common in chaotic models), one could obtain all the information about the signal by looking at the higher-order cumulants of the observations $\{Z(t)\}$. For example, if one wishes to obtain the dimensions of the state $x_{t+1} = f(x_t)$ in the presence of noise, canonical correlations of the higher powers of Z_t, say, Z_t^2, may be useful. It is similar to the approach followed by Broomhead and King[7] where they performed covariance analysis for extracting dimensions of the state. It may be interesting to point out that the procedure suggested by Broomhead and King[7] for estimating the dimensions is similar to the techniques proposed by Subba Rao[29] for estimating the order of a linear autoregressive model in the presence of noise. This technique is called Principal Component Analysis (in the presence of noise, it is called canonical factor analysis) which is a widely used technique in multivariate analysis for reducing the dimensions of the vector. The dimension of the "state" of the vector does in fact correspond to the number of non-zero eigenvalues of the sample covariance matrix of the vector, and the variance of the noise can be estimated by using the smallest eigenvalues of the covariance matrix (see Subba Rao[29]). This technique recently has been extended to nonlinear models by Subba Rao and da Silva[30] for estimating the order of the bilinear models (in situations when noise is both present and not present). Subba Rao and da Silva[30] have proposed a higher-order canonical correlations approach for determining the order of bilinear systems.

We now consider two specific examples.

Let $x(t)$ satisfy an autoregressive model of order p of the form

$$x(t) + a_1 x(t-1) + \ldots + a_p x(t-p) = e(t),$$

where $\{e(t)\}$ is a sequence of independent, identically distributed random variables. The object is to estimate $(a_1, a_2, \ldots, a_p)$ when a sample of $\{Z(t)\}$ is available. It is possible to estimate these parameters of the signal using third-order moments of $\{z(t)\}$ (for details, see Parzen[19]).

The other example we consider is the estimation of the frequencies of the series $\{x(t)\}$, when

$$x(t) = \sum_{i=1}^{\kappa} A_i \cos(\omega_i t + \phi_i),$$

where the independent random phases $\{\phi_i\}$ are distributed uniformly over the interval $(-\pi, \pi)$ and the colored noise $N(t)$ is Gaussian and stationary.

For a given sample $(Z(1), Z(2), \ldots, Z(m))$, the phases can be considered to be deterministic, and it has been shown by Subba Rao and Gabr[28] that the modulus of bispectrum of $Z(t)$ will show clear peaks corresponding to the frequencies of the signal.

BISPECTRUM AND CHAOS Trajectories generated by some nonlinear deterministic difference equations (or differential equations) look random (stochastic), and this phenomenon is referred to as "chaos." It is interesting to note that only nonlinear equations (not all nonlinear equations) produce this phenomenon. So it is interesting to discriminate between deterministic chaos and random process. It is well known that second-order covariances (or second-order spectra) do not provide adequate information of these nonlinear models (and this feature is similar to bilinear models) and as such they cannot be used for discrimination purposes (see Eubank and Farmer[8] and Brock[4]). Since second-order spectra are not adequate, it is necessary to calculate the higher-order spectra, and the simplest higher-order spectrum is bispectrum. In this paper, our object is to report the preliminary empirical results we obtained when we calculated bispectra from the samples generated from three well-known chaotic models. These results have yet to be confirmed theoretically.

TENT MAP Let the time series $\{x(t)\}$ satisfy the model

$$x(t+1) = \begin{cases} \frac{x(t)}{a} & 0 \le x(t) \le a, \\ \frac{(1-x(t))}{1-a} & a \le x(t) \le 1, \end{cases}$$

where $0 \le a \le 1$. If $x(0)$ is uniformly distributed over the interval $(0,1)$, then $x(t)$, for all t, is uniformly distributed over the interval $(0,1)$. Sakai and Tokomaru[22] have shown that the second-order covariances of the tent model similar to those of the linear model

$$x(t+1) = (2a-1)x(t) + e(t+1),$$

where $\{e(t)\}$ are independent, identically distributed. We have generated 250 observations with $a = 0.24$ and $x(0) = 0.6$, as shown in Figure 3. The spectrum and bispectrum are calculated from the sample. The spectrum looks like the spectrum of a linear model confirming Sakai and Tokomaru,[22] and the normalized bispectral density function (2.4) is plotted in Figure 4. It is clear that the process cannot be linear, and also we see a ridge along the line $\omega_1 + \omega_2 = 1.2\pi$ which seems to be common in chaotic models.

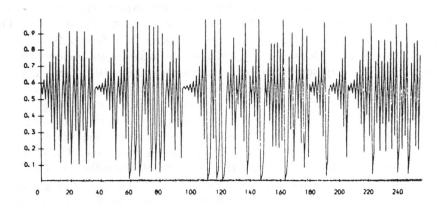

FIGURE 3 Sample from tent map.

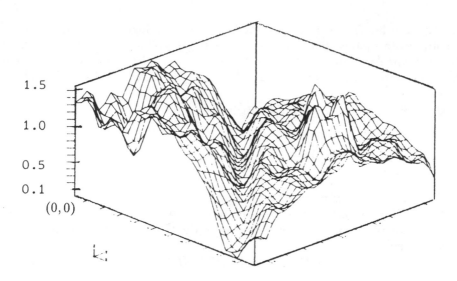

1.5

1.0

0.5

0.1

(0,0)

FIGURE 4 Normalized bispectrum (modulus) of tent map.

LOGISTIC MAP Let the series be generated from the model $x(t + 1) = ax(t)(1 - x(t)), (t = 1, 2, \ldots, 500)$. It is well known that we observe interesting trajectories when $3 < a \leq 4$. More unstable trajectories (or chaos) seem to occur for $a \geq 3.5$. In order to see this we have calculated the bispectrum for various values of "a" using the product windows with the truncation point $M = 32$ as described earlier. The modulus of the bispectrum $\mid f(\omega, \omega) \mid$ for several values of a from 3.0 to 3.9 are plotted in Figure 5. For values of $a, 3 \leq a \leq 3.4$, the values of the modulus of the bispectrum are very, very small (the values are smaller than 10^{-7} (see Figures 7 and 9). Though we find a ridge along the line $\omega_1 + \omega_2 = \pi$ in the bispectra, they are not significant because their values are very small. When $a = 3.5$, we find three dominant peaks, one at the frequency corresponding to the period-four units (this is when period doubling starts[24]). The bispectral values obtained when $a > 3.5$ are several times larger than the corresponding values when $a < 3.5$. It is also interesting to compare these values with the values we obtained for $a < 3.4$. These are several times bigger. As the value of a increases, we find several ridges in the modulus of the bispectrum, a phenomenon which we have observed consistently in chaotic models.

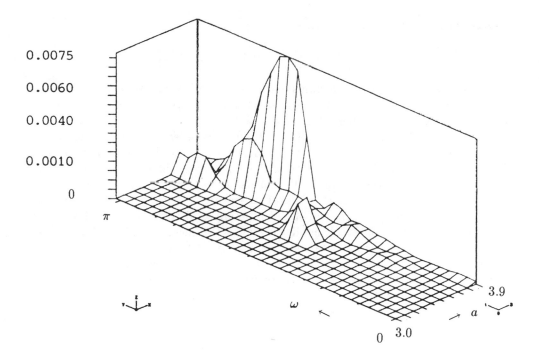

FIGURE 5   Plot of $| f(\omega, \omega) |$ against a of the logistic map.

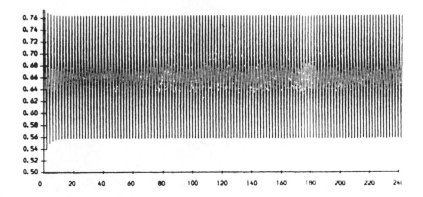

FIGURE 6   Sample from logistic map with $a = 3.0$.

HÉNON MAP Now consider the map

$$x(t) = 0.3x(t-2) - 1.4x^2(t-1) + 1 \qquad (t = 1, 2, \ldots, 512).$$

The plot of the data and the modulus of the bispectrum are given in Figures 20 and 21, respectively. The shape of the bispectrum is similar to the shapes we observed for the logistic map for values of $a = 3.5$ to 3.9.

From these calculations, we feel that the higher-order spectra can distinguish between nonlinear deterministic stable models and nonlinear deterministic models which produce "chaos." On the basis of this empirical evidence, we can conclude that Canadian lynx data and sunspot numbers are *not* chaotic, i.e., they are nonlinear and nondeterministic. Another important feature is that for the chaotic models, the "energy" in the spectrum and the bispectrum is distributed in the higher frequency range, and for the stable nonlinear models the energy is in the low frequency range (for example, sunspot data and Canadian lynx data).

In the following, we briefly consider one nonlinear model introduced recently in time series, and study its usefulness for forecasting purposes.

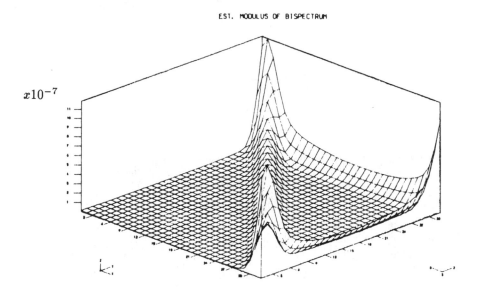

FIGURE 7 Bispectrum of sample (logistic) with $a = 3.0$.

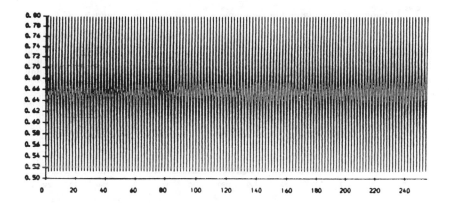

FIGURE 8   Sample from logistic map with $a = 3.1$.

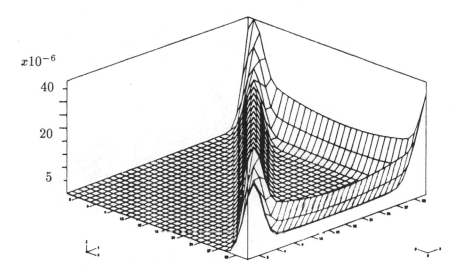

FIGURE 9   Bispectrum of sample (logistic) with $a = 3.1$.

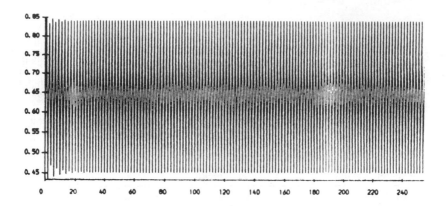

FIGURE 10   Sample from logistic map with $a = 3.4$.

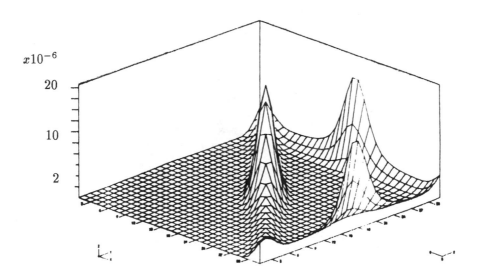

FIGURE 11   Bispectrum of sample (logistic) with $a = 3.4$.

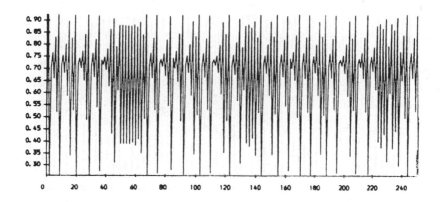

FIGURE 12   Sample from logistic map with $a = 3.5$.

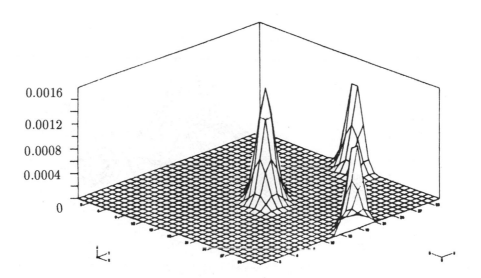

FIGURE 13   Bispectrum of sample (logistic) with $a = 3.5$.

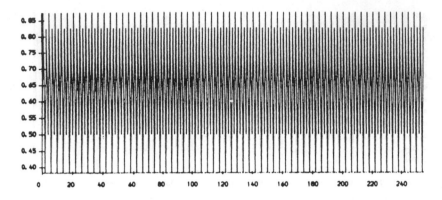

FIGURE 14   Sample from logistic map with $a = 3.7$.

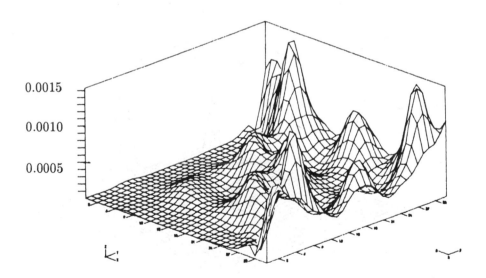

FIGURE 15   Bispectrum of sample (logistic) with $a = 3.7$.

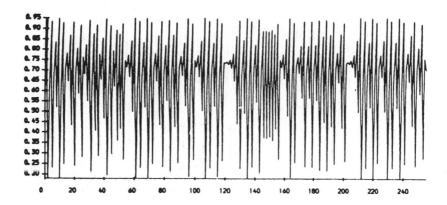

FIGURE 16   Sample from logistic map with $a = 3.8$.

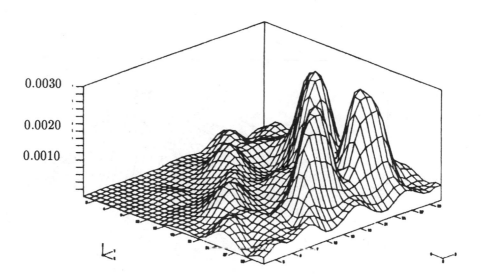

FIGURE 17   Bispectrum of sample (logistic) with $a = 3.8$.

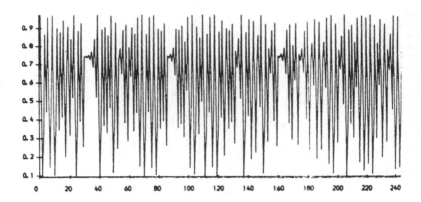

FIGURE 18   Sample from logistic map with $a = 3.9$.

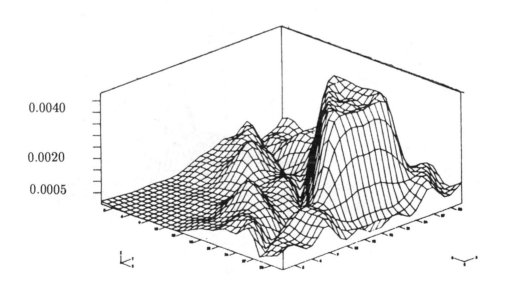

FIGURE 19   Bispectrum of sample (logistic) with $a = 3.9$.

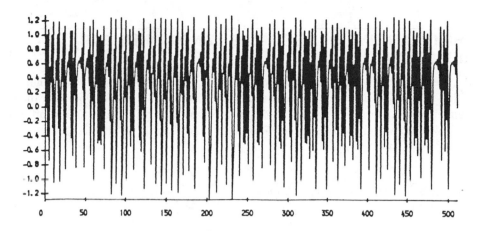

FIGURE 20 Sample from Hénon map.

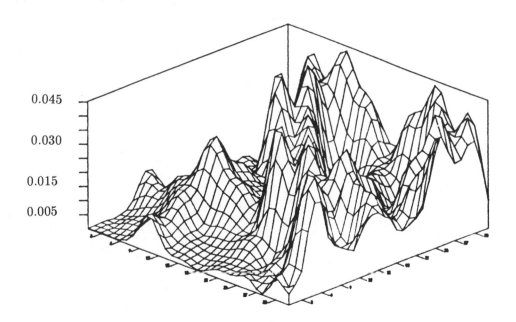

FIGURE 21 Bispectrum of sample from Hénon map.

4. BILINEAR TIME-SERIES MODEL AND NONLINEAR PREDICTION

Once we decide that the series is nonlinear, it is important to see whether we can find a finite-parameter nonlinear model to describe the series. One such model (Granger and Andersen,[10] Subba Rao[26] and Subba Rao and Gabr[27]) is a bilinear model, whose analytic properties have been extensively studied. Let $x(t)$ satisfy the difference equation

$$x(t) + \sum_{j=1}^{p} a_j x(t-j) = e(t) + \sum_{j=1}^{q} b_j e(t-j) + \sum_{i=1}^{P}\sum_{j=1}^{Q} a_{ij} x(t-i)e(t-j). \quad (4.1)$$

The above model is called the bilinear model and is denoted by $BL(p,q,P,Q)$. It is linear in $x(t), e(t)$ but not jointly. For $p = P, q = 0, Q = 1$, Subba Rao[26] has shown that one can write an equivalent state-space form and then it is easy to evaluate the moments of the process $x(t)$. The solution of the equation can be written in the form of Volterra series. In Figure 22 and 23, we show plots of the observations generated from the bilinear models. It may be observed that when the coefficients of the nonlinear part of the model tend towards the nonstationary region, the trajectories generated by these bilinear models may produce behavior

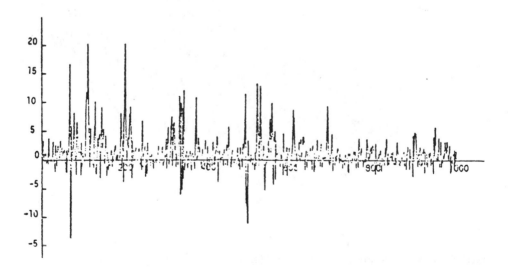

FIGURE 22   Bilinear series from $x(t) = 0.4x(t-1)e(t-1) + e(t)$.

similar to "chaotic" models. The theoretical properties of the bilinear models have been extensively investigated (see Brockwell and Liu,[3] Subba Rao,[26] Bhaskara Rao et al.,[1] Tuan and Tran,[32] and Terdik and Subba Rao[31]). The parameters of the model are estimated by minimizing the least-squares criterion $\sum e^2(t)$, and this minimization is done using iterative techniques.[27] We now consider a real example.

SUNSPOT NUMBERS We consider the first 221 observations (from the year 1700 to 1920) for fitting the bilinear model, and then calculate one-step-ahead predictions for the subsequent 35 years. The following model is fitted to the data[27]

$$
\begin{aligned}
x(t) & -1.5x(t-1)+0.7x(t-2)-0.12x(t-9) \\
& = 6.8-0.146x(t-2)e(t-1)+0.01x(t-8)e(t-1) \\
& \quad -0.01x(t-1)e(t-3)-0.01x(t-4)e(t-3) \\
& \quad +0.004x(t-2)e(t-1)+0.002x(t-3)e(t-2)+e(t).
\end{aligned}
\tag{4.2}
$$

The adequacy of the fit is checked by testing for the independence (and normality) of the residuals. The mth step ahead forecast is calculated from the conditional expectation

$$
\hat{X}(t_0, m) = E(x(t_0 + m) \mid x(s), \quad s \le t_0).
$$

These forecasts together with the true values are given (for $m = 1$) in Table 2. This model seems to have the smallest mean sum of squares, and also error variances calculated up to five steps compared to other nonlinear models (see Table 6.1 of Subba Rao and Gabr[27]).

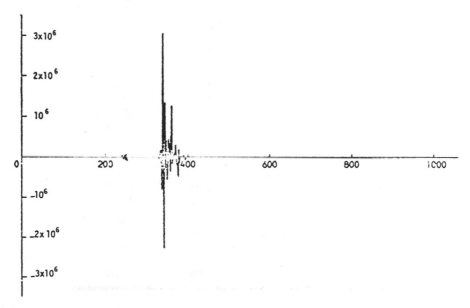

FIGURE 23   Bilinear series from $x(t) = 0.8x(t-1) - 0.4x(t-2) + 0.6x(t-1)e(t-1) + 0.7x(t-2)e(t-1) + e(t)$.

TABLE 2 One-step-ahead prediction of sunspot numbers. X(I) is the actual sunspot number; SBL is the number predicted by the bilinear model in the text; the others are auto-regressive models described elsewhere.[27]

I	X(I)	SBL	FULL AR	SUBSET AR
222	26.1	24.55	24.60	19.07
223	14.2	11.43	13.13	18.69
224	5.8	10.08	13.74	11.39
225	16.7	9.90	9.96	13.12
226	44.3	37.52	35.25	32.82
227	63.9	75.67	67.83	68.32
228	69.0	77.86	74.57	74.11
229	77.8	73.93	67.11	67.14
230	64.9	75.04	72.20	71.42
231	35.7	57.03	49.87	48.73
232	21.2	29.34	16.12	17.56
233	11.1	9.37	14.05	14.27
234	5.7	.68	11.49	11.28
235	8.7	10.94	13.28	14.23
236	36.1	21.84	24.12	23.90
237	79.7	63.85	57.50	57.25
238	114.4	105.82	98.80	97.95
239	109.6	106.99	114.89	115.36
240	88.8	85.10	86.85	85.86
241	67.8	58.95	60.36	60.34
242	47.5	38.76	44.48	44.04
243	30.6	24.79	27.11	29.44
244	16.3	20.28	17.92	19.95
245	9.6	16.20	14.69	15.51
246	33.2	16.58	21.63	21.55
247	92.6	73.04	59.19	59.94
248	151.6	140.92	119.53	120.44
249	136.3	148.47	157.61	158.32
250	134.7	121.87	104.42	103.54
251	83.9	87.77	104.82	106.93
252	69.4	50.79	45.94	41.80
253	31.5	30.73	40.88	49.52
254	13.9	7.01	11.57	9.15
255	4.4	14.83	4.20	11.58
256	38.0	7.65	22.63	18.31
$\hat{\sigma}_e^2(1)$		123.77	190.89	214.07

5. LONG-RANGE DEPENDENT BILINEAR MODEL

A time series $\{x(t)\}$ is said to be short-range memory model if its autocovariances $R(s)$ satisfy the condition $\sum \mid R(s) \mid < \infty$. All the standard stationary linear models satisfy this condition. A process $\{x(t)\}$ for which $\sum \mid R(S) \mid = \infty$ is said to be long-range dependent. This definition of the long-range dependence in terms of the second-order covariances is because of the fact that the process so far considered in the literature is assumed to be Gaussian. If the process is not Gaussian, this condition has to be replaced by conditions involving higher-order cumulants which takes care of nonlinear dependence. Recently there is an enormous interest in this type of model because of its application in hydrology, meteorology, economics, and many related fields. Also these models are closely related to fractals. One way of generating the long-range dependent Gaussian time series is from the model[11,15]

$$x(t) = (1 - B)^{-\delta} e(t) , \delta \in (0, \frac{1}{2}) , \tag{5.1}$$

and $\{e(t)\}$ is a Gaussian white noise. The spectral density function of this process is of the form

$$f(\lambda) = \frac{\sigma^2}{2\pi} (4 \sin^2 \frac{\lambda}{2})^{-\delta} \text{ and as } \lambda \to 0 , f(\lambda) \to c\lambda^{-2\delta} .$$

The other types of linear long-range memory models are generated from the models of the form

$$(1 + a_1 B + \ldots + a B_p^p)(1 - B)^{\delta} x(t) = (1 + b_1 B + \ldots + b_q B^q) e(t) ,$$

where $\delta \in (-1/2, 1/2)$ and the usual assumptions on the polynomials $Z^p + a_1 Z^{p-1} + \ldots + a_p$ and $Z^q + b_1 Z^{q-1} + \ldots + b_q$ are imposed. The condition that $\delta \in (-1/2, 1/2)$ ensures both stationarity and invertibility of the model. Recently, attempts have been made to extend the definition of long-range dependence to non-Gaussian situations but mainly restricted to linear models with non-Gaussian variables $\{e(t)\}$. Here, our object is to extend this definition to nonlinear models. Since we know that bilinear models have covariances similar to the linear models, and they decay to zero exponentially, it is natural to generate long-range nonlinear models through these bilinear models; and the covariances and the third-order moments of these processes will decay at hyperbolic rate in contrast to bilinear models. We can define two types of nonlinear models (for further details see Subba Rao, Terdik, and Bhaskara Rao.[34]). For example, they can be generated as follows:

Let $w(t) = (1 - B)^{-\delta} e(t)$, where $e(t)$ is a Gaussian white noise. We can write for $\delta \in (-1/2, 1/2), w(t) = \sum_{j=0}^{\infty} \psi_j e(t - j)$, where $\psi_j = \Gamma(j + \delta)/\Gamma(\delta)\Gamma(j + 1)$, as $j \to \infty, j^{1-\delta}\psi_j \to 1/\Gamma(\delta)$. Let $x(t)$ satisfy

$$x(t) = ax(t - 1) + bx(t - 1)w(t - 1) + w(t) .$$

We can define the above as long-range bilinear model $LRBL(1,0,1,1)$. The solution of the above bilinear model can be written in the Volterra series form,[27] the first term of which will be linear in $w(t)$, second will be quadratic in $w(t)$, and so on. Equivalently, we can write a Wiener-Ito representation. For example, when we take first two terms of Volterra expansion only, we have

$$x(t) = x_1(t) + x_2(t),$$

where $x_1(t)$ and $x_2(t)$ satisfy the difference equations

$$x_1(t) = ax_1(t-1) + w(t), \mid a \mid < 1$$
$$= \sum_j a^j w(t-j),$$
$$x_2(t) = ax_2(t-1) + bx_1(t-1)w(t-1)$$
$$= \sum_j a^j bx_1(t-j-1)w(t-j-1).$$

We can show that the covariances and higher-order moments decay to zero at a much slower rate (hyperbolic rate) than the usual bilinear model. The other nonlinear model is of the form

$$x(t) = ax(t-1) + bx(t-1)e(t-1) + e(t),$$

and

$$y(t) = (1-B)^{-\delta}x(t).$$

For this model, the covariances decay at the same rate as the usual bilinear model and the spectrum and the bispectrum will have singularities in the neighborhood of the origin.

The properties and the usefulness of the above models will be discussed in a later paper.

ACKNOWLEDGMENTS

I wish to thank Drs. M. Casdagli, S. Eubank, and D. Farmer for inviting me to present the paper at the conference which helped me to learn a lot on "chaos." The last section of this paper was a result of the inspiration provided by the first two weeks of the meeting on Program on Time Series organized by the Institute of Mathematics and Application, Minneapolis, Minnesota, July 2–27, 1990, and the author is very grateful to the organizers for the invitation and to NSF for financial assistance to attend this meeting and the excellent facilities offered. Lastly, I must thank Dr. Jingsong Yuan for the computations reported in this paper.

REFERENCES

1. Bhaskara Rao, M., T. Subba Rao, and A. M. Walker. "On the Existence of Some Bilinear Time Series Models." *J. Time Ser. Anal.* **4(2)** (1983): 95–110.

2. Brillinger, D. R., and M. Rosenblatt. *Asymptotic Theory of Estimates of the rth Order Spectra in Spectral Analysis of Time Series*, edited by B. Harris. New York: Wiley, 1967.

3. Brock, W. A. "Causality, Chaos, Explanation and Prediction in Economics and Finance." In *Beyond Belief: Readomness, Prediction, and Explanation in Science*, edited by J. Casti and A. Karlquist. Boca Raton, FL: CRC Press, 1991.

4. Brock, W. A. "Distinguishing Random and Deterministic Systems: Abridged Version." *J. Econ. Theor.* **40** (1986): 168–195

5. Brockett, P. L., M. Hinich, and D. Paterson. "Bispectral Based Tests for the Detection of Gaussianity and Linearity in Time Series." *J. Amer. Statist. Assoc.* **83** (1988): 657–664.

6. Brockwell, P. J., and J. Liu. "On the General Bilinear Time Series Model." *J. Appl. Prob.* **25** (1988): 553–564.

7. Broomhead, D. S., and G. P. King. "Extracting Qualitative Dynamics from Experimental Data." *Physica* **20D** (1986): 217–236.

8. Eubank, S., and D. Farmer. "An Introduction to Chaos and Randomness." In *1989 Lectures in Complex Systems*, edited by E. Jen, 70–190. Santa Fe Institute Studies in the Sciences of Complexity, Lect. Vol. II. Redwood City, CA: Addison-Wesley, 1989.

9. Godfrey, M. D. "An Explorationary Study of the Bispectrum of Economic Time Series." *Appl. Statist.* **14** (1965): 48–69.

10. Granger, C. W. J., and A. P. Andersen. "An Introduction to Bilinear Time Series Models." Göttingen: Vandenhoeck and Ruprecht, 1978.

11. Granger, C. W. J., and R. Joyeux. "An Introduction to Long Memory Time Series Models and Fractional Differencing." *J. Time Ser. Anal.* **1** (1980): 15–29.

12. Hasselman, K., W. Munk, and G. Macdonald. *Bispectra of Ocean Waves in Time Series Analysis*, edited by M. Rosenblatt, 125–139. New York: John Wiley, 1963.

13. Hinich, M. J. "Testing for Gaussianity and Linearity of a Stationary Time Series." *J. Time Ser. Anal.* **3** (1982): 169–178.

14. Holland, K. N., K. S. Lii, and M. Rosenblatt. In *Bispectra and Energy Transfer Grid Generated Turbulence in Development in Statistics*, edited by P. R. Krishnaiah, 123–155. New York: Academic Press, 1979.

15. Hosking, J. R. M. "Fractional Differencing." *Biometrika* **68** (1981): 165–176.

16. Huber, P. J., B. Kleiner, T. H. Gasser, and G. Dumermuth. "Statistical Methods for Investigating Phase Relations in Stationary Stochastic Processes." *IEEE Trans. Auto Electroacoust* (1971): 78–86.

17. Lii, K. A., M. Rosenblatt, and C. Van Atta. "Bispectral Measurements in Turbulence." *J. Fluid Mech.* **77** (1976): 46–52.
18. Lii, K. S., and M. Rosenblatt. "Deconvolution and Estimation of Transfer Function Phase and Coefficients for Non-Gaussian Linear Processes." *Ann. Statist.* **10** (1982): 1195–1208.
19. Parzen, E. "Time Series Analysis for Models for Signal Plus White Noise." *Spectral Analysis of Time Series*, edited by B. Harris, 233–257. **20**, 1966.
20. Priestley, M. B. *Spectral Analysis and Time Series*, Vol. 1. New York: Academic Press, 1981.
21. Rosenblatt, M., and J. W. Van Ness. "Estimation of the Bispectrum." *Ann. Math. Statist.* **36** (1965): 1120–1136.
22. Sakai, H., and H. Tokomaru. *Autocorrelation of a Certain Chaos*. IEEE Trans. Acoust. Spectra and Signal Processing, ASSP - 2F, 588–590, 1980.
23. Sato, T., K. Sasaki, and Y. Nakamura. "Real Time Bispectral Analysis of Gear Noise and its Applications to Contactless Diagnosis." *J. Acoust. Soc. Am.* **2** (1977): 382–387.
24. Stewart, I. *Does God Play Dice*. Oxford: Basil Blackwell, 1989.
25. Subba Rao, T., and M. Gabr. "A Test for Linearity of Stationary Time Series." *J. Time Ser. Anal.* **1, 2** (1980): 145–158.
26. Subba Rao, T., G. Terdik and M. Bhaskara Rao. "On the Long-Range Bilinear Models." 1991 (under preparation).
27. Subba Rao, T., and M. Gabr. *An Introduction to Bispectral Analysis and Bilinear Time Series Models*. Lecture Notes in Statistic, vol. 24. Berlin: Springer-Verlag, 1984.
28. Subba Rao, T., and M. Gabr. "The Estimation of the Bispectral Density Function and the Detection of Periodicities in a Signal." *J. Mult. Anal.* **27** (1988): 457–472.
29. Subba Rao, T. "Canonical Factor Analysis and Stationary Time Series Models." *Sankhya. Ser. B.* **38** (1976): 256–271.
30. Subba Rao, T., and M. G. A. da Silva. *Identification of Bilinear Time Series Models*. Submitted for publiction, 1990.
31. Terdik, G., and T. Subba Rao. "On Wiener-Ito Representation and the Best Linear Predictors for Bilinear Time Series." *J. Appl. Prob.* **26** (1989): 274–286.
32. Tuan, Phan Dinh, and Tat Tran Lan. "On the First Order Bilinear Time Series Model." *J. Appl. Prob.* **18** (1981): 617–627.
33. Van Atta, C. W. "Inertial Range Bispectrum in Turbulence." *Phys. Fluids* **22, 8** (1979): 1440–1442.

Section III: Dynamical Systems

Henry D. I. Abarbanel
Department of Physics, Institute for Nonlinear Science, and Marine Physical Laboratory, Scripps Institution of Oceanography, University of California at San Diego, Mail Code R-002, La Jolla, CA 92093–0402

Local and Global Lyapunov Exponents on a Strange Attractor

Predictability is limited by positive Lyapunov exponents in chaotic systems, so knowing what the Lyapunov exponents are for an observed system is critical for establishing the limits on how much one can predict, and thus influence or control, about that system. We describe new methods for determining global Lyapunov exponents from data alone including some tools for determining how many of the exponents are "real" and how many artifacts arise from choosing too large an embedding dimension for one's data. We also introduce the notion of *local Lyapunov exponents* which describe orbit instabilities a fixed number of steps ahead rather than an infinite number of steps ahead. These local exponents approach their asymptotic value, the familiar global Lyapunov exponents, as a power of the number of steps. All fluctuations around the mean also approach zero as a power which is characteristic of the invariant distribution of points on the attractor.

1 INTRODUCTION

In nonlinear systems, predictability is limited by the presence of positive Lyapunov exponents when the system is chaotic. Indeed, it is often stated that the existence of positive Lyapunov exponents and the accompanying sensitivity to initial conditions is the essential aspect of chaos for real-world processes. Actually computing the Lyapunov exponents for a given system can be quite a tricky operation, and in the past the algorithms given in the literature have either relied on knowing the explicit differential equations or maps which define the dynamics of the system or have been limited to the determination of the largest Lyapunov exponent, when it is positive. In this talk we will cover the following topics:

- Global Lyapunov Exponents—Revisited

  - Definition
  - Significance
  - Computing Them from Data
  - Examples—Hénon Map, Ikeda map, Lorenz Attractor, Mackey-Glass Time Delay Equation
  - The Role of Noise

- Local Lyapunov Exponents—Visited

  - Definition
  - Significance
  - Distribution Across an Attractor
  - Examples from Computer Work—Hénon Map, Ikeda Map, Lorenz 63 Attractor, and Lorenz 84 Attractor

Not all the results from each example will be given, but they can be found in the papers referred to in the bibliography.[1,3] After discussing these topics we will indicate future projects worth pursuing.

2 GLOBAL LYAPUNOV EXPONENTS

Phase-space reconstruction is so widely discussed both at this workshop and elsewhere,[4,13,21] that we shall simply assume that we have been given scalar data $x(n) = x(t_0 + n\Delta t)$ and from it made multivariate data by time delays

$$\mathbf{y}(n) = [x(n), x(n+T), \dots, x(n+T(d-1))], \tag{1}$$

and we leave to other sources the ways in which one chooses the time delay T and the embedding dimension d. We shall address later what to do if the embedding dimension is too large, but that is coming up.

Lyapunov exponents are defined by reference to an orbit $\mathbf{y}(1) \rightarrow \mathbf{y}(2) \rightarrow \cdots \rightarrow$ $\mathbf{y}(N)$ governed by a map $\mathbf{F}(\mathbf{x})$ from R^d to itself:

$$\mathbf{y}(n+1) = \mathbf{F}(\mathbf{y}(n)). \tag{2}$$

A small perturbation, $\mathbf{w}(l)$ to this orbit at the point $\mathbf{y}(l)[\mathbf{y}(l) \rightarrow \mathbf{y}(l) + \mathbf{w}(l)]$, evolves via the tangent space map

$$\mathbf{w}(k) = \mathbf{DF}(\mathbf{y}(k-1))\mathbf{w}(k-1), \tag{3}$$

where

$$\mathbf{DF}(\mathbf{x})_{ab} = \frac{\partial F_a(\mathbf{x})}{\mathbf{x}_b} \tag{4}$$

is the Jacobian of the map. This means the perturbation at step l evolves into $\mathbf{w}(l+L)$ which is L steps along the orbit, and we have

$$\begin{aligned} \mathbf{w}(l+L) &= [\mathbf{DF}(\mathbf{y}(l+L-1)) \cdot \mathbf{DF}(\mathbf{y}(l+L-2)) \cdots \mathbf{DF}(\mathbf{y}(l))]\mathbf{w}(l) \\ &= \mathbf{DF}^L(l)\mathbf{w}(l). \end{aligned} \tag{5}$$

Lyapunov exponents are defined by the Oseledec Multiplicative Ergodic Theorem[4,17,18,19] which takes $\mathbf{DF}^L(\mathbf{x})$ and forms the matrix

$$\mathbf{OSL}(\mathbf{x}) = \{\mathbf{DF}^L(\mathbf{x}) \cdot [\mathbf{DF}^L(\mathbf{x})]^T\}^{1/2L}, \tag{6}$$

then considers the limit of this as $L \rightarrow \infty$. Oseledec proves that the eigenvalues $\exp[\lambda_i]$, $i = 1, 2, \ldots, d$ of $\mathbf{OSL}(\mathbf{x})$ exist as $L \rightarrow \infty$ and are independent of $\mathbf{x}$. This means that the Lyapunov exponents λ_i are invariants of the dynamics. The eigendirections of $\mathbf{OSL}(\mathbf{x})$ for large L also exist, but are dependent on $\mathbf{x}$. We will have use of these eigendirections later.

Because they are independent of the position $\mathbf{x}$ on the orbit from which the composition of matrices in $\mathbf{DF}^L$ was begun, the λ_i can be used to classify the attractor. Even more critically for the subject of this workshop, they define the limits of predictability for any model of the dynamics. In one sense this is obvious since the largest exponent, λ_1, governs the size of essentially any perturbation to the orbit. When that perturbation grows to order unity, predictability is lost. In a more precise sense, the sum of the positive Lyapunov exponents is, by Pesin's formula,[4,18] the Kolmogorov-Sinai entropy which dictates, in a strict information theoretic sense, the rate of loss of predictability. The computation of λ_i becomes important since the system defines its own predictability by them.

If we are given the dynamics as a map

$$\mathbf{y}(n+1) = \mathbf{F}(\mathbf{y}(n)), \tag{7}$$

or a flow

$$\frac{d\mathbf{y}(t)}{dt} = \mathbf{F}(\mathbf{y}(t)), \tag{8}$$

then there are clear and standard methods for evaluating the λ_i.[2,9] The only tricky part is that the matrix $\mathbf{DF}^L(\mathbf{x})$ becomes quite ill conditioned since $\lambda_1 \geq \lambda_2 \geq \ldots \geq \lambda_d$, and the eigenvalues of DF^L are $\exp[L\lambda_1], \exp[L\lambda_2], \ldots, \exp[L\lambda_d]$. Various standard linear algebra routines found in the popular LINPAC or CLAMS libraries can handle this issue.

To evaluate the λ_i from data, however, is trickier. Basically we do not know either the correct embedding dimension for the phase space (if we are observing scalar data $x(n)$ only), nor do we know the dynamics which maps $\mathbf{y}(n) \mapsto \mathbf{y}(n+1)$, so we do not know the Jacobian matrices $\mathbf{DF}(\mathbf{y}(n))$.

Earlier work[5,20,23] on this subject made the essential suggestion that to construct the Jacobian matrices, one should look not only at the points on the orbit $\mathbf{y}(n)$ but also at the neighbors in state space or phase space of these points. If one knew how a whole cluster of points distributed in phase space near the orbit mapped forward, then one could estimate the partial derivatives in space which are required in $\mathbf{DF}(\mathbf{x})$. Call the difference vector between the orbit point, $\mathbf{y}(n)$, and its rth neighbor, $\mathbf{y}^r(n)$, $\mathbf{z}^r(n) = \mathbf{y}^r(n) - \mathbf{y}(n)$. Then we need to know how $\mathbf{z}^r(n)$ maps into $\mathbf{z}^r(n+1)$. Earlier work using this idea implemented it by assuming a *local linear* map from time n to time $n+1$:

$$z^r(n+1)_a = M_{ab}(\mathbf{y}(n))\mathbf{y}_b(n), \tag{9}$$

and identified the mapping matrix $M(x)$ with $\mathbf{DF}(\mathbf{x})$.

One feature of the use of local linear mappings was that only the largest Lyapunov exponent λ_l was reliably determined by the procedure. We have investigated the idea of making *local polynomial* maps from neighborhood to neighborhood as a means of relieving the Jacobian $\mathbf{DF}(\mathbf{x})$ of the double burden of mapping from neighborhood in highly structured attractors and also reproducing the correct Lyapunov exponents. The mathematical problem addressed here is that of estimating from data presented at finite $\mathbf{z}(n)$ the properties of the $\mathbf{z} \to \mathbf{z}$ map as $\mathbf{z} \to 0$, namely the derivatives of the map evaluated on the orbit. A polynomial map addresses two important issues in determining the required Jacobian:

- The attractor may be quite convoluted and curved within a single time step of the given data for the chosen value of sampling rate Δt. The polynomial allows the local map to account for this curvature. The attractor may also involve significant curvature from time step to time step, and this is taken care of as well.
- Small errors in the evaluation of $\mathbf{DF}(\mathbf{x})$ will lead to enormous errors in the determination of the λ_i for $i = 2, 3, \ldots, d$, since the matrices involved are so ill conditioned. Using local polynomial maps allows us to accurately determine the linear term in the map.

We represent the local neighborhood-to-neighborhood map as

$$\begin{aligned} z^r(n+1) = {} & M(n) \cdot z^r(n) + Q(n) \cdot z^r(n)z^r(n) \\ & + C(n) \cdot z^r(n)z^r(n)z^r(n) + \ldots \end{aligned} \tag{10}$$

We determine the linear $[M(n)]$, quadratic $[Q(n)]$, cubic $[C(n)]$, and higher-order tensors by a least-squares fit to the residuals in this expression, and we do it at each time step. Then we throw away terms that are higher order than linear and construct $\mathbf{DF}^L$ by multiplication of L terms $M(n)$. The eigenvalues of the resulting product of matrices are determined by the QR decomposition procedure outlined in Eckmann and Ruelle[4] and Eckmann et al.[4,5]

An example[3] of this procedure is shown in Table 1 where results from calculations on the Lorenz attractor[12] are shown. In arriving at these results, we took two long data sets of the $x(n)$ degree of freedom from the Lorenz system and created three vectors by time lag embedding. In constructing Table 1(a) we used data with five digits of accuracy and in Table 1(b), data with nine digits of accuracy. In Table 1(b) we also determined the distance between points required to establish who is and who is not a neighbor in a large dimension to studiously avoid errors in identifying neighbors. In each table we give the results of our computations for the use

TABLE 1 Lyapunov exponents for the Lorenz system

Order of Polynomial	True Values $\lambda_1 = 1.51$	$\lambda_2 = 0$	$\lambda_3 = 22.5$
(a)			
50,000 data points $d_L = 3,\ d_G = 3$ 5 digits of accuracy and analyzed using $d = 3$ for varying orders of the polynomial fit.			
Linear	1.4504	-0.0057123	-13.999
Quadratic	1.5027	-0.046041	-19.448
Cubic	1.5121	0.0069641	-22.925
Quartic	1.5561	0.032219	-23.465
(b)			
20,000 data points $d_L = 3,\ d_G = 3$ 9+ digits of accuracy and analyzed for mapping orders one through five.			
Linear	1.549	-0.09470	-14.31
Quadratic	1.519	-0.02647	-20.26
Cubic	1.505	-0.005695	-22.59
Quartic	1.502	-0.002847	-22.63
Quintic	1.502	-0.000387	-22.40

of linear, quadratic, cubic, and quartic maps. For the higher accuracy data, we also evaluated the results for quintic maps.

It is clear that after cubic maps were used, very little improvement was achieved in the determination of the λ_i whose true values are indicated at the top of the table. The loss of accuracy in the bottom row of Table 1(b) is from the use of a large order map with inaccurate data, and the degradation is due to the numerical errors in the least-squares algorithm for the large matrices involved. This problem is absent in the more accurate data. In these results the accuracy of the data was used as a proxy for noisy data, and we shall return directly to the issue of noise in a moment.

The next issue we address concerns the matter of the embedding dimension. We embedded the Lorenz attractor data in three dimensions because of our and others' experience that this was an adequate size for the embedding. However, suppose we had decided to embed the $x(n)$ Lorenz data in $d = 4$? From the point of view of dynamics, nothing bad has happened. From the point of view of Lyapunov exponent determinations, we suddenly have four exponents, not three. One must be spurious!

If we simply compute with four-dimensional vectors and use cubic maps, say, then the four exponents which come out of the calculation are $\lambda_1 = 18.04$, $\lambda_2 = 1.502$, $\lambda_3 = -0.00055$, and $\lambda_4 = -22.77$. One of these is false, and clearly it is the largest exponent which is an order of magnitude larger than the true largest exponent!

To discriminate against this spurious large exponent, we used the other piece of information in the Oseledec matrix $\mathbf{OSL}(\mathbf{x})$, namely the eigendirections associated with the Lyapunov exponents. Our reasoning was that if there is a true positive exponent, then stretching will occur along that direction in the neighborhood of any point $\mathbf{x}$. This direction changes from point to point along the attractor, but is easy to establish at any given $\mathbf{x}$. If there is stretching, then there will be many points of the data along that eigendirection. If the number of points along that eigendirection is not large, then the (positive) exponent is a fake.

To determine in a quantitative fashion the "thickness" along an eigendirection of $DF^L(\mathbf{x})$, we made a least-squares fit of the data transverse to this direction, call it $L_i(\mathbf{x})$. So we fit all the points in the neighborhood of $\mathbf{x}$ by a polynomial of desired order with the $d - 1$-dimensional manifold described by a polynomial and taken tangent to $\mathbf{L}_i(\mathbf{x})$. The residual in this fit is a measure of the "thickness." In Table 2 we show the result of this procedure when we use an embedding dimension of four for data from the Lorenz attractor and use linear through quartic polynomial neighborhood-to-neighborhood maps. The data thicknesses TH_i clearly show an enormously thin result for the largest exponent and for the smallest exponent. The latter is good since it is negative. The former announces the presence of a spurious exponent since the attractor should be "thick" in that direction. In the other positive direction and the neutral direction, the attractor is acceptably thick. We then toss out the spurious, very large exponent as an artifact of our choosing too large an embedding dimension.

TABLE 2 Lyapunov exponents and thicknesses for Lorenz system[1]; $d_L = 4, d_G = 7$.

Order of Polynomial Fit; Lyapunov Exponents & Thicknesses			
1	$\lambda_1 = 1.936$	$\lambda_2 = 0.8019$	$\lambda_3 = -1.137$	$\lambda_4 = -13.44$
	$TH_1 = 0.4666$	$TH_2 = 1.083$	$TH_3 = 1.111$	$TH_4 = 0.1614$
2	$\lambda_1 = 4.364$	$\lambda_2 = 1.401$	$\lambda_3 = -0.6559$	$\lambda_4 = -20.57$
	$TH_1 = 0.00256$	$TH_2 = 0.4116$	$TH_3 = 0.4539$	$TH_4 = 0.00116$
3	$\lambda_1 = 18.04$	$\lambda_2 = 1.502$	$\lambda_3 = -0.00055$	$\lambda_4 = -22.77$
	$TH_1 = 1.89E - 07$	$TH_2 = 0.1288$	$TH_3 = 0.0910$	$TH_4 = 2.7E - 04$
4	$\lambda_1 = 26.96$	$\lambda_2 = 1.503$	$\lambda_3 = -0.00484$	$\lambda_4 = -22.55$
	$TH_1 = 2.46E - 10$	$TH_2 = 0.0656$	$TH_3 = 0.0656$	$TH_4 = 5.57E - 05$

[1] Lyapunov exponents and thicknesses of the attractor along the corresponding Lyapunov direction vectors for the Lorenz system. The calculation was done with 20,000 data points. There is one spurious exponent. For second-order polynomials and above, the spurious exponent separates from the true ones and can be identified by its extremely small thickness value.

We have found similar behavior for our work on the Ikeda Map[11] and the Mackey-Glass[14] time-delay equation. The reports on this work are to be found in the references.[3]

In the first table we used accuracy of the input data as a proxy for the addition of noise to the signal whose exponents we want. To address this question directly, we added noise of some level to the data directly and computed the Lyapunov exponents as a function of the amplitude of $R = $ Noise/Signal. In Figure 1 we show the Lyapunov exponents which come out of our procedure as a function of R when we choose an embedding dimension of $d = 3$. Clearly for R very small, there is no problem, and the largest and next largest exponents are quite robust. For $R \approx 10^{-4}$, however, the smallest exponent becomes quite unreliable, and by the time $R \approx 10^{-2}$, it is all over for accurate determination of the exponents from noisy data.

As ever, along with bad news comes good news. In Figure 2 we show the same calculation with the choice of embedding dimension $d = 4$. Now there is one spurious exponent, and for R very small we see that appearing as a false positive exponent, as we have discussed. Now as we increase R, the fake exponent wanders all over the place, and exposes its falsity by its extreme sensitivity to noise even when those noise levels are miniscule. This is easy to understand since there is no data in the fake directions associated with the fake exponents; they are an artifact of

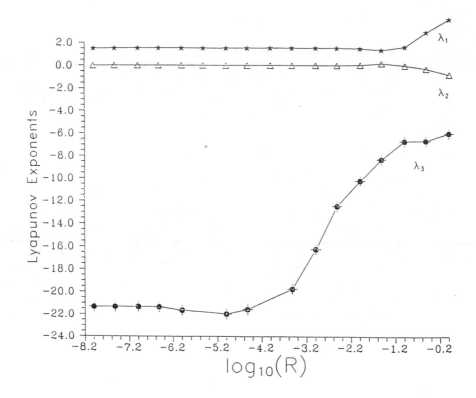

FIGURE 1 The effect of external noise on the determination of Lyapunov exponents for the Lorenz system. In this figure the dimension is $d = 3$. On the horizontal axis is the $\log_{10}$ of $R = $ (Noise Level)/(Signal Level).

our numerical procedure and bad choice of embedding dimension. Noise, however, respects no dynamics and populates, in a random fashion, directions in the larger phase space which are not explored by the dynamics. This causes major changes in the locations of neighbors as perceived by our algorithm and thus in the values of $\mathbf{DF}^L(\mathbf{x})$ we use. So noise added to clean data can indicate the presence of false exponents.

3 LOCAL LYAPUNOV EXPONENTS

The λ_i we have discussed so far are the eigenvalues of $\mathbf{OSL}(\mathbf{x})$ as $L \to \infty$; that is L quite large. Thus we learn about the average or global instability of perturbations to an orbit from these exponents. We may not be really interested in predictability for $L \to \infty$, however. Indeed, if we were interested in weather, we would be interested in predictability of our models for 5 days or 5 weeks, and not for 10^5 days ≈ 300 years or 10^5 weeks $\approx 20,000$ years. We have explored[1] the idea of evaluating the eigenvalues of $\mathbf{OSL}(\mathbf{x})$ for *finite* L and considering

- how the exponents and the variation about the mean of these exponents depend on L, and
- the distribution of the exponents around the attractor.

The eigenvalues of $\mathbf{OSL}(x)$ are written as $\exp[\lambda_i(\mathbf{x}, L)]$, $i = 1, 2, \ldots, d$, and their dependence on the location in phase space $\mathbf{x}$ and the composition length along the

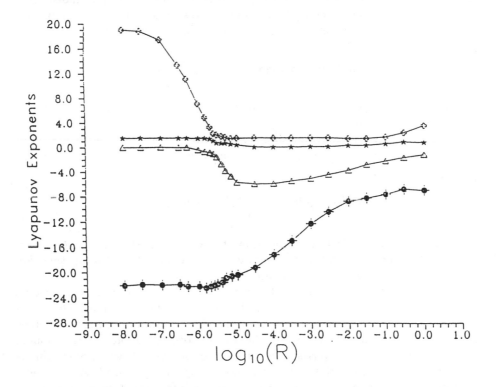

FIGURE 2 The effect of external noise on the determination of Lyapunov exponents for the Lorenz system. In this figure the dimension is $d = 4$. The spurious exponent wanders from about $+19$ to nearly -6 as the noise level is varied. Note that the exponents do not cross each other but prefer to switch roles as they become close. In the $d = 3$ case, the correct exponents are more robust against the addition of noise. On the horizontal axis is the $\log_{10}$ of $R = $ (Noise Level)/(Signal Level).

attractor is made explicit. Oseledec's theorem states that as $L \to \infty$, the $\lambda_i(\mathbf{x}, L)$ become independent of $\mathbf{x}$. If we average the $\lambda_i(\mathbf{x}, L)$ over the invariant distribution

$$\rho(\mathbf{x}) = \lim_{N \to \infty} \frac{1}{N} \sum_{k=1}^{N} \delta^d(\mathbf{x} - \mathbf{y}(k)), \tag{11}$$

then by the usual argument we have an invariant under the action of the dynamics $\mathbf{x} \mapsto \mathbf{F}(\mathbf{x})$. This we call

$$\overline{\lambda}_i(L) = \int d^d x \rho(\mathbf{x}) \lambda_i(\mathbf{x}, L). \tag{12}$$

In the limit that L becomes large, this is just λ_i since $\lambda_i(\mathbf{x}, L)$ is independent of $\mathbf{x}$ in that limit. The rate at which $\overline{\lambda}_i(L)$ approaches this limit is of interest to us. We are also interested in the moments about the mean $\overline{\lambda}_i(L)$ which we call

$$\sigma_i(p, L) = \int d^d x \rho(\mathbf{x}) [\lambda_i(\mathbf{x}, L) - \overline{\lambda}_i(L)]^p, \tag{13}$$

and each of these moments must tend to 0 as L is increased. Again this is a result of the Oseledec theorem.

Form the probability distribution of exponents by counting the frequency of their taking specific values as we proceed around the attractor; this distribution is

$$P_i(\lambda, L) = \lim_{N \to \infty} \frac{1}{N} \sum_{k=1}^{N} \delta\big(\lambda_i(\mathbf{y}(k), L) - \lambda\big), \tag{14}$$

as one can prove directly from the definition of $\rho(\mathbf{x})$. This is the natural definition of the distribution of local exponents. In practice, of course, the delta functions are replaced by sharp peaks of finite resolution.

The implication of the Osedelec theorem for $P_i(\lambda, L)$ is that it will become narrower and narrower as L is increased and will be concentrated on λ_i. The width of this distribution is important for the predictability of the system $L\Delta t$ time steps ahead of any point on the attractor. The width gives us a quantitative handle on how much the exponent varies as we move around the attractor. Since this is directly connected to our ability to predict how small changes in phase space will grow or contract relative to an orbit, this is just the information we wish to have.

We have investigated these properties of $P_i(\lambda, L)$ and report on them now.

3.1 NUMERICAL EXAMPLES

Our first example is drawn from the two-dimensional Hénon map of the plane to itself[10]

$$x(n+1) = 1.0 + y(n) - ax(n)^2,$$
$$y(n+1) = bx(n),$$
(15)

with $a = 1.4$ and $b = 0.3$. The Jacobian matrix $\mathbf{DF(x)}$ has a constant determinant of $\log(b) = -1.204$, so both the global and local Lyapunov exponents satisfy $\lambda_1 + \lambda_2 = \log(b)$.

To examine the local Lyapunov exponents of order L, $\lambda_i(\mathbf{x}, L)$, we generated a data set with 15,000 points on the Hénon attractor after letting all transients die out. We then computed the average of the local exponents for $L = 5, 10, \ldots, 1250$. In Figure 3 we show on the same plot the distributions, normalized to unity, of the larger local Lyapunov exponent for $L = 10$, 100, and 1000. The width of the distributions clearly shrinks, and the peak centers closely around the value 0.412 with variations at $L = 1000$ of ± 0.013. This is completely consistent with previous

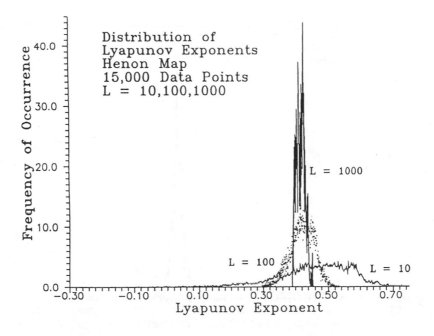

FIGURE 3 The distribution of the larger Lyapunov exponent for the Hénon map for composition lengths of $L = 10, 100$, and 1000. Each distribution is normalized to unity and is from a sample of 15,000 points on an orbit of the Hénon map.

calculations of λ_1. For λ_2 the picture is the same with the average tending to -1.62 ± 0.013, as it must. In each case the limit is approached as $1/\sqrt{L}$. More precisely the best fit to the RMS value of the variation about the mean for the Hénon exponents is $0.45/L^{0.52}$ using the data from $L = 10$ to $L = 1250$. Discarding the point at $L = 10$, we find the exponent in this variation with L to change to 0.48 instead of 0.52. It is safe to claim that the actual exponent is approximately 0.5. This is borne out by the fit to the decrease for $\overline{\lambda}_1(L)$ to its $L \to \infty$ value. For the larger exponent this is $\overline{\lambda}_1(L) \approx 0.412 + (0.047/L^{0.5})$.

We have also examined the dependence on L for the exponents of the model of Lorenz.[12] This is the familiar three-dimensional model extracted from the study of convection in the lower atmosphere

$$\dot{x}(t) = \sigma\big(y(t) - x(t)\big)$$
$$\dot{y}(t) = -x(t)x(t) + rx(t) - y(t) \qquad (16)$$
$$\dot{z}(t) = x(t)y(t) - bz(t),$$

and the values $a = 16.0$, $b = 4.0$, and $r = 45.92$ are taken.

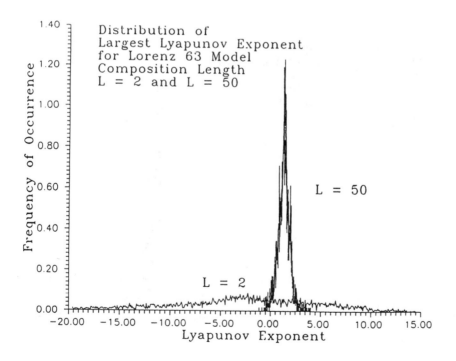

FIGURE 4 The distribution of the largest Lyapunov exponent for the Lorenz 63 model for composition lengths of $L = 2$ and $L = 50$. Each distribution is normalized to unity.

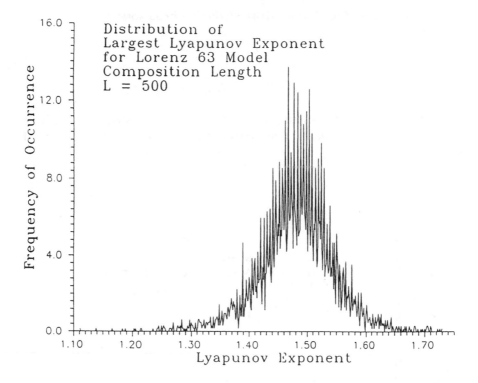

FIGURE 5 The distribution of the largest Lyapunov exponent for the Lorenz 63 model for a composition length of $L = 500$. The distribution is normalized to unity; note the difference in scale from Figure 4.

In Figure 4 we show the distribution of the largest exponent of the Lorenz model for composition lengths $L = 2$ and $L = 50$. In Figure 5 is the same for $L = 500$. Note the change in scale between these figures which clearly demonstrates the narrowing of the distribution. Also note that the distributions are not very smooth and for smallish L, they are quite broad!

In the next two figures (Figure 6 and Figure 7), we show how the two of the moments $\sigma_i(p, L)$ for $p = 2, 4$ behave as a function of L. We actually show the square root of $\sigma_1(2, L)$ and the fourth root of $\sigma_1(4, L)$. Each approaches its limit, 0, as it should, and each does so as an inverse power of L. This power approach to the limits is characteristic of the systems we have examined, and we have found that the specific power depends on the system and on the moment considered. In other words, we have found as a rule the following behaviors:

$$\overline{\lambda}_i(L) = \lambda_i + \frac{c_i}{L^{\nu_i}}$$

$$\sigma_i(p, L) = \frac{C_i(p)}{L^{\nu_i(p)}}, \tag{17}$$

with c_i and $C_i(p)$ some constants, and the powers dynamically determined. It may be that $\nu_i(p) \approx p\xi_i$ with the ξ_i independent of p, but we are not quite sure of that yet.

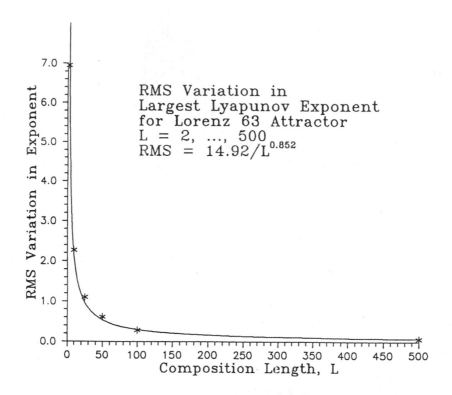

FIGURE 6   The RMS variation of the largest of the Lyapunov exponents of the Lorenz 63 model about its mean: $\sqrt{\sigma_1(2, L)}$. The variation in L is $\sqrt{\sigma_1(2, L)} \approx 14.92L^{0.85}$. The stars mark calculated points and the solid line is a best fit with a power-law dependence.

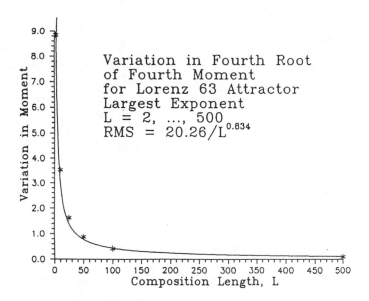

FIGURE 7 The variation of the fourth root of $\sigma_1(4, L)$ for the larger of the exponents from the Lorenz 63 model. $[\sigma_1(4, L)]^{0.25} \approx 20.26L^{-0.83}$. The stars mark calculated points and the solid line is a best fit with a power-law dependence.

The implication of these results are at least these:

- The widths of the distribution of exponents, as indicated by the variation about the means, decrease slowly to zero for large L. There may well be significant variation around an attractor of Lyapunov exponents in any realistic setting where one wants to predict a few steps ahead.

- The approach to the limit λ_i, the global exponent, of the averaged local Lyapunov $\overline{\lambda}_i(L)$ is also slow, but if we can establish that $\overline{\lambda}_i(L)$ behaves as a constant plus a power correction, we may be able to establish the limit using a finite step in L data by extrapolation.

- Since the $\overline{\lambda}_i(L)$ and the $\sigma_i(p, L)$ or, for that matter, the $P_i(\lambda, L)$ are invariants; each of them may be used to characterize the dynamics and classify it. This allows us to go beyond the use of just the $L \to \infty$ limit, the λ_i for this purpose. Of course, the distributions themselves have implications for predictability which we have indicated.

4 CONCLUSIONS

There are two items which I want to touch on in this conclusion. One is a combination of the topics mentioned which is an important matter for further research, and the other is an attempt to address some questions raised at the time I actually gave the talk at the workshop of which these are the proceedings. There are comments on earlier work by Nese[15,16] and Vastano and Moser[22] which are made in the published papers, and I direct the reader to these for those references.

4.1 A RESEARCH ISSUE

The combination of the ideas in this talk will provide a useful means of extracting the local Lyapunov exponents from data. The idea is very simple to state. Use the methods of Abarbanel, Brown, and Bryant to accurately construct local Jacobians from data using local polynomial maps. With these numerical Jacobians, $\mathbf{DF}(\mathbf{x})$, we can compose them to make $\mathbf{DF}^L(\mathbf{x})$ and from this extract $\lambda_i(\mathbf{x}, L)$ and $\overline{\lambda}_i(L)$ and the various moments. It seems a straightforward and useful exercise to develop the algorithms for this and demonstrate their use on both laboratory and simulated data.

4.2 EARLIER WORK—COMMENTS AND OBSERVATIONS

During the talk at this workshop, I was asked by Peter Grassberger about the relationship between the local Lyapunov exponents here and those defined by Fujisaka[6,7] and by Grassberger and Procaccia.[8]

Fujisaka defines local Lyapunov exponents for only one-dimensional maps, but this is a small point. He considers the map $x(k + 1) = f(x(k))$, and defines the quantity

$$
\begin{aligned}
\Lambda(k) &= \frac{1}{L} \sum_{j=1} L \log[|f'(x(j + k))|] \\
&= \log\left\{ \prod_{j=1}^{L} |f'(x(k + j))|^{1/L} \right\},
\end{aligned}
\tag{18}
$$

which is clearly the same as our $\lambda(x(k), L)$ for one dimension. He now *holds L fixed* calling it one "chaos-period," which is a concept I do not understand at all and refer you to his first paper for its introduction. He then studies various properties of $\Lambda(k)$ averaged over $x(0)$ using the usual invariant density. He concludes that there is a kind of diffusion of perturbations to an orbit with distances growing as $\sqrt{k}$ and a diffusion constant related to the dynamics. While he has identified more or less the same object, his interpretation of it and his use of the idea is really quite different from what we have presented here. Nonetheless, it seems quite an interesting paper.

Grassberger and Procaccia are concerned with the evolution of small deviations from an orbit $\Delta_i(t)$—they use continuous time which is not an important issue; that is, their t is our L—and write a presumed differential equation for $\log[|\Delta_i(t)|]$ of the form

$$\frac{d\log[|\Delta_i(t)|]}{dt} = \lambda_i + \eta_i(t), \tag{19}$$

where they take $\eta_i(t)$ to be zero-mean, white, Gaussian noise:

$$\begin{aligned}\langle\eta_i(t)\rangle &= 0,\\ \langle\eta_i(t)\eta_j(t')\rangle &= Q_{ij}\delta(t-t'),\end{aligned} \tag{20}$$

with the tensor Q_{ij} a set of constants. The averages $\langle\bullet\rangle$ are over the invariant density. These assumptions lead to the statement that the RMS value of the local Lyapunov exponent should behave as $1/\sqrt{t}$ for large times.

One could relax their assumptions in the following way. Explicitly recognize that the distances $\Delta_i(t)$ depend on the location $\mathbf{x}$ on the attractor where one starts, so write $\Delta_i(\mathbf{x}, t)$, and then also write the "noise" term as $\eta_i(\mathbf{x}, t)$. The averages involved are over $\mathbf{x}$ so they can be t dependent. Now write $\log[|\Delta_i(\mathbf{x}, t)|]$ as $\lambda_i(\mathbf{x}, t)t$ and the solution of the equation for $\Delta_i(\mathbf{x}, t)$ becomes

$$\lambda(\mathbf{x}, t) = \lambda_i + \frac{1}{t}\int_0^t d\tau\,\eta_i(\mathbf{x}, \tau), \tag{21}$$

which, upon averaging over $\mathbf{x}$ using the invariant density $\rho(\mathbf{x})$, is in our notation

$$\overline{\lambda}_i(t) = \lambda_i + \frac{1}{t}\int_0^t d\tau\,\eta_i(\tau), \tag{22}$$

where now $\eta_i(t)$ is the phase-space average of the fluctuating term driving the local perturbations from an orbit. It should not be taken to be zero, but should vanish in the limit as $t \to \infty$. Our numerical experiments indicate that, in fact, for large t, $\eta_i(t)$ behaves as $t^{-\nu_i}$.

Similarly one can relax the assumption that $\langle[\eta_i(\mathbf{x}, t) - \eta_i(t)][\eta_j(\mathbf{x}, t') - \eta_j(t')]\rangle$ is a delta function. Indeed, the numerical experiments indicate it falls as a power of time which is dynamically determined. The relationship between that power and the fall off of the moments about the mean we have reported above is easy to establish, and I leave it to the reader.

Given the numerical observations we have reported here and those further ones in the papers referred to, we see that the Grassberger-Procaccia formalism is a transcription to a stochastic differential equation language and then to a Fokker-Planck-like equation of the numerical observations set out above. It could well be a useful way to state the results, though the non-Gaussianity of the variations of the $\eta(\mathbf{x}, t)$ which are observed may make the Fokker-Planck equation cumbersome. This point might disappear if, as we have conjectured, the $\nu_i(p)$ scale in p as $\nu_i(p) \approx p\xi_i$. It is important to note, however, that the assumption that the noise is zero-mean, Gaussian, and white is not obvious for dynamics and needs to be carefully checked for each system.

ACKNOWLEDGMENTS

We thank the members of INLS for numerous discussions on this subject and Professor D. Thomson of Penn State for making the thesis of Nese available to us. Also Dr. John Vastano kindly sent us a preprint of his work with Moser, and we appreciate that. We also thank the organizers of this Workshop: Martin Casdagli, Stephen Eubank, J. Doyne Farmer, and Peter Grassberger for the opportunity to participate. The support from the Santa Fe Institute and NATO is much appreciated. This work was supported in part by the U.S. Department of Energy, Office of Basic Energy Sciences, Division of Engineering and Geosciences, under Contract N00014-89-D-0142 D0#15.

REFERENCES

1. Abarbanel, H. D. I., R. Brown, and M. B. Kennel. "Variation of Lyapunov Exponents on a Strange Attractor." Submitted to the *Journal for Nonlinear Science*, September, 1990.
2. Benettin, G., L. Galgani, A. Giorgilli, and J.-M. Strelcyn. *Meccanica* **15** (1980): 9.
3. Brown, R., Bryant, P., and H. D. I. Abarbanel. "Computing the Lyapunov Spectrum of a Dynamical System from Observed Time Series." *Phys. Rev.* **A43** (1991): 2787. We refer to this paper as BBA. A shorter version of this paper appears in *Phys. Rev. Lett.* **65** (1990): 1523.
4. Eckmann, J.-P. and D. Ruelle. "Ergodic Theory of Chaos and Strange Attractors." *Rev. Mod. Phys.* **57** (1985): 617.
5. Eckmann, J.-P., S. 0. Kamphorst, D. Ruelle, and S. Ciliberto. "Lyapunov Exponents from Time Series." *Phys. Rev. A* **34** (1986): 4971. We refer to this work as EKRC.
6. Fujisaka, H. "Statistical Dynamics Generated by Fluctuations of Local Lyapunov Exponents." *Prog. Theor. Phys.* **70** (1983): 1264.
7. Fujisaka, H. 'Theory of Diffusion and Intermittency in Chaotic Systems." *Prog. Theor. Phys.* **71** (1984): 513.
8. Grassberger, P., and I. Procaccia. "Dimensions and Entropies of Strange Attractors from a Fluctuating Dynamics Approach." *Physica D* **13D** (1984): 34.
9. Greene, J. M., and J. S. Kim. "The Calculation of Lyapunov Spectra." *Physica D* **24** (1987): 213–225.
10. Hénon, M. "A Two-Dimensional Mapping with a Strange Attractor." *Commun. Math. Phys.* **50** (1976): 69.
11. Ikeda, K. *Opt. Commun.* **30** (1979): 257.
12. Lorenz, E. N. "Deterministic, Nonperiodic Flow." *J. Atmos. Sci.* **20** (1963): 130.

13. Mane, R. "On the Dimension of the Compact Invariant Sets of Certain Nonlinear Maps." In *Dynamical Systems and Turbulence, Warwick 1980*, edited by D. Rand and L. S. Young. Lecture Notes in Mathematics, vol. 898, 230. Berlin: Springer, 1981.
14. Mackey, M. C., and L. Glass. *Science* **197** (1977): 287.
15. Nese, Jon M. "Quantifying Local Predictability in Phase Space." *Physica D* **35** (1989): 237–250.
16. Nese, Jon M. "Predictability of Weather and Climate in a Coupled-Ocean Atmosphere Model: A Dynamical Systems Approach." Ph.D. thesis, Department of Meteorology, Pennsylvania State University, August, 1989; and references therein.
17. Oseledec, V. I. "A Multiplicative Ergodic Theorem. Lyapunov Characteristic Exponents for Dynamical Systems." *Trudy Mosk. Mat. Obsc* (Moscos Math. Soc.) **19** (1968): 197.
18. Pesin, Ya. B. "Lyapunov Characteristic Exponents and Smooth Ergodic Theory." *Usp. Mat. Nauk* (Russian Math. Surveys) **32** (1977): 55–112.
19. Ruelle, D. "Ergodic Theory of Differentiable Dynamical Systems." Technical Paper No. 50, Publications Mathématiques of the Institut des Hautes Etudes Scientifiques, 1979, 27–58.
20. Sano, M., and Y. Sawada. *Phys Rev. Lett.* **55** (1985): 1082.
21. Takens, F. "Detecting Strange Attractors in Turbulence." In *Dynamical Systems and Turbulence, Warwick 1980*, edited by D. Rand and L. S. Young. Lecture Notes in Mathematics, vol. 898, 366. Berlin: Springer, 1981.
22. Vastano, J. A., and R. D. Moser. "Lyapunov Exponent Analysis and the Transition to Chaos in Taylor-Couette Flow." Preprint, Center for Turbulence Research, Stanford University and NASA-Ames Research Center, May 10, 1990.
23. Wolf, A., J. B. Swift, H. L. Swinney, and J. A. Vastano. "Deterimining Lyapunov Exponents from a Time Series." *Physica* **16D** (1985): 285.

Thomas P. Meyer and Norman H. Packard
Physics Department and the Center for Complex Systems Research, Beckman Institute,
University of Illinois at Urbana-Champaign, 405 North Mathews Avenue, Urbana, IL 61801

Local Forecasting of High-Dimensional Chaotic Dynamics

We use a genetic learning algorithm to learn patterns in data produced by a
high-dimensional chaotic attractor. The learned patterns are relationships
between a region of the attractor and the future behavior of orbits passing
through this region. The learned patterns give an accurate local profile of
the attractor, and provide good forecasts.

1. INTRODUCTION

We consider the problem of forecasting a complex time series. Generally this problem is hindered by three difficulties: (i) dynamical noise due to the presence of chaos; (ii) observational and stochastic noise, uncorrelated noise coupled to the system; and (iii) statistical fluctuations due to small amounts of data. In this paper, we focus on the first difficulty, and briefly comment on the other two.

Using a standard technique in dynamical-systems data analysis, we will begin by embedding the time series data in a d-dimensional space, $\mathcal{X}$.[8] We will consider the case where d is large in order to get a maximally deterministic map from $\vec{x}$ to y for as much of the input space as possible.

Typically, parts of the data set are predictable and others not. Geometrically, this means that some regions of $\mathcal{X}$ will be predictable and others not. Our goal is to discover these predictable regions of $\mathcal{X}$. We can cut the space into regions an infinite number of ways. We will use a version of the genetic algorithm to search through the set of possible regions.

The genetic algorithm is a machine-learning technique developed largely by Holland.[4] It it particularly useful when the solution sought after lies in a high-dimensional, very complex space. A version of the genetic algorithm has been specially tailored to look through the set of regions of a discrete data space[9]; in this paper, we employ a generalization adapted to data having continuous values.

We consider only a single time series; the method is straightforwardly generalizable to multivariate series, with a corresponding increase in computational expenditure. Furthermore, the method may also address many other problems besides forecasting.[9]

2. REPRESENTATION

We will consider a finite data set of size N, which comes from a continuous variable that is sampled at discrete time intervals,

$$\{\xi^1, \xi^2, \dots, \xi^N\},$$

and the case where the sampling has high resolution, so that we can consider the sampled data values ξ^t as continuous rather than discrete.

From this sequence, we can construct a sequence of points in a corresponding d-dimensional state space[8] $\mathcal{X}$,

$$\{\vec{x}^d, \vec{x}^{d+1}, \dots, \vec{x}^N\},$$

where each $\vec{x}^t = (\xi^{t-d+1}, \dots, \xi^t)$ is a set of past values in the time series. From the data, we are interested in determining the map from the past values $(\xi^{t-d+1}, \dots, \xi^t)$ to the future value at a time t' later, $\xi^{t+t'}$.

$$\boxed{\xi^{t-d+1}, \xi^{t-d+2}, \dots, \xi^t} \longrightarrow \boxed{\xi^{t+t'}}$$

Equivalently, we wish to construct a map between $\vec{x}^t$ and $y^t \equiv \xi^{t+t'}$.

To learn this map, we construct a set of input-output vectors that embody the graph of the mapping we need to learn,

$$\{(\vec{x}^d, y^d), \dots, (\vec{x}^{N-t'}, y^{N-t'})\}.$$

The values of each $(\vec{x}, y)$ pair are obtained from a measurement which can be discrete or continuous; in this paper we will consider the continuous case. The discrete case is discussed in Packman.[9]

Commonly used forecasting approaches include factor analysis, auto-correlation, and linear or nonlinear regression techniques.[1] Although these types of approaches work well for systems that have simple probabilistic or deterministic global structure, they generally fail for systems with complex temporal dependencies.

Typically, chaotic systems have extremely complex dependencies among past and future states. In part, this complexity comes from a non-uniform spreading rate of nearby trajectories as they move through the system's state space. The non-uniform spreading rates lead directly to areas having different degrees of predictability. It is precisely this fact which we wish to exploit. Whenever the system is found within a region with a low cumulative spreading rate, its future iterates can be predicted with a higher degree of accuracy.

2.1 CONDITIONAL SETS

In order to exploit the inherent non-uniformity of the system's predictability, we construct a state-space magnifying glass, which will allow us to look in greater detail at a particular region, while ignoring everything which is not within our field of vision. We do this by imposing restrictions on the past values, of the form $(a < x_i < b)$, and considering only the cases in which the conjunction of all restrictions are met. A typical condition C might be the conjunction

$$C = \{(5.5 < x_2 < 7.5) \ \wedge \ (35 < x_4 < 41) \ \wedge \ (-3.2 < x_9 < 4.1)\}.$$

This conjunction specifies a "hyperslab" of state space, which we will refer to as the *conditional set* $\mathcal{X}_C$. $\mathcal{X}_C$ will typically contain only a fraction of the data in our complete set.

Analysis using conditional sets, which place us in regions of interest of the state space, allows us to measure local properties of a dynamical system. Once we are confined to a local region, we can perform a series of measurements on this subsystem to determine its attributes. This conditional set will be deemed good or "fit," if the probability distribution $P(y \mid \vec{x} \in \mathcal{X}_C)$ has a small standard deviation, i.e., the predictability of y for $\vec{x} \in \mathcal{X}_C$ is high. This distribution is estimated from the y values for which the corresponding $\vec{x} \in \mathcal{X}_C$. Predictability of $\mathcal{X}_C$ is quantified by the amount of information contained about y^t given that $\vec{x}^t$ was found in $\mathcal{X}_C$, i.e., $-\log \sigma$, where σ is the standard deviation of $P(y \mid \vec{x} \in \mathcal{X}_C)$. The term $-\log \sigma$ is the information with respect to a flat distribution. Since we have some information about the *a priori* distribution of y (estimated from the entire collection of y values), we must measure information with respect to this. This is given by $-\log \sigma/\sigma_o$, where σ_o is the width of the unconditional distribution $P(y)$.

---

[1] See any standard statistics text, for example, Pfaffenberger and Patterson.[11]

The estimation of predictability may be flawed by lack of data. For example, consider the case where $\mathcal{X}_C$ contains only one data point. Because the width of the distribution estimated from a single data point is zero, the information as measured goes to infinity. In general, low statistics will cause an overestimation of predictability. We will thus modify our measure of predictability, which is the estimated information in $P(y \mid \vec{x} \in \mathcal{X}_C)$,[7,12,13,5] to give us a *fitness function*

$$\mathcal{F}(\mathcal{C}) = -\log \frac{\sigma}{\sigma_o} - \frac{\alpha}{N}.$$

$\mathcal{F}(\mathcal{C})$ is the *fitness* of a conjunction of conditions. The first term in $\mathcal{F}(\mathcal{C})$ measures the amount of information contained in $P(y \mid \vec{x} \in \mathcal{X}_C)$. The second term ensures the statistics of the numbers used in calculating the first term are reliable. The number of points in our subset is denoted by N, and α is simply a constant, which can be adjusted to meet the goals of a particular application. A thorough analysis of the proper form of the second term can be found in Richards.[13]

2.2 SEARCH PROCESS

Each conjunction $\mathcal{C}$ places us inside a region $\mathcal{X}_C$. Our objective is to search through the space of all possible $\mathcal{C}$ to discover which ones correspond to the regions $\mathcal{X}_C$ that contain the points with the highest predictability.

This space of all conjunctions of conditions is in fact infinite. For this reason, it is impossible in general to be assured of ever finding the best $\mathcal{C}$, where best is defined as the set which maximizes the fitness $\mathcal{F}$. What we must accept instead is a reasonably thorough escapade through the space of conditions, with an algorithm trained to detect where optima of $\mathcal{F}$ may rest. Because of the assumed high complexity of this space, a simple hill-climbing approach will not be able to paint an accurate portrait of the fitness landscape. The search process must be able to obtain a global picture. This can be achieved by a *population* of searchers. But still this may not be enough. Throwing a handful of searchers onto a landscape and telling each to hill climb will result in a handful of local maxima, but the result will not contain any global information if the landscape is complex. To gain global information, the searchers must interact dynamically with each other and convey information about their neighborhood to the other members. They must then process this information and modify their behavior accordingly.

2.3 LEARNING PROCESS

We now introduce our genetic algorithm, which has been specifically designed for this task. The genetic algorithm works with a population, each member of which is described with a set of genes $(g_1, \ldots, g_n)$ which we will call a genome $\vec{g}$. Each genome is assigned a fitness, $\mathcal{F}$, which is used to distinguish good gene combinations from bad.

In this application, a gene g_i is equivalent to a condition on one of the d-dimensions, and a genome $\vec{g}$ is equivalent to the conjunction $\mathcal{C}$. The population of genomes $\mathcal{G} \equiv \{\vec{g}\}$ changes with time, and new members of the population are formed from the previous generation through selection and modification of the genes. The genomes are modified in ways that are analogous to modification of biological genomes; the maps that take old genomes to new genomes are called *genetic operators*.

We use four types of mutation operators and one type of crossover operator. These operators act upon the current members of $\mathcal{G}$, introducing new genomes to the population. These new genomes replace the less-fit members of the previous generation.

- Mutation A adds a new condition to the conjunction:

$$\{(3.2 < x_6 < 5.5) \wedge (0.2 < x_8 < 4.8)\} \longrightarrow$$
$$\{(3.2 < x_6 < 5.5) \wedge (0.2 < x_8 < 4.8) \wedge (3.4 < x_9 < 9.9)\}.$$

- Mutation B does just the opposite:

$$\{(3.2 < x_6 < 5.5) \wedge (0.2 < x_8 < 4.8) \wedge (3.4 < x_9 < 9.9)\} \longrightarrow$$
$$\{(3.2 < x_6 < 5.5) \wedge (3.4 < x_9 < 9.9)\}.$$

- Mutation C takes an existing condition, and symmetrically either broadens or shrinks the acceptable range:

$$\{(\mathbf{3.2} < x_6 < \mathbf{5.5}) \wedge (3.4 < x_9 < 7.9)\} \longrightarrow$$
$$\{(\mathbf{3.9} < x_6 < \mathbf{4.8}) \wedge (3.4 < x_9 < 7.9)\}.$$

- Mutation D either increases or decreases the average of the acceptable values of x_i while keeping the range constant:

$$\{(3.9 < x_6 < 4.8) \wedge (\mathbf{3.4} < x_9 < \mathbf{7.9})\} \longrightarrow$$
$$\{(3.9 < x_6 < 4.8) \wedge (\mathbf{4.7} < x_9 < \mathbf{9.2})\}.$$

- Crossover takes two existing members of $\mathcal{G}$ and exchanges approximately half of the d constraints on $\vec{x}$ to create two new genomes.

$$\left\{\begin{array}{l} \{(3.9 < x_6 < 4.8) \wedge (3.4 < x_7 < 7.9) \wedge (7.7 < x_9 < 8.2)\} \\ \{(0.3 < x_1 < 1.7) \wedge (7.7 < x_3 < 8.2)\} \end{array}\right\} \longrightarrow$$
$$\left\{\begin{array}{l} \{(7.7 < x_3 < 8.2) \wedge (3.4 < x_7 < 7.9)\} \\ \{(0.3 < x_1 < 1.7) \wedge (3.9 < x_6 < 4.8) \wedge (7.7 < x_9 < 8.2)\} \end{array}\right\}$$

The *genetic evolution* of a population $\mathcal{G}$ is given by the following sequence of events:

0. Initialize the population, typically with a set of random genomes $\mathcal{G}$.
1. Calculate the fitness of all genomes using the fitness function.
2. Order the population by fitness.
3. Refit the genomes as a function of the population, as described below.
4. Discard a fraction of the population with low fitness, and replace the deleted members with alterations of the remaining population, using the genetic operators.
5. A generation is completed. Go to step 1 and repeat.

The necessary interactions for a global search process are present in steps two, three, and four. Steps two and four together perform the task of removing hapless searches from the process and relocating them to (hopefully) more useful areas guided by the fitter members.

Step three maintains diversity among the population members and inhibits them from congregating onto the same peak. This is accomplished by penalizing less-fit $\mathcal{C}$'s if they contain any of the same points already present in a fitter member. This is justifiable in that if a vector $\vec{x}^{\,t}$ is present in a conditional set, we already have a prediction for it. We gain little predictability through its presence in the less-fit member. Note that through this process, the presence of genomes alters the structure of the fitness landscape. When many similar or equivalent genomes are present, they are in competition for the same territory, and it becomes less desirable to be of this "species." The fittest survive, and the others die out, analogous to the overcrowding phenomena occurring in nature.

3. RESULTS

For our subject dynamical system, we chose the Mackey-Glass equation, introduced as a model for blood flow, and subsequently popularized in the nonlinear field due to its richness in structure[6,3]

$$\frac{dx}{dt} = \frac{ax_\tau}{1 + x_\tau^c} - bx,$$

where $x(t)$ is the state variable, $x_\tau(t) = x(t - \tau)$, and a, b, c, and τ are constants.

The Mackey-Glass equation is in principle infinite dimensional, in the sense that a future value depends on a *continuum* of past values. It has been shown, however, that the dynamics display a rich variety of attractors whose dimension increases roughly linearly (with the exception of low-dimensional windows) with the delay time τ.[3] We will thus vary τ to create example time series with varied complexities.

For parameter values of $a = 0.2$, $b = 0.1$, and $c = 10$, the system becomes chaotic above $\tau = 16.8$, where the dimensionality is just over three. For large values

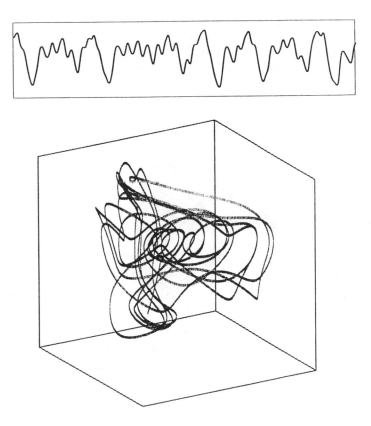

FIGURE 1 $\tau = 30$ seconds. Top: x^t versus time graph plotted for 800 seconds. Bottom: reconstruction plot x^t versus $x^{t-\tau}$ versus $x^{t-2\tau}$ for the same time period.

of τ the dimension of the attractor grows roughly monotonically with τ.[3] We have examined the attractor for τ values of 30 and 150, where the dimensionality has been calculated to be about 3.6 and 10 respectively.[3] The orbits are plotted as 3-dimensional reconstructions and as x versus t graphs in Figures 1 and 2.

The metric entropy for this system with $\tau = 30$, as computed by summing the positive Lyapunov exponents[10]

$$h_\mu = \sum \lambda_i^+,$$

is about 0.01. This means each time interval of approximately 100 seconds will on average double any initial uncertainty of the position of the trajectory in state space. With this in mind, we presented our learning algorithm with a series of past values, taking readings every second, and hid the evolution for 100, 200, and 400 seconds. The dimensionality of the state space was set to 50. (This corresponds to observing

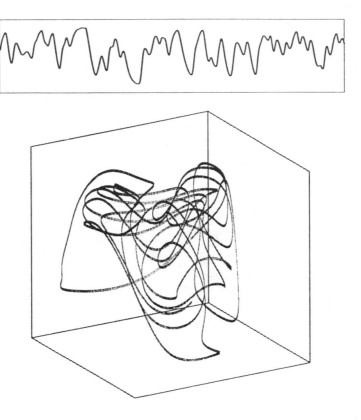

FIGURE 2 $\tau = 150$ seconds. Top: x^t versus time graph plotted for 1000 seconds. Bottom: reconstruction plot x^t versus $x^{t-\tau}$ versus $x^{t-2\tau}$ for the same time period.

the position of the trajectory once per second for 50 seconds.) The objective was to find conditions on the past values which would place the trajectory into a volume of state space for which the cumulative spreading rate along the direction $y^t = \xi^{t+t'}$ is minimal.

The results are shown in Figure 1. Plotted are all vectors $\vec{x}$ which are contained in $\mathcal{X}_C$, and their subsequent evolution up to and past the prediction time $t + t'$. A vertical line has been drawn at prediction time $t + t'$. The shaded regions correspond to the "unobservability window." Experimentally, this may correspond to the time necessary to make adjustments or take other actions based on observations. The area prior to this region is the time when we make our observations. A vertical line in this region corresponds to a condition on the past state of the trajectory. The conjunction of conditions taken together focus our attention to the subset of trajectories that fall within $\mathcal{X}_C$. The search process can be thought of as an automated approach for finding locally optimal reconstruction delay variables. While there is some divergence during the period of the unobservability window, the trajectories clearly collapse down to a very well determined region at $t + t'$.

While there is some divergence during the period of the unobservability window, the trajectories clearly collapse down to a very well determined region at $t + t'$.

The results for the trials with delay of $\tau = 150$ are shown in Figure 4. Here the intermediate divergence and subsequent collapse is much more striking. So much so that the authors originally rejected these results, concluding the algorithm had a loose enough rein to allow it to select trajectories that coincidentally happened to be near each other at the appointed time. It wasn't until after rigorous testing that the authors assented to their vigor. Appendix A contains the values for all of the conditional sets used in this paper. The reader is encouraged to examine the evolution of these sets, as it is truly a wonderfully unexpected phenomenon.

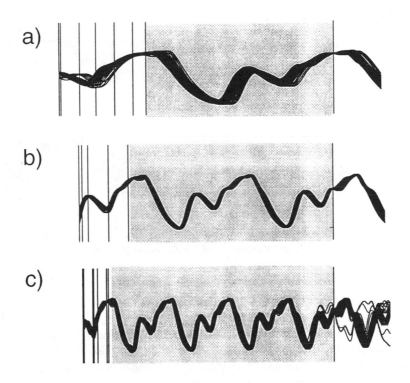

FIGURE 3   Figures a, b, and c show the evolution of $\vec{x}^{\,t} \in \mathcal{X}_C$ from Eq. (1) with $\tau = 30$, and $t' = 100, 200,$ and 400 seconds respectively. The vertical lines to the left of the gray region denote the location of the conditions in C. The vertical line to the right of the gray region denotes the prediction time in the future The gray area extends across the intermediate unobservable region.

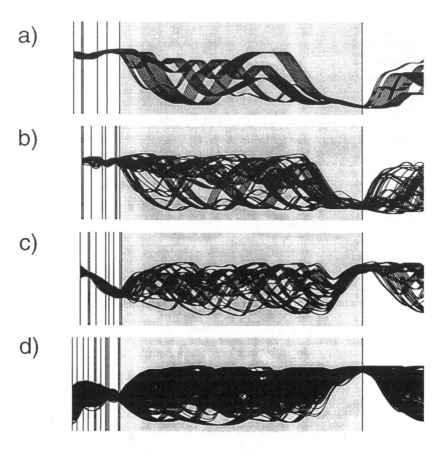

FIGURE 4 The four best $\mathcal{X}_C$'s and the evolution of points $\vec{x}^t \in \mathcal{X}_C$ for data from equation (1) with $\tau = 150$. (See previous caption for description of graphs.)

Farmer[3] has previously reported the surprising result that for the Mackey-Glass equation, as the delay time τ increases, the metric entropy of the system remains approximately constant, although the dimensionality increases linearly with time. Our results reflect this to some extent, in that the predictability for the two τ values were similar, although in the case $\tau = 150$, more conditions were needed to specify the past state of the system.

Since it is known that this system has a low-dimensional attractor, we were hopeful to construct a low-dimensional portrait of the system. Our learning algorithm did just that. We have found that the dimensionality depends upon location in state space; i.e., the attractor is multifractal. In some regions, two dimensions were enough to characterize the system, and in other regions, more were needed. The number of conditions necessary to maximize the predictability from a region places an upper bound on the local dimensionality of the system.

3.1 STATISTICAL SIGNIFICANCE

A question which must be asked is "How significant are the conditions which were learned?" Certainly if one were allowed to select a subset of events from a process— even a random process—a sufficiently large number of times, he will eventually come up with a set having an arbitrarily small standard deviation, even if there is no actual causality. How can we tell we haven't done the same? The simplest answer lies in performing what is called an "out of sample" test. This simply means than you train on one data set and test on another. If your predictive power is equivalent across data sets, one can be confident that the algorithm has indeed found some structure, and not simply a statistical fluke. This was done for the results presented here, and indeed they were robust. It should be noted that this process merely turns a first-order anomaly into a second-order one. Simply stated, if your "fluke" happens one time in a million, it becomes one in a trillion that it will pass this two-pronged test.

Unfortunately, we don't always have the capability of generating a second data set on which to train. This is often the case when the data is generated by experimental or naturally occurring physical systems. If this is the case, we can only do a comparison and get a relative measure of confidence. The procedure is to randomly shuffle the data set to create a new independently and identically distributed (IID) data set.[2] This data set will have the same overall distribution, but all correlations will have been removed. Applying the learning algorithm independently to a number of such data sets will yield a distribution of values of the maximum predictability found for each one. The predictability found in each of these cases is due *entirely* to statistical fluctuations discovered by the learning algorithm. If the conditions corresponding to the correctly ordered data are not significantly better than the average over the IID sets, you may conclude that you were simply "learning the noise." The drawback of using this method for evaluating the statistical significance of the learned patterns is simply that this process can be *extremely* computationally expensive.

4. CONCLUSION

We have introduced a genetic learning algorithm which is designed to differentiate regions of attractors. The criterion used for differentiation was the net information this region contained about the systems behavior at a particular future time. This information is measured using only the data in the region. For different distances into the future, different regions led to maximal predictability.

This algorithm searches through a high-dimensional (frequently more than 100) state space to find which past values best quantify the target value. From the numerous available dimensions, the algorithm chooses how many dimensions and which combinations have the most predictable dynamics for orbits passing through a region. This can be thought of as an automated approach to discovering the best *local* reconstruction coordinates. The number of dimensions necessary to determine the target value may be a measure of the local embedding dimension of the observed attractor.

One possible generalization of the current technique is to consider arbitrary conditions rather than the conjunctions used here. The inclusion of disjunctions would allow arbitrary subsets of $\mathcal{X}$ to be selected, instead of just hyperslabs. This modification would entail a new genetic code and new genetic operators, but otherwise the genetic algorithm would remain the same.

5. ACKNOWLEDGMENTS

The authors have benefited from conversations with W. Brock, J. Breeden, B. Nieswand, L. Rendell, and F. Richards. This research was supported in part by the National Science Foundation, grant number PHY 86-58062, and by a Shell Faculty Award.

APPENDIX A

Figure 3 conditional sets ($\tau = 30$):

a. One hundred time steps into the future:

$$
C = \begin{cases}
(x_{01} > 0.801) \wedge (x_{02} < 0.941) \wedge \\
(x_{12} > 0.667) \wedge (x_{21} < 0.945) \wedge \\
(x_{31} < 1.161) \wedge (x_{41} > 1.236) \wedge \\
(x_{48} > 1.273)
\end{cases} \longrightarrow y = 1.29 \pm 0.007.
$$

b. Two hundred time steps into the future:

$$
C = \begin{cases}
(x_{01} < 0.671) \wedge (x_{04} > 0.570) \wedge \\
(x_{10} < 0.968) \wedge (x_{30} > 0.769) \wedge \\
(x_{49} > 1.159)
\end{cases} \longrightarrow y = 1.06 \pm 0.008.
$$

c. Four hundred time steps into the future:

$$
C = \begin{cases}
(x_{01} < 0.927) \wedge (x_{03} > 0.687) \wedge \\
(x_{19} < 0.747) \wedge (x_{21} < 1.151) \wedge \\
(x_{28} > 0.825) \wedge (x_{44} > 1.257) \wedge \\
(x_{47} > 1.272)
\end{cases} \longrightarrow y = 1.06 \pm 0.024.
$$

Figure 4 conditional sets ($\tau = 150$):

a. Best C for $t' = 150$:

$$
C = \begin{cases}
(x_{20} > 1.122) \wedge (x_{25} < 1.330) \wedge \\
(x_{26} > 1.168) \wedge (x_{35} < 1.342) \wedge \\
(x_{41} > 1.304) \wedge (x_{49} > 1.262)
\end{cases} \longrightarrow y = 0.18 \pm 0.014.
$$

b. Second best C for $t' = 150$:

$$
C = \begin{cases}
(x_{25} < 1.330) \wedge (x_{26} > 1.177) \wedge \\
(x_{31} > 1.127) \wedge (x_{38} > 1.156) \wedge \\
(x_{40} < 1.256) \wedge (x_{46} > 1.194) \wedge \\
(x_{47} < 1.311) \wedge (x_{49} > 1.070)
\end{cases} \longrightarrow y = 0.27 \pm 0.019.
$$

c.   Third best $\mathcal{C}$ for $t' = 150$:

$$\mathcal{C} = \begin{cases} (x_{24} > 0.992) \wedge (x_{29} < 1.150) \wedge \\ (x_{30} > 1.020) \wedge (x_{34} < 1.090) \wedge \\ (x_{40} < 0.951) \wedge (x_{42} > 0.599) \wedge \\ (x_{45} > 0.591) \wedge (x_{49} < 0.763) \wedge \\ (x_{50} > 0.576) \end{cases} \longrightarrow y = 1.22 \pm 0.024.$$

d.   Fourth best $\mathcal{C}$ for $= 150$:

$$\mathcal{C} = \begin{cases} (x_{19} < 0.967) \wedge (x_{22} < 1.049) \wedge \\ (x_{26} > 0.487) \wedge (x_{29} < 1.066) \wedge \\ (x_{33} > 0.416) \wedge (x_{34} < 1.008) \wedge \\ (x_{37} < 1.331) \wedge (x_{40} < 0.941) \wedge \\ (x_{41} > 0.654) \wedge (x_{42} > 0.262) \wedge \\ (x_{48} > 0.639) \wedge (x_{49} < 0.814) \end{cases} \longrightarrow y = 1.34 \pm 0.034.$$

It is possible that as a conditional set evolves, the addition of new dimensions causes a previously useful condition to become superfluous. Indeed, the authors suspect this to be the case for many of the examples presented here. If one wished to use the number of conditions contained in a set as a measure of local dimension, an "Occom's Razor" should first be applied to the set to remove any vestigial conditions.

REFERENCES

1. Brock, W. "Simple Technical Trading Rules and the Stochastic Properties of Stock Returns." Preprint, 1990.
2. Farmer, J. Doyne. "Chaotic Attractors of an Infinite-Dimensional Dynamical System." *Physica* **4D** (1981).
3. Holland, J. H. *Adaptation in Natural and Artificial Systems: An Introductory Analysis with Applications to Biology, Control, and Artificial Intelligence.* Ann Arbor: Michigan University Press, 1975.
4. Li, W. "Mutual Information Functions Versus Correlation Functions." *J. Stat. Phys.* **60(5/6)** (1990): 823–837.
5. Mackey, M. C., and L. Glass. *Science* **197** (1977): 287.
6. Meyer, T. P., F. C. Richards, and N. H. Packard. "A Learning Algorithm for the Analysis of Complex Spatial Data." *Phys. Rev. Lett.* **63** (1989).
7. Packard, N. J. Crutchfield, D. Farmer, R. Shaw. "Geometry from a Time Series." *Phys. Rev. Lett.* **45** (1980): 712.
8. Packard, Norman H. "A Genetic Learning Algorithm for the Analysis of Complex Data." *Complex Systems* October (1990).
9. Pesin, Y. B. *Uspekhi Mat. Nauk* **32** (1977): 55.
10. Pfaffenberger, R., and J. Patterson. *Statistical Methods.* Irwin, 1988.
11. Richards, F. C., T. P. Meyer, and N. H. Packard. "Extracting Cellular Automaton Rules Directly from Experimental Data." *Physica* **D 45** (1990): 189.
12. Richards, Fred C. "Learning Two-Dimensional Spatial Dynamics From Experimental Data." Ph.D. thesis, University of Illinois at Urbana-Champaign, 1991.
13. Shaw, R. "Strange Attractors, Chaotic Behavior, and Information Flow." *Z. Naturforschung* **36a** (1981): 80.

Martin Casdagli
Santa Fe Institute, 1660 Old Pecos Trail, Santa Fe, New Mexico 87501

A Dynamical Systems Approach to Modeling Input-Output Systems

Motivated by practical applications, we generalize theoretical results on the nonlinear modeling of autonomous dynamical systems to input-output systems. The inputs driving the system are assumed to be observed, as well as the outputs. The underlying dynamics coupling inputs to outputs is assumed to be deterministic. We give a definition of chaos for input-output systems, and develop a theoretical framework for state-space reconstruction and modeling. Most of the results for autonomous deterministic systems are found to generalize to input-output systems with some modifications, even if the inputs are stochastic.

1. INTRODUCTION

There has been much recent interest in the nonlinear modeling and prediction of time-series data, as evidenced by this conference proceedings volume. For approaches motivated by deterministic chaos, see Casdagli,[2] Crutchfield,[5] Farmer,[7,8] and references therein. For approaches motivated by nonlinear stochastic models see Tong.[15] In both these approaches, it is assumed that a time series $x(t)$ is obtained from observations of an autonomous dynamical system, possibly perturbed by unobserved forces or noise. By contrast, when modeling *input-output systems*, in addition to an observed *output* time series $x(t)$, there is also available an observed *input* time series $u(t)$; see Figure 1.

Modeling input-output systems is appropriate in many applications. For example, in vibration testing, a mechanical device may be subjected to a controlled random forcing to test its robustness in a simulated environment. Also, in scientific experiments, one may be interested in the response of a system to various forms of stimulus. There are many other disciplines, for example, meteorology and economics, in which pairs of input-output time series from input-output systems are available for analysis.

In this paper we will develop a theory for the nonlinear modeling of input-output systems based on deterministic dynamics. This deterministic approach assumes that the input-output time series arises from a finite-dimensional dynamical system

$$\frac{ds}{dt} = f(s(t), u(t)) \tag{1}$$
$$x(t) = h(s(t)), \tag{2}$$

where $s(t) \in \Re^d$ denotes a d-dimensional state, and for simplicity we take $u(t) \in \Re$ to be a scalar input, $f : \Re^d \times \Re \to \Re^d$ to be a smooth flow, and $h : \Re^d \to \Re$ to be a scalar measurement function. It is then natural to attempt to model and forecast the behavior of the input-output system with a nonlinear deterministic model of the form

$$x(t) = P(x(t - \tau), x(t - 2\tau), .., x(t - m\tau), u(t), u(t - \tau), .., u(t - (l - 1)\tau)), \tag{3}$$

where P is a nonlinear function fitted to the input-output time series data. This deterministic approach has been applied by Hunter to a variety of practical examples.[11]

Of course, the deterministic system (1) and (2) is only an approximation to reality. First, it is assumed that only d independent modes of the system are excited to a significant amplitude, where in practice d is reasonably small. Second, it is assumed that effects due to unobserved sources of noise and measurement errors are small enough to be ignored. If these assumptions are violated, a stochastic approach to nonlinear modeling may be more appropriate, and it is natural to include noise terms in the above equations. For a stochastic approach to the nonlinear modeling of input-output systems, see Billings et al.[1]

FIGURE 1 Conceptual model of a single input-single output system.

Under the above deterministic assumptions, we are interested in the following theoretical questions. First, how should chaos be defined and quantified for the system (1)? If the input time series $u(t)$ is periodic, this question trivially reduces to that of autonomous systems by considering the appropriate time-T Poincaré map. However, we are mostly interested in the case where $u(t)$ is a random time series. If $u(t)$ is random, then $x(t)$ is random, so neither time series *by itself* is chaotic. Hence, this must be a question about the structure of the *pair* of time series $u(t), x(t)$. We will address this question in section 2. Second, if we wish to construct a nonlinear deterministic model of the form (3), how many lags m and l should be chosen for a system of dimension d? In the case of autonomous systems, this question has been addressed by Takens[14]; see also Sauer et al.[13] We are also interested in how accurate such a model is likely to be as a function of the length of the time series available to construct it, and the dimension d of the underlying dynamical system. In the case of autonomous systems, this question has been addressed by Farmer and Sidorowich[7]; see also Casdagli.[2] We will investigate these questions for input-output systems in section 3. Finally, we summarize the conclusions in section 4.

2. CHAOS IN INPUT-OUTPUT SYSTEMS

In this section we give a definition of chaos for input-output systems. We also investigate the usefulness of this definition in quantifying predictability for a numerical example. For simplicity we will assume that time is discrete, so that the system (1) and (2) is replaced by

$$s_{n+1} = f(s_n, u_n) \qquad (4)$$
$$x_{n+1} = h(s_{n+1}). \qquad (5)$$

All of the results in this section generalize naturally to continuous-time input-output systems by using results about Lyapunov exponents in continuous-time autonomous systems.[6,16]

2.1 DEFINITION OF THE LARGEST LIAPUNOV EXPONENT

Suppose the system (4) is initialized at two slightly different states, and in both cases is subjected to the *same* sequence of inputs u_n. Then if the system states diverge exponentially in time, we say the input-output system is chaotic, and the rate of divergence is given by the largest Liapunov exponent. The largest Liapunov exponent quantifies the degree to which the system is predictable in the long term, *assuming that the input time series is always observed*. This observability assumption is satisfied in many of the applications mentioned in the introduction. For example, in vibration testing, one may only be able to observe the state of the system at rare intervals in the past due to measurement problems, and desire to predict the present state of the system given a sequence of inputs to the system. The assumption of having the input sequence available is also relevant for problems of reducing noise on output sequences observed in the past. Note that if the input time series is random and *unobserved*, the system is unpredictable even in the short term. We now make the above notions more precise.

We will only be concerned with the largest Liapunov exponent, which is defined as follows. Let $Df(s, u)$ denote the derivative of f at s with u held constant. Let Df^T denote the matrix product

$$Df^T(s_0, u_0, .., u_{T-1}) = \prod_{i=0}^{T-1} Df(s_i, u_i) \tag{6}$$

where the s_i are generated from Eq. (4) starting from a given initial state s_0. Then the largest Liapunov exponent λ_1 is defined by Eq. (7), where ds is an arbitrary initial vector, and $\| \cdot \|$ denotes the Euclidean norm.

$$\lambda_1 = \lim_{T \to \infty} \frac{1}{T} \log(\| Df^T ds \| / \| ds \|) \tag{7}$$

If $\lambda_1 > 0$, we say the system is *chaotic*.

The above definition of the Lyapunov exponent λ_1 at first sight depends on the initial state s_0, the sequence of inputs $u_0, u_1, \ldots$, and the tangent vector ds. However, suppose that the sequence of inputs is drawn from a realization of a stationary random or deterministic process. Then in the case of random inputs, by multiplicative ergodic theorems (see Eckmann and Ruelle[6]), the limit Eq. (7) exists with probability one, and depends only on the ergodic invariant measure to which the initial state s_0 is attracted. This invariant measure is often unique, for example in the case of Gaussian inputs. In the case of deterministic inputs, although there are no general theorems, it is observed numerically that the limit Eq. (7) exists, and depends only on the basin of attraction in which s_0 lies.

The above definition coincides with the definition of Liapunov exponents for randomly driven dynamical systems with unobserved inputs.[6] In this case, the input is assumed to be small, and models noise perturbations. However, if the inputs are observed, the largest Liapunov exponent may be used to describe the divergence of trajectories *even for large amplitude inputs* as follows. Let s_0 and s_0' denote two close initial conditions, subjected to the same sequence $u_0, \ldots, u_{T-1}$ of inputs. Then the divergence of trajectories will be described for moderate T by

$$s_T - s_T' \approx Df^T(s_0, u_0, \ldots, u_{T-1})(s_0 - s_0'). \tag{8}$$

Hence, using Eq. (7) we obtain

$$\| s_T - s_T' \| \approx e^{\lambda_1 T} \| s_0 - s_0' \| . \tag{9}$$

We now illustrate the above ideas with the randomly driven Ikeda map where f is taken to be

$$f(x, y, u) = 1 + a(x \cos t - y \sin t) + u, a(x \sin t + y \cos t), \tag{10}$$

where $t = 0.4 - 6.0/(1 + x^2 + y^2)$, $a = 0.7$, and the inputs u_n are independently identically distributed (IID) Gaussians with variance η^2. Figure 2 illustrates the dependence of the Liapunov exponent λ_1 on the noise level η. The Liapunov exponent was computed numerically using 10^5 iterates for each value of η, with the QR algorithm described in Eckmann[6] to avoid overflow. Observe that a smooth transition from chaotic to non-chaotic behavior occurs at $\eta \approx 0.95$. Unlike autonomous systems, it is impossible to locate this transition by inspection of the invariant measure, which in randomly driven systems is always smooth.

In the above example, even if η lies in the non-chaotic region, the time series s_i and u_i appear irregular. This is illustrated in Figure 3 for the non-chaotic case $\eta = 1.2$. Figure 3(a) illustrates a realization $u_1, \ldots, u_{100}$ of a time series of IID Gaussian inputs. Figure 3(b) illustrates how the invariant measure for the input-output system is filled out by the iterates $s_i = (x_i, y_i)$, for $i = 1, \ldots, 10000$. Also shown in Figure 3(b) is the fractal invariant measure in the autonomous case $\eta = 0$. The random inputs perturb the system so that the states s_i fill out a more disperse, smooth invariant measure. Figure 3(c) illustrates the output time series $x_1, \ldots, x_{100}$. The output time series considered by itself must be modeled by a stochastic process. However, if the input time series is also observed, the output time series may be modeled deterministically using an input output model. Although the output time series appears irregular, it is in fact the output of a non-chaotic, deterministic input-output system.

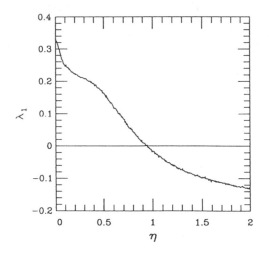

FIGURE 2   Dependence of the Liapunov exponent λ_1 on the noise level η for the randomly driven Ikeda map.

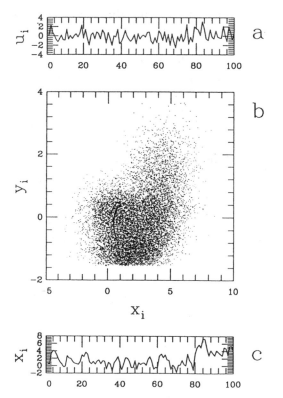

FIGURE 3   Input-output time series for a non-chaotic randomly driven Ikeda map in the case $\eta = 1.2$. (a) Input time series u_i. (b) Invariant measures for $\eta = 1.2$ and $\eta = 0$. (c) Output time series x_i.

2.2 FLUCTUATIONS IN DIVERGENCE RATES

It is well known in autonomous dynamical systems that the divergence of trajectories fluctuates to a large extent depending on the regions of the state space visited, so that Eq. (9) is only a crude approximation; for example, see Nese.[12] An investigation of the limitations of Eq. (9) is important, as it is the divergence of trajectories which directly limits predictability. A more accurate approximation to the divergence of trajectories may be obtained as follows. Suppose that the initial state s_0 is known to lie in a ball $B_\epsilon(s_0')$ of radius ϵ about the state s_0'. Then s_T must lie in the region $f^T(B_\epsilon(s_0'))$. Now assume that T is small enough so that f^T may be approximated by its linearization $Df^T(s_0', u_0, \ldots, u_{T-1})$. Then the region $f^T(B_\epsilon(s_0'))$ is approximately an ellipsoid, with a semi-major axis of length $\| Df^T \| \epsilon$, where $\| M \|$ denotes the largest singular value[1] of the matrix M. In this way we obtain the estimate (11) of the rate of divergence of trajectories, which depends on the specific values of $s_0', u_0, \ldots, u_{T-1}$.

$$\| s_T - s_T' \| \approx \| Df^T(s_0', u_0, \ldots, u_{T-1}) \| \epsilon \tag{11}$$

We now illustrate this numerically with the randomly driven Ikeda map. Figure 4 corresponds to a chaotic case with $\eta = 0.5$ and $\lambda_1 \approx 0.164$. Three different pairs of initial conditions s_0, s_0' were chosen with $\| s_0 - s_0' \| = \epsilon = 1.4 \times 10^{-4}$, and subjected to three different sequences of random inputs. The solid lines are plots of $\| s_T - s_T' \|$

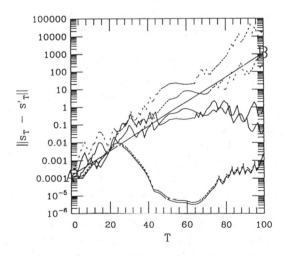

FIGURE 4 Fluctuations in divergence rates for a chaotic randomly driven Ikeda map, with noise level $\eta = 0.5$. Direct computations of the divergence of 3 pairs of trajectories are represented by solid lines. The dashed lines represent the approximation of Eq. (11). The line AB represents the approximation of Eq. (9).

[1] The largest singular value of a matrix M is equal to the square root of the largest eigenvalue of the matrix $M^\dagger M$, where $\dagger$ denotes the transpose of a matrix.

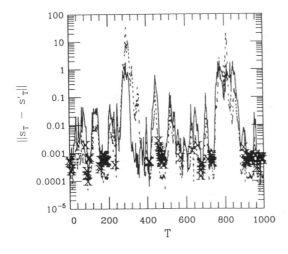

FIGURE 5 Fluctuations in divergence rates for a non-chaotic randomly driven Ikeda map, with noise level $\eta = 1.2$, and dynamic noise added.

against T. The dashed lines are plots of $\| Df^T (s'_0, u_0, \ldots, u_{T-1}) \| \epsilon$ against T. The line AB represents the anticipated rate of divergence corresponding to a largest Liapunov exponent $\lambda_1 \approx 0.164$. Observe that there are considerable fluctuations between the divergence of the three pairs. In fact, one of the pairs appears to be following a non-chaotic path over the times T considered. Observe that the dashed lines give an excellent approximation to the divergence of trajectories, but that the line AB only gives a crude approximation to the divergence of trajectories. This shows that the Liapunov exponent λ_1 only gives an average[2] rate of divergence of trajectories; a more accurate analysis requires computations of the state-dependent matrices $Df^T (s'_0, u_0, \ldots, u_{T-1})$.

Finally, we consider a non-chaotic case with $\eta = 1.2$. Since the largest Liapunov exponent is negative, $\| s_T - s'_T \|$ converges to zero as T increases with probability one (but note that s_T does not converge to a fixed point or periodic orbit, as would be the case for an autonomous system). To make the problem more complicated, unobserved "dynamic" noise of standard deviation 10^{-4} was added to the system during the iterations, setting a noise floor. Figure 5 shows the divergence of one pair of trajectories, with the same conventions in Figure 4, except that the matrix Df^T is reset to the identity whenever its largest singular value is less than one. This convention simulates the noise floor, and is marked with an "X" in Figure 5. Observe that although the system is non-chaotic, there are several fluctuations where $\| s_T - s'_T \| \approx 1$, so that the system is essentially unpredictable at these times T, even when the noise level is only of order 10^{-4}. However, the times when the system becomes unpredictable may be anticipated from computations of Df^T, as

---

[2] We found numerically that λ_1 approximately describes the geometric average of the divergence of trajectories at short times T. In the case of one-dimensional dynamical systems, this is a consequence of the additive ergodic theorem, but does not hold exactly in higher-dimensional dynamical systems.

illustrated by the dashed lines. Note that this phenomenon cannot occur in non-chaotic autonomous systems, as it requires a non-trivial invariant measure.

The ideas of this section have been applied by Hunter in the case of continuous time randomly driven input-output systems. Insights were gained into phenomena observed in modeling and forecasting input-output time series from a randomly driven Duffing equation and a randomly driven beam moving in a double well potential.[11]

3. STATE-SPACE RECONSTRUCTION AND MODELING

The results of section 2 show that the input-output systems (1) and (4) are predictable in the short term, assuming that both the states $s(t)$ and the inputs $u(t)$ are observed over time with reasonable accuracy. In practice, prediction would then be accomplished either from knowledge of the dynamics f, or by fitting a non-parametric model to Eq. (1) or (4), as is done in the autonomous case.[2,5,7] If the system is non-chaotic, long-term prediction is then possible by iterating the model, though there may be occasional intervals of unpredictability, as shown in Figure 5.

In many applications, it is more likely that a scalar time series of outputs $x(t)$ is observed, as in Eqs. (2) and (5). This complicates the problem, and just as in the autonomous case, a multidimensional state space must be reconstructed. A natural choice is to try a reconstruction of the form (3) for modeling. Indeed, just such a choice has been used in the control theory literature. For example in the stochastic modeling approach of Billings et al.,[1] P is chosen to be a polynomial, and m and l are chosen by trial and error. By contrast, in a deterministic modeling approach, we recommend using local approximation to construct P, and have a theory of how m and l depend on the state-space dimension d. In this section we investigate how many lags m and l should be chosen for a system of dimension d. We then investigate how accurate the model (3) is likely to be as a function of the length of the time series available to construct it. We will assume that time is discrete, until subsection 3.3.

3.1 STATE-SPACE RECONSTRUCTION

The idea behind state-space reconstruction is that a long enough sequence of observations coded in the *delay vector*

$$v_n = x_n, \ldots, x_{n-m+1}, u_n, \ldots, u_{n-l+1} \tag{12}$$

should uniquely and smoothly determine the unobserved d-dimensional state s_n of the system (4) and (5). If this can be achieved, it follows that there exists a smooth function $P : \Re^{m+l} \to \Re$ satisfying Eq. (13) for all v_n, and it is sensible to estimate P from time-series data.

$$x_{n+1} = P(v_n) \tag{13}$$

In the case of autonomous dynamical systems, it is a consequence of a theorem of Takens that if $m > 2d$, then such a smooth function P exists, for a generic set of functions f and h defining the underlying dynamics and measurement function.[14] This result has recently been strengthened by Sauer et al., so that "generically" can be replaced by "prevalent" (which essentially means full measure), and the state-space dimension d can be replaced by the attractor's box counting dimension; investigations are also made into what happens when $m \leq 2d$ and when other more general forms of reconstruction are used.[13]

In the case of input-output systems, we will argue below that, subject to genericity conditions on f and h, then if $m > 2d$ and $l > 2d$, a globally smooth function P exists satisfying Eq. (13) for almost all input sequences. Moreover, if $m = l = d+1$, an almost everywhere smooth function P exists satisfying Eq. (13). As a corollary, it follows that for non-chaotic input-output systems with $\lambda_1 < 0$, the input time series alone can be used to determine the future outputs arbitrarily accurately. This result is obtained by iterating the model (13), and observing that the outputs x_i for $i > n + \lambda_1^{-1} \log \epsilon$ become independent of $x_n, \ldots, x_{n-m+1}$ to accuracy proportional to ϵ, and thus essentially depend only on the input sequence. By contrast, the output time series alone can be used to determine the future outputs only if the inputs come from a deterministic process.

Rather than giving a rigorous mathematical treatment, we will give a heuristic argument here. We believe that this argument may be made rigorous by straightforward generalizations of the theorems for the autonomous case. Define the map $\Phi : \Re^d \times \Re^{m-1} \to \Re^m$ by

$$\Phi(s, u_{n-1}, \ldots, u_{n-m+1}) =$$
$$\ldots, h\big(f(f(f(s, u_{n-m+2}), u_{n-m+1}))\big), h\big(f(s, u_{n-m+1})\big), h(s) \tag{14}$$

The map Φ is well defined and smooth. To obtain a good state-space reconstruction, m must be chosen large enough so that there is a unique solution for s to the nonlinear equation

$$\Phi(s, u_{n-1}, \ldots, u_{n-m+1}) = x_n, \ldots, x_{n-m+1} \tag{15}$$

in terms of the x_i and u_i, which depends smoothly on the x_i and u_i. The solution for s identifies the unobserved state s_{n-m+1}. If this can be achieved for all integers n, it follows that a smooth function P exists satisfying Eq. (13) by substituting $s = s_{n-m+1}$ into Eq. (4) and iterating.

It is clear that for Eq. (15) to have a unique solution for s, we must in general have $m > d$. This is because if $m = d$, then Eq. (15) constitutes d simultaneous nonlinear equations for the d unknown components of s, and in general there are several different solutions for s. To break this degeneracy, suppose we take $m = d+1$. Then, for generic f and h, the extra simultaneous equation for s is expected to pick out the unique solution, unless the solution lies on a "bad" subset Σ^{d-1} of $\Re^d$ of dimension $d-1$. The location of Σ^{d-1} depends on the inputs u_i as well as the functions f and h. The situation is illustrated geometrically for $d = 2$ in Figure 6. As m is increased by one, the dimension of the bad set Σ is generically decreased by one, until when $m > 2d$ there is no bad set at all.

So far the argument has paralleled that for the autonomous case. However, there is an additional complication that arises with the inputs when $m > 2d$. In the case that $m = 2d + 1$, it is expected that if the inputs lie on a "bad" subset U of $\Re^{2d}$ of dimension $2d - 1$, then a bad subset Σ of dimension zero will be induced on $\Re^d$ so that Eq. (15) will not have a unique solution for s. Since the set U generically has measure zero, we will ignore it. If m is increased further, the bad sets U will still have dimension $2d - 1$ in $\Re^{m-1}$, and do not in general disappear.

The above argument has concentrated on the uniqueness of the solution s to Eq. (15). To address the issue of the smooth dependence of the solution s on the x_i and u_i requires an application of the implicit function theorem. This says that there is smooth dependence if the matrix $D\Phi$ for the derivative of Φ with respect to s has full rank at the solution for s. The matrix $D\Phi$ generically has full rank if m is chosen large enough. It turns out that the conditions on m derived above to ensure uniqueness of solutions are generically strong enough to also ensure full rank, hence smoothness. The arguments parallel those for the autonomous case, which can be found clearly expressed in Sauer et al.[13]

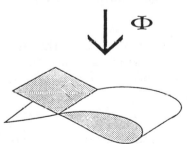

FIGURE 6   An illustration of the map Φ with the inputs u_i held fixed, for $d = 2$ and $m = 3$. There is a one-dimensional "bad" subset Σ of the state space $\Re^2$, which gets mapped by Φ to the one-dimensional self-intersection shown. If the vector of outputs $x_n, \ldots, x_{n-m+1}$ lies on the self intersection $\Phi(\Sigma)$, then there is not a unique solution for s to Eq. (15).

In the case of autonomous dynamical systems, a theory of state-space reconstruction has been developed which applies when there are low levels of observational noise on the scalar output time series.[3,4] This theory can be generalized to multivariate time series, and input-output time series.[9] The theory quantifies the "goodness" of a state-space reconstruction in terms of formulae involving the noise level, the measurement function, the information flow between variables, and the state-space reconstruction technique used. By studying a variety of examples, insights can be gained into the limitations imposed on a reconstruction technique by observational noise, and how to minimize such effects.

3.2 MODELING AND SCALING LAWS

The results of subsection 3.1 show how a good choice of lags m and l for the model P of (13) depends on the underlying state-space dimension d. We now consider the problem of estimating P non-parametrically from an input-output time series of length N, when the functions f and h are unknown. Suppose that a local approximation technique is used to construct an estimate $\hat{P}_N$ for P. We measure the accuracy of the model $\hat{P}_N$ by the RMS *prediction error* $\sigma(\hat{P}_N)$ defined by

$$\sigma(\hat{P}_N) = \left(\langle (x_{n+1} - \hat{P}_N(v_n))^2 \rangle_n \right)^{1/2}, \tag{16}$$

where the RMS average is taken over a typical orbit of vectors v_n in the reconstructed state space.

One of the advantages of local approximation is that a *scaling law* can be derived which shows how the predictor error $\sigma(\hat{P}_N)$ scales with N. In the autonomous case, it was conjectured by Farmer and Sidorowich that the prediction error scales with N according to

$$\sigma(\hat{P}_N) = O(N^{-q/D}), \quad N \to \infty \tag{17}$$

where D is the information dimension[3] of the attractor, and q is the order of the local approximation technique ($q = 1$ for step function approximation, and $q = 2$ for local linear approximation).[7]

In the case of input-output systems, a scaling law such as Eq. (17) is also expected to hold, where D is the dimension of the subspace of $\Re^{m+l}$ that the reconstructed states v_n are constrained to lie in. In the case of randomly driven input-output systems, D is given by Eq. (18), where we have taken $l = m \geq d+1$.

$$D = d + m. \tag{18}$$

This can be seen as follows. First, because of randomness, there are no constraints on the m inputs. Second, for any fixed values of the inputs, the map Φ of Eq. (14) constrains the m outputs to lie on a d-dimensional subspace of $\Re^m$. See Figure 6 for

[3] It can be shown that the RMS average must in general be replaced by a geometric average for this scaling law to be true for multifractal attractors.[4]

an illustration in the case $d = 2$. Moreover, if the system is randomly driven, then the outputs will not collapse onto an attractor of dimension less than d. Equation (18) follows by combining these two observations.

It follows from Eq. (18) that the lower m can be held subject to retaining continuity properties, the more accurate the model $\hat{P}_N$ will be, according to the scaling law (17). In subsection 3.1 we showed that if $m = d+1$, then a smooth map P exists almost everywhere. If local approximation is used to construct $\hat{P}_N$, then these discontinuities are rarely expected to have an effect. The effects of the discontinuities can only be reduced at the expense of increasing m. In autonomous systems, this is not such a problem, since the dimension D of the reconstructed attractor is independent of the number of lags m if $m > D$. By contrast, for randomly driven input-output systems, we recommend taking $l = m = d+1$, so that $D = 2d+1$. This result implies that models for randomly driven input-output systems are expected to be much less accurate than those for autonomous systems, since for autonomous systems we have $D \le d$.

We now illustrate the above ideas with the randomly driven Ikeda map, taking $\eta = 0.5$, the measurement function $h(x, y) = x$, and $l = m = 3$, so that $D = 5$. We computed predictive models $\hat{P}_N$ using both step function approximation (with the nearest neighbor used to make the prediction) and local linear approximation (using the $2(2m+1) = 14$ nearest neighbors to fit a locally affine map). We approximated the predictor error $\sigma(\hat{P}_N)$ using

$$\sigma^2(\hat{P}_N) \approx \frac{1}{100} \sum_{n=N}^{N+99} (x_{n+1} - \hat{P}_N(v_n))^2 . \tag{19}$$

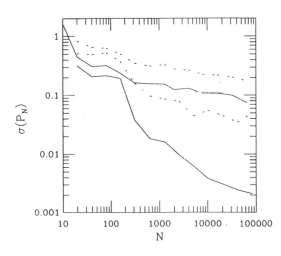

FIGURE 7   Dependence of the predictor error on the number of data points N, for a randomly driven Ikeda map with $\eta = 0.5$, using both step function and local linear approximation. The two dashed lines were obtained using the RMS average (19). The lower curve is for local linear approximation. The two solid lines were obtained using a geometric average. The lower curve is for local linear approximation. The geometric averages obey the scaling law $N^{-q/5}$ reasonably accurately, where $q = 1$ for step function approximation and $q = 2$ for local linear approximation.

This approximation amounts to an out of sample RMS measure of the accuracy of the one-step predictive model $\hat{P}_N$. The results are shown in Figure 7. The dashed lines represent the RMS average (19), and the solid lines represent a geometric average. The anticipated scaling law (17) is observed to be reasonably well obeyed for the geometric average, which is more robust to occasional bad predictions.

3.3 CONTINUOUS TIME

In the case of continuous time systems, we assume that time-series data u_i and x_i are obtained by sampling every τ units of time. In the case of autonomous systems, it is well known that results for discrete time immediately apply to continuous time by considering the time-τ map f^τ, which is defined to be the map evolving a state $s(t)$ to $s(t + \tau)$. By uniqueness theorems for solutions to differential equations, the map f^τ is a well defined and smooth map on the state space $\Re^d$. The properties of the sampled time series are then completely determined by f^τ.

In the case of input-output systems, the above approach is complicated by the input term $u(t)$. A time-τ map f^τ can still be defined by

$$s((n+1)\tau) = f^\tau \left(s(n\tau), u[n\tau, (n+1)\tau] \right). \tag{20}$$

The notation of Eq. (20) is intended to convey that the map f^τ depends on the values of the inputs $u(t)$ in the *entire interval* $n\tau \le t \le (n+1)\tau$. In principle the arguments of subsection 3.1 can be applied to the map f^τ, except that the delay vector v_n must be defined in terms of the inputs $u(t)$ by

$$v_n = x_n, \ldots, x_{n-m+1}, u[(n+1)\tau, (n-l+1)\tau]. \tag{21}$$

However, in practice the input values $u(t)$ are only observed at discrete intervals of time. If the inputs arise from a deterministic continuous time process, then a knowledge of the sampled input time series $u_n, \ldots, u_{n-l+1}$ is often enough to uniquely constrain $u(t)$ over the entire interval $(n+1)\tau \le t \le (n-l+1)\tau$. The results of subsections 3.1 and 3.2 then go through unchanged. On the other hand, if the inputs arise from a stochastic continuous time process, information on $u(t)$ is in general lost. The results of Sections 3.1 and 3.2 then only hold approximately. The approximation will be good if the sequence $u_n, \ldots, u_{n-l+1}$ constrains $u(t)$ over the entire interval $(n+1)\tau \le t \le (n-l+1)\tau$ to a narrow range of possibilities. Whether this will be the case depends on the nature of the stochastic process and the length of the sampling interval τ.

4. CONCLUSIONS

We now summarize the results of the above theoretical investigations, and comment on the implications for the deterministic modeling of input-output systems arising in applications.

First, we have given a definition of chaos for input-output systems. This definition was motivated by the fact that if the input sequence driving the dynamical system is observed, then the output sequence, which considered *by itself* may be stochastic, can become deterministic and even non-chaotic when considered *together with* the input sequence. If the input-output system is non-chaotic, then the outputs can be predicted far into the future, assuming all the inputs, and the underlying dynamics, are known. Unlike the autonomous case, non-chaotic input-output systems can be hard to distinguish from chaotic input-output systems, since the output time series can in either case appear irregular.

Second, we have developed a theoretical framework for the deterministic modeling of input-output systems when only a scalar output time series is available in addition to the input time series, and the underlying dynamics is unknown. This framework is based on a generalization of Takens' embedding theorem from autonomous to input-output systems. We have also derived scaling laws describing how the accuracy of local models depends on the length of the input-output time series available, and the dimension of the underlying dynamics. For a discrete-time randomly driven input-output system of state-space dimension d, we recommend using a local linear model with $d + 1$ lags on both input and output, which has an accuracy of order $N^{-2/(2d+1)}$ for making short-term predictions. This result implies that the modeling of randomly driven input-output systems is expected to be much less accurate than for autonomous systems. Additional problems have been identified for continuous-time randomly driven input-output systems which do not occur for autonomous systems.

We have tested the above theoretical results on a simple example: the randomly driven Ikeda map. Applications to real data from vibration testing and meteorology have been made by Hunter.[11] In these applications some more sophisticated state-space reconstruction techniques were used to reduce the effects of observational noise. The reconstruction techniques used included generalizations of the local principal components technique proposed by Casdagli et al.,[3] to input-output systems. As expected by the above theoretical results, in these applications it was found essential to use both input and output time series to construct an accurate model. The theoretical results outlined above have also provided insights into the number of lags required in modeling vibration testing systems, where estimates for the underlying state-space dimension are sometimes known, and into the limitations of the application of such models if the system is chaotic or of high dimension.

There appears to be increasing interest within the dynamical systems community in input-output systems. For example, Hübler has investigated the question of how to design input forces which optimally probe, or control experimental dynamical systems.[10] In this paper we have concentrated on theoretical issues of how to construct deterministic models given the inputs and outputs for the input-output system. Given the potentially wide range of applications of input-output dynamical systems, it will be interesting to see whether other theoretical approaches will be developed and applied in the future.

ACKNOWLEDGMENTS

I would like to thank Norman Hunter and Doyne Farmer for useful conversations. I am grateful to partial support from the National Institute of Mental Health under grant 1-R01-MH47184-01.

REFERENCES

1. Billings, S. A., K. M. Tsang, and G. R. Tomlinson. "Application of the Narmax Method to Nonlinear Frequency Response Estimation." In *Proceedings of the 6th International IMAC Conference*, 1987.
2. Casdagli, M. "Nonlinear Prediction of Chaotic Time Series." *Physica D* **35** (1989): 335–356.
3. Casdagli, M., S. Eubank, J. D. Farmer, and J. Gibson. "State Space Reconstruction in the Presence of Noise." *Physica D* **51** (1991): 52.
4. Casdagli, M. "Nonlinear Forecasting, Chaos and Statistics." In *Modeling Complex Phenomena*, edited by L. Lam. New York: Springer-Verlag. To appear.
5. Crutchfield, J. P., and B. S. McNamara. "Equations of Motion from a Data Series." *Complex Systems* **1** (1987): 417–452.
6. Eckmann, J. P., and D. Ruelle. "Ergodic Theory of Chaos and Strange Attractors." *Rev. Mod. Phys.* **57** (1985): 617.
7. Farmer, J. D., and J. J. Sidorowich. "Predicting Chaotic Time Series." *Phys. Rev. Lett.* **59(8)**: 845–848.
8. Farmer, J. D., and J. J. Sidorowich. "Exploiting Chaos to Predict the Future and Reduce Noise." In *Evolution, Learning and Cognition*, edited by Y. C. Lee. Singapore: World Scientific, 1988.
9. He, X. Personal communication, 1991
10. Hubler, A., and E. Luscher. "Resonant Stimulation and Control of Nonlinear Oscillators." *Naturwissenschaften* **76** (1989): 67–69.
11. Hunter, N. F. "Application of Nonlinear Time Series Models to Driven Systems." This volume.
12. Nese, J. M. "Quantifying Local Predictability in Phase Space." *Physica D* **35** (1989): 237-250.

13. Sauer, T., J. A.Yorke, and M. Casdagli. "Embedology." *J. Stat. Phys.*, to appear.
14. Takens, F. "Detecting Strange Attractors in Fluid Turbulence." In *Dynamical Systems and Turbulence*, edited by D. Rand and L.-S. Young. Berlin: Springer-Verlag, 1981.
15. Tong, H. "Nonlinear Time Series Analysis: A Dynamical Systems Approach." Oxford: Oxford University Press, 1990.
16. Wolf, A., J. Swift, H. Swinney and J. Vastano. *Physica D* **16** (1985): 285.

Wallace E. Larimore
Adaptics, Inc., 40 Fairchild Drive, Reading, MA 01867

Identification and Filtering of Nonlinear Systems Using Canonical Variate Analysis

States for a nonlinear time series are constructed directly from a nonlinear canonical variate analysis (CVA) of the past and future of the process. Such states can be computed sequentially by solution of the maximal correlation problem. A state-space innovations representation for the Markov process is given in terms of the canonical variable states. The problem of choosing a state for the nonlinear system that has minimal state order is developed leading to the additional requirement of independence, that is, an independent canonical variate analysis (ICVA). Optimal normalizing transformations are developed and related to mutual information and independence of the canonical variables. Computational algorithms are developed for determination of the canonical variable states, state-space model fitting, and construction of nonlinear stochastic filters. The performance of the computational procedures are demonstrated on simulated data of the Lorenz chaotic attractor, a multiple equilibria nonlinear system, including process excitation noise. From observation of only one of the three states of the Lorenz attractor, the full dynamics of the system are determined. The filtered state estimate is accurate, and the identified nonlinear system has the same nonlinear character as the true process including chaos and multiple equilibria.

1. INTRODUCTION

In the CVA approach, the approximation problem is formulated directly in terms of finding an optimal approximation of a specified state order. The states are to be determined as linear combinations of basis functions expressed as nonlinear functions of the past observations of the process. The states are computed in a CVA that determines the linear combinations of the nonlinear functions of the past that have predictive value for the future evolution of the process. The computation is fundamentally a linear procedure. Even in the Hilbert space of all possible nonlinear functions of the past, the solution to the CVA problem is shown to be the result of an iteration of the projections on subspaces of nonlinear functions—a linear procedure for each iteration. It is shown that the optimal selection of states for a given order k, has the same first $k-1$ states as the solution for the case of finding only $k-1$ states. Thus, the problem solution for state order k contains in canonical form the solution for all lower orders and the determination of the kth state given the solution for the $k-1$ states can efficiently proceed in an iterative fashion. Since there is no *a priori* model structure imposed on the problem, the procedure will detect any significant model structure contained in the observed data.

For reasons of space, proofs have been deleted and some of the discussions are abbreviated. More details are contained in Larimore[16] including numerous references to the related literature. Early developments in this approach to the nonlinear problem are contained in Larimore,[14,15] while a tutorial review of the present results for the linear problem are contained in Larimore.[18]

Sections 2 through 4 develop the nonlinear CVA problem in a Hilbert space setting of nonlinear functions. This is applied to nonlinear systems in section 5. The resulting nonlinear CVA does not result in minimal state order which requires the additional condition of independence rather than orthogonality as discussed in section 6. This is further related to optimal normalizing transformations and mutual information in sections 7 and 8. Sections 9 and 10 demonstrate the nonlinear CVA procedure on simulated data from the Lorenz attractor.

2. HILBERT SPACE OF NONLINEAR FUNCTIONS

The extension of CVA to nonlinear systems requires a more general setting such as a Hilbert space of nonlinear functions. Let $X = (x_1, \ldots, x_m)^T$ and $Y = (y_1, \ldots, y_n)^T$ be two sets of random variables defined on a probability space Ω with respect to the probability measure μ and let P denote the induced probability measure on R^M and R^N respectively. The random variables X and Y will play the role of the past p_t and future f_t defined in section 5 offering a simplicity in the notation where the notion of time t is not required in the present section.

For a given positive integer r, consider the space $\mathcal{F}_X^r$ and $\mathcal{F}_Y^r$ of all Borel measurable r-dimensional vector functions $f(X) = (f_1(X), \ldots, f_r(X))^T$ and $g(Y) = (g_1(Y), \ldots, g_r(Y))$ satisfying

$$E f(X) = 0, \quad E g(Y) = 0 \tag{1}$$

where $E(\ )$ denotes expectation. The space is a linear vector space on which we define the inner product

$$\langle f, g \rangle - tr E f(X) g^T(Y) = E \sum_{i=1}^{r} f_i(X) g_i^T(Y) = tr \Sigma_{fg} \tag{2}$$

where, for zero-mean random functions as in Eq. (1), the covariance matrix notation $\Sigma_{fg} = E f(X) g(Y)^T$ is used. The pseudonorm is given by

$$\| f \| = \langle f, f \rangle^{1/2} \tag{3}$$

$\| \cdot \|$ is not a norm since $\| f \| = 0$ may hold when $f(x) \neq 0$ for x in a set A with probability $P_x(A) = 0$. The spaces $\mathcal{F}_X^r$ and $\mathcal{F}_Y^r$ are separable Hilbert spaces under the inner product.

Projection operators on the Hilbert space can be expressed in terms of conditional expectations and are shown to be the solution of an optimal prediction problem. Such a prediction problem is central to the nonlinear CVA problem discussed in the following sections.

Consider two random variables u and v defined on a probability space with respect to a probability measure P. Then the conditional expectation $E(v|u)$ of v with respect to u is a random variable for each value of u and satisfies

$$\int_A E(v|u) dP = \int_A v dP \quad \text{for every } A \in \mathcal{A}_u \tag{4}$$

where $\mathcal{A}_u$ is the least sigma algebra of measurable sets of Ω for which u is measurable.

The concept of projection on a subspace provides the optimal solution to linear problems. Let $P_{\mathcal{K}}$ denote the operator of orthogonal projection on a subspace $\mathcal{K}$. Then, for any $h \in \mathcal{H}$,

1. $h_k = P_{\mathcal{K}} h$ if and only if for some $h_k \in \mathcal{K}$ and for all $g \in \mathcal{K}$, $\langle h, g \rangle = \langle h_k, g \rangle$.
2. $h_k = P_{\mathcal{K}} h$ if and only if for some $h_k \in \mathcal{K}$

$$\min_{g \in \mathcal{K}} \| h - g \| = \| h - h_k \| . \tag{5}$$

The conditional expectation has optimal properties as a projection operator in the Hilbert space $\mathcal{F}$.

THEOREM 1 (Optimal Projection): The optimal projection $P_{\mathcal{F}_u} v$ of the random variable v on the Hilbert subspace $\mathcal{F}_u$ of nonlinear functions of the vector of random variables u is the conditional expectation

$$P_{\mathcal{F}_u} v = E(v|u) \,. \tag{6}$$

3. MULTIVARIATE NONLINEAR CVA

In this section, a nonlinear generalization to the linear CVA problem[7] is developed.

First, consider the problem where f and g are fixed functions and we wish to find the nonlinear function $\hat{g}(f(X))$ such that the relative prediction error

$$\| g(Y) - \hat{g}(f(X)) \|_{\Sigma_{gg}^\dagger} \doteq E\{[g(Y) - \hat{g}(f(X))]^T \Sigma_{gg}^\dagger [g(Y) - \hat{g}(f(X))]\} \tag{7}$$

is minimum where ($\dagger$) denotes the pseudoinverse operation. As shown in the previous section, the solution is given by the conditional expectation projection operator

$$\hat{g} = E(g|f) \,. \tag{8}$$

Now with the optimal prediction $\hat{g}$ given by Eq. (8), consider the r-rank nonlinear prediction problem of finding an r-dimensional nonlinear function $f(X)$ of X and r-dimensional nonlinear function $g(Y)$ of Y so as to minimize Eq. (7). Specifically consider the following minimization problem.

RANK R NONLINEAR PREDICTION PROBLEM: For a given positive integer r, find r-dimensional vector functions $f(X)$ and $g(Y)$ minimizing the relative prediction error

$$\min_{(f,g)} \| g(Y) - \hat{g}(f(X)) \|_{\Sigma_{gg}^\dagger} \tag{9}$$

where $\hat{g} = E(g|f)$.

Since $f(X)$ enters the problem only through $\hat{g}(f(X))$, it is sufficient to include the nonlinearity of $\hat{g}$ in the function $f(X)$. In this case for a particular $g(Y)$, the optimal prediction is given by

$$\hat{g}(f(X)) = \Sigma_{gf} \Sigma_{ff}^\dagger f(X) \tag{10}$$

and the prediction error is

$$\| g(Y) - \hat{g}(f(X)) \|_{\Sigma_{gg}^\dagger} = tr \Sigma_{gg}^\dagger [(\Sigma_{gg} - \Sigma_{gf}) \Sigma_{ff}^\dagger \Sigma_{fg}] \,. \tag{11}$$

The above measure is invariant to linear transformations of both f and g. Thus, we are free to impose the constraints

$$\Sigma_{ff} = \Sigma_{gg} = I_r \tag{12}$$

so that f and g are each orthonormal sets of functions. Then the prediction error reduces to

$$tr(I_r - \Sigma_{gf}\Sigma_{fg}) = r - tr(\Sigma_{gf}\Sigma_{fg}) = r - \sum_{i,j=1}^{r} \langle f_i(X), g_j(Y) \rangle^2 \tag{13}$$

which is the sum of the squares of the elements of the covariance matrix Σ_{fg}.

The optimal prediction problem then reduces to finding f and g to solve the maximization problem

$$\max_{\Sigma_{ff} = \Sigma_{gg} = I_r} tr(\Sigma_{gf}\Sigma_{fg}). \tag{14}$$

Note that in the univariate case of $r = 1$, the problem is precisely the maximal correlation problem to be discussed in the next section.

One further simplification occurs by doing a singular value decomposition (SVD) on the covariance matrix Σ_{fg}

$$\Sigma_{fg} = UDV^T, \quad D = Diag(d_1, \geq \ldots \geq d_k > 0, \ldots, 0) \tag{15}$$

Then the prediction error reduces to

$$r - \sum_{i=1}^{k} d_i^2 \tag{16}$$

which is the sum of the squared canonical correlations d_i. The optimal prediction problem is then given by the solution of

$$\max_{\substack{\Sigma_{ff} = \Sigma_{gg} = I_r \\ \Sigma_{fg} = Diag}} \sum_{i=1}^{r} d_i^2. \tag{17}$$

For a given fixed R, let f^R and g^R denote the R-dimensional functions maximizing Eq. (17) for $r = R$. Then it is easily shown that the maximum of Eq. (17) is achieved for all $r < R$ with f^r and g^r as the first r components of f^R and g^R respectively and with D^r the first r components of D^R.[17] Thus the problem is transformed to the sequence of R one-dimensional problems for $r = 1, \ldots, R$. In particular, this is summarized in the following theorem.

THEOREM 2 (Sequential Selection): The vector functions f and g giving an optimal solution to the nonlinear prediction problem (9) are given sequentially by the following procedure: For each r, find the pair of functions (f_r, g_r) such that they are orthogonal to the previously selection functions $f^{r-1} = (f_1, \ldots, f_{r-1})^T$ and $g^{r-1} = (g_1, \ldots, g_{r-1})^T$ respectively and maximize the correlation, i.e., such that

$$d_r = \max_{\substack{\Sigma_{f^{r-1}f_r} = \Sigma_{g^{r-1}g_r} = 0 \\ \Sigma_{f_r f_r} = \Sigma_{g_r g_r} = 1}} \Sigma_{f_r g_r} . \tag{18}$$

4. MAXIMAL CORRELATION AND PROJECTION OPERATORS

The solution to the nonlinear CVA problem was reduced in the previous section to the problem of finding scalar functions $f(X)$ and $g(Y)$ of the respective sets X and Y of random variables such that the correlation is maximized. In this section, properties of the maximal correlation are investigated and related to projection operators on Hilbert spaces. We follow primarily the development of Renyi[20] (see also Csaki and Fischer).[3,4]

Consider random vectors $X = (x_1, \ldots, x_m)$ and $Y = (y_1, \ldots, y_n)$. The *maximal correlation* of X and Y is defined as

$$\rho^*(X, Y) = \sup_{f, g} \rho(f(X), g(Y)) = \sup_{\substack{f, g \\ \| f \| = 1 \\ \| g \| = 1}} E[f(X)g(Y)] \tag{19}$$

where f and g run over all Borel measurable functions with zero mean; i.e., $Ef = Eg = 0$, and $\rho(f, g)$ is the correlation coefficient given in this case by $E[f(X)g(Y)]$. The maximal correlation satisfies the following properties:

- $\rho^*(X, Y)$ is defined for any pair of random vectors X and Y, neither of them being a constant with probability 1.
- $\rho^*(X, Y) = \rho^*(Y, X)$.
- $0 \le \rho^*(X, Y) \le 1$.
- $\rho^*(X, Y) = 0$ if and only if X and Y are *stochastically independent*.
- $\rho^*(X, Y) = 1$ if there is a *strict dependence* between X and Y; i.e., $f(X) = g(Y)$ for some non-zero Borel measurable functions f or g. The converse requires some additional conditions.
- *Invariance.* Under 1-1 onto Borel-measurable transformations f and g, $\rho^*(f(X), g(Y)) = \rho^*(X, Y)$.
- If the joint distribution of X and Y is normal, then $\rho^*(X, Y)$ is the maximum canonical correlation, i.e., the sup is achieved by considering only *linear functions f and g*.

When the maximum is attained for some pair of functions f and g, then certain operator equations involving projections are satisfied. These operator equations are used below to establish the existence of functions f and g attaining the maximum.

Consider the situation when the optimal solution exists to the CVA problem (14). Then there exist $f_0(X)$ and $g_0(Y)$ such that

$$E(g_0(Y)|X) = \Sigma_{g_0 f_0} f_0(X). \tag{20}$$

recalling the normalization (12), and similarly

$$E(f_0(X)|Y) = \Sigma_{f_0 g_0} g_0(Y). \tag{21}$$

Furthermore by taking the conditional expectation of Eq. (21) with respect to X and expressing the result in terms of projections operators, we have

$$P_{\mathcal{F}_X} P_{\mathcal{F}_Y} f_0 = E(E(f_0(X)|Y)|X) = \Sigma_{f_0 g_0} \Sigma_{g_0 f_0} f_0(X) \tag{22}$$

and similarly

$$P_{\mathcal{F}_Y} P_{\mathcal{F}_X} g_0 = E(E(g_0(Y)|X)|Y) = \Sigma_{g_0 f_0} \Sigma_{f_0 g_0} g_0(Y). \tag{23}$$

Thus, the optimal solutions f_0 and g_0 are eigenvectors of a successive projection operator. The function f_0 is projected successively on the function space $\mathcal{F}_Y$ and $\mathcal{F}_X$ and similarly for g_0. The eigenvectors f_0 and g_0 have the common eigenvalue $\Sigma_{g_0 f_0}^2$ which is equal to the squared maximal correlation.

To determine the existence of an optimal solution f_0 and g_0 to Eq. (22) and Eq. (23), we study the pair of operator equations

$$Af = D^2 f \tag{24}$$
$$Bg = D^2 g \tag{25}$$

where A and B are the operators defined on the left-hand side of Eqs. (22) and (23), respectively. We wish to determine under what conditions there exist solutions f_0 and g_0 to these operator equations with D equal to the maximal correlation.

The existence of the maximum is insured by results from the theory of operators on Hilbert spaces. We will show that the operators A and B are bounded, self-adjoint and continuous under the assumption that the distributions are regular as defined below. Such operators must then have eigenfunction solutions f and g attaining the maximum eigenvalue D^2.

Renyi[20] establishes the following theorem:

THEOREM 3: The operators A and B are bounded self-adjoint linear operators such that the maximum eigenvalue is $(\rho^*(X,Y))^2$, the squared maximal correlation.

To show that A and B are completely continuous, some restrictions on the distribution of the random variables are required.

DEFINITION The dependence between the vector random variables X and Y is *regular* if the joint distribution $P(X,Y)$ is absolutely continuous with respect to the direct product P_{X*Y} of the marginal distributions P_X and P_Y where the direct product distribution is defined by

$$P_{X*Y}(x \in X \text{ and } y \in Y) = P_X(x \in X)P_Y(y \in Y). \tag{26}$$

If the dependence between X and Y is regular, then the probability density $k(x,y)$ of the joint distribution $P(X,Y)$ relative to the product distribution $P(X)P(Y)$ exists as a Radon-Nikodyn derivative satisfying

$$P_{X*Y}(C) = \int \int_C k(x,y)dP_X(x)dP_Y(y) \tag{27}$$

for every Borel set C in the product space $X*Y$.

If X and Y are regular, then the conditional expectation operator has a useful expression as an integral operator with kernel $k(x,y)$[4]

$$E(f(X)|Y) = \int f(x)k(x,y)dP(x). \tag{28}$$

The continuity of the operators A and B is shown in the following theorem.

THEOREM 4: If the dependence between X and Y is regular and $Ek^2(x,y)$ is finite, then the operators A and B are completely continuous.

The projection relationships (20) and (21) can be used in an algorithm to compute functions f and g maximizing the correlation. The alternating conditional expectation algorithm[1,2] begins with any non-zero functions and alternates between the two projection relationships. For a general class of functions, the algorithm converges. The ACE algorithm has been used extensively on sample data in multivariate regression in many variables.

5. NONLINEAR MARKOV PROCESSES

In this section, various aspects and representations of Markov processes are developed. The fundamental properties of the state of a Markov process are reviewed. Given a state for a Markov process, the development of the state-space innovations representation is immediate.

A fundamental concept in the CVA approach is the *past* and *future* of a process. Suppose that data are given consisting of observed outputs y_t and observed inputs u_t at time points labeled $t = 1, \ldots, N$ that are equally spaced in time. Associated with each time t is a past vector p_t consisting of the past outputs and inputs occurring prior to time t as well as a future vector f_t consisting of outputs at time t or later, specifically

$$p_t = (y_{t-1}^T, y_{t-2}^T, \ldots, u_{t-1}^T, u_{t-2}^T, \ldots)^T, \quad f_t = (y_t^T, y_{t+1}^T, \ldots)^T. \tag{29}$$

For simplicity, consider first purely stochastic processes with no observed deterministic input to the system. A fundamental property of a nonlinear, strict sense Markov process of finite state order is the existence of a finite-dimensional state x_t which is a nonlinear function of the past p_t

$$x_t = C_t(p_t) \tag{30}$$

with $C_t(\cdot)$ a nonlinear function. The state x_t has the property that the conditional probability of the future f_t conditioned on the past p_t is identical to that of the future f_t conditioned on the finite-dimensional state x_t so

$$P\{f_t|p_t\} = P\{f_t|x_t\}. \tag{31}$$

Thus, only a finite amount of information from the past is relevant to the future evolution of the process.

To extend this concept to processes involving deterministic controls or inputs, the effects of future inputs must first be removed from the future outputs. Let v_t denote the future inputs $(v_t^T, u_{t+1}^T, \ldots)$ and consider the conditional random variable $f_t|v_t$. Then the process is a *controlled Markov process* of order k if there exists a k-order state such that the conditional distribution of $f_t|v_t$ given the past p_t is identical to the conditional distribution of $f_t|v_t$ given the state x_t so

$$P\{(f_t|v_t)|p_t\} = P\{(f_t|v_t)|x_t\}. \tag{32}$$

This is equivalent of the statement that

$$P\{f_t|(v_t, p_t)\} = P\{f_t|(v_t, x_t)\}. \tag{33}$$

Now suppose that the state x_t is given from CVA of the past and future as in the previous sections with $X = p_t$ and $Y = f_t$, and we wish to obtain the generally nonlinear state equations describing the state evolution and observed output from the observed inputs and unobserved disturbances. First we define, for a given selection for the state x_t of the process, the *innovations* process which is the error in the optimal nonlinear prediction $E(y_t|x_t, u_t)$ of the process y_t from the state x_t and the input u_t given by

$$\nu_t = y_t - E(y_t|x_t m u_t). \tag{34}$$

Then the following shows that the vector (ν_t, u_t, x_t) is a state at time $t+1$ and thus the state evolution can be obtained as a nonlinear function of these variables.[15]

THEOREM 5: Suppose that the joint and marginal densities among p_t, f_t, v_t, u_t, and y_t are non-zero. Then the state at time $t+1$ is a function of x_t, u_t, and y_t, and the state evolves as

$$x_{t+1} = \phi(x_t, u_t, \nu_t) \tag{35}$$

where the innovation process ν_t is an orthogonal increment process defined by

$$y_t = \mu_t(x_t, u_t) + \nu_t \tag{36}$$

where $\mu_t(x_t, u_t)$ is the projection of y_t on $\mathcal{F}_{(x_t, u_t)}$.

The importance of this is in the evolution of the state equations. Let x_{t+1} be a minimal order state. Then from the above, the variables (ν_t, u_t, x_t) generate the subspace $\mathcal{F}_{(\nu_t, u_t, x_t)}$ containing the state x_{t+1} so that x_{t+1} can be found by projection

$$x_{t+1} = E(x_{t+1} | (\nu_t, u_t, x_t)) = \phi_t(\nu_t, u_t, x_t) \tag{37}$$

using the conditional expectation operator $E(\cdot | \cdot)$.

6. MINIMAL STATE RANK

CVA provides a very useful procedure for construction of states for a nonlinear process. Such a state vector, however, is not of minimal order. Orthogonality is not sufficient to exclude redundancy among the canonical variables. What is required is independence rather than orthogonality. The construction of independent canonical variables is discussed in this section. First, several other approaches to minimal realization are discussed.

The problem is to construct a state-space realization where the order of the state vector is minimal. One approach to minimal realization involves the use of the Rank Theorem of differential topology. The Rank Theorem of differential topology is a generalization of the inverse function theorem that states conditions under which a differentiable map can be transformed locally into a linear map by a smooth change of coordinates of the domain and range variables. The rank theorem gives a local result; that is, for a given point it guarantees the existence of a mapping from the past to a set of variables equal in number to the rank of the past/future function and that agrees with the state in a neighborhood of the given point. Unfortunately, in general this state cannot be extended to a global state. Under additional restrictions, the global minimal realization has the same rank as the local realizations, but not in general.

The problem of existence and uniqueness of minimal realizations of nonlinear systems has been investigated extensively for deterministic nonlinear systems. Under suitable conditions where the state-space system has various analytic properties,[5,8,21] there exists a minimal realization that is globally observable and controllable and is unique up to isomorphism. While such results are of great theoretical value, they give little guidance for construction of minimal realizations in practice.

Now we turn to the problem of minimal realization in the context of CVA. Suppose that two canonical variables g_1 and g_2 are such that there is no functional redundancy between them in the sense that for any functions e and f, $e(g_1)$ and $f(g_2)$ are uncorrelated. This is equivalent to the statement that the maximal correlation is zero, i.e., $\rho^*(g_1, g_2) = 0$, which from section 4 is the case if and only if g_1 and g_2 are stochastically independent random variables. Thus we seek a stochastically independent canonical variate analysis (ICVA); i.e., replace the requirement of orthogonality with that of stochastic independence.

This would require that the canonical variables are mutually independent rather than just orthogonal. In particular, in the notation of section 4, the two sets X and Y of random variables are independent if and only if $\rho^*(X, Y) = 0$. Thus for an *independent CVA*, in the Sequential Selection Theorem 2, replace the orthogonality condition $\langle g^{(r-1)}, g_r \rangle = 0$ with the mutual independence condition $\rho^*(g^{(r-1)}, g_r) = 0$.

Thus, a procedure for constructing a set of random variables independent of another set is required. In particular, we wish to show that for any sets X and Y of random variables, there exists a set $Z(X, Y)$ of random variables such that X and Z are mutually independent and span the same space as X and Y. One version of such a theorem is given as follows:

THEOREM 6: If the density $k(x, y)$ of the joint distribution $P_{X,Y}(x, y)$ with respect to the product $P_X(x) P_Y(y)$ of the marginals exists, and is continuous and non-zero, then there exists a transformation $Z(X, Y)$ such that the map: $(X, Y) \to (X, Z)$ is 1–1 and X and Z are mutually independent.

For the construction of independent canonical variables, at each step of the CVA procedure, the subspace independent of the previous canonical variables would need to be constructed. In fact, this is given simply by the variables Z above as stated in the following theorem:

THEOREM 7: The random variables Z independent of X generate the subspace $\mathcal{M}$ of the function space $\mathcal{F}_{(X,Y)}$ that is independent of the variables X, i.e., $\mathcal{M} = \mathcal{F}_Z$.

In the proofs of these theorems, conditional and marginal distributions provide a means of constructing independent random variables. From the equivalence of mutual independence and zero maximal correlation, it may be possible to use the orthogonality of all functions of the two sets to define implicit algorithms such as the ACE algorithm for constructing mutually independent random variables. More work is needed on efficient algorithms for construction of independent canonical variables particularly from observational data.

A further issue to be addressed in future work is the minimal rank issue. Conditions under which the transformation to mutually independent random variables is a 1–1 transformation need to be established. The minimal rank of the state space will be guaranteed if any transformation to a set of mutually independent random variables has an inverse transformation to the original variables which holds except on a set of probability zero. This condition also enters into the discussion in the next section.

7. OPTIMAL NORMALIZING TRANSFORMATIONS

The canonical variables have a more general optimality property than maximal correlation. It is shown that the canonical variables are the result of transforming the original variables to variables that are closest to normality. The measure of closeness to normality involves an entropy measure. Thus a first-order approximation of the distribution of the canonical variables would be the normal distribution.

A starting point that suggests a much more basic relationship follows.[11,12]

THEOREM 8: If X and Y are two jointly normal random variables, then the maximal correlation occurs for *linear* transformations $g(X)$ and $h(Y)$, and if the maximal correlation is positive, then strictly nonlinear transformations will strictly decrease the correlation.

The result does not generalize to the multivariate case for the reasons discussed in the previous section—statistically or functionally dependent variables may be orthogonal in the case of nonlinear transformation of the variables. In this section, the term linear canonical variables and linear CVA will mean the usual CVA considering only linear functions of the random variables. The strongest multivariate result appears to be that given in Lancaster.[11]

THEOREM 9: Let X and Y be jointly normal random vectors, and let the nonlinear transformations $g_i(X)$ and $h_j(Y)$ be recursively defined so that $E(g_i g_j) = E(h_i h_j) = 0$ for $i < j$. Then g_1 and h_1 have maximal correlation if they are respectively the first pair of linear canonical variables. If for $i > 1$ we have $\rho_i > \rho_1^2$, then the maximal correlation of g_i and h_i are given respectively by the ith pair of linear canonical variables.

The condition $\rho_i > \rho_1^2$ is sufficient to insure that nonlinear functions that are orthogonal to the previously defined canonical variables will not have large enough correlation. If, however, the condition of orthogonality is replaced by that of independence or equivalently zero maximal correlation, then the following multivariate generalization is obtained.

THEOREM 10: Let X and Y be jointly normal random vectors of dimensions k and ℓ respectively, and let the nonlinear transformations $g_j(X)$ and $h_j(Y)$ be recursively defined such that $\rho^*(g_i, g_j) = \rho^*(h_i, h_j) = 0$ for $i < j$. Then the functions g_j and h_j have maximal correlation if they are respectively the jth pair of linear canonical variables c_j and d_j. For $\rho_j > 0$, g_j and h_j are strictly linear functions respectively of $c_1, \ldots, c_k$ and $d_1, \ldots, d_\ell$.

With the added requirement of independence among the canonical variables, the univariate result of Theorem 8 is generalized to the multivariate case by Theorem 10. Thus, in the case of joint normality, of all possible transformations the independent canonical variables relevant to prediction are normally distributed and the corresponding prediction problem among the canonical variables is linear.

From this section it is clear that independence of the canonical variables is a necessary condition for removal of redundancy. That is, if independence of the canonical variables is not required, then particular examples are easily constructed where redundancy is present.

8. MUTUAL INFORMATION AND APPROXIMATION

In this section, a broader interpretation of canonical variables for random variables with arbitrary probability distributions is obtained. This is important since the correlation coefficient as a measure of dependence appears to be very simplistic and to miss much of the complexity of arbitrary distributions.

The previous section addresses the case where there exists a 1–1 transformation to a set of jointly normal random variables. But what about the more general case where there may not exist such a transformation. The correlation coefficient has an interpretation in terms of mutual information and approximating normal distributions. As noted by Gelfand and Yaglom,[6] the mutual information between two Gaussian random variables c and d is

$$E\left\{\ln\left(\frac{p(c,d)}{p(c)p(d)}\right)\right\} = -\frac{1}{2}\ln(1 - \rho_{cd}^2) \tag{38}$$

so that maximizing the correlation is tantamount to maximizing the mutual information.

To generalize this procedure to nonlinear processes requires a number of new concepts and results. The linear case depends completely on the Gaussian assumption which is violated in the nonlinear case. Also, in the non-Gaussian case the expression for the mutual information no longer holds. However, the basic computational procedure of the CVA method is minimizing the function (38) of the correlation coefficient based upon second moments of the variables, which is computationally direct and efficient. Thus, it is an attractive computational procedure if reasons for its use in nonlinear processes are developed.

Consider arbitrary random variables c and d with some arbitrary joint density function $p(c,d)$. Consider the normal densities $n(c,d)$, $n(c)$, and $n(d)$ using the first- and second-order moments μ_c, μ_d, and σ_{cc}, σ_{dd}, σ_{cd} respectively of the true distribution of c and d. For simplicity of notation, the means μ_c and μ_d will be assumed to be zero which will not affect the results below. Now consider the expected mutual information of the normal densities with expectation E_p taken with respect to the true density $p(c, d)$

$$E_p \left\{ \ln \left(\frac{n(c, d)}{n(c)n(d)} \right) \right\} = E_p \left\{ \ln \left(\frac{n(d|c)}{n(d)} \right) \right\}$$

$$= -\frac{1}{2} E_p \left\{ \ln 2\pi (\sigma_{dd} - \sigma_{dc}\sigma_{cc}^{-1}\sigma_{cd}) \right.$$

$$+ \frac{(d - \sigma_{dc}\sigma_{cc}^{-1}c)^2}{(\sigma_{dd} - \sigma_{dc}\sigma_{cc}^{-1}\sigma_{cd})} - \ln 2\pi\sigma_{dd} - \frac{d^2}{\sigma_{dd}} \right\} \qquad (39)$$

$$= -\frac{1}{2} \ln \left(1 - \frac{\sigma_{cd}^2}{\sigma_{cc}\sigma_{dd}} \right)$$

$$= -\frac{1}{2} \ln (1 - \rho_{cd}^2).$$

Thus the function (39) of ρ_{cd} has the interpretation as the average over the true density of the mutual information of the hypothesized normal densities $n(c, d)$, $n(c)$, and $n(d)$ based upon the true covariances. The presence of the true density p in the expectation may at first appear to be a problem since it is often not encountered in information arguments; however, in terms of the statistical inference problem, it will play the role of the "truth" or data. As with relative or Kullback information (see Kullback,[10] pp. 6; Larimore[13]), consider a relative measure of approximation in terms of the *expected relative mutual information* between the true density $p(c, d)$ and an approximating normal density $n(c, d)$ expressed as

$$\int p(c, d) \left\{ \ln \frac{p(c, d)}{p(c)p(d)} - \ln \frac{n(c, d)}{n(c)n(d)} \right\} dcdd. \qquad (40)$$

Now suppose that $c(x)$ and $d(y)$ are transformations of the two univariate random variables x and y respectively. The first term in the integrand is the expected mutual information which is invariant if c and d are 1–1 transformations. Thus Eq. (40) is a function only of the second term. From Eq. (39), minimizing the measure (40) in the case that c and d are 1–1 is equivalent to maximizing the correlation coefficient ρ_{cd}.

Now we consider the case where the transformations c and d are multivariate transformations of the random vectors x and y respectively. For the first term in Eq. (40) to be constant, the transformations c and d must be 1–1. This suggests yet another generalization of CVA under the additional constraint that the transformations c and d be 1–1. Suppose that in addition the transformations c and d are required to produce pairs of random variables (c_i, d_i) that are independent of (c_j, d_j) for $i \neq j$, as developed in section 7. Then the first term of Eq. (39) would be a constant and the second term would decompose into the sum of terms Eq. (40)

involving each of the pairs of components c_i and d_i of the transformed variables, i.e.,

$$E_p \left\{ \ln \left(\frac{n(c, d)}{n(c)n(d)} \right) \right\} = -\frac{1}{2} \sum_i \ln[1 - \rho^2(c_i, d_i)]. \tag{41}$$

Now if the pairs (c_i, d_i) are sequentially chosen as independent canonical variables, then the measure Eq. (41) would be maximized incrementally for each additional pair (c_i, d_i). Since $\ln(1 - \rho^2)$ is monotone in ρ, this is equivalent to the sequential selection procedure in Theorem 2 of minimizing the trace of $\sum_{fg}$ with the additional requirement of independence.

In the literature of maximal correlation and nonlinear canonical variables (see section 4), there appears to be no consideration of the multivariate case where orthogonality is replaced by independence or equivalently zero maximal correlation. From several points of view, this appears to be the natural nonlinear generalization for the multivariate case:

- In the case of the multivariate normal distribution where orthogonality is equivalent to independence, it is independence that generalizes to the nonlinear case, not orthogonality (see Theorem 10).
- Independence removes the functional and statistical redundancy present in the nonlinear case as discussed in section 6. This provides the basis for studying the selection of a minimal order state.
- The mutual information measure expressed in terms of canonical correlations generalizes to the multivariate case if the canonical variables are independent.

The computational aspects of the independent CVA appear to be significantly greater since it involves the construction of independent random variables rather than orthogonal ones.

9. SIMULATION RESULTS

In the remaining sections of the paper, only orthogonal CVA rather than independent CVA is considered. Further work remains for the development of efficient algorithms for the independent CVA.

The Markov process considered is the Lorenz attractor[19] with process excitation noise. The differential equations are discretized with $\Delta t = 0.01$, and white process noise is added to the state equations so that the discrete time equations used for simulation become

$$x_{t+1}^{(1)} = x_t^{(1)} + \Delta t \sigma \left(x_t^{(2)} - x_t^{(1)} \right) + n_t^{(1)}; \tag{42}$$

$$x_{t+1}^{(2)} = x_t^{(2)} + \Delta t \left[\rho x_t^{(1)} - x_t^{(2)} - x_t^{(1)} x_t^{(3)} \right] + n_t^{(2)}; \tag{43}$$

$$x_{t+1}^{(3)} = x_t^{(3)} - \Delta t \left[\beta x_t^{(3)} + x_t^{(1)} x_t^{(2)} \right] + n_t^{(3)}. \tag{44}$$

The values of the parameters used in the simulation are $\sigma = 10$, $\rho = 28$, and $\beta = 8/3$ which results in the much studied chaos of the system. The noise covariance matrix of the white process excitation noise $(n_t^{(1)}, n_t^{(2)}, n_t^{(3)})^T$ used in the simulation is $10^{-2} \times I_3$ with I_3 as the three-dimensional identity matrix. The presence of process excitation noise provides a much more difficult identification problem since the process no longer is exactly predictable given exact arithmetic. Most studies of identification of chaos consider only the presence of additive white noise which can be reduced by simple averaging of the observations. The time correlation introduced by the nonlinear process dynamics presents a much more difficult problem for identification.

For system identification, the measurement observation data are $y_t = x_t^{(1)}$, the first component of the discretized Lorenz process observed at 1000 time points, which is shown in Figure 1. Note that the presence of the noise on the process is very noticeable. Below it is shown that the entire three-dimensional dynamics of the process can be reconstructed from the measured first component.

The measurements y consisting of only the first component $x^{(1)}$ of the Lorenz attractor are used to compute nonlinear functions of the past as basis functions for canonical variate analysis. The past p_t consists of functions that are powers and products of up to degree three in the first three lags $(y_{t-1}, y_{t-2}, y_{t-3})$ of the measurements y so that functions of the form

$$f_{i_1, i_2, i_3}(y_{t-1}, y_{t-2}, y_{t-3}) = y_{t-1}^{i_1} y_{t-2}^{i_2} y_{t-3}^{i_3} \quad \text{for } i_1 + i_2 + i_3 \leq 3 \qquad (45)$$

are considered. There are 20 such basis functions. The future f_t is the vector of outputs up to 20 lags into the future so

$$f_t = (f_t, \ldots, f_{t+20})^T . \qquad (46)$$

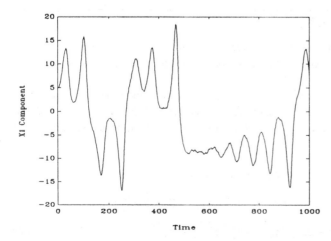

FIGURE 1 Data used in state reconstruction and model identification.

TABLE 1
Canonical Correlations

Index	Canonical Correlation
1	0.9999
2	0.9746
3	0.9043
4	0.6062
5	0.3022
6	0.1782
7	0.1626
8	0.1539
9	0.1309
10	0.0969
11	0.0940
12	0.0827
13	0.0686
14	0.0581
15	0.0461
16	0.0149
17	0.0102
18	0.0041
19	0.0011

A canonical variate analysis of sample covariances of the past and future are given in Table 1. Note that the canonical correlations drop until a floor is hit at 0.1782, and from this point on the canonical correlations fall off slowly. This is typical behavior of sample canonical correlations and most likely the canonical correlations less than or equal to 0.1782 are not statistically significant. This suggests that there are five statistically significant canonical variables. In the discussion below, the canonical state is chosen as the first five canonical variables.

10. STATE RECONSTRUCTION AND MODEL IDENTIFICATION

While the canonical states have optimal properties for embedding the dynamics of the observations, they, in general, do not directly relate to the states of a process that may be of interest for a particular purpose. In the present case, it is important to assess the ability of the canonical states c to predict the full three-dimensional motion of the Lorenz attractor state x.

To obtain an estimate of the original state x of the Lorenz attractor, an approximate nonlinear transformation is constructed from the canonical states c to x. The transformation $h(c)$ is constructed by polynomial regression on polynomials in the states c up to degree 6. The estimated state $\hat{x}$ in the original three-dimensional coordinates of the Lorenz attractor is

$$\hat{x} = h(c). \tag{47}$$

Phase plane plots of pairs of the components of the Lorenz attractor state x are shown in Figure 2 for original "true" states (solid line) and the reconstructed state estimates $\hat{x}$ (dashed line). Note that components $x^{(1)}$ and $x^{(2)}$ are estimated accurately in that the reconstructed trajectories based upon CVA provided by the regression of the canonical states on the Lorenz states gives an accurate average estimate. The estimate of the component $x^{(3)}$ is much noisier, but on the average provides a good estimate of the true value of $x^{(3)}$. The exception is for values of $x^{(3)}$ below about 20 that appear to have been distorted. There are several reasons for this distortion:

- From Eq. (42), the variable $x^{(3)}$ is not directly observed by $y_t = x_t^{(1)}$ but only indirectly through $x_{t-1}^{(2)}$.
- From Eq. (43), the variable $x^{(3)}$ is related to $x^{(2)}$ through the term $x_{t-1}^{(1)} x_{t-1}^{(3)}$ so the effect of $x^{(3)}$ is proportional to the magnitude of $x^{(1)}$.

The distortion in the phase plane plots is seen to be most severe for $x^{(1)}$ and $x^{(3)}$ small and for $x^{(2)}$ and $x^{(3)}$ small. Either of these conditions will cause poor identifiability in those regions of the trajectory from the discussion above. More precise state reconstruction will require more data. Another alternative is to measure more components of the process. Apart from this, however, the state estimate provides a very faithful reconstruction of the process although it is somewhat noisy. The noisy estimate is to be expected since only one of the components of the process was used for measurement data.

An approach to improved accuracy in the selection of the canonical variables is the direct construction of the nonlinear functions that define the transformation to canonical variables. This approach has been developed in the literature of nonlinear regression in high-dimensional spaces and is solved, in particular, by the alternating conditional expectation (ACE) algorithm.[1] The ACE algorithm constructs a spline function for the transformation that adaptively determines the knot locations so as to minimize the statistical error of the solution. This approach appears to be one of the most promising to pursue in future research.

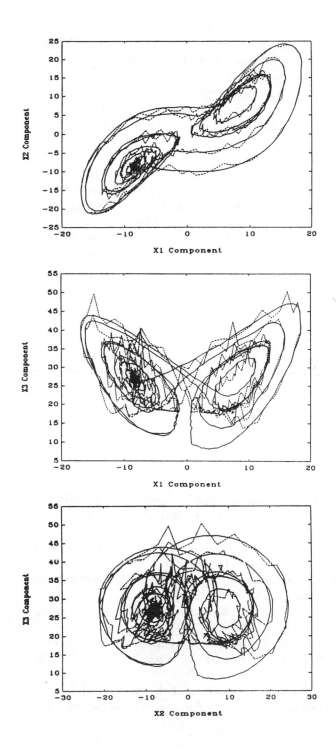

FIGURE 2 Phase space of true vs. reconstructed process.

For model identification, Theorem 5 provides a state-space innovations model of a Markov process. If the state for the process is given, then it is only necessary to determine the nonlinear output function μ, innovations process ν_t, and the state transition function ϕ by nonlinear regression.

The function μ is considered as a sixth-degree polynomial in the canonical state c_t. The error is the innovation process ν_t. Then the transition function ϕ is considered as a sixth-degree polynomial in the canonical state and innovations. The polynomial coefficients are found by nonlinear regression of the canonical state c_t one step ahead on power product terms in c_t and ν_t.

To investigate the dynamics of the resulting identified state space-model, data was simulated using the model with no process excitation noise. Phase-plane plots of pairs of the components of the canonical state variables c_t were computed for the reconstructed states and for states simulated from the identified model. The trajectories constructed from the observed data had the same dynamical character as those simulated from the identified model. Details are contained in Larimore.[16]

ACKNOWLEDGMENT

This work was supported by the Air Force Office of Scientific Research under contract with Computational Engineering, Inc., presently a part of Coleman Research Corporation.

REFERENCES

1. Brieman, L., and J. H. Friedman. "Estimating Optimal Transformations for Multiple Regression and Correlation." *J. Amer. Stat. Assoc.* **80** (1985): 580–619.
2. Brieman, A. "Remarks on Functional Canonical Variates, Alternating Least Squares Methods and ACE." *Ann. Stat.* **18(3)** (1990): 1032–1069.
3. Csaki, P., and J. Fischer. "Contributions to the Problem of Maximal Correlation." *Publ. Math. Inst. Hung. Acad. Sci.* **5** (1960): 325–337.
4. Csaki, P., and J. Fischer. "On the General Notion of Maximal Correlation." *Magyar Tud. Akad. Mat. Kutato Int. Kozl.* **8** (1963): 27–51.
5. Gauthier, J. P., and G. Bornard. "Global Realizations of Analytic Input-Output Mappings." *SIAM J. Control and Optimization* **24(3)** (1986): 509–521.
6. Gelfand, I. M., and A. M. Yaglom. "Calculation of the Amount of Information about a Random Function Contained in Another Such Function." *Amer.*

Math. Soc. Trans. Series 2 **12** (1959): 199–236. Originally appeared in *Usp. Mat. Nauk* **12** (1956): 3–52.

7. Hotelling, H. "Relations Between Two Sets of Variates." *Biometrika* **28** (1936): 321–377.

8. Jakubczyk, B. "Existence and Uniqueness of Realizations of Nonlinear Systems." *SIAM J. Control and Optimization* **24(2)** (1986).

9. Koyak, R. A. "On Measuring Internal Dependence in a Set of Random Variables." *Annals of Statistics* **15(3)** (1987): 1215–1228.

10. Kullback, S. *Information Theory and Statistics.* New York: Dover, 1960.

11 Lancaster, H. O. "Kolmogorov's Remark on the Hotelling Canonical Correlations." *Biometrika* **53** (1966): 585–588.

12. Lancaster, H. O. *The Chi-Squared Distribution.* New York: John Wiley, 1969.

13. Larimore, W. E. "Predictive Inference, Sufficiency, Entropy, and an Asymptotic Likelihood Principle." *Biometrika* **70** (1983): 175–81.

14. Larimore, W. E. "Identification of Nonlinear Systems Using Canonical Variate Analysis." *Proc. 26th IEEE Conference on Decision & Control*, Los Angeles, CA, 1987.

15. Larimore, W. E. "Generalized Canonical Variate Analysis of Nonlinear Systems." *Proceedings of the 27th IEEE Conference on Decision and Control*, December 7-9, 1988, Austin, TX, Vol. 3, 1720–5.

16. Larimore, W. E. "System Identification and Filtering of Nonlinear Controlled Markov Processes by Canonical Variate Analysis." Final Report for Air Force Office of Scientific Research, Computational Engineering, Inc., 1989.

17. Larimore, W. E. "A Unified View of Reduced Rank Multivariate Prediction Using a Generalized Singular Value Decomposition." Submitted for publication, 1989.

18. Larimore, W. E. "Canonical Variate Analysis for System Identification, Filtering, and Adaptive Control." *Proc. 29th IEEE Conference on Decision and Control*, Honolulu, Hawaii, December, Vol. 1, 596–604.

19. Lorenz, E. N. "Deterministic Nonperiodic Flow." *J. Atmospheric Sciences* **20** (1963): 130–41.

20. Renyi, A. "On Measures of Dependence." *Acta. Math. Acad. Sci. Hungar.* **10** (1959): 441–451.

21. Sussman, H. J. "Existence and Uniqueness of Minimal Realizations of Nonlinear Systems." *Math. Systems Theory* **10** (1977): 263–284.

Aviv Bergman,* Peter Grassberger,† and Thomas P. Meyer‡
*SRI International Artificial Intelligence Center, 333 Ravenswood Ave., Menlo Park, CA 94025, and Department of Biological Sciences, Stanford University, Stanford, CA 95305; †Physics Department, University of Wuppertal, D-5600 Wuppertal 1, Germany; and ‡CCSR, University of Illinois, Urbana-Champaign, IL 61801

Forecasting Probabilities with Neural Networks

We apply neural networks to forecasting symbol sequences. More precisely, we forecast probabilities of forthcoming symbols, using sample sequences as training inputs in a back-propagation learning algorithm. Specific features of our method are the use of entropy as a cost function and a $T \to \infty$ "annealing" schedule which allows us to converge to optimal forecasts in spite of conflicting inputs. Optimal forecasts also give us estimates for the Shannon entropy of the sequences and, in principle, algorithms for optimal coding.

1. INTRODUCTION

The problem we pose is that of estimating probabilities on the basis of observed frequencies. More specifically, we consider a sequence of discrete symbols s_n chosen from a finite alphabet. For simplicity we assume that the alphabet is binary, i.e., the sequence is composed of bits $s_n \in \{0, 1\}$. In principle, this represents no restriction since any sequence can be encoded in binary form. In practice, however, a binary coding might not be convenient. In such cases, our methods can be extended to arbitrary k-ary alphabets rather straightforwardly as discussed below.

Nonlinear Modeling and Forecasting, SFI Studies in the Sciences of Complexity, Proc. Vol. XII, Eds. M. Casdagli & S. Eubank, Addison-Wesley, 1992

In a simple forecasting scheme, one might be satisfied with predicting the most likely next symbol. But in many cases this would not be sufficient. Even if one of the symbols (say, $s = 0$) were always the more likely, the probability to observe the other symbol might vary considerably, and this might be of relevance to the observer. Thus, we want instead to estimate the *probabilities* for the next symbol to be 0 or 1. Forecasting not a single step but several steps ahead would again lead to a rather straightforward generalization, but we shall, for simplicity, discuss only the one-step case.

To make the problem well posed, we assume that the sequence is drawn randomly from a stationary distribution. This means that the s_i are realizations of random variables S_j, and that the block probabilities

$$P_k(s_1 \ldots s_k) = \text{prob}(S_{N+1} = s_1, \ldots S_{N+k} = s_k) \tag{1}$$

are independent of N for all integers k and satisfy the standard Kolmogorov consistency relations. Let us assume that the partial sequence $(s_1, s_2, \ldots, s_n)$ is already known, and we want to find the conditional probability for the next symbol to be 1 which we denote as

$$P_{n+1} = P(S_{n+1} = 1 \mid s_1, s_2, \ldots, s_n), \quad n = 1, 2, \ldots . \tag{2}$$

This is the optimal forecast, and it has to be distinguished from our actual forecast which might be less than optimal for some reasons. We call the actual (in general, non-optimal) forecast p_{n+1}.

We imagine the forecasting to be a process where we start with zero knowledge, forecast $p_1 \in [0, 1]$, are told the actual value $s_1 \in \{0, 1\}$ by a "supervisor," then forecast p_2, are again told the actual symbol by the supervisor, and so on. Notice that in this way, if we succeed in making the optimal forecasts every time, we will have, at the end, reconstructed the entire sequence with the minimal amount of information (the supervisor could transmit only the needed information, for example, by supervising the forecasts of several sequences simultaneously; here, we shall not deal with the important problem of how the supervisor codes his information optimally). This shows that the problems of optimal coding and of optimal forecasting are closely related, a fact stressed by Rissanen.[13] Therefore, the method described below is relevant, not only for forecasting but also for data compression.

If the sequence is a Markov chain of known order K, asymptotically optimal forecasting is in principle easy. We can simply make a histogram of all "words" of length $K + 1$, and estimate from it the block probabilities $P_{K+1} = P_{K+1}(s_1 \ldots s_K, 1)/ P_K(s_1 \ldots s_K)$. For small training sets one has to be careful with avoiding errors in estimating the block probabilities, but this is not a serious problem for small K and long sequences.

A more elegant (and more efficient) forecasting scheme is based on Lempel-Ziv-type codes.[20] There, the sequence is broken up into "words" which are simple extensions of previous words, so that a coding sequence consists of the numbers of these words and their extensions. As one can easily see, the efficiency of this method relies on the fact that at every time n, each of the possible words will appear with essentially the same probability. Thus, the natural forecasting problem associated with Lempel-Ziv codes is that of entire words, not of single symbols.

Problems both with the histogram method and with Lempel-Ziv-type codings arise mainly if the sequence is either not Markovian, or of very high (and possibly unknown) order K. This is exactly the situation we are interested in. It arises, e.g., in written natural languages where phonetic, grammatic, and syntactic constraints give correlations over a vast range. While the single-letter entropy in written English (i.e., the Shannon entropy of a "monkey English" in which only the character frequencies are correct, but not any correlations) is ≈ 4.5 bits/character, the true entropy seems to be at most 1 bit/character.[4,7] Similar long-range correlations are seen in symbol sequences generated by complex dynamical systems. Indeed, the range of correlations has been suggested as a complexity measure in such cases.[3,6,18]

The problems with the histogram method are that for large K, we must store 2^K entries (this alone might already be a problem, unless most of these entries are zero), and we need enough data to estimate 2^K probabilities. If all 2^K probabilities were, indeed, completely independent, this would be unavoidable and no method could improve on this. We expect, however, that these probabilities are not independent. The lack of Markovian structure (with order $< K$) just tells us that they do not *factorize*, and no more. But we do not, in general, have any *a priori* guess what other structure there might be, and, thus, we cannot use it in the histogram method.

The problem with Lempel-Ziv algorithms is similar, in that they only take into account specific structures in the observed sequence. In this case these structures are not factorizing probabilities but exact repeats. Lempel-Ziv codes are efficient if all usable structure in the data is in the form of repeats of long "words." If these words are interrupted by insertions, contain "point mutations," or follow grammatical rules which cannot be "discovered" by the algorithm due to lack of data or lack of memory, then Lempel-Ziv algorithms are not optimal (though they would become optimal for infinite data and infinite memory).

We propose that neural networks could represent a way out of this impasse. Though they are typically much slower than both methods mentioned above, they could have enough flexibility to encode the relevant information about the past only, and could, thus, perform well with much smaller memory and on much smaller data sets.

As we shall see, this is indeed the case. We use a rather conventional three-layer feed-forward network which incorporates a number of novel or at least unusual features. We shall describe our implementation in the next section, and discuss the obtained results in section 3.

2. THE NETWORK

As we said, the network is implemented as a rather conventional feed-forward network. It has K linear input nodes (K defines, thus, the width of the window on which the forecast is based), one layer of M hidden nodes, and one output node. While the inputs are binary, the output is a real number and is interpreted as the forecasted probability P_n, i.e., our best approximation for $P_n = \text{prob}(s_n = 1)$. The threshold functions at the hidden and at the output nodes are chosen as

$$g(x) = \frac{1}{(1 + e^{-x})}. \tag{3}$$

In the simplest version, back propagation (i.e., gradient descent)[14] is used as our learning algorithm.

Special care must be taken regarding the cost function. Traditionally, feed-forward networks are used for learning deterministic input-output relations, and the value of the output should, after training, agree with the desired output provided by the training examples. This is not so in our case. Here, each training example consists of a pair (s, s_n) where s is the input window $(s_{n-K} \ldots s_{n-1})$ and the "output" s_n is a binary digit, while the corresponding output of the network is the real number $p_n \in [0, 1]$. We claim that the unique optimal cost function, is in this case,[1,8,16]

$$e = -\sum_{n=1}^{N} [s_n \log p_n + (1 - s_n) \log(1 - p_n)]. \tag{4}$$

Since the correct probabilities for blocks of length $K+1$ are $P_{K+1}(s_1 \ldots s_{K+1})$, and the network has a memory of only K time steps back, the expectation value of e is

$$\langle e \rangle = -\sum_{s_1 \ldots s_K} [P_{K+1}(s_1 \ldots s_K, 1) \log p_n + P_{K+1}(s_1 \ldots s_K, 0) \log(1 - p_n)]. \tag{5}$$

Using well-known results from information theory, this is easily seen to have a unique minimum with

$$\min_{p_n} \langle e \rangle = h_K. \tag{6}$$

Here h_K is the Kth-order Shannon block entropy,

$$h_K = -\sum_{s_1 \ldots s_{K+1}} [P_{K+1}(s_1 \ldots s_{K+1}) \log P_{K+1}(S_{K+1} \mid s_1 \ldots s_{K+1})]. \tag{7}$$

This minimum is reached, if and only if

$$p_{n+1} = P(s_{n+1} = 1 \mid s_1, s_2, \ldots, s_n) \forall n, \tag{8}$$

i.e., if we have hit the correct forecasts for all n. If the forecasts are not exact, the difference $\langle e \rangle - h_K$ is just the Kullback-Leibler relative entropy between P and p.

In the limit $K \to \infty$, one has $h_K \to h$, and, thus, the average cost function is just the entropy if the forecast is optimal.

Notice that if the steps in the gradient descent corresponding to back propagation are finite, the true minimum can never be reached since the cost of each individual training example is not going to a minimum (only the *average* cost does so). In conventional applications (with 0/1 outputs), a similar effect arises if the input-output relation is not always consistent. Here, we cannot, of course, speak of inconsistencies since we never can exclude any output (estimated probabilities should never be exactly 0 or 1), but the effect is similar. In order to come as close as possible to the exact optimum, we, thus, used an "annealing" scheme where the step size goes towards 0 for large times. This is formally similar to simulated annealing in a Monte Carlo simulation, but towards *infinite* temperature; compare with Sompolinsky et al.[17] We found that the precise annealing schedule was not very important, and finally chose to decrease the step size as $1\sqrt{n}$ for large n.

As an alternative to Eq. (4), one might be tempted to use, e.g.,

$$e' = \frac{1}{2} \sum_{n=1}^{N} (s_n - p_n)^2 \,. \tag{9}$$

Though this seems simpler than Eq. (4), the back-propagation equations derived from it are actually more complicated. As noted also in Solla,[16] the back-propagation equations simplify when using Eq. (4), since

$$\frac{\partial e}{\partial p_n} = \frac{(p_n - s_n)}{p_n(1 - p_n)} \tag{10}$$

and, thus, if $p_n = 1/(1 + e^{-x})$,

$$\frac{\partial e}{\partial x} = p_n - s_n \,, \tag{11}$$

while using Eq. (9) yields the more cumbersome $\partial e/\partial x = (p_n - s_n)p_n(1 - p_n)$. More important than this technical disadvantage, the average value of e' after learning is not related to the Shannon entropy but to the order-2 Renyi entropy.[12] Since the latter does not satisfy all Khinchin axioms for a satisfactory information measure, we must expect suboptimal performance, in general.

We should point out that already a network with $K = M = 1$ is able to make exact forecasts if the sequence is a first-order Markov chain. This is easy to prove analytically. For higher-order Markov chains, this is no longer true, unless the number of nodes in the hidden layer is blown up to $\approx 2^K$. The latter would, of course, not be interesting, since the essential advantage of the network would be lost. Indeed, the histograms mentioned above can be viewed as linear networks with essentially this many nodes.

Let us finally discuss the necessary modifications for the case of a k-ary alphabet. Instead of a single probability, P_{n+1}, we have now k forecasts

$$P^i_{n+1} = P(s_{n+1} = i \mid s_1, s_2, \ldots, s_n), \quad i = 1, 2, \ldots k, \quad n = 1, 2, \ldots . \qquad (12)$$

subject to the constraint $\sum_{i=1}^{k} P^i_n = 1$. Our estimates for these are P^i_{n+1}. The cost function whose average value gives the entropy is[1]

$$e = \sum_n e_n, \quad e_n = -\sum_i \delta_{s_n,i} \log p^i_n . \qquad (13)$$

where $\delta_{s_n,i}$ is a Kronecker delta. A suitable network will now have $k-1$ output nodes, each giving a number $q_r (r = 1, 2, \ldots, k-1)$ between 0 and 1. The first is just the estimated probability for $s_n = 1$, i.e., $q_1 = p^1_n$. The rth output q_r is the estimated conditional probability for $s_n = r$, conditioned on $s_n \geq r$, i.e.,

$$p^r_n = (1 - q_1) \ldots (1 - q_{r-1}) q_r . \qquad (14)$$

Inserting this into Eq. (13), we can write e_n as a sum of terms involving $\log q_r$ and $\log(1 - q_r)$ very similar to Eq. (4). More generally, we could also use any other hierarchical grouping of the symbols. We, then, interpret q_r as the conditional probability for a symbol in group r to appear, conditioned on appearance of its parent group.

The input nodes in the case of a k-ary alphabet can be made such that for each input s_m there are $k-1$ nodes (or k nodes, if slight redundancy causes no storage problems), with the ith node having input "1" if and only if $s_m = i$.

RESULTS AND DISCUSSION

Up to now, we tested the network on the following examples:

a. the logistic map $x' = 1 - ax^2$, with a binary partition at $x = 0$ ($s = 0$ if $x < 0, s = 1$ if $x > 0$), at $a = 1.71$;

b. the same map at $a = 1.8$;

c. the Hénon map $(x', y') = (1 + by - ax^2, x)$, with $a = 1.4, b = .3$ and with a partition at $y = 0$;

---

[1] In Hopfield,[8] indeed, a generalization of Eq. (13) was used. Assume that $\tilde{P}^i_{n,a}$ is a conditional probability for $s_n = i$ with some (not necessarily stationary) probability measure compatible with all observations available to the forecaster. Two possible choices are $\tilde{P}^i_{n,a} = P^i_n$ and $\tilde{P}^i_{n,b} = \delta_{s_n,i}$. In Hopfield,[8] the cost function was $e_n = -\sum_i \tilde{P}^i_n \log p^i_n$. With the second choice $\tilde{P}^i_{n,b}$ this gives Eq. (13). We thank A. Lapedes for a discussion clarifying this point.

d. the logistic map at $a = 1.7499$, where it is very strongly intermittent (there is a Pomeau-Manneville intermittency point[11] at $a = 1.75$);

e. sequences obtained by iterating a one-dimensional cellular automaton (rule "30" in Wolfram's notation) for a small and fixed number of times, starting from an input with rather low entropy; and

f. bit strings obtained from ASCII codes of English texts.

(A-C) In the first examples, it was, of course, easy to produce as much data as we wanted. Thus, in our first tests we did not follow the common usage in networks learning deterministic input-output relations, where the training examples are used repeatedly. Instead, in these tests we used each training example only once. In this way we mainly investigate how efficiently the network can *store* the learned information, as compared, e.g., to a histogram. In subsequent tests we checked the performance when the training examples were presented repeatedly. In this way, we test more stringently the ability of the network to generalize.

The first three examples gave excellent results. In all cases, the entropies achieved with networks having K hidden nodes (and K input nodes) agreed within less than 1% with the Kth-order block entropies. This is very encouraging but maybe not too surprising, as these symbol strings do not show very strong long-range correlations. Thus, the forecasts are not too hard to learn. We should point out, however, that even here our positive results are far from trivial. Networks with m real parameters did far better than histograms with m entries (i.e., with window size $\log_2 m$).

It is common usage to initialize feed-forward neural nets with randomly chosen weights. In this way, one breaks the permutation symmetry among the internal nodes which would prevent convergence. In our case, with the same number of input and intermediary nodes, another initialization proved at least as efficient. In this, we set originally $w_{ij} = \delta_{ij}$, where the input to the jth node in the intermediate layer is $\sum_i w_{ij} s_i$. Thus, we break the symmetry in the intermediate layer by associating each internal node with one input symbol.

(D) The first "hard" problem was the intermittent example (d). Surprisingly, the program did very well even there. Results are shown in Figure 1. One might object that this is not too surprising since the structure of the long-range correlations is rather regular (long repeats 011011011011...), but, e.g., a network with $20 + 20$ nodes gave already results which would have been hard to obtain from histograms.

(E) The next example (proposed first in Lindgren[9]) is essentially the hardest artificial test case we could think of. Rule 30 has ...000... as an invariant state, but an isolated "1" creates a triangle which seems completely random inside.[19] Moreover, the time evolution is subjective, implying that the entropy of the input string is preserved at every time step. Finally, completely random strings form an invariant state, and nearly any configuration seems to be attracted to this state, suggesting that the evolution is mixing with uniform invariant measure. Assume, now, one starts the iteration with a string of known low entropy. Due to the mixing, any local measurement will find after sufficiently many iterations that the entropy is 1 bit/symbol. After t iterations, one needs a window of width $> 2t$ to recover the correct value of the entropy.

In our simulations, we started with random input strings with a ratio 1:19 of 1's and 0's. This gives an entropy $h = 0.286$ bits. After $t = 3$ iterations, a network with $K = M = 8$ gave an entropy ~10% higher than h_8, but this was reduced to a satisfactory error of ~1% by going over to $M = 12$. Definitely the limits of our present method were reached with $t = 20$. While, there, our largest feasible histogram ($K = 17$) gave $h_{17} = .413$ bits, our largest feasible network ($K = M = 30$) gave ~ .68 bits.

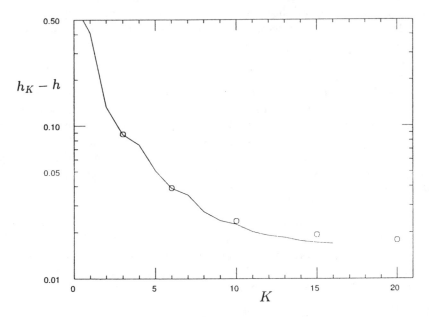

FIGURE 1 Entropy estimates based on windows of length K, using histograms (continuous line) resp. networks with $K + K$ nodes (circles).

(F) The written text analyzed consisted of several Shakespeare plays obtained from the NeXT library, a collection of Associated Press articles, and a portion of a German Bible. In all cases, the ASCII text was converted to binary in a straightforward manner, with no attempt at compression (even the vestigal eighth bit was left intact). Each of the Shakspearean plays consisted of approximately 1 million bits in this form. The APA articles totalled 1.8 Mbits in length and the Bible 6.7 million. The number of input bits K in the neural network was varied from 10 to 60 (K was set equal to M in all cases). The text file studied in most detail was "King Lear." For the largest network used in our simulations ($K = 60$), the network converged to a value of approximately 0.37 bits/bit or, equivalently, 2.96 bits/character. For $K = 10, 20, 30$, and 40, this text was compressed to respectively 4.73, 3.66, 3.35, and 3.16 bits per character. This trend indicates that a larger network would have further increased compression capability. The resource limiting the size of the network was CPU time needed for the learning process. This is in direct contrast to the method of compiling histograms, in which case the limited resource is memory.

The other plays showed very nearly the same compressibility. For $N = K = 30$, we obtained 3.31, 3.35, and 3.34 bits/character. The apparent entropy of the APA articles was slightly higher (3.40). The German text was the most compressible; its entropy, based again on a 30-bit input window, was estimated to be 2.96 bits/character. When the window size was increased to 50, this dropped to 2.64.

It was conjectured that we could further decrease the length by initially compressing the data manually into 5-bit "words." Since a 5-bit word allows us only 32 different characters, we must alter the text slightly. The most harmless way immediately obvious to the authors was the following: (1) change all uppercase characters to their lowercase equivalent; (2) replace all tabs and carriage returns by blanks; and (3) let everything else (interpunctuations, special characters) equal "other." We now have a 28-letter alphabet in which each character can be mapped into a unique 5-bit word. Of course, we cannot uniquely reconstruct the original text from this mapping, but the advantage we gain is that a 40-bit input window will now contain eight characters, whereas before it could only contain five. For the APA collection, the network could only reduce the entropy per bit to 0.752 bits. The corresponding entropy/character is 3.75, considerably higher than for the raw data. This is surprising since the preprocessed data contained less information than the original text. If it is not a special problem of our network, it means that the discarded information was important for predicting some of the retained structures.

Another interesting question is whether using one compression scheme and then another can further increase compressibility. To address this question, the text (this time "Macbeth") was first compressed using the UNIX "compress" command, and the network was then trained on the compressed version. The original compression reduced the length to 3.73 bits/character, and the network, using $K = 40$ was able to reduce this an additional 25% to 2.8 bits/character. This is considerably better than results obtained from the uncompressed data.

4. CONCLUSIONS

It seems that our neural networks do very well where other methods also work, but all available methods have difficulties when dealing with the same hard problems. Notice, however, that the difficulties they encounter are very different: while the limitation in the histogram was purely due to storage (the same would be true for Lempel-Ziv-based forecasts), the limitation in the network case was CPU time. It seems quite likely that CPU time can be reduced considerably by using more efficient updating algorithms. While back propagation uses steepest gradient descent, it could, e.g., be much more efficient to use conjugate gradient descent methods. Other improvements could be achieved by adjusting the topology of the network systematically to the problem at hand.[5,10]

Apart from this, we think that the connections between forecasting, data compression, and neural networks shown in the present work are, by themselves, very interesting.

Finally, we should mention that the entropy estimates obtained using neural networks are always upper bounds (like Lempel-Ziv estimates), provided, of course, that the sequences used for training are typical. This would not be true for estimates obtained from histograms, if the frequencies were estimated naively as ratios of observed frequencies. In this case, the algorithm would misinterpret all fluctuations as real structures, and underestimate entropies. Upper estimates from histograms could only be obtained if Laplace's successor rule[2] is used, but then the overestimation can be very substantial for low statistics.

ACKNOWLEDGMENTS

We are very indebted to Jean-Pierre Nadal, Alan Lapedes, and Sara Solla for most valuable discussions.

REFERENCES

1. Baum, E. B., and F. Wilczek. In *Proceedings of IEEE Conference on Neural Information Processing Systems*. Denver: IEEE, 1987
2. Bulmer, M. G. *Principles of Statistics*. New York: Dover, 1979
3. Chaitin, G. "Towards a Mathematical Definition of 'Life.'" In *The Maximum Entropy Principle*, edited by R. D. Lefine and M. Tribus. Cambridge: MIT Press, 1979
4. Cover, T. M., and R. C. King. *IEEE Trans. Inform. Th.* **24** (1978): 413.
5. Golea, M., and M. Marchand. *Europhys. Lett.* **12** (1990): 205.

6. Grassberger, P. *Int. J. Theor. Phys.* **25** (1986): 907.
7. Grassberger, P. *IEEE Trans. Inform. Th.* **35** (1989): 669.
8. Hopfield, J. *Proc. Nat. Acad. Sci.* **84** (1987): 8429.
9. Lindgren, K. Gothenburg thesis, 1989.
10. Nadal, J. P. *Int. J. Neural Syst.* **1** (1989): 55.
11. Pomeau, Y., and P. Manneville. *Commun. Math. Phys.* **74** (1980): 189.
12. Renyi, A. *Probability Theory.* Amsterdam: North-Holland, 1970
13. Rissanen, J. *Stochastic Complexity in Statistical Inquiry.* Singapore: World Scientific, 1989
14. Rumelhart, D. E., G. Hinton, and R. J. Williams. *Nature* **323** (1986): 533.
15. Shaw, R. *The Dripping Faucet as a Model Chaotic System.* Santa Cruz, CA: Aerial Press, 1984
16. Solla, S. A., E. Levin, and M. Fleisher. *Complex Systems* **2** (1988): 625.
17. Sompolinsky, H., N. Tishby, and H. S. Seung. *Phys. Rev. Lett.* **65** (1990): 1683.
18. van Emden, M. H. *An Analysis of Complexity.* Amsterdam: Mathematical Centre Tracts, 1975
19. Wolfram, S. *Adv. Appl. Math.* **7** (1986): 123.
20. Ziv, J., and A. Lempel. *IEEE Trans. Inform. Th.* **24** (1978): 530.

James P. Crutchfield
Physics Department, University of California, Berkeley, California 94720;
Internet: chaos@gojira.berkeley.edu.

Semantics and Thermodynamics

Inferring models from given data leads through many different changes in representation. Most are subtle and profitably ignored. Nonetheless, any such change affects the semantic content of the resulting model and so, ultimately, its utility. A model's semantic structure determines what its elements mean to an observer that has built and uses it. In the search for an understanding of how large-scale thermodynamic systems might themselves take up the task of modeling and so evolve semantics from syntax, the present paper lays out a constructive approach to modeling nonlinear processes based on computation theory. It progresses from the microscopic level of the instrument and individual measurements, to a mesoscopic scale at which models are built, and concludes with a macroscopic view of their thermodynamic properties. Once the computational structure of the model is brought into the analysis, it becomes clear how a thermodynamic system can support semantic information processing.

Nonlinear Modeling and Forecasting, SFI Studies in the Sciences of Complexity,
Proc. Vol. XII, Eds. M. Casdagli & S. Eubank, Addison-Wesley, 1992   317

NONLINEAR MODELING: FACT OR FICTION?

These ambiguities, redundancies, and deficiencies recall those attributed by Dr. Franz Kuhn to a certain Chinese encyclopedia entitled *Celestial Emporium of Benevolent Knowledge*. On those remote pages it is written that animals are divided into (a) those that belong to the Emperor, (b) embalmed ones, (c) those that are trained, (d) suckling pigs, (e) mermaids, (f) fabulous ones, (g) stray dogs, (h) those that are included in this classification, (i) those that tremble as if they were mad, (j) innumerable ones, (k) those drawn with a very fine camel's hair brush, (l) others, (m) those that have just broken a flower vase, (n) those that resemble flies from a distance.

J. L. Borges, "The Analytical Language of John Wilkins," page 103.[5]

What one intends to do with a model colors the nature of the structure captured by it and determines the effort used to build it. Unfortunately, such intentions most often are not directly stated, but rather are implicit in the choice of representation. To model a given time series, should one use (Fourier) power spectra, Laplace transforms, hidden Markov models, or neural networks with radial basis functions?

Two problems arise. The first is that the choice made might lead to models that miss structure. One solution is to take a representation that is complete: a sufficiently large model captures the data's properties to within an error that vanishes with increased model size. The second, and perhaps more pernicious, problem is that the limitations imposed by such choices are not understood *vis à vis* the underlying mechanisms. This concerns the appropriateness of the representation.

The basis of Fourier functions is complete. But the Fourier model of a square wave contains an infinite number of parameters and so is of infinite size. This is not an appropriate representation, since the data is simply described by a two-state automaton.[1] Although completeness is a necessary property, it simply does not address appropriateness and should not be conflated with it.

Nonlinear modeling, which I take to be that endeavor distinguished by a geometric analysis of processes represented in a state space, offers the hope of describing more concisely and appropriately a range of phenomena hitherto considered random. It can do this since it enlarges the range of representations and forces an appreciation, at the first stages of modeling, of nonlinearity's effect on behavior. Due to this, nonlinear modeling necessarily will be effective.

From the viewpoint of appropriateness, however, nonlinear modeling is an ill-defined science: discovered nonlinearity being the product largely of assumptions made by and resources available to the implementor, and not necessarily a property of the process modeled. There is, then, a question of scientific principle that

---

[1]It is an appropriate representation, though, of the response of free space to carrying weak electromagnetic pulses. Electromagnetic theory is different from the context of modeling *only* from given data. It defines a different semantics.

transcends its likely operational success: How does nonlinearity allow a process to perform different classes of computation and so exhibit more or less complex behavior? This is where I think nonlinear modeling can make a contribution beyond engineering concerns. The contention is that incorporating computation theory will go some distance to basing modeling on first principles.

COMPUTATIONAL MECHANICS

The following discussion reviews an approach to these questions that seeks to discover and to quantify the intrinsic computation in a process. The rules of the inference game demand ignorance of the governing equations of motion. Each model is to be reconstructed from the given data. It follows in the spirit of the research program for chaotic dynamics introduced under the rubric of "geometry from a times series,"[34] though it relies on many of the ideas and techniques of computation and learning theories.[2,21]

I first set up the problem of modeling nonlinear processes in the general context in which I continually find it convenient to consider this task.[10] This includes delineating the effect the measurement apparatus has on the quality and quantity of data. An appreciation of the manner in which data is used to build a model requires understanding the larger context of modeling; namely, given a fixed amount of data, what is the best explanation? Once an acceptable model is in hand, there are a number of properties that one can derive. It becomes possible to estimate the entropy and complexity of the underlying process and, most importantly, to infer the nature of its intrinsic computation. Just as statistical mechanics explains macroscopic phenomena as the aggregation of microscopic states, the overall procedure of modeling can be viewed as going from a collection of microscopic measurements to the discovery of macroscopic observables, as noted by Jaynes.[23] The resulting model summarizes the relation between these observables. Not surprisingly its properties can be given a thermodynamic interpretation that captures the combinatorial constraints on the explosive diversity of microscopic reality. This, to my mind, is the power of thermodynamics as revealed by Gibbsian statistical mechanics.

The following sections are organized to address these issues in just this order. But before embarking on this, a few more words are necessary concerning the biases brought to the development.

The present framework is "discrete unto discrete." That is, I assume the modeler starts with a time series of quantized data and must stay within the limits of quantized representations. The benefit of adhering to this framework is that one can appeal to computation theory and to the Chomsky hierarchy, in particular, as giving a complete spectrum of model classes.[21] By complete here I refer to a procedure that, starting from the simplest, finite-memory models and moving toward the universal Turing machine, will stop with a finite representation at the least powerful computational model class. In a few words that states the overall inference methodology.[11] It addresses, in principle, the ambiguity alluded to above of selecting the wrong modeling class. There will be somewhere in the Chomsky hierarchy an optimal representation which is finitely expressed in the language of the least powerful class.

Finally, note that this framework does not preclude an observer from employing finite precision approximations of real-valued probabilities. I have in mind here using arithmetic codes to represent or transmit approximate real numbers.[4] That is, real numbers are algorithms. It is a mistake, however, to confuse these with the real numbers that are a consequence of the inference methodology, such as the need at some point in time to solve a Bayesian or a maximum entropy estimation problem, as will be done in a later section. This is a fact since an observer constrained to build models and make predictions within a finite time, or with infinite time but access to finite resources, cannot make use of such infinitely precise information. The symbolic problems posed by an inference methodology serve rather to guide the learning process and, occasionally, give insight when finite manipulations of finite symbolic representations lead to finite symbolic answers.

FUZZY β-INSTRUMENTS

The universe of discourse for nonlinear modeling consists of a process P, the measuring apparatus $\mathcal{I}$, and the modeler itself. Their relationships and components are shown schematically in Figure 1. The goal is for the modeler, taking advantage of its available resources, to make the "best" representation of the nonlinear process. This section concentrates on the measuring apparatus. The modeler is the subject of a later section. The process, the object of the modeler's ultimate attention, is the unknown, but hopefully knowable, variable in this picture. And so there is little to say, except that it can be viewed as governed by stochastic evolution equations

$$\vec{X}_{t+\Delta t} = \vec{F}(\vec{X}_t, \vec{\xi}_t, t) \tag{1}$$

where $\vec{X}_t$ is the configuration at time t, $\vec{\xi}_t$ some noise process, and $\vec{F}$ the governing equations of motion.[2] The following discussion also will have occasion to refer to the process's measure $\mu(\vec{X})$ on its configuration space and the entropy rate $h_\mu(\vec{X})$ at which it produces information.

The measuring apparatus is a transducer that maps $\vec{X}_t$ to some accessible states of an instrument $\mathcal{I}$. This instrument has a number of characteristics, most of which should be under the modeler's control. The primary interaction between the instrument and the process is through the measurement space $\mathcal{R}^D$ which is a

---

[2]I explicitly leave out specifying the (embedding) dimension of the process. This is a secondary statistic that is estimated[12] as a topological property of the model, *not* something intrinsic to the present view of the process.

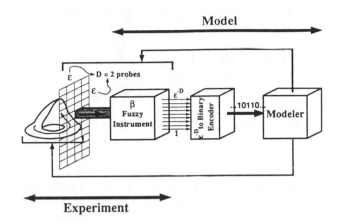

FIGURE 1 The Big Channel. The flow of information (measurements) on the shortest time scales is from the left, from the underlying process, to the right toward the modeler. The latter's task is to build the "best" representation given the available data set and computational resources. On longer time scales the modeler may modify the measuring apparatus and vary the experimental controls on the process. These actions are represented by the left-going arrows. Notice that from the modeler's perspective there is a region of ambiguity between the model and the experiment. The model includes the measuring apparatus since it instantiates many of the modeler's biases toward what is worth observing. But the experiment also includes the measuring apparatus since it couples to the process. Additionally, the apparatus is itself a physical device with its own internal dynamics of which the modeler may be unaware or incapable of controlling.

projection $\mathcal{P}$ of $\vec{X}_t$ onto (say) a Euclidean[3] space whose dimension is given by the number D of experimental probes. The instrument's resolution ϵ in distinguishing the projected states partitions the measurement space into a set $\Pi_\epsilon(D) = \{\pi_i : \pi_i \subset \mathcal{R}^D, i = 0, \ldots, \epsilon^{-D}\}$ of cells. Each cell π_i is the equivalence class of projected states that are indistinguishable using that instrument. The instrument represents the event of finding $\mathcal{P}(\vec{X}_t) \in \pi_i$ by the cell's label i. With neither loss of generality nor information, these indices are then encoded into a time-serial block code. As each measurement is made, its code is output into the data stream. In this way, a time series of measurements made by the instrument becomes a string, the data stream s available to the modeler. This is a discretized set of symbols $s = \ldots s_{-4} s_{-3} s_{-2} s_{-1} s_0 s_1 s_2 s_3 s_4 \ldots$ where, in a single measurement made by the modeler, the instrument returns a symbol $s_t \in \mathbf{A}$ in an alphabet $\mathbf{A}$ at time index $t \in \mathbf{Z}$. Here I take a binary alphabet $\mathbf{A} = \{0, 1\}$.

[3] If measuring p phases, for example, then the associated topology would be $\mathcal{R}^{D-p} \times \mathcal{T}^p$.

This gives the overall idea, but it is, in fact, a gross simplification. I will discuss two important elements that are left out: the instrument temperature and the cell dwell time.

As described, the measurement partition $\Pi_\epsilon(D)$ is "crisp." Each partition cell is associated with an indicator function that maps the state $\vec{x} = \mathcal{P}(\vec{X}) \in \mathcal{R}^D$ onto a symbolic label for that element depending on whether the state is or is not in the domain of that indicator function. But no real instrument implements a crisp measurement partition. There are errors in the assignment of a state $\vec{x}$ to a cell and so an error in the resulting symbol. There are two kinds of errors that one might consider.

The first is a classification error in which the cell is misidentified with the projected state $\vec{x} = \mathcal{P}(\vec{X}_t)$ independent of its location within the measurement cell. If the error rate probability is taken to be p, then the instrument's effective temperature $T_{\text{inst}} = (k_{\text{Boltzmann}}\beta_{\text{inst}})^{-1}$ is simply $\beta_{\text{inst}} = \log_2(1-p)/p$. This is not a very realistic view of classification error. Physical devices, such as analog-to-digital converters, fail more in correct classification near the cell boundaries since they cannot implement exact decision thresholds. In this case, error is not uniform over the partition cells.

One solution to this follows the spirit of fuzzy logic which suggests that the cell indicator function be generalized to a membership function that decays outside a cell.[43] An example of a fuzzy instrument that accounts for this somewhat realistically is to convolve the boundaries of the cells in the crisp partition $\Pi_\epsilon(D)$ with a Fermi-Dirac density. The membership function then becomes

$$\pi_\epsilon^{\beta_{\text{inst}}}(\vec{X}) \propto \frac{1}{e^{\beta_{\text{inst}}(||\mathcal{P}(\vec{X})-\vec{x}_\pi||-\epsilon/2)} + 1}$$

where β_{inst} is the fuzzy partition's inverse temperature and $\vec{x}_\pi \in \mathcal{R}^D$ is the cell's center in the measurement space. At zero temperature the crisp partition is recovered, $\Pi_\epsilon^\beta(D) \xrightarrow[\beta\to\infty]{} \Pi_\epsilon(D)$. At sufficiently high temperatures, the instrument outputs random sequences uncorrelated with the process within the cell.

The algebra of fuzzy measurements will not be carried through the following. I will simply leave behind at this point knowledge of the fuzzy partition. The particular consequences for doing this correctly, though, will be reported elsewhere. The main result is that when done in this generality, the ensuing inference process is precluded from inferring too much and too precise a structure in the source.

The second element excluded from the Big Channel concerns the time $\vec{X}_t$ spends in each partition cell. To account for this there should be an additional time series that gives the cell dwell time for each state measurement. Only in special circumstances will the dwell time be constant when the partition is a uniform coarse-graining. When ergodicity can be appealed to, the average dwell time τ can be used. In any case, it is an important parameter and one that is readily available, but often unused.

The dwell time suggests another instrument parameter, the frequency response, or, more properly dropping the Fourier modeling bias, the instrument's dynamic response. On short time scales the instrument's preceding internal states can affect its resolution in determining the present state and the dwell time. In the simplest case, there is a shortest time below which the instrument cannot respond. Then passages through a cell that are too brief will not be detected or will be misreported.

All of these detailed instrument properties can be usefully summarized by the information acquisition rate $\dot{I}$. In its most general form, it is given by the information gain of the fuzzy partition Π_ϵ^β with respect to the process's asymptotic distribution $\mu(\vec{X})$ projected onto the measurement space. That is,

$$\dot{I}(\tau,\beta,\epsilon,D) - \tau^{-1}H\left(\Pi_\epsilon^\beta(D)|\mathcal{P}(\mu(\vec{X}))\right) \tag{3}$$

where $H(\{P\}|Q)$ is the averaged information gain of distributions $\{P\}$ with respect to Q. Assuming ignorance of the process's distribution allows some simplification and gives the measurement channel capacity

$$\dot{I}(\tau,\beta,\epsilon,D) = \tau^{-1}H(\Pi_\epsilon^\beta(D))$$
$$= \dot{I}(\tau,\infty,\epsilon,D) - \tau^{-1}H(\pi_\epsilon^\beta(D)) \tag{4}$$

$$\text{where } \dot{I}(\tau,\infty,\epsilon,D) = \tau^{-1}\log_2\|\Pi_\epsilon(D)\| = -\tau^{-1}D\log_2\epsilon$$

and where $H(\pi_\epsilon^\beta(D))$ is the entropy of a cell's membership function and $\|\Pi_\epsilon(D)\|$ is the number of cells in the crisp partition. At high temperature

$$H(\pi_\epsilon^\beta(D))\xrightarrow[\beta\to 0]{}\log_2\|\Pi_\epsilon^\beta(D)\|$$

and the information acquisition rate vanishes, since each cell's membership function widens to cover the measurement space. At low temperature $H(\pi_\epsilon^\beta(D))\xrightarrow[\beta\to\infty]{} 0$ and the crisp instrument is recovered.

THE MODELER

Beyond the instrument, one must consider what can and should be done with information in the data stream. Acquisition of, processing, and inferring from the measurement sequence are the functions of the modeler. The modeler is essentially defined in terms of its available inference resources. These are dominated by storage capacity and computational power, but certainly include the inference method's efficacy, for example. Delineating these resources constitutes the barest outline of an observer that builds models. Although the following discussion does not require further development at this abstract a level, it is useful to keep in mind since particular choices for these elements will be presented.

The modeler is presented with s, the bit string, some properties of which were just given. The modeler's concern is to go from it to a useful representation. To do this the modeler needs a notion of the process's effective state and its effective equations of motion. Having built a model representing these two components, any residual error or deviation from the behavior described by the model can be used to estimate the effective noise level of the process. It should be clear when said this way that the noise level and the sophistication of the model depend directly on the data and on the modeler's resources. Finally, the modeler may have access to experimental control parameters. And these can be used to aid in obtaining different data streams useful in improving the model by (say) concentrating on behavior where the effective noise level is highest.

The central problem of nonlinear modeling now can be stated. Given an instrument, some number of measurements, and fixed *finite* inference resources, how much computational structure in the underlying process can be extracted?

LIMITS TO MODELING

Before pursuing this goal directly, it will be helpful to point out several limitations imposed by the data or the modeler's interpretation of the data.

In describing the data stream's character, it was emphasized that the individual measurements are only indirect representations of the process's state. If the modeler interprets the measurements as the process's state, then it is unwittingly forced into a class of computationally less powerful representation. This consists of finite Markov chains with states in **A** or in some arbitrarily selected state alphabet.[4] This will become clearer through several examples used later on. It is important at this early stage to not over-interpret the measurements' content as this might limit the quality of the resulting models.

The instrument itself obviously constrains the observer's ability to extract regularity from the data stream and so it directly affects the model's utility. The most basic of these constraints are given by Shannon's coding theorems.[40] The instrument was described as a transducer, but it also can be considered to be a communication channel between the process and the modeler. The capacity of this channel is $\dot{I} = \tau^{-1} H(\Pi_\epsilon^\beta(D))$. As $\beta \to \infty$ and if the process is deterministic and has entropy $h_\mu(\vec{X}) > 0$, a theorem of Kolmogorov's says that this rate is maximized for a given process if the crisp partition $\Pi_\epsilon(D)$ is generating.[26] This property requires infinite sequences of cell indices to be in a finite-to-one correspondence with the process's states. A similar result was shown to hold for the classes of process of interest here: deterministic, but coupled to an extrinsic noise source.[13] Note that the generating partition requirement necessarily determines the minimal number D of probes required by the instrument.

For an instrument with a crisp generating partition, Shannon's noiseless coding theorem says that the measurement channel must have a capacity higher than the process's entropy

$$\dot{I} \geq h_\mu(\vec{X}) \,. \tag{5}$$

---

[4] As done with hidden Markov models.[18,37]

If this is the case, then the modeler can use the data stream to reconstruct a model of the process and, for example, estimate its entropy and complexity. These can be obtained to within error levels determined by the process's extrinsic noise level.

If $\dot{I} < h_\mu(\vec{X})$, then Shannon's theorem for a channel with noise says that the modeler will not be able to reconstruct a model with an effective noise level less than the equivocation $h_\mu(\vec{X}) - \dot{I}$ induced by the instrument. That is, there will be an "unreconstructable" portion of the dynamics represented in the signal.

These results assume, as is also done implicitly in Shannon's existence proofs for codes, that the modeler has access to arbitrary inference resources. When these are limited there will be yet another corresponding loss in the quality of the model and an increase in the apparent noise level. It is interesting to note that if one were to adopt Laplace's philosophical stance that all (classical) reality is deterministic and update it with the modern view that it is chaotic, then the instrumental limitations discussed here are the general case. And apparent randomness is a consequence of them.

THE EXPLANATORY CHANNEL

A clear statement of the observer's goal is needed, beyond just estimating the best model. Surely a simple model is to be desired from the viewpoint of understandability of the process's mechanism and as far as implementation of the model in (say) a control system is concerned. Too simple a model, though, might miss important structure, rendering the process apparently stochastic and highly unpredictable when it is deterministic, but nonlinear. The trade-off between model simplicity and large unpredictability can be explained in terms of a larger goal for the modeler: to explain to another observer the process's behavior in the most concise manner, but in detail as well. Discussion of this interplay will be couched in terms of the explanatory channel of Figure 2.

Before describing this view, it is best to start from some simple principles. To make contact with existing approaches and for the brevity's sake, the best model will be taken to be the most likely. If one had access to a complete probabilistic description of the modeling universe, then the goal would be to maximize the conditional probability $Pr(\mathbf{M}|\mathbf{s})$ of the model $\mathbf{M}$ given the data stream $\mathbf{s}$. This mythical complete probablistic description $Pr(\mathbf{M}, \mathbf{s})$ is not available, but an approximation can be developed by factoring it using Bayes' rule[22]

$$Pr(\mathbf{M}|\mathbf{s}) = \frac{Pr(\mathbf{s}|\mathbf{M})P(\mathbf{M})}{\sum_{M \in \mathcal{M}} P(\mathbf{s}|\mathbf{M})}. \tag{6}$$

There are several comments. First and foremost, all of these probabilities are conditioned on the choice of model class $\mathcal{M}$. Second, all of the terms on the right-hand side refer to a single data stream $\mathbf{s}$. Third, $Pr(\mathbf{s}|\mathbf{M})$ is the probability that a model $\mathbf{M} \in \mathcal{M}$ produces the given data. Thus, candidate models are considered to be generators of data. With sufficient effort, then, $Pr(\mathbf{s}|\mathbf{M})$ can be estimated. Finally, the normalization $Pr(\mathbf{s}) = \sum_{M \in \mathcal{M}} Pr(\mathbf{s}|\mathbf{M})$ depends only on the given data and so can be dropped since it is a constant when maximizing $Pr(\mathbf{M}|\mathbf{s})$ over the model class.

Shannon's coding theorem established that an event of probability p can be optimally represented by a code with length $-\log_2 p$ bits.[40] The search for the most likely explanation is tantamount to constructing the shortest code $\hat{\mathbf{X}}$ for the data $\mathbf{s}$. The length of the optimal code is then $||\hat{\mathbf{X}}|| = -\log_2 Pr(\mathbf{M}|\mathbf{s})$. Using the above Bayesian decomposition of the likelihood, it follows that

$$||\hat{\mathbf{X}}|| \propto -\log_2 Pr(\mathbf{s}|\mathbf{M}) - \log_2 Pr(\mathbf{M}) . \tag{7}$$

The resulting optimization procedure can be described in terms of the explanatory channel of Figure 2. There are two observers $\mathbf{A}$ and $\mathbf{B}$ that communicate an explanation $\mathbf{X}$ via a channel. The input to this explanatory channel, what the modeler $\mathbf{A}$ sees, is the data stream $\mathbf{s}$; the output, what $\mathbf{B}$ can resynthesize given the explanation $\mathbf{X}$, will be denoted $\mathbf{s}'$.

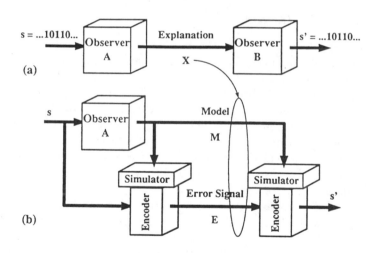

FIGURE 2   The Explanatory Channel. The upper portion (a) of the figure illustrates the bare channel: observer $\mathbf{A}$ communicates an explanation $\mathbf{X}$ to observer $\mathbf{B}$. The lower portion (b) shows in more detail the two subchannels making up the explanatory channel: the model channel transmits model $\mathbf{M}$ and the error channel transmits an error signal $\mathbf{E}$. $\mathbf{M}$ is first built by observer $\mathbf{A}$ and then loaded into $\mathbf{A}$'s simulator. It is also transmitted to $\mathbf{B}$ which loads into its simulator. $\mathbf{A}$ develops the error signal $\mathbf{E}$ as the deviation of the measurements in the data stream $\mathbf{s}$ from those predicted by simulating the model. Only these deviations are transmitted, and at less precision than the original individual measurements. $\mathbf{B}$ is to resynthesize a data stream $\mathbf{s}'$ by simulating the model and when that is not predictive, to use information from the error signal. $\mathbf{X}$ then *explains* $\mathbf{s}$ if $\mathbf{s}' = \mathbf{s}$.

As shown in Figure 2(b) $\mathbf{X}$ is transmitted over two subchannels. The first is the modeling channel along which a model $\mathbf{M}$ is communicated. The second is the error channel along which an error signal $\mathbf{E}$ is transmitted. $\mathbf{E}$ is that portion of s unexplained by the model $\mathbf{M}$.

There are two criteria for a good explanation:

1. $\mathbf{X}$ must explain s. That is, $\mathbf{B}$ must be able to resynthesize the original data: $\mathbf{s'} = \mathbf{s}$.
2. The explanation must be as short as possible. That is, the length $||\mathbf{X}|| = ||\mathbf{M}|| + ||\mathbf{E}||$ in bits of $\mathbf{X}$ must be minimized.

The efficiency of an explanation or, equivalently, of the model depends on the compression ratio

$$\mathcal{C}(\mathbf{M}, \mathbf{s}) = \frac{||\mathbf{X}||}{||\mathbf{s}||} = \frac{||\mathbf{M}|| + ||\mathbf{E}||}{||\mathbf{s}||} . \tag{8}$$

This quantifies the efficacy of an explanation employing model $\mathbf{M}$. $\mathcal{C}$ is then a cost function over the space $\mathcal{M}$ of possible models. The optimal model $\hat{\mathbf{M}}$ minimizes this cost

$$\mathcal{C}(\hat{\mathbf{M}}, \mathbf{s}) = \inf_{\mathbf{M} \in \mathcal{M}} \mathcal{C}(\mathbf{M}, \mathbf{s}) . \tag{9}$$

Figure 3 illustrates the basic behavior of the cost function $\mathcal{C}$ over the model space $\mathcal{M}$. There are two noteworthy extremes. When the model is trivial, $\mathbf{M} = \emptyset$, it predicts nothing about the data stream. In this case, the entire data stream must be sent along the error channel. $||\mathbf{E}|| = ||\mathbf{s}||$ is as large as it can be, but the model is small $||\mathbf{M}|| = 0$. The second, complementary extreme occurs when the data stream is taken as its own model $\mathbf{M} = \mathbf{s}$. No information needs to be sent on the error channel since the model explains all of the data stream. In this case, $||\mathbf{M}|| = ||\mathbf{s}||$ and $||\mathbf{E}|| = 0$. This is the overfitting regime: model parameters come to fit each measurement.

The view provided by the explanatory channel turns on the existence of an optimal code for a given information source. The semantics is decidedly different, though, in that it interprets the code as consisting of two parts: the model and the error. This is the main innovation beyond Shannon's theory. It apparently was given its first modern articulation by Kemeny[24] as an implementation of Ockham's dictum that "diversity should not be multiplied beyond necessity." It has been put on a more rigorous foundation by Rissanen[38] who adapted the Kolmogorov-Chaitin-Solomonoff algorithmic theory of inductive inference[29] to the needs of universal coding theory, and by Wallace and Freeman[41] in the domain of statistical classification.

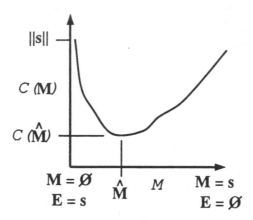

FIGURE 3   Model optimality. A schematic illustration of the features of the cost function $\mathcal{C}(\mathbf{M}, s)$ over the model class $\mathcal{M}$. The topology of the latter is extremely important and is by no means one dimensional. The right portion of the graph is the region of overfitting: so many parameters are used in the model that they begin to directly reflect individual measurements. The left portion of the graph is the region of high apparent randomness: the model captures so little of the data that there is large prediction error.

The notion of the explanatory channel might seem nonetheless to be a bit of an abstraction as far as modeling nonlinear processes is concerned. It was implemented, in effect, in a software system for reconstructing the equations of motion from dynamical system time series.[12] The system contained a symbolic dynamical system interpreter as the simulation portion of the channel. The error signal was determined as the deviation of the input trajectory from the deterministic dynamic. Initially, it was averaged by assuming the deviations to be IID Gaussian variables. Optimal models were then selected by minimizing the model entropy[5] which consisted of a model complexity term and a prediction error term. In this view, the precision of the error signal along different directions in the tangent space to the dynamic is modulated by the spectrum of the associated local Lyapunov characteristic exponents.

OPTIMAL INSTRUMENTS

The quantitative search for an optimal model extends to criteria for building optimal instruments. In one view at least, the instrument is part of the model. There are two basic principles that are easily summarized

1. Use all the data, and
2. Nothing but the data.

[5]A variant of Akaike's Boltzmann Information Criterion for model order selection.[1]

Formally, these translate into the following criteria for instruments.

1. Maximize the conditional entropy $h_\mu(s|\mathcal{I}(\tau, \epsilon, \beta, D))$ of the data over the space of instruments. As will be seen later, h_μ is readily estimated using the reconstructed model $\hat{M}(s|\mathcal{I}(\tau, \epsilon, \beta, D))$.
2. Minimize the complexity of the reconstructed machine

$$\inf_{M \in \mathcal{M}} \|M(s|\mathcal{I}(\tau, \epsilon, \beta, D))\|. \tag{9}$$

With sufficiently large data sets, prediction errors dominate over model size and only the first optimization need be performed. In this regime, early results demonstrated that in accordance with Kolmogorov's theorem, the maxima are attained at generating partitions. Going somewhat beyond the theorem, they also showed that the dependence near the maxima was smooth. Later results showed that the order of the conditional entropy maximum is determined by, and so is an indication of, the smoothness of the equations of motion.[10] For finite and especially small data sets, however, the model size plays a significant role. In that regime the criteria are optimized simultaneously over the space of instruments. Exactly how this is done to select the optimal instrument $\hat{\mathcal{I}}$ will be left for discussion elsewhere.

The overall picture here is a formalism for implementing the Baconian scientific algorithm of experimentation and refinement. In the drive to understand and predict more of the process, the modeler updates the instrument. An improved model allows the instrument to be modified to remove discovered regularity from the measurements before the information is put into the data stream. In this way, over long times the instrument as transducer provides an increasingly more informative data stream that in principle narrows in on behavior that is less well modeled. One consequence of the coding theoretic view is that, as the instrument takes into account more and more regularity, the resulting data stream from it looks more and more like noise. Concomitantly, the residual regularity requires ever larger inference resources to extract.

Such a high level view of inductive inference is all very well and good; especially in light of the rather large number of parameters that appear. There is one problem, however, that goes to the heart of its coding theoretic premises. This is the almost complete lack of attention to the functional properties of the reconstructed models. It is exactly these properties that have scientific value. Furthermore, that value is independent of the amount of data used to find the model. This problem is reflected in the formalism's ignorance of the topological and metric properties of the model class and range of classes. The claim in the following is that these can be accounted for more directly with a measure of complexity and an investigation of computational properties of individual models. To address these the next section begins to focus on a particular class of models. Once their inference algorithm is outlined and some basic properties described, the discussion examines their utility and semantic content.

COMPUTATION FROM A TIME SERIES

On what sort of structure in the data stream should the models be based? If the goal is prediction, as the preceding assumed, then a natural object to reconstruct from the data series is a representation of the instantaneous state of the process. Unfortunately, as already noted, individual measurements are only indirect representations of the process's state. Indeed, the instrument simply may not supply data of adequate quality in order to discover the true states independent of the amount of data. So how can the process's "effective" states be accessed?

The answer to this turns on a generalization of the "reconstructed states" introduced, under the assumption that the process is a continuous-state dynamical system, by Packard et al.[34] The contention there was that a single time series necessarily contained all of the information about the dynamics of that time series. The notion of reconstructed state was based on Poincaré's view of the intrinsic dimension of an object.[36] This was defined as the largest number of successive cuts through the object resulting in isolated points. A sphere in three dimensions by his method is two dimensional since the first cut typically results in a circle and then a second cut, of that circle, isolates two points. Packard et al. implemented this using probability distributions conditioned on values of the time series' derivatives. This was, in fact, an implementation of the differential geometric view of the derivatives as locally spanning the graph of the dynamic.

In this reconstruction procedure a state of the underlying process is identified once the conditional probability distribution is peaked. It was noted shortly thereafter that in the presence of extrinsic noise, a number of conditions is reached beyond which the conditional distribution is no longer sharpened.[13] And, as a result, the process's state cannot be further identified. The width of the resulting distribution then gives an estimate of the effective extrinsic noise level and the minimum number of conditions first leading to this situation, an estimate of the effective dimension.

The method of time derivative reconstruction gives the key to discovering states in discrete times series.[6] For discrete time series a state is defined to be the set of subsequences that render the future conditionally independent of the past.[7] Thus, the observer identifies a state at different times in the data stream as being in identical conditions of ignorance about the future. The set of future subsequences following from a state is called its **morph**.

For this definition of state, several reconstruction procedures have been developed. In brief, the simplest method consists of three steps. In the first all length D subsequences in the data stream are represented as paths in a depth D binary "parse" tree. In the second, the morphs are discovered by associating them with the distinct depth $L = D/2$ subtrees found in the parse tree down to depth $D/2$.

---

[6] The time delay method appears not to generalize.

[7] This notion of state is widespread; appearing in various guises in early symbolic dynamics, ergodic, and automata theories. It is the basic notion of state in Markov chain theory.[14]

The number of morphs is then the number of effective states. In the final step, the state-to-state transitions are found by looking at how each state's associated subtrees map into one another on the parse tree.[11,14,15]

This procedure reconstructs from a data stream a "topological" machine: the skeleton of states and allowed transitions. There are a number of issues concerning statistical estimation, including error analysis and probabilistic structure, that need to be addressed.[16] But this outline suffices for the present purposes. The estimated models are referred to as ϵ-machines in order to indicate their dependence not only on measurement resolution, but also indirectly on all of the instrumental and inferential parameters discussed so far.

ϵ-MACHINES

The product of machine reconstruction is a set of states that will be associated with a set $\mathbf{V} = \{v\}$ of vertices and a set of transitions associated with a set $\mathbf{E} = \{e : e \sim v \underset{s}{\to} v', v, v' \in \mathbf{V}, s \in \mathbf{A}\}$ of labeled edges. Formally, the reconstruction procedure puts no limit on the number of machine states inferred. Indeed, in some important cases the number is infinite, such as at phase transitions.[15] In the following $\mathbf{V}$ will be a finite set and the machines "finitary." One depiction of the reconstructed machine M is as a labeled directed graph $G = \{\mathbf{V}, \mathbf{E}\}$. Examples will be seen shortly. The full probabilistic structure is described by a set of transition matrices

$$\mathcal{T} = \{T^{(s)} : (T^{(s)})_{vv'} = p_{v \underset{s}{\to} v'}, v, v' \in \mathbf{V}, s \in \mathbf{A}\} \tag{11}$$

where $p_{v \underset{s}{\to} v'}$ denotes the conditional probability to make a transition to state v' from state v on observing symbol s.

A stochastic machine is a compact way of describing the probabilities of a possibly infinite number of measurement sequences. The probability of a given sequence $\mathbf{s}^L = s_0 s_1 s_2 \ldots s_{L-1}, s_i \in \mathbf{A}$, is recovered from the machine by the telescoping product of conditional transition probabilities

$$p(\mathbf{s}^L) = p_{v_0} p_{v_0 \underset{s_0}{\to} v_1} p_{v \underset{s_1}{\to} v_2} \cdots p_{v_{L-1} \underset{s_{L-1}}{\to} v_n} . \tag{12}$$

Here v_0 is the unique start state. It is the state of total ignorance, so that, at the first time step, $p_{v_0} = 1$. The sequence $v_0, v_1, v_2, \ldots, v_{L-1}, v_L$ consists of those states through which the sequence drives the machine. To summarize, a machine is the set $M = \{\mathbf{V}, \mathbf{E}, \mathbf{A}, \mathcal{T}, v_0\}$.

Several important statistical properties are captured by the stochastic connection matrix

$$T = \sum_{s \in \mathbf{A}} T^{(s)} \tag{13}$$

where $(T)_{vv'} = p_{v \to v'}$ is the state to state transition probability, unconditioned by the measurement symbols. By construction every state has an outgoing transition. This is reflected in the fact that T is a stochastic matrix: $\sum_{v' \in \mathbf{V}} p_{vv'} = 1$. It should be clear that by dropping the input alphabet transition labels from the machine, the detailed, call it "computational," structure of the input data stream has been lost. All that is retained in T is the state transition structure and this is a Markov chain. The interesting fact is that Markov chains are a proper subset of stochastic finitary machines. Examples later on will support this contention. It is at exactly this step of unlabeling the machine that the "properness" appears.

The stationary-state probabilities $\vec{p}_{\mathbf{V}} = \{p_v : \sum_{v \in \mathbf{V}} p_v = 1, v \in \mathbf{V}\}$ are given by the left eigenvector of T

$$\vec{p}_{\mathbf{V}} T = \vec{p}_{\mathbf{V}} . \tag{14}$$

The entropy rate of the Markov chain is then

$$h_\mu(T) = -\sum_{v \in \mathbf{V}} p_v \sum_{v' \in \mathbf{V}} p_{v \to v'} \log_2 p_{v \to v'} . \tag{15}$$

This measures the information production rate in bits per time step of the Markov chain. Although the mapping from input strings to the chain's transition sequences is not in general one-to-one, it is finite-to-one. And so, the Markov chain entropy rate is also the entropy rate of the original data source

$$h_\mu(M) = -\sum_{v \in \mathbf{V}} p_v \sum_{v' \in \mathbf{V}} \sum_{s \in \mathbf{A}} p_{v \underset{s}{\to} v'} \log_2 p_{v \underset{s}{\to} v'} . \tag{16}$$

The complexity[8] quantifies the information in the state-alphabet sequences

$$C_\mu(M) = H(\vec{p}_{\mathbf{V}}) = -\sum_{v \in \mathbf{V}} p_v \log_2 p_v . \tag{17}$$

It measures the amount of memory in the process. For completeness, note that there is an edge complexity that is the information contained in the asymptotic edge distribution $\vec{p}_{\mathbf{E}} = \{p_e = p_v p_{v \underset{s}{\to} v'} : v, v' \in \mathbf{V}, e \in \mathbf{E}, s \in \mathbf{A}\}$

$$C_\mu^e(M) = -\sum_{e \in \mathbf{E}} p_e \log_2 p_e . \tag{18}$$

These quantities are not independent. Conservation of information at each state leads to the relation

$$C_\mu^e = C_\mu + h_\mu . \tag{19}$$

[8] Within the reconstruction hierarchy this is actually the finitary complexity, since the context of the discussion implies that we are considering processes with a finite amount of memory. However, I have not introduced this restriction in unnecessary places in the discussion. The finitary complexity has been considered before in the context of generating partitions and known equations of motion.[19,31,42]

And so, there are only two independent quantities when modeling a process as
a stochastic finitary machine. The entropy h_μ, as a measure of the diversity of
patterns, and the complexity C_μ, as a measure of memory, have been taken as the
two elementary coordinates with which to analyze a range of sources.[15]

There is another set of quantities that derive from the skeletal structure of the
machine. Dropping all of probabilistic structure, the growth rate of the number of
sequences it produces is the topological entropy

$$h = \log_2 \lambda(T_0) \tag{20}$$

where $\lambda_0(T_0)$ is the principle eigenvalue of the connection matrix $T_0 = \sum_{s \in \mathbf{A}} T_0^{(s)}$.
The latter is formed from the labeled matrices

$$\left\{ T_0^{(s)} : \left(T_0^{(s)} \right)_{vv'} = \begin{cases} 1 & p_{v \xrightarrow{s} v'} > 0 \\ 0 & \text{otherwise} \end{cases} s \in \mathbf{A} \right\}. \tag{21}$$

The state and transition topological complexities are

$$\begin{aligned} C &= \log_2 \|\mathbf{V}\|; \\ C^e &= \log_2 \|\mathbf{E}\|. \end{aligned} \tag{22}$$

In computation theory, an object's complexity is generally taken to be the size
in bits of its representation. The quantities just defined measure the complexity of
the reconstructed machine. As will be seen in the penultimate section, when these
entropies and complexities, both topological and metric, are integrated into a single
parametrized framework, a thermodynamics of machines emerges.

COMPLEXITY

It is useful at this stage to stop and reflect on some properties of the models that
I have just described how to reconstruct. Consider two extreme data sources. The
first, highly predictable, produces a stream of 1s; the second, highly unpredictable,
is an ideal random source of binary symbols. The parse tree of the predictable
source is a single path of 1s. And there is a single subtree, at any depth. As a result
the machine has a single state and a single transition on $s = 1$: a simple model of
a simple source. For the ideal random source, the parse tree, again to any depth, is
the full binary tree. All paths appear in the parse tree since all binary subsequences
are produced by the source. There is a single subtree, of any morph depth at all
parse tree depths: the full binary subtree. And the machine has a single state with
two transitions; one on $s = 1$ and one on $s = 0$. A simple machine, even though the
source produces the widest diversity of binary sequences.

A simple gedanken experiment serves to illustrate how complexity is a measure
of a machine's memory capacity. Consider two observers $\mathbf{A}$ and $\mathbf{B}$, each with the
same model $\mathbf{M}$ of some process. $\mathbf{A}$ is allowed to start machine $\mathbf{M}$ in any state and
uses it to generate binary strings that are determined by the edge labels of the

transitions taken. These strings are passed to observer **B** which traces their effect through its own copy of **M**. On average how much information about **M**'s state can **A** communicate to **B** via the binary strings? If the machine describes (say) a period-three process, e.g., it outputs strings like $101101101\ldots$, $011011011\ldots$, and $110110110\ldots$, it has $\|\mathbf{V}\| = 3$ states. Since **A** starts **M** in different states, **B** can learn only the information of the process's phase in the period-three cycle. This is $\log_2 \|\mathbf{V}\| = 1.584\ldots$ bits of information about the process's state, if **A** chooses the initial states with equal probability. However, if the machine describes an ideal random binary process, by definition **A** can communicate no information to **B**: there is no structure in the sequences to use for this purpose. This is reflected in the fact, as already noted above, that the corresponding machine has a single state and its complexity is $\log_2 1 = 0$. In this way, a process's complexity is the amount of information that someone controlling its start state can communicate to another.

These examples serve to highlight one of the most basic properties of complexity, as I use the term. Both predictable and random sources are simple in the sense that their models are small. Complex processes in this view have large models. In computational terms, complex processes have, as a minimum requirement, a large amount of memory as revealed by many internal states in the reconstructed machine. Most importantly, that memory is structured in particular ways that support different types of computation. The sections below on knowledge and meaning show several consequences of computational structure.

In the most general setting, I use the word "complexity" to refer to the amount of information contained in observer-resolvable equivalence classes. For finitary machines, the complexity is measured by the quantities labeled above by C. This notion has been referred to as the "statistical complexity" in order to distinguish it from the Chaitin-Kolmogorov complexity,[9,27] the Lempel-Ziv complexity,[28] Rissanen's stochastic complexity,[38] and others[45,44] which are all equivalent in the limit of long data streams to the process's Kolmogorov-Sinai entropy $h_\mu(\vec{X})$. If the instrument is generating and $\mu(\vec{X})$ is absolutely continuous, these quantities are given by the entropy rate of the reconstructed machine, Eq. (16).[7] Accordingly, I use the word "entropy" to refer to such quantities. They measure the diversity of sequences a process produces. Implicit in their definitions is the restriction that the modeler must pay computationally for each random bit. Simply stated, the overarching goal is exact description of the data stream. In the modeling approach advocated here, the modeler is allowed to flip a coin or to sample the heat bath to which it may be coupled. "Complexity" is reserved in my vocabulary to refer to a process's structural properties, such as memory and other types of computational capacity.

This is not the place to review the wide range of alternative notions of "complexity" that have been discussed more recently in the physics and dynamics literature. The reader is referred to the comments and especially the citations elsewhere.[14,15] It is important to point out, however, that the notion defined here does not require knowledge of the equations of motion, the prior existence of exact conditional probabilities, Markov or even generating partitions of the state space, continuity and differentiability of the state variables, nor the existence of periodic orbits. Furthermore, the approach taken here differs from those based on the construction of universal codes in the emphasis on the model's structure. That emphasis brings it into direct contact with the disciplines of stochastic automata, formal language theory, and thermodynamics.

Finally, statistical complexity is a highly relative concept that depends directly on the assumed model class. In the larger setting of hierarchical reconstruction, it becomes the finitary complexity since it measures the number of states in a finite-state machine representation. But there are other versions appropriate, for example, when the finitary complexity diverges.[11]

CAUSALITY

There are a few points that must be brought out concerning what these reconstructed machines represent. First, by the definition of future-equivalent states, the machines give the minimal information dependency between the morphs. In this respect, they represent the causality of the morphs considered as events. The machines capture the information flow within the given data stream. If state **B** follows state **A**, then **A** is a cause of **B** and **B** is one effect of **A**. Second, machine reconstruction produces minimal models up to the given prediction error level. This minimality guarantees that there are no other events (morphs) that intervene, at the given error level, to render **A** and **B** independent. In this case, we say that information flows from **A** to **B**. The amount of information that flows is given by the mutual information $\mathbf{I(A;B) = H(B) - H(B|A)}$ of observing state-event **A** followed by state-event **B**. Finally, time is the natural ordering captured by machines. An ϵ-machine for a process is then the minimal causal representation reconstructed using the least powerful computational model class that yields a finite complexity.

KNOWLEDGE RELAXATION

The next two sections investigate how models can be used by an observer. An observer's knowledge $\mathcal{K}_P$ of a process P consists of the data stream, its current model, and how the information used to build the model was obtained.[9] Here the latter is given by the measuring instrument $\mathcal{I} = \{\Pi_\epsilon^\beta(D), \tau\}$. To facilitate interpretation and calculations, the following will assume a simple data acquisition discipline with uniform sampling interval τ and a time-independent zero-temperature measurement partition Π_ϵ. Further simplification comes from ignoring external factors, such as what the observer intends or needs to do with the model, by assuming that the observer's goal is solely optimal prediction with respect to the model class of finitary machines.

[9] In principle, the observer's knowledge also consists of the reconstruction method and its various assumptions. But it is best to not elaborate on this here. These and other unmentioned variables are assumed to be fixed.

The totality of knowledge available to an observer is given by the development of its $\mathcal{K}_P$ at each moment during its history. If we make the further assumption that by some agency the observer has at each moment in its history optimally encoded the available current and past measurements into its model, then the totality of knowledge consists of four parts: the time series of measurements, the instrument by which they were obtained, and the current model and its current state. Stating these points so explicitly helps to make clear the upper bound on what the observer can know about its environment. Even if the observer is allowed arbitrary computational resources, given either finite information from a process or finite time, only a finite amount of structure can be inferred.

An ϵ-machine is a representation of an observer's model of a process. To see its role in the change in $\mathcal{K}_P$, consider the situation in which the model structure is kept fixed. Starting from the state v_0 of total ignorance about the process's state, successive steps through the machine lead to a refinement of the observer's knowledge as determined by a sequence of measurements. The average increase in $\mathcal{K}_P$ is given by a diffusion of information throughout the model. The machine transition probabilities, especially those connected with transient states, govern how the observer gains more information about the process with longer measurement sequences.

A measure of information relaxation on finitary machines is given by the time-dependent finitary complexity

$$C_\mu(t) = H(\vec{p}_\mathbf{V}(t)) \tag{23}$$

where $H(P) = -\sum_{p_i \in P} p_i \log_2 p_i$ is the Shannon entropy of the distribution $P = \{p_i\}$ and

$$\vec{p}_\mathbf{V}(t+1) = \vec{p}_\mathbf{V}(t)T \tag{24}$$

is the probability distribution at time t beginning with the initial distribution $p_\mathbf{V}(0) = (1, 0, 0, \ldots)$ concentrated on the start state. The latter distribution represents the observer's state of total ignorance of the process's state, i.e., before any measurements have been made, and correspondingly $C_\mu(0) = 0$. $C_\mu(t)$ is simply (the negative of) the Boltzmann H-function in the present setting. And we have the analogous result to the H-theorem for stochastic ϵ-machines: $C_\mu(t)$ converges monotonically when $\vec{p}_\mathbf{V}(t)$ is sufficiently close to $\vec{p}_\mathbf{V} = \vec{p}_\mathbf{V}(\infty) : C_\mu(t) \underset{t \to \infty}{\to} C_\mu$. That is, the time-dependent complexity limits on the finitary complexity. Furthermore, the observer has the maximal amount of information about the process, i.e., the observer's knowledge is in equilibrium with the process when $C_\mu(t+1) - C_\mu(t)$ vanishes for all $t > t_{\text{lock}}$, where t_{lock} is some fixed time characteristic of the process.

For finitary machines there are two convergence behaviors for $C_\mu(t)$. These are illustrated in Figure 4 for three processes: one P_3 which is period 3 and generates $(101)^*$, one P_2 in which only isolated zeros are allowed, and one P_1 that generates 1s in blocks of even length bounded by 0s. The first behavior type, illustrated by P_3 and P_2, is monotonic convergence from below. In fact, the asymptotic approach occurs

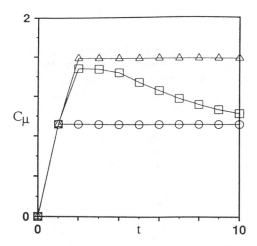

FIGURE 4 Temporal convergence of the complexity $C_\mu(t)$ for a period-three process P_3 (triangles), a Markovian process P_2 whose support is a subshift of finite type (circles), and a process P_1 that generates blocks of even numbers of 1s surrounded by 0s (squares).

in finite time. This is the case for periodic and recurrent Markov chains, where the latter refers to finite-state stochastic processes whose support is a subshift of finite type (SSFT). The convergence here is over-damped.

The second convergence type, illustrated by P_1, is only asymptotic; convergence to the asymptotic state distribution is only at infinite time. There are two subcases. The first is monotonic increasing convergence; the conventional picture of stochastic process convergence. The second subcase P_1 is nonmonotonic convergence. In this case, starting in the condition of total ignorance leads to a critically damped convergence with a single overshoot of the finitary complexity. With other initial distributions, oscillations, i.e., underdamped convergence, can be seen. Exact convergence is only at infinite time. This convergence type is associated with machines having cycles in the transient states or, in the classification of symbolic dynamics, with machines whose support is a strictly Sofic[32] system (SSS).[10] For these, at some point in time the initial distribution spreads out over more than just the recurrent states. $C_\mu(t)$ can then be larger than C_μ. Beyond this time, it converges from above. Much of the detailed convergence behavior is determined, of course, by T's full eigenvalue spectrum. The interpretation just given, though, can be directly deduced by examining the reconstructed machine's graph G. One aspect which is less immediate is that for SSSs the initial distribution relaxes through an infinite number of Cantor sets in sequence space. For SSFTs there is only a finite number of Cantor sets.

[10] SSS shall also refer, in context, to stochastic Sofic systems.[25]

TABLE 1 β-locking times for the periodic P_3, isolated
0s P_2, and even 1s P_1, processes. Note that for the latter the locking time is substantially longer and depends
on β. For the former two, the locking times indicate
the times at which asymptotic convergence has been
achieved. The observer knows the state of the underlying process with certainty at those locking times. For
P_1, however, at $t = 17$ the observer is partially phase-
locked with knowledge of 99% of the process's state
information.

Locking Times at 1% Level	
Process	Locked at time
Period 3	2
Isolated 0s	1
Even 1 blocks	17

This structural analysis indicates that the ratio

$$\Delta C_\mu(t) = \frac{|C_\mu - C_\mu(t)|}{C_\mu} \tag{25}$$

is largely determined by the amount of information in the transient states. For
SSSs this quantity only asymptotically vanishes since there are transient cycles
in which information persists for all time, even though their probability decreases
asymptotically. This leads to a general definition of (chaotic or periodic) phase and
phase locking. The phase of a machine at some point in time is its current state.
There are two types of phase of interest here. The first is the process's phase and
the second is the observer's phase which refers to the state of the observer's model
having read the data stream up to some time. The observer has β-locked onto the
process when $\Delta C_\mu(t_{\text{lock}}) < \beta$. This occurs at the locking time t_{lock} which is the
longest time t such that $\Delta C_\mu(t) = \beta$. When the process is periodic, this notion of
locking is the standard one from engineering. But it also applies to chaotic processes
and corresponds to the observer knowing what state the process is in, even if the
next measurement cannot be predicted exactly.

These two classes of knowledge relaxation lead to quite different consequences
for an observer even though the processes considered above all have a small number
of states (2 or 3) and share the same single-symbol statistics: $Pr(s = 1) = 2/3$
and $Pr(s = 0) = 1/3$. In the over-damped case, the observer knows the state of
the underlying process with certainty after a finite time. In the critically damped
situation, however, the observer has only approximate knowledge for all times. For

example, setting $\beta = 1\%$ leads to locking times shown in Table 1. Thus, the ability of an observer to infer the state depends crucially on the process's computational structure, viz. whether its topological machine is an SSFT or an SSS. The presence of extrinsic noise and observational noise modify these conclusions systematically.

It is worthwhile to contrast the machine model of P_1 with a model based on histograms, or look-up tables, of the same process. Both models are given sufficient storage to exactly represent the length-3 sequence probability distribution. They are then used for predictions on length-4 sequences. The histogram model will store the probabilities for each length-3 sequence. This requires 8 bins each containing an 8-bit approximation of a rational number: 3 bits for the numerator and 5 for the denominator. The total is 67 bits which includes an indicator for the most recent length-3 sequence. The machine model (see Figure 5) must store the current state and five approximate rational numbers, the transition probabilities, using 3 bits each: one for the numerator and two for the denominator. This gives a model size of 17 bits.

Two observers, each given one or the other model, are presented with the sequence 010. What do they predict for the event that the fourth symbol is $s = 1$? The histogram model predicts

$$Pr(1|010) \approx Pr(1|10) = \frac{Pr(101)}{Pr(10)} = \frac{2/18}{2/9} = \frac{1}{2}, \tag{26}$$

whereas the machine model predicts

$$Pr(1|010) = p_{\mathbf{C} \rightarrow \mathbf{A}} = 0. \tag{27}$$

The histogram model gives the wrong prediction. It says that the fourth symbol is uncertain when it is completely predictable. A similar analysis for the prediction of measuring $s = 1$ having observed 011 shows the opposite. The histogram model predicts $s = 1$ is more likely $p_{s=1} = 2/3$; when it is, in fact, not predictable at all $p_{s=1} = 1/2$. This example is illustrative of the superiority of stochastic machine models over histogram and similar look-up table models of time-dependent processes. In fact, there are processes with finite memory for which no finite-size sequence histogram will give correct predictions.

In order to make the physical relevance of SSSs and their slow convergence more plausible, the next example is taken from the Logistic map at a Misiurewicz parameter value. The Logistic map is an iterated mapping of the unit interval

$$x_{n+1} = f_r(x_n) = rx_n(1 - x_n), r \in [0, 4], x_0 \in [0, 1]. \tag{28}$$

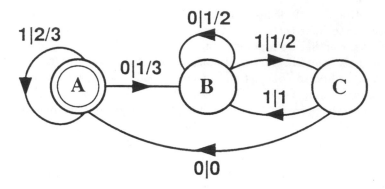

FIGURE 5 The even system generates sequences $\{\ldots 01^{2n}0 \ldots : n = 0, 1, 2 \ldots\}$ of 1s of even length, i.e., even parity. There are three states $\mathbf{V} = \{A, B, C\}$. The state A with the inscribed circle is the start state v_0. The edges are labeled $s|p$ where $s \in \mathbf{A}$ is a measurement symbol and $p \in [0, 1]$ is a conditional transition probability.

The control parameter r governs the degree of nonlinearity. At a Misiurewicz parameter value, the chaotic behavior is governed by an absolutely continuous invariant measure. The consequence is that the statistical properties are particularly well behaved. These parameter values are determined by the condition that the iterates $f^n(x_c)$ of the map's maximum $x_c = 1/2$ are asymptotically periodic. The Misiurewicz parameter value r' of interest here is the first root of $f_{r'}^4(x_c) = f_{r'}^5(x_c)$ below that at $r = 4$. Solving numerically yields $r' \approx 3.9277370017867516$. The symbolic dynamics is produced from the measurement partition $\Pi_{1/2} = \{[0, x_c], [x_c, 1]\}$. Since this partition is generating, the resulting binary sequences completely capture the statistical properties of the map. In other words, there is a one-to-one mapping between infinite binary sequences and almost all points on the attractor.

Reconstructing the machine from one very long binary sequence in the direction in which the symbols are produced gives the four-state machine $M_{r'}^{\rightarrow}$ shown in Figure 6. The stochastic connection matrix is

$$T = \begin{pmatrix} 0.636 & 0.364 & 0.000 & 0.000 \\ 0.724 & 0.000 & 0.276 & 0.000 \\ 0.000 & 0.000 & 0.000 & 1.000 \\ 0.000 & 0.521 & 0.479 & 0.000 \end{pmatrix} . \tag{29}$$

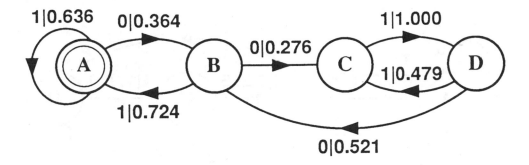

FIGURE 6 The machine $M_{r'}^{\rightarrow}$ reconstructed by parsing in forward presentation order a binary sequence produced using a generating partition of the Logistic map at a Misiurewicz parameter value.

Reconstructing the machine from the same binary sequence in the opposite direction gives the reverse-time machine $M_{r'}^{\leftarrow}$ shown in Figure 7. Its connection matrix is

$$T = \begin{pmatrix} 0.636 & 0.364 & 0.000 & 0.000 \\ 0.000 & 0.000 & 0.276 & 0.724 \\ 0.000 & 0.000 & 0.000 & 1.000 \\ 0.000 & 1.000 & 0.000 & 0.000 \end{pmatrix}. \tag{30}$$

Notice that $M_{r'}^{\leftarrow}$ has a transient state and three recurrent states compared to the four recurrent states in $M_{r'}^{\rightarrow}$. This suggests the likelihood of some difference in complexity convergence. Figure 8 shows that this is the case by plotting $C_\mu(M_{r'}^{\rightarrow}, t)$ and $C_\mu(M_{r'}^{\leftarrow}, t)$ for positive and negative times, respectively. Not only do the convergence behaviors differ in type, but also in the asymptotic values of the complexities: $C_\mu(M_{r'}^{\rightarrow}) \approx 1.77$ bits and $C_\mu(M_{r'}^{\leftarrow}) \approx 1.41$ bits. This occurs despite the fact that the entropies must be and are the same for both machines: $h(M_{r'}^{\rightarrow}) = h(M_{r'}^{\leftarrow}) \approx 0.82$ bits per time unit and $h_\mu(M_{r'}^{\rightarrow}) = h_\mu(M_{r'}^{\leftarrow}) \approx 0.81$ bits per time unit. Although the data stream is equally unpredictable in both time directions, an observer learns about the process's state in two different ways and obtains different amounts of state information. The difference

$$\Delta C_{\leftrightarrows}^{\rightarrow} = C_\mu(M_{r'}^{\rightarrow}) - C_\mu(M_{r'}^{\leftarrow}) \approx 0.36 \text{ bits} \tag{31}$$

is a measure of the computational irreversibility of the process. It indicates the process is not symmetric in time from the observer's viewpoint. This example serves to distinguish machine reconstruction and the derived quantifiers, such as complexity, from the subsequence-based measures, such as the two-point mutual information and the excess entropy.

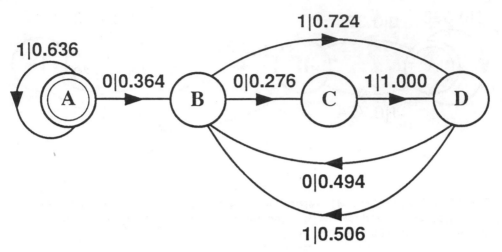

FIGURE 7   The machine $M_{r\vec{l}}$ reconstructed by parsing in reverse presentation order a binary sequence produced using a generating partition of the Logistic map at a Misiurewicz parameter value.

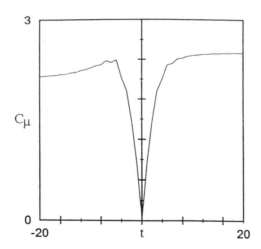

FIGURE 8   What the observer sees, on average, in forward and reverse lag-time in terms of the complexity convergence $C_\mu(t)$ for $M_{r\vec{l}}$ and $M_{r\vec{l}}$. Data for the latter are plotted on the negative lag time axis. Note that not only do the convergence characteristics differ between the two time directions, but the asymptotic complexity values are not equal.

MEASUREMENT SEMANTICS OF CHAOS

Shannon's communication theory tells one how much information a measurement gives. But what is the meaning of a particular measurement? Sufficient structure has been developed up to this point to introduce a quantitative definition of an observation's meaning. Meaning, as will be seen, is intimately connected with hierarchical representation. The following, though, concerns meaning as it arises when crossing a single change in representation and not in the entire hierarchy.[11]

A universe consisting of an observer and a thing observed has a natural semantics. The semantics describes the coupling that occurs during measurement. The attondant meaning derives from the dual interpretation of the information transferred at that time. As already emphasized, the measurement is, first, an indirect representation of the underlying process's state and, second, information that updates the observer's knowledge. The semantic information processing that occurs during a measurement thus turns on the relationship between two levels of representation of the same event.

The meaning of a message, of course, depends on the context in which its information is made available. If the context is inappropriate, the observation will have no basis with which to be understood. It will have no meaning. If appropriate, then the observation will be "understood." And if that which is understood—the content of the message—is largely unanticipated, then the observation will be more significant than a highly likely, "obvious" message.

In the present framework, context is set by the model held by the observer at the time of a measurement. To take an example, assume that the observer is capable of modeling using the class of stochastic finite automata. And, in particular, assume the observer has estimated a stochastic finite automaton[11] and has been following the process sufficiently long to know the current state with certainty. Then at a given time the observer measures symbol $s \in \mathbf{A}$. If that measurement forces a disallowed transition, then it has no meaning other than that it lies outside of the contexts (morphs) captured in the current model. The observer clearly does not know what the process is doing. Indeed, formally the response is for the observer to reset the machine to the initial state of total ignorance. If, however, the measurement is associated with an allowed transition, i.e., it is anticipated, then the degree $\Theta(s)$ of meaning is

$$\Theta(s) = -\log p_{\substack{\longrightarrow v \\ s}} . \tag{32}$$

[11] Assume also that the estimated machine is deterministic in the sense of automata theory: the transitions from each state are uniquely labeled: $p_e = p(v, v', s) = p_v p_{v \underset{s}{\longrightarrow} v'}$. This simplifies the discussion by avoiding the need to define the graph indeterminacy as a quantitative measure of ambiguity.[14] Ambiguity for an observer arises if its model is a stochastic nondeterministic automaton.

Here $\underset{s}{\longrightarrow} v$ denotes the machine state $v \in \mathbf{V}$ to which the measurement brings the observer's knowledge of the process's state. $p_{\underset{s}{\longrightarrow} v}$ is the corresponding morph's probability which is given by the associated state's asymptotic probability. The meaning itself, i.e., the content of the observation, is the particular morph to which the model's updated state corresponds. In this view a measurement selects a particular pattern from a palette of morphs. The measurement's meaning is the selected morph[12] and the amount of meaning is determined by the latter's probability.

To clarify these notions, let's consider as an example a source that produces infinite binary sequences for the regular language[21] described by the expression $(0+11)^*$. Assume further that the choice implied by the "+" is made with uniform probability. An observer given an infinite sequence of this type reconstructs the stochastic finite machine shown in Figure 5. The observer has discovered three morphs: the states $\mathbf{V} = \{$ A, B, C $\}$. But what is the meaning of each morph? First, consider the recurrent states B and C. State B is associated with having seen an even number of 1s following a 0; C with having seen an odd number. The meaning of B is "even" and C is "odd." Together the pair $\{$B, C$\}$ recognize the parity of the data stream. The machine as a whole accepts strings whose substrings of the form $01 \ldots 10$ have even parity of 1s. What is the meaning of state A? As long as the observer's knowledge of the process's state remains in state A, there has been some number of 1s whose parity is unknown, since a 0 must be seen to force the transition to the parity state B. This state, a transient, serves to synchronize the recurrent states with the data stream. This indicates the meaning content of an individual measurement in terms of the state to which it and its predecessors bring the machine.

Before giving a quantitative analysis, the time dependence of the state probabilities must be calculated. Recall that the state probabilities are updated via the stochastic connection matrix

$$\vec{p}_{\mathbf{V}}(t+1) = \vec{p}_{\mathbf{V}}(t) \begin{pmatrix} \frac{2}{3} & \frac{1}{3} & 0 \\ 0 & \frac{1}{2} & \frac{1}{2} \\ 0 & 1 & 0 \end{pmatrix} \tag{33}$$

where $\vec{p}_{\mathbf{V}}(t) = (p_A(t), p_B(t), p_C(t))$ and the initial distribution is $\vec{p}_{\mathbf{V}}(0) = (1, 0, 0)$. The time-dependent state probabilities are found, using the z-transform, to be

$$p_A(t) = \left(\frac{2}{3}\right)^t \qquad\qquad t = 0, 1, 2, \ldots$$

$$p_B(t) = 2\left(\frac{2}{3}\right)^t - 2^{1-t} \qquad t = 0, 1, 2, \ldots \tag{34}$$

$$p_C(t) = \begin{cases} \left(\frac{2}{3}\right)^t - 2^{1-t} & t = 1, 2, 3, \ldots \\ 0, & t = 0 \end{cases}$$

[12]I simplify here. The best formal representation of meaning at present uses the set-theoretic structure that the machine induces over the set of observed subsequences. This in turn is formulated via the lattice theory[3] of machines.[20]

TABLE 2 Observer's Semantic Analysis of Parity Source in Figure 5

Observer in state	Measures Symbol	Interprets Meaning as	Degree of Meaning (bits)	Amount Information (bits)
A	1	Unsynchronized	∞	0.585
A	0	Synchronize	0.585	1.585
B	1	Odd number of 1s	1.585	1
B	0	Even number of 1s	0.585	1
C	1	Even number of 1s	0.585	0
C	0	Confusion: lose sync, reset to start state	0	∞

Any time a disallowed transition is forced the current state is reset to the start state and $p_V(t)$ is reset to the distribution representing total ignorance which is given by $\vec{p}_V(0)$.

What then is the quantitative degree of meaning of particular measurements? Let's consider all of the possibilities: all possible contexts, i.e., current states, and all possible measurements. t steps after a reset, the observer is

1. In the sync state and measures $s = 1$: $\Theta_{\text{sync}}^t(1) = -\log_2 p_{1 \to A} = t(\log_2 3 - 1)$;

2. In the sync state and measures $s = 0$: $\Theta_{\text{sync}}^t(0) = -\log_2 p_{0 \to B} = -\log_2 p_B(t)$;

   e.g., $\Theta_{\text{sync}}^1(0) = \log_2 3 \approx 1.584$ bits;

3. In the even state and measures $s = 1$: $\Theta_{\text{even}}^t(1) = -\log_2 p_{1 \to C} = -\log_2 p_C(t)$,

   $t > 1$; e.g., $\Theta_{\text{even}}^2(1) = \log_2 6 \approx 2.584$ bits;

4. In the even state and measures $s = 0$: $\Theta_{\text{even}}^t(0) = -\log_2 p_{0 \to B} = -\log_2 p_B(t)$;

   e.g., $\Theta_{\text{even}}^2(0) = 1 + 2\log_2 3 - \log_2 7 \approx 1.372$ bits;

5. In the odd state and measures $s = 1$: $\Theta_{\text{odd}}^t(1) = -\log_2 p_{1 \to B} = -\log_2 p_B(t)$;

   e.g., $\Theta_{\text{odd}}^3(1) = 2 + 3\log_2 3 - \log_2 3 - \log_2 37 \approx 1.545$ bits; or

6. In the odd state and measures $s = 0$, a disallowed transition. The observer resets the machine: $\Theta_{\text{odd}}^t(0) = -\log_2 p_{0 \to A} = -\log_2 p_A(0) = 0$.

In this scheme states B and C cannot be visited at time $t = 0$ nor state C at time $t = 1$.

Assuming no disallowed transitions have been observed, at infinite time $\vec{p}_V = (0, 2/3, 1/3)$ and the degrees of meaning are, if the observer is

1. In the sync state and measures $s = 1$: $\Theta_{\text{sync}}(1) = -\log_2 p_{1 \to A} = \infty$;

2. In the sync state and measures $s = 0$: $\Theta_{\text{sync}}(0) = -\log_2 p_{\underset{0}{\to}\text{B}} = \log_2 3 - 1 \approx$ 0.584 bits;

3. In the even state and measures $s = 1$: $\Theta_{\text{even}}(1) = -\log_2 p_{\underset{1}{\to}\text{C}} = \log_2 3 \approx 1.584$ bits;

4. In the even state and measures $s = 0$: $\Theta_{\text{even}}(0) = -\log_2 p_{\underset{0}{\to}\text{B}} = \log_2 3 - 1 \approx$ 0.584 bits;

5. In the odd state and measures $s = 1$: $\Theta_{\text{odd}}(1) = -\log_2 p_{\underset{1}{\to}\text{B}} = \log_2 3 - 1 \approx 0.584$ bits; or

6. In the odd state and measures $s = 0$, a disallowed transition. The observer resets the machine: $\Theta_{\text{odd}}(0) = -\log_2 p_{\underset{0}{\to}\text{A}} = -\log_2 p_{\text{A}}(0) = 0$.

Table 2 summarizes this analysis for infinite time. It also includes the amount of information gained in making the specified measurement. This is given simply by the negative binary logarithm of the associated transition probability.

Similar definitions of meaning can be developed between any two levels in a reconstruction hierarchy. The example just given concerns the semantics between the measurement symbol level and the stochastic finite automaton level.[11] Meaning appears whenever there is a change in representation of events. And if there is no change, e.g., a measurement is considered only with respect to the population of other measurements, an important special case arises.

In this view Shannon information concerns degenerate meaning: that obtained within the same representation class. Consider the information of events in some set E of possibilities whose occurrence is governed by arbitrary probability distributions $\{P, Q, \ldots\}$. Assume that no further structural qualifications of this representation class are made. Then the Shannon self-information $-\log p_e$, $p_e \in P$, gives the degree of meaning $-\log_2 p_{\underset{e}{\to}e}$ in the observed event e with respect to total ignorance. Similarly, the information gain $I(P; Q) = \sum_{e \in E} p_e \log_2 p_e / q_e$ gives the average degree of "meaning" between two distributions. The two representation levels are degenerate: both are the events themselves. Thus, Shannon information gives the degree of meaning of an event with respect to the set E of events and not with respect to an observer's internal model; unless, of course, that model is taken to be the collection of events as in a histogram or look-up table. Although this might seem like vacuous re-interpretation, it is essential that general meaning have this as a degenerate case.

The main components of meaning as defined above, should be emphasized. First, like information it can be quantified. Second, conventional uses of Shannon information are a natural special case. And third, it derives fundamentally from the relationship *across* levels of abstraction. A given message has different connotations depending on an observer's model and the most general constraint is the model's level in a reconstruction hierarchy. When model reconstruction is considered to be a time-dependent process that moves up a hierarchy, then the present discussion suggests a concrete approach to investigating adaptive meaning in evolutionary systems: emergent semantics.

In the parity example above, I explicitly said what a state and a measurement "meant." Parity, as such, is a human linguistic and mathematical convention, which has a compelling naturalness due largely to its simplicity. A low-level organism, though, need not have such a literary interpretation of its stimuli. Meaning of (say) its model's states, when the state sequence is seen as the output of a preprocessor,[13] derives from the functionality given to the organism, as a whole and as a part of its environment and its evolutionary and developmental history. Said this way, absolute meaning in nature is quite a complicated and contingent concept. Absolute meaning derives from the global structure developed over space and through time. Nonetheless, the analysis given above captures the representation level-to-level origin of "local" meaning. The tension between global and local entities is not the least bit new to nonlinear dynamics. Indeed, much of the latter's subtlety is a consequence of their inequivalence. Analogous insights are sure to follow from the semantic analysis of large hierarchical processes.

MACHINE THERMODYNAMICS

The atomistic view of nature, though professed since ancient times, was largely unsuccessful until the raw combinatorial complication it entailed was connected to macroscopic phenomena. Founding thermodynamics on the principles of statistical mechanics was one of, if not the major, influence on its eventual acceptance. The laws of thermodynamics give the coarsest constraints on the microscopic diversity of large many-particle systems. This same view, moving from microscopic dynamics to macroscopic laws, can be applied to the task of statistical inference of nonlinear models. And so it is appropriate after discussing the "microscopic" data of measurement sequences and the reconstruction of "mesoscopic" machines from them, to end with a discussion at the largest scale of description: machine thermodynamics. This gives a concise description of the structure of the infinite set of infinite sequences generated by a machine and also of their probabilities. It does this, in analogy with the conventional thermodynamic treatment of microstates, by focusing on different subsets of allowed sequences.

The first step is the most basic: identification of the microstates. Consistent with machine reconstruction's goal to approximate a process's internal states, microstates in modeling are the individual measurement subsequences.[14] Consider the set $sub_L(\mathbf{s})$ of all length L subsequences occurring in a length N data stream $\mathbf{s}$. The probability of a subsequence $\omega \in sub_L(\mathbf{s})$ is estimated by $p_\omega \approx N^{-1}N_\omega$, where N_ω is the number of occurrences of ω in the data stream. The connection with the

---

[13] This preprocessor is a transducer version of the model that takes the input symbols and outputs strings in the state alphabet **V**.

[14] Going from individual measurements in a data stream to subsequences is a change in representation from the raw data to the parse tree, a hierarchical data structure.

physical interpretation of thermodynamics follows from identifying a microstate's energy with its self-information

$$U_\omega = -\log_2 p_\omega \, . \tag{35}$$

That is, improbable microstates have high energy. Energy macrostates are then given by grouping subsequences of the same energy U into subsets $\{\omega: U_\omega = U, \omega \in \mathrm{sub}_L(s)\}$. At this point there are two distributions: the microstate distribution and an induced distribution over energy macrostates. Their thermodynamic structure is captured by the parametrized microstate distribution

$$p_\omega(\beta) = \frac{2^{-\beta U_\omega}}{Z_\beta(L)} \tag{36}$$

where β accentuates or attenuates a microstate's weight solely according to its energy. This is the same role (inverse) temperature plays in classical thermodynamics. The partition function

$$Z_\beta(L) = \sum_{\omega \in \mathrm{sub}_L(s)} 2^{-\beta U_\omega} = \sum_{\omega \in \mathrm{sub}_L(s)} p_\omega^\beta \tag{37}$$

gives the total signal space volume of the distribution $P_\beta(L) = \{p_\omega(\beta): \omega \in \mathrm{sub}_L(s)\}$. Starting from this point statistical mechanics explains thermodynamic properties as constraints on how this volume changes under various conditions.

From these definitions an extensive, system-size-dependent thermodynamics follows directly. For example, given an infinitely long data stream s, the average total energy in all length L sequences is

$$U(L) = \sum_{\omega \in \mathrm{sub}_L(s)} p_\omega(\beta) U_\omega = Z_\beta^{-1} \sum_{\omega \in \mathrm{sub}_L(s)} U_\omega 2^{-\beta U_\omega} ;$$
$$U(L) = -Z_\beta^{-1} \sum_{\omega \in \mathrm{sub}_L(s)} p_\omega^\beta \log_2 p_\omega . \tag{38}$$

And the thermodynamic entropy is given by

$$S(L) = H(P_\beta(L)) = -\sum_{\omega \in \mathrm{sub}_L(s)} p_\omega(\beta) \log_2 p_\omega(\beta) \tag{39}$$

where $H(P_\beta(L))$ is the Shannon information of the microstate distribution.

These are definitions of the extensive, L-dependent thermodynamic parameters for a closed system thermally coupled to its environment. The total energy U exists in several forms. The most important of which is the thermal energy TS, where S is the thermodynamic entropy and T is the temperature. The remaining "free" energy is that which is stored via a reversible process and is retrievable by one. For a closed and nonisolated system, it is the Helmholtz free energy F. The fundamental equation expressing energy conservation is then

$$U = F + TS. \tag{40}$$

In modeling, an observer is put into contact with the process and attempts, by collecting measurements and estimating models, to come to "inferential" equilibrium by finding the optimal model. The above thermodynamics describes the situation where the information in the data stream exists in two forms. The first is that which is randomized and the second is that responsible for the deviation from equilibrium. The thermodynamic analog of the Helmholtz free energy is

$$F(L) = -\beta^{-1} \log_2 Z_\beta(L). \tag{41}$$

It governs the amount of nonrandom information in the ensemble described by $P_\beta(L)$ at the given temperture β^{-1}.

There are three temperature limits of interest in which the preceding thermodynamics can be simply described.

1. Equilibrium, $\beta = 1$: The original subsequence distribution is recovered: $p_\omega = p_\omega(1)$ and $Z_1 = 1$. All of the information is "thermalized" $U_1 = \beta^{-1}S_1$ and the Helmholtz free energy vanishes $F_{\beta=1} = 0$.
2. Infinite temperture, $\beta = 0$: All microstates are "excited" and are equally probable: $p_\omega(0) = Z_0^{-1}$, where the partition function is equal to the total number of microstates: $Z_0 = ||\text{sub}_L(s)||$. The effective signal space volume is largest in this limit. The average energy is just the sum of the microstate energies: $U_0 = Z_0^{-1} \sum U_\omega$. The entropy simply depends of the multiplicity of microstates $S_0 = \log_2 ||\text{sub}_L(s)||$. The free energy diverges.
3. Zero temperature, $\beta = \infty$: The least energetic, or most probable, microstate ω_∞ dominates: $p_\omega = \delta_{\omega\omega_\infty}$, the signal space volume is the smallest $Z_\infty = 0, U_\infty = U_{\omega_\infty}$, and the entropy vanishes $S_\infty = 0$.

The goal for an observer is to build a model that reproduces the observed data stream, including the probability structure of the latter. In thermodynamic terms, the model should minimize the Helmholtz free energy. This is what machine reconstruction produces: a stochastic automaton that is in inferential equilibrium with the given data. How it does this is described elsewhere.[16] The following will cover the basic methods for this, using them to investigate the thermodynamic structure of a machine's invariant subsequences and distributions.

Dividing each of the extensive quantities by the volume L yields thermodynamic densities. And upon taking the thermodynamic limit of the densities, the asymptotic growth rates of the extensive parameters are obtained. These growth rates are intensive. They can be directly computed from the reconstructed machine. In a sense the machine itself is an intensive thermodynamic object: the effective computational equations of motion.

To obtain the intensive thermodynamics from a given stochastic machine M with $T = \{T^{(s)}: s \in \mathbf{A}\}$, a new set $\{T_\beta^{(s)}: s \in \mathbf{A}\}$ of parametrized transition matrices are defined by

$$\left(T_\beta^{(s)}\right)_{vv'} = e^{-\beta I_{v \underset{s}{\longrightarrow} v'}} \tag{42}$$

where $I_{v \underset{s}{\longrightarrow} v'} = -\log_2 p_{v \underset{s}{\longrightarrow} v'}$ is the information obtained on making the transition from state v to state v' on symbol s. Note that as the parameter β is varied the transition probabilities in the original machine are given different weights while the overall "shape" of the transitions is maintained. This is the intensive analog of the effect β has on the extensive distribution P_β above.

Many thermodynamic properties are determined by the parametrized connection matrix

$$T_\beta = \sum_{s \in \mathbf{A}} T_\beta^{(s)} . \tag{43}$$

There are two quantities required from this matrix. Its principal eigenvalue

$$\lambda_\beta = \sup_{i=0,\dots,\|\mathbf{V}\|-1} \{\lambda_i : {}_i\vec{p}_\mathbf{V} T_\beta = {}_i\vec{p}_\mathbf{V} \lambda_i\} \tag{44}$$

and the associated right eigenvector $\vec{r}_\mathbf{V}$

$$T_\beta \vec{r}_\mathbf{V} = \lambda_\beta \vec{r}_\mathbf{V} . \tag{45}$$

Note, however, that T_β is not a stochastic matrix. In fact,

$$\sum_{v' \in \mathbf{V}} (T_\beta)_{vv'} \begin{cases} \geq 1 & \beta < 1; \\ = 1 & \beta = 1; \\ \leq 1 & \beta > 1. \end{cases} \tag{46}$$

It does not directly describe, for example, the probabilities of the subset of sequences that are associated with the relative transition weightings at the given β, except, of course, at $\beta = 1$.

There is, however, an "equilibrium" machine, whose stochastic connection matrix is denoted $\mathcal{S}_\beta$, that produces the same sequences but with the relative weights given by T_β. Recall that the state of macroscopic equilibrium is determined by one of the variational principles:

1. At given total entropy, the equilibrium state minimizes the energy;
2. At given total energy, it maximizes the thermodynamic entropy.

Using the latter entropy representation, the equilibrium machine is that with maximal thermodynamic entropy subject to the constraints imposed by T_β. That is, all of the nonzero edge probabilities are allowed to vary. $\mathcal{S}_\beta$ describes the process over the allowed subsequences which is in thermodynamic equilibrium at the given temperature. It is found using Shannon's entropy maximization formula[35,40]

$$S_\beta = \frac{D^{-1}(\vec{r}_\mathbf{v}) T_\beta D(\vec{r}_\mathbf{v})}{\lambda_\beta} \tag{47}$$

where $D(\vec{v})$ is a diagonal matrix with the components of $\vec{v}$ on the diagonal. Since this is a stochastic matrix, its principal eigenvalue is unity. However, the associated left eigenvector $\vec{p}_\mathbf{v}$,

$$\vec{p}_\mathbf{v} S_\beta = \vec{p}_\mathbf{v} , \tag{48}$$

when normalized in probability gives the asymptotic state distribution.

The entropy rate, as seen in a previous section, is

$$h_\mu(S_\beta) = - \sum_{v \in \mathbf{V}} p_v \sum_{v' \in \mathbf{V}} p_{v \to v'} \log_2 p_{v \to v'} \tag{49}$$

where $p_{v \to v'} = (S_\beta)_{vv'}$. The β-complexities are given by

$$C_\beta = - \sum_{v \in \mathbf{V}} p_v \log_2 p_v$$
$$C_\beta^e = - \sum_{e \in \mathbf{E}} p_e \log_2 p_e \tag{50}$$

where $p_{e=(v,v')} = p_v p_{v \to v'}$. The metric ($\beta = 1$) and topological ($\beta = 0$) quantities are directly recovered. That is,

$$h = h_0 \text{ and } C = C_0;$$
$$h_\mu = h_1 \text{ and } C_\mu = C_1. \tag{51}$$

The relation

$$C_\beta^e = C_\beta + h_\mu(S_\beta) \tag{52}$$

again constrains the entropy rate and the complexities.

Physically speaking $I_{v \to v'} = - \log_2 p_{v \to v'}$ plays the role of an interaction energy between two states and β is related to the inverse temperature. Although the same support, i.e., set of sequences and topological machine, exists at all temperatures, varying β accentuates the measure of the process S_β over different paths in the machine or, equivalently, over different subsets of sequences. One subset's weight changes relative to others as dictated by T_β's elements.

In the limit of long sequences, the partition function's growth rate is governed by the maximal eigenvalue λ_β of the machine matrix T_β. That is,

$$Z_\beta(L) \underset{L \to \infty}{\propto} \lambda_\beta^L . \tag{53}$$

The machine Helmholtz free-energy density becomes

$$F = -\lim_{L \to \infty} \frac{1}{L} \beta^{-1} \log_2 Z_\beta(L)$$
$$F = -\beta^{-1} \log_2 \lambda_\beta \tag{54}$$

and the thermodynamic entropy density is

$$S = k_B h_\mu(\mathcal{S}_\beta) \tag{55}$$

where k_B is Boltzmann's constant. Using the basic thermodynamic relation between these, the total energy density is then readily computed by noting

$$U = F + TS$$
$$U = \beta^{-1}(h_\mu(\mathcal{S}_\beta) - \log_2 \lambda_\beta) \tag{56}$$

using the identification $\beta^{-1} = k_B T$.

In the entropy representation, the function $S(U)$, computed from Eqs. (55) and (56), determines the thermodynamic "potential" along an arc $\mathcal{S}_\beta$ in the model space $\mathcal{M}$ of consistent stochastic machines. Consistent machines are those having the same set of allowed sequences as those observed in the data stream. At each fixed β the equilibrium machine is estimated via Eq. (47). Here equilibrium refers to a closed and isolated system specified by a fixed temperature and so a fixed average energy U. In contrast, the graph of $S(U)$ concerns a closed, but nonisolated system in contact with an energy reservoir at temperature $\beta^{-1} = \partial U/\partial S$. It gives the entropies and energies for the family of machines $\mathcal{S}_\beta$. The equilibrium machine occurs at $\beta = 1$ where the free energy vanishes and all of the unconstrained information is "thermal" or randomized.

This is the thermodynamic analog of a cost function like that over model space $\mathcal{M}$ as shown in Figure 3. It is not the same, however, since (i) $S(U)$ is computed in the thermodynamic limits of a long data stream and long sequence length and (ii) it represents two different optimizations, one at each temperature and the other over all temperatures. This is the view of statistical estimation developed in large deviation theory.[8,17] It suggests a rather different appreciation of the Sinai-Ruelle-Bowen thermodynamic formalism[6,33,39] for invariant measures of dynamical systems as a foundation for nonlinear modeling.

Independent of this modeling interpretation, Eqs. (55) and (56) give a direct way to study macroscopic properties of the sequences produced by a stochastic machine. In particular, the shape of $S(U)$ determines the variation in entropy and energy within subsets of sequences that are invariant under the process. It indicates rather directly the range of likely fluctuations in observed sequences. Two examples will serve to illustrate these points.

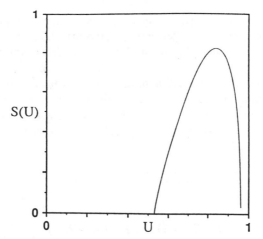

FIGURE 9 The fluctuation spectrum, thermodynamic entropy density $S(U)$ versus internal energy density U, for the machine $M_{r'}^{\rightarrow}$.

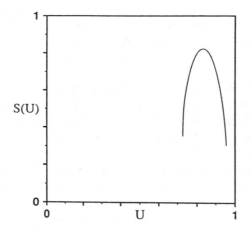

FIGURE 10 The fluctuation spectrum, thermodynamic entropy density $S(U)$ versus internal energy density U, for the machine $M_{r'}^{\leftarrow}$.

Figures 9 and 10 show the "fluctuation" spectra, the thermodynamic entropy density $S(U)$ versus the energy density U, for the Misiurewicz machines $M_{r'}^{\rightarrow}$ and $M_{r'}^{\leftarrow}$. Notice the rather large difference in character of the two spectra. This is another indication of the computational irreversibility of the underlying process. The topological entropies, found at the spectra's maxima $h = S(U{:}\beta = 0)$, and the metric entropies, found at the unity slope point on the curves $h_\mu = S(U:\beta = 1)$, are the same, despite this. The energy extremes U_{min} and U_{max}, as well as the thermodynamic entropies $S(U_{min})$ and $S(U_{max})$, differ significantly due to the irreversibility.

By way of ending this section, a final thermodynamic analogy will be mentioned. One of the first experimentally accessible measures of complexity was the excess entropy.[13] The total excess entropy $F_\beta(L)$ is a coarse measure of the average amount of memory in a measurement sequence above and beyond its randomized information. It is defined as follows

$$F_\beta(L) = H_\beta(L) - h_\beta L \tag{57}$$

where

$$H_\beta(L) = (1 - \beta)^{-1} \log_2 Z_\beta(L) \tag{58}$$

is the total Renyi entropy and $h_\beta = (1-\beta)^{-1} \log_2 \lambda_\beta$ is the Renyi entropy rate. This was referred to as the free information[14] since it is easily seen to be analogous to a free energy. The free information is the Legendre transform of the Renyi entropy H_β that replaces its length dependence with an intensive parameter h_β. If subsequence length L is again associated with the volume V, a thermodynamic pressure can be associated with h_β. Finally, since the free information is an approximation of the finitary complexity,[14] the latter is also seen to be a type of free energy.

A more detailed development of machine thermodynamics can be found elsewhere.[16] The preceding outline hopefully serves to indicate a bit of its utility and interest.

To summarize, the thermodynamic analysis suggests that the total information in a data stream, as extracted by machine reconstruction and as interpreted by an observer with a model, exists in two forms: one thermal and the other associated with information processing. The first, randomness in the data, is measured by the thermodynamic entropy. The second, structure in the data that causes it to deviate from simple thermal equilibrium, is that available to do mathematical work. This work might consist of communicating from a process to an observer; this is information transmission in space. It also can be available as static memory, which is information transmission across time. Most usefully, it can be available to do genuine computation and to support thereby semantic information processing.

SCIENCE AS DATA COMPRESSION?

Thinking back to the explanatory channel, these considerations lead me to disagree with the philosophical premise implicit in the universal coding theory approach to nonlinear modeling. While I accept the mathematics and use the optimization criteria, its own semantics appears wanting. Science is not data compression. The structure of models is ultimately more important than their use in encoders and decoders for the efficient encapsulation of experience. In the limit of large data streams and positive entropy processes, i.e., the realm of universal coding theory, the model is essentially ignored and prediction error dominates. At the end of the day, though, good models are independent of the amount of data used to originally infer them.[15] This point was emphasized in the preceding by the analysis of the effects their computational structure had on knowledge relaxation and on their semantic structure. Even these naked, mathematical objects, with which one typically does not associate meaning, do imply a semantic structure for the act of measurement. And it is this semantics that gives models their scientific value.

The preceding discussion, though only an outline, attempted to put these issues in a sufficiently large arena so that they can stand on their own. At the beginning there are dynamical systems whose diverse and complicated phenomenology has rapidly become better understood. They enrich our view of natural phenomena, though they do not necessarily deepen it. The contrast between their often simple specification and their creation of apparent complexity leads to computational mechanics. Computation theory in this development appears as the theory *par excellence* of structure. During the 1960s it gave the foundation for a theory of randomness. But that success should not blind us to the pressing need for constructive measures of complexity for physical, chemical, biological, and economic systems that go beyond randomness. Descriptions of complexity need not always pay for randomness. This is as true of statistical inference applied to nonlinear modeling as it is of thermodynamic and evolutionary systems. Indeed, it is one of the primary lessons of nonlinear dynamics that effective randomness is cheap and easily regenerated. Concomitantly, it also shows that ideal randomness is just that: an ideal that is expensive and, in principle, impossible to objectively obtain. Fortunately, nature does not seem to need it. Often only randomness effective for the task at hand is required.

This tension between randomness and order, the result of which is complexity, has always been a part of the problem domain of thermodynamics. Indeed, phase transitions and, especially, critical phenomena are the primary evidence of nature's delicate balance between them. Given this observation, the question now presents itself to nonlinear modeling, What types of computation are supported by physical systems at phase transitions, at the interface between order and chaos?[16] Away from "critical" processes, classical thermodynamics forms a solid basis on which to build nonlinear modeling. To the extent Gibbsian statistical mechanics is successful, so too will optimal modeling be. Though, as I just mentioned, there is much to question within this framework. Having described the analogy between thermodynamics and optimal modeling, another deeper problem suggests itself.

Classical thermodynamics foundered in its description of critical phenomena due to its confusion of the (observable) average value of the order parameter with

---

[15]To describe the behavior of a thermodynamic system it suffices to communicate the equations of state, approximate macroscopic parameters, and possibly the force laws governing the microscopic constituents. Exact description is not only undesirable, but well nigh impossible.

[16]A first, constructive answer can be found elsewhere.[15]

its most likely value. So, too, the universal coding theoretic association of the optimal model with the most likely Eq. (7) can fail for processes with either low entropy or near phase transitions. This will be especially exaggerated for "critical" processes that exhibit fluctuations on all scales. In these cases, fluctuations dominate behavior and averages need not be centered around the most likely value of an observable. This occurs for high-complexity processes, such as those described by stochastic context-free and context-sensitive grammars,[15,30] since they have the requisite internal computational capacity to cause the convergence of observable statistics to deviate from the Law of Large Numbers.

Having told this modeling story somewhat briefly, I hope it becomes at least a little clearer how the view of microscopic processes offered by statistical mechanics needs to be augmented. The examples analyzed demonstrate that the computational structure and semantic content of processes are almost entirely masked and cannot be articulated within the conventional framework. But it is exactly these properties that form the functional substrate of learning and evolutionary systems. The claim here is that an investigation of the intrinsic computation performed by dynamical systems is a prerequisite for understanding how physical systems might spontaneously take up the task of modeling their nonlinear environments. I believe the engineering by-products of this program for forecasting, design, and control, will follow naturally.

> The greatest sorcerer [writes Novalis memorably] would be the one who bewitched himself to the point of taking his own phantasmagorias for autonomous apparitions. Would not this be true of us?

> J.L. Borges, "Avatars of the Tortoise," page 115.[5]

ACKNOWLEDGMENTS

I would like to express my appreciation for discussions with Karl Young and Jim Hanson. Many thanks are due to the Santa Fe Institute, where the author was supported by a Robert Maxwell Foundation Visiting Professorship, for the warm hospitality during the writing of the present review. Funds from NASA-Ames University Interchange NCA2–488 and AFOSR contract AFOSR-91-0293, also contributed to this work.

REFERENCES

1. Akaike, H. "An Objective Use of Bayesian Models." *Ann. Inst. Statist. Math.* **29A** (1977): 9.
2. Angluin, D., and C. H. Smith. "Inductive Inference: Theory and Methods." *Comp. Surveys* **15** (1983): 237.
3. Birkhoff, G. *Lattice Theory.* Providence, RI: American Mathematical Society, 1967.
4. Blahut, R. E. *Principles and Practice of Information Theory.* Reading, MA: Addison-Wesley, 1987.
5. Borges, J. L. *Other Inquisitions 1937–1952.* New York: Simon and Schuster, 1964.
6. Bowen, R. *Equilibrium States and the Ergodic Theory of Anosov Diffeomorphisms.* Lecture Notes in Mathematics, vol. 470. Berlin: Springer-Verlag, 1975.
7. Brudno, A. A. "Entropy and the Complexity of the Trajectories of a Dynamical System." *Trans. Moscow Math. Soc.* **44** (1983): 127.
8. Bucklew, J. A. *Large Deviation Techniques in Decision, Simulation, and Estimation.* New York: Wiley-Interscience, 1990.
9. Chaitin, G. "On the Length of Programs for Computing Finite Binary Sequences." *J. ACM* **13** (1966): 145.
10. Crutchfield, J. P. "Noisy Chaos." Ph.D. Thesis, University of Califomia, Santa Cruz, 1983. Published by University Microfilms International, Michigan, (800) 521-0600, Order No. 84-09371.
11. Crutchfield, J. P. "Reconstructing Language Hierarchies." In *Information Dynamics*, edited by H. A. Atmanspracher and H. Scheingraber, 45. New York: Plenum, 1991.
12. Crutchfield, J. P., and B. S. McNamara. "Equations of Motion from a Data Series." *Complex Systems* **1** (1987): 417.
13. Crutchfield, J. P., and N. H. Packard. "Symbolic Dynamics of Noisy Chaos." *Physica* **7D** (1983): 201.
14. Crutchfield, J. P., and K. Young. "Inferring Statistical Complexity." *Phys. Rev. Lett.* **63** (1989): 105.
15. Crutchfield, J. P., and K. Young. "Computation at the Onset of Chaos." In *Entropy, Complexity, and the Physics of Information*, edited by W. H. Zurek, 223. Santa Fe Institute Studies in the Sciences of Complexity, Vol. VIII. Redwood City, CA: Addison-Wesley, 1990.
16. Crutchfield, J. P., and K. Young. "ε-Machine Spectroscopy." Preprint, 1991.
17. Ellis, R. S. *Entropy, Large Deviations, and Statistical Mechanics*, Grundlehren der mathematischen Wissenschaften, Vol. 271. New York: Springer-Verlag, 1985.
18. Fraser, A. "Chaotic Data and Model Building." In *Information Dynamics*, edited by H. Atmanspracher and H. Scheingraber, 125. New York: Plenum, 1991.

19. Grassberger, P. "Toward a Quantitative Theory of Self-Generated Complexity." *Intl. J. Theo. Phys.* **25** (1986): 907.

20. Hartmanis, J., and R. E. Stearns. *Algebraic Structure Theory of Sequential Machines.* Englewood Cliffs, NJ: Prentice Hall, 1966.

21. Hopcroft, J. E., and J. D. Ullman. *Introduction to Automata Theory, Languages, and Computation.* Reading, MA: Addison-Wesley, 1979.

22. Howson, C., and P. Urbach. *Scientific Reasoning: The Bayesian Approach.* La Salle, IL: Open Court, 1989.

23. Jaynes, E. T. "Where Do We Stand On Maximum Entropy?" In *Delaware Symposium on the Foundations of Physics*, Vol. 1. Berlin: Springer-Verlag, 1967.

24. Kemeny, J. G. "The Use of Simplicity in Induction." *Phil. Rev.* **62** (1953): 391.

25. Kitchens, B., and S. Tuncel. "Finitary Measures for Subshifts of Finite Type and Sofic Systems." *Memoirs of the AMS* **58**, no. 338 (1985).

26. Kolmogorov, A. N. "A New Metric Invariant of Transient Dynamical Systems and Automorphisms in Lebesgue Spaces." *Dokl. Akad. Nauk. SSSR* **119** (1958): 861. (Russian) *Math. Rev.* **21(2035a)**.

27. Kolmogorov, A. N. "Three Approaches to the Concept of the Amount cf Information." *Prob. Info. Trans.* **1** (1965): 1.

28. Lempel, A., and J. Ziv. "On the Complexity of Individual Sequences." *IEEE Trans. Info. Th.* **IT-22** (1976): 75.

29. Li, M., and P. M. B. Vitanyi. "Kolmogorov Complexity and Its Applications." Technical Report CS-R8901, Centruum voor Wiskunde en Informatica, Universiteit van Amsterdam, 1989.

30. Li, W. "Generating Non-Trivial Long-Range Correlations and $1/f$ Spectra by Replication and Mutation." Santa Fe Institute Working Paper No. 91-01-002, 1991.

31. Lindgren, K., and M. G. Nordahl. "Complexity Measures and Cellular Automata." *Complex Systems* **2** (1988): 409.

32. Marcus, B. "Sofic Systems and Encoding Data." *IEEE Trans. Info. Th.* **31** (1985): 366.

33. Oono, Y. "Large Deviation and Statistical Physics." *Prog. Theo. Phys.* **99** (1989): 165.

34. Packard, N. H., J. P. Crutchfield, J. D. Farmer, and R. S. Shaw. "Geometry From a Time Series." *Phys. Rev. Lett.* **45** (1980): 712.

35. Parry, W., and S. Tuncel. *Classification Problems in Ergodic Theory.* London Mathematical Society Lecture Notes Series, Vol. 67. London: Cambridge University Press, 1982.

36. Poincaré, H. *Science and Hypothesis.* New York: Dover Publications, 1952.

37. Rabiner, L. R. "A Tutorial on Hidden Markov Models and Selected Applications." *IEEE Proc.* **77** (1989): 257.

38. Rissanen, J. "Stochastic Complexity and Modeling." *Ann. Stat.* **14** (1986): 1080.

39. Ruelle, D. *Thermodynamic Formalism.* Reading, MA: Addison-Wesley, 1978.

40. Shannon, C. E., and W. Weaver. *The Mathematical Theory of Communication*. Champaign-Urbana: University of Illinois Press, 1962.
41. Wallace, C. S., and F. P. Freeman. "Estimation and Inference by Compact Coding." *J. R. Statist. Soc. B* **49** (1987): 240.
42. Wolfram, S. "Computation Theory of Cellular Automata." *Comm. Math. Phys.* **96** (1984): 15.
43. Zadeh, L. A. *Fuzzy Sets and Applications: Selected Papers*. New York: Wiley, 1987.
44. Ziv, J. "Complexity and Coherence of Sequences." In *The Impact of Processing Techniques on Communications*, edited by J. K. Skwirzynski, 35. Dordrecht: Nijhoff, 1985.
45. Zurek, W. H. "Thermodynamic Cost of Computation, Algorithmic Complexity, and the Information Metric." *Nature* **341** (1989): 119.

Matthew Koebbe and Gottfried Mayer-Kress
Department of Mathematics, University of California at Santa Cruz, Santa Cruz, CA 95064

Use of Recurrence Plots in the Analysis of Time-Series Data

Recovering information about the recurrent behavior of a non-stationary time series typically requires a large data set and extensive off-line analysis. However, the technique of creating a recurrence plot can be used to graphically represent global recurrent structures. Characteristic elements of recurrent patterns include line segments of various orientations. We will classify some of these structures by examining several types of computer-generated time series. We will then identify some of the structures in recurrence plots of experimentally derived time series. We will see how geometrical properties of these plots also provide insight into local rates of divergence and dimensional complexity. We conclude with a classification of the dynamics corresponding to various line segment orientations.

INTRODUCTION

Dealing with the nonlinear dynamics of biological data, one is at the same time encouraged and disappointed with developments over the past ten years: whereas diffeomorphic invariants such as Lyapunov exponents and dimensions have suggested new insights into biological processes, results have also been equivocal due to the assumptions that the data series are autonomous and that their lengths are much longer than the characteristic times of the system in question. By and large, such assumptions cannot be supported.[3,6,7]

In order to study these erratic time series, Eckmann, Kamphorst, and Ruelle introduced a new graphical tool in 1987, which they called a recurrence plot.[1] In their paper, they begin with a time-ordered sequence of vectors in $\Re^n$ and let $x(i)$ denote the ith vector in this sequence. Defining $\mathcal{Z}_m = \{0, 1, 2, ..., m-1\}$, they define a function from an $n \times n$ array onto $\mathcal{Z}_2$ according to the following rule: a value of 1 is assigned to the (i, j)th element of the array if and only if $x(j) \in B(x(i), r(i))$, where $B(x(i), r(i))$ is the ball of radius $r(i)$ centered at $x(i)$. If an array element value of 1 is made to correspond to the color black and 0 denotes white, then this colored array is known as a recurrence plot. Their recurrence plots tend to be nearly symmetric with respect to the diagonal $i = j$ because, in general, if $x(i)$ is close to $x(j)$, then $x(j)$ is close to $x(i)$. (Notice that the plot is not necessarily symmetric as the authors did not require that $r(i) = r(j)$.) Large-scale diagonal line segments parallel to $i = j$ indicate periodic behavior which is not always visible in the original data. At the same time, the average length of small-scale line segments parallel to $i = j$ were claimed to be proportional to the inverse of the largest positive Lyapunov exponent. Additionally, drift in the signal is recognized by the overall reduction of recurrences away from the main diagonal.

In order to extend the original concept, we require that $r(i) = r(j)$ and modify the definition of a recurrence plot in order to eliminate the resulting redundancies. (For previous efforts in this direction see Mayer-Kress[8] and Zbilut et al.[13]) The relationship between the appearance of line segments of various orientations and the dynamics underlying the time series is determined by examining elementary data sets. These same line segments are also observed in experimental data. Their stability with respect to increasing embedding dimension is also examined. We see that the horizontal line segments are among the most stable and, in addition, are the geometrical feature which provides local divergence rate information. We complete our analysis with a discussion of the manner in which the creation of recurrence plots inspires an efficient algorithm for calculating the local dimensional complexity.

DEFINITIONS

It is the time-ordered sequence of vectors obtained by reconstruction from our given data series which permits us to understand how geometrical features of our recurrence plots give us insight into the dynamics underlying this data series. In order to develop this relationship, we begin with a time-ordered sequence of real numbers, $\{d_i\}_{i=1}^{N}$, where $N \in \mathcal{N}$ is the total number of data points ($N < \infty$), and we think of this data sequence as being a discretized one-dimensional projection of an integral curve of some dynamic in $\Re^n$, where n is a positive integer. We use the now widely known technique of reconstruction[11] to generate $\{v_i\}_{i=1}^{N_v}$, a time-ordered sequence of vectors in $\Re^{n_0}$, where N_v is the total number of vectors and $n < n_0$.

(By the Taken's Embedding Theorem, the dynamics of the reconstruction will be diffeomorphic to the original generating dynamics if the embedding dimension is twice the original dimension plus one.[12]) Consider the distance $\delta(i,j)$ between a vector $\vec{v}_i$ at time i and a vector $\vec{v}_j$ at time j, $\delta : \mathcal{Z}_m \times \mathcal{Z}_n \longrightarrow \Re^+$ with:

$$\delta(i,j) = \|\vec{v}_i - \vec{v}_{(i+j)}\|, \tag{1}$$

where $m + n < N$. Next, we define a function $\rho : \Re^1 \rightarrow \mathcal{Z}_k$, where k is given by the resolution of the chosen color map. (Colors are used to represent recurrence function values for enhanced visual recognition.) In the original paper by Kamphorst, Eckmann, and Ruelle,[1] this function was taken to be ($k = 2$):

$$\rho(\delta(i,j)) = \begin{cases} 0, & \text{if } \delta(i,j) > c \\ 1, & \text{if } \delta(i,j) < c \end{cases} \tag{2}$$

where $c \in \Re^1$ is a suitably chosen (see below) cutoff value. At this point, we define a recurrence plot to be the graph of the composition of these two functions

$$\rho \circ \delta : \mathcal{Z}_m \times \mathcal{Z}_n \longrightarrow \mathcal{Z}_k$$

($\rho \circ \delta$ is the recurrence function referred to above.) We note that, as defined here, the (i,j)th element of the $m \times n$ recurrence plot array takes on a value of 1 if the vectors $\vec{v}_i$ and $\vec{v}_{(i+j)}$ are close (or recurrent) and a value of 0, otherwise. This is known as the black-and-white color map ($k = 2$.)

In order to provide a more concrete realization of these ideas, we define the diameter, $d \in \Re^+$, of a reconstruction to be $d = \max \{\|\vec{v}_i - \vec{v}_j\|, i,j \in \{1, ..., N_v\}$ and $i \neq j\}$. Ten percent of the diameter is a typical cutoff value, c. Though choosing a different function ρ clearly affects the recurrence plot, for clarity of the ensuing discussion, we will choose this function to be as given in Eq. (2) with c assuming a value between $0.1d$ and $0.28d$. For the rest of the discussion, vectors $\vec{v}_i$ and $\vec{v}_j$ will be called close if $\|\vec{v}_i - \vec{v}_j\| < c$ and far, otherwise.

Note that with δ defined as above, an $m \times n$ recurrence plot gives us insight into the dynamics underlying the first $m + n$ vectors. We can define $\delta(i,j)$ more generally, however:

$$\delta(i,j) = \|\vec{v}_{(i+c_l)} - \vec{v}_{[(i+c_l)+(j+r_l)]}\|, \tag{3}$$

where $c_l \in \mathcal{Z}$ allows us to study the $m + n$ vectors following $\vec{v}_{c_l}$. The integer r_l is to be interpreted as a delay factor and permits us to study longer recurrences. In particular, with non-zero c_l and r_l, a recurrence plot defined using Eq. (3) gives information about the recurrences between the sets of vectors $\{\vec{v}_{(c_l+1)}, ..., \vec{v}_{(c_l+n)}\}$ and $\{\vec{v}_{(c_l+r_l+2)}, ..., \vec{v}_{(c_l+r_l+m+n)}\}$. Fixing $i_0 \in \mathcal{Z}$ and letting $r_l = 0$, we note that the m array elements $d(i_0,j) = \|\vec{v}_{i_0} - \vec{v}_{(i_0+r_l+j)}\|, (j = 1, 2, ..., m)$ contain information about the recurrences between the i_0th vector and the next m vectors in the time series. By letting $r_l \in \mathcal{Z}^+$, we can study the recurrences between the vector $\vec{v}_{(i_0+1)}$ and the set of vectors $\{\vec{v}_{(i_0+r_l+1)}, \vec{v}_{(i_0+r_l+2)}, \cdots, \vec{v}_{(i_0+r_l+m)}\}$, i.e., between the i_0th vector and any set of m vectors. Lastly, we can incorporate the concept of a delay in a yet more general fashion. We can define

$$\delta(i,j) = \|\vec{v}_{(c_l+i)} - \vec{v}_{[(c_l+i)+(r_d+1)*j)]}\| . \tag{4}$$

In this way, the delay r_d allows us to study recurrences over an even greater domain within the reconstruction sequence. This is typically accomplished, however, by filtering the data before any reconstruction takes place (Discarding every other data point from the original time series has the same effect as letting $r_d = 1$ in Eq. (4).)

Having noted that we can generalize the definition of the distance function, δ, in order to permit a study of the data series at different time scales, we will lose no generality during the rest of our analysis by assuming that $r_l = c_l = 0$. Let us turn our attention to the connection between recurrence plots and the dynamics underlying the original time series by studying the geometrical features which are revealed in the recurrence plots of a selection of elementary time series.

EXAMINATION OF ELEMENTARY TIME SERIES

As mentioned above, throughout this section we will assume that δ and ρ are as defined in Eqs. (1) and (2).

SQUARE WAVE

This time series is simply an alternating sequence of $l \in \mathcal{Z}^+$ values of 0 and l values of 1, giving a period of $2l$. Two distinct geometrical features are seen in the $\Re^1$ recurrence plot of this data, as shown in Figure 1. (The prefix $\Re^n$-indicates that the original time series is embedded in $\Re^n$. All embeddings use a delay of 1, unless otherwise noted.)

We will begin our classification with the vertical line segments of length l which are clearly visible in this plot. These line segments represent array elements

$$\{(i,j),(i,j+1),(i,j+2),\dots,(i,j+l-1)\} ,$$

which, in turn, correspond to the distances

$$\{\delta(i,j) = \|\vec{v}_i - \vec{v}_{(i+j)}\| ,$$
$$\delta(i,j+1) = \|\vec{v}_i - \vec{v}_{(i+j+1)}\| ,$$
$$\vdots$$
$$\delta(i,j+l-1) = \|\vec{v}_i - \vec{v}_{(i+j+l-1)}\| .$$

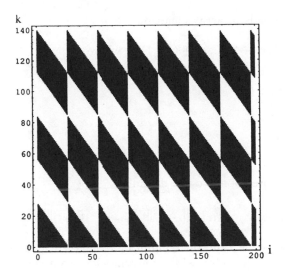

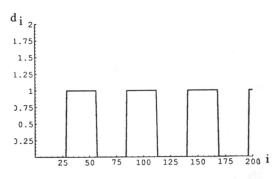

FIGURE 1 A recurrence plot of the square-wave data described in the text embedded in $\Re^1$. We used the l_1 norm in defining δ via Eq. (1) for all the figures. We also used a 140 × 200 array and the black-and-white color map in all the figures. The diamond shapes are most usefully interpreted as repeated sequences of shifted vertical line segments and shifted line segments with a slope of -1. The first 200 data values, connected by straight lines, are displayed below.

Thus, these vertical segments of length l represent a vector which remains close to a set of l successive vectors forward in time. This behavior is obvious in our square wave example, but can now be generally recognized. Looking at this particular example more closely, we notice that these line segments are l array elements in length, that they occur in l successive columns (though shifted down by one array element each time), and that they are separated by l successive columns before repeating. Since this pattern repeats throughout the plot, we can conclude that an arbitrary data point will alternate between being close to l successive data values and being far from l successive values.

An equally noticeable line-segment orientation appearing in Figure 1 is a slope of -1. Consider such a line segment passing through the array element (s_0, t_0) and having length l. This segment corresponds to array elements

$$\{(s, \phi(s)): \ s = s_0, s_0 + 1, s_0 + 2, ..., s_0 + (l-1)\}, \ \text{where}$$
$$\phi(s) = t_0 - (s - s_0)$$
$$= -s + (s_0 + t_0).$$

These elements can also be expressed as

$$\{(i,j), (i+1, j-1), (i+2, j-2), \ldots, (i+l-1, j-(l-1))\},$$

which makes it more clear that they correspond to the distances

$$\{\|\vec{v}_i - \vec{v}_{(i+j)}\|, \|\vec{v}_{(i+1)} - \vec{v}_{(i+j)}\|, \|\vec{v}_{(i+2)} - \vec{v}_{(i+j)}\|, \ldots, \|\vec{v}_{(i+l-1)} - \vec{v}_{(i+j)}\|\}.$$

So these line segments represent the same type of dynamics as the vertical line segments do in that they correspond to a vector which remains close to a set of

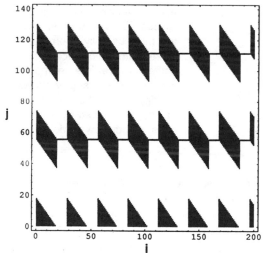

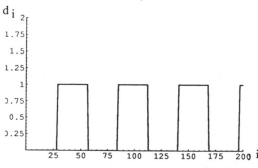

FIGURE 2 A recurrence plot of the same time series used in Figure 1 embedded in $\Re^{10}$. Since we are using the l_1 norm and $c = .1d$, two components of a vector pair being far from each other results in the vectors themselves being far from one another. Thus, the diamond-shaped regions have shrunk, exposing the horizontal line segments which reflect the periodicity of the time series.

l successive earlier vectors. l of these line segments also occur successively, and alternate with l empty diagonal segments, producing large-scale diamond-shaped regions in our plot (see Figure 1.)

Another commonly seen line-segment orientation is more easily recognized as this data set is embedded in higher dimensions. Figure 2 is an $\Re^{10}$ recurrence plot of the same data used in Figure 1.

Here we notice horizontal line segments which cross the entire plot every $2l$ rows. A horizontal line segment of length l which passes through the array element (s_0, t_0) corresponds to array elements

$$\{(s, \phi(s)) : s = s_0, s_0 + 1, s_0 + 2, \ldots, s_0 + (l-1)\}, \text{ where}$$
$$\phi(s) = t_0 .$$

or

$$\{(i, j), (i+1, j), (i+2, j), \ldots, (i+l-1, j)\}$$

which correspond to the distances

$$\{\|\vec{v}_i - \vec{v}_{(i+j)}\|, \|\vec{v}_{(i+1)} - \vec{v}_{(i+j+1)}\|, \|\vec{v}_{(i+2)} - \vec{v}_{(i+j+2)}\|, \ldots, \|\vec{v}_{(i+l-1)} - \vec{v}_{(i+j+l-1)}\|\} .$$

Thus, these line segments correspond to the portions of the data which result in reconstructed vectors coming close to one another and remaining close for an interval of time. In particular, evenly spaced, vertically repeating, horizontal line segments correspond to periodic recurrence. We'll look further into this phenomenon in the next section.

TRIANGLE WAVE

The recurrence plot of this data set again shows the horizontal line segments which we saw in the last data set. Here, we also see periodically repeating line segments with a negative slope which is less than -1. In particular, the diagonal line segments shown in Figure 3 represent a sequence of array elements

$$\{(s, \phi(s)), s = s_0, s_0 + 1, s_0 + 2, \ldots, s_0 + (l-1)\}, \text{ where}$$
$$\phi(s) = t_0 - 2(s - s_0)$$
$$= -2s + (t_0 - 2s_0) .$$

or

$$\{(i, j), (i+1, j-2), (i+2, j-4), \ldots, (i+l-1, j-2(l-1))\}$$

which correspond to the distances

$$\{\|\vec{v}_i - \vec{v}_{(i+j)}\|, \|\vec{v}_{(i+1)} - \vec{v}_{(i+j-1)}\|, \|\vec{v}_{(i+2)} - \vec{v}_{(i+j-2)}\|, \ldots, \|\vec{v}_{(i+l-1)} - \vec{v}_{(i+j-(l-1))}\|\} ,$$

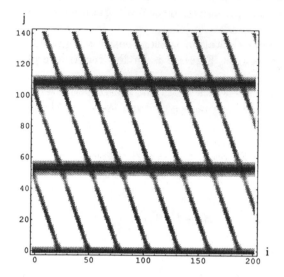

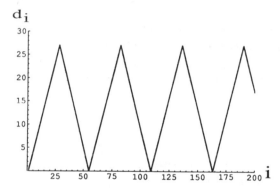

FIGURE 3 A recurrence plot of the triangle wave whose first 200 data values are shown below. Data is embedded in $\Re^5$ with a delay of 1. We note that three colors are visible in this plot—white indicates vector pairs which are far from one another, grey represents vector pairs which are separated by $.2d$, and black indicates vector pairs which are recurrent within $.1d$.

i.e., the sequence of vectors $\{\vec{v}_k\}_{k=i}^{i+l-1}$ is close to the sequence $\{\vec{v}_k\}_{k=i+j}^{i+j-(l-1)}$! We see this line-segment orientation any time there is a sequence of vectors which appear inverted in time elsewhere in the reconstruction.

OSCILLATING SAW-TOOTH WAVE

Since this data set is not so typical, we mention that it consists of n equally spaced increasing values followed by $3n$ equally spaced increasing values (whose range of increase is the same as the range of the previous n increasing values). In the recurrence plot of this data, shown in Figure 4, we see another line-segment orientation—line segments with positive slopes. These line segments represent (in this case) a sequence of array elements

$$\{(s, \phi(s)) : \ s = s_0, s_0 + 1, s_0 + 2, \ldots, s_0 + (l - 1)\}, \text{ where}$$
$$\phi(s) = t_0 + 2(s - s_0)$$
$$= 2s + (t_0 - 2s_0).$$

or

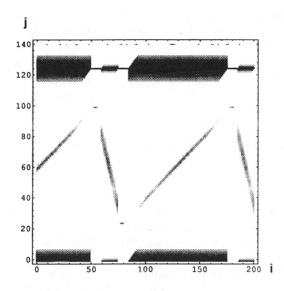

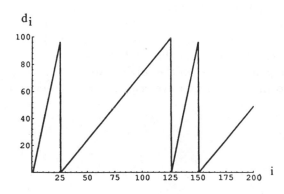

FIGURE 4 A recurrence plot demonstrating the line segments with a positive slope which indicates a time dilation between two segments of the data. This data has been embedded in $\Re^{10}$ with a delay of 1 and a cutoff $c = .1d$. Again, we would see that the diagonal line segments would correspond to closer recurrences in $\Re^5$. Note that since this data is periodic, the time dilation is accompanied by a time contraction which is represented by the line segments with a slope (strictly) between −1 and 0.

$$\{(i, j), (i + 1, j + 2), (i + 2, j + 4), \ldots, (i + l - 1, j + 2(l - 1))\}$$

which correspond to the distances

$$\{\|\vec{v}_i - \vec{v}_{(i+j)}\|, \|\vec{v}_{(i+1)} - \vec{v}_{(i+j+3)}\|, \|\vec{v}_{(i+2)} - \vec{v}_{(i+j+6)}\|, \ldots, \|\vec{v}_{(i+l-1)} - \vec{v}_{(i+j+3(l-1))}\|.\}$$

Thus, this line segment with a slope of positive two represents the presence of a time dilation between two segments of the reconstruction; the sequence of vectors $\{\vec{v}_k\}_{k=i}^{i+l-1}$ is, successively, close to every third vector in the sequence $\{\vec{v}_k\}_{k=i+j}^{i+j+3(l-1)}$.

RELAXATION OSCILLATOR

Since we will be referring to these ideas throughout the rest of the paper, we should precisely define what we mean by time inversion, dilation, and contraction. A sequence of vectors $\{\vec{v}_i\}_{i=0}^n$ is said to be inverted in time elsewhere in the data if there exists a sequence of vectors $\{\vec{v'}_i\}_{i=0}^n$ such that $\vec{v}_i$ is close to $\vec{v'}_{n-i}$ for all $i = 0$ to n. A sequence of vectors $\{\vec{v}_i\}_{i=0}^n$ is said to be dilated with respect to time elsewhere in the data if there exists a sequence $\{\vec{v'}_i\}_{i=0}^{d \cdot n}, d \in \mathcal{Z}^+$ such that $\vec{v}_i$ is close to $\vec{v'}_{d \cdot i}$ for all $i = 0$ to n. A contraction is defined similarly.

Now, as we increase the embedding dimension, let us examine the stability of the dynamics corresponding to the line-segment orientations we have seen. An $\Re^1$ recurrence plot of this time series shows all of the above-described geometrical features if the period of the sine wave is made to decrease as its amplitude increases (for example, $\{100j\sin(2\pi j/(125 - j))\}_{j=0}^{86}$), as seen in Figure 5.

Notice that only the line segments of slope zero and slope -2 (a portion of the time series is inverted elsewhere in the data) impose a constraint on successive vectors in the two segments of the reconstruction which are recurrent, and it is these orientations which are most persistent as the embedding dimension is increased.

The vertical line segments and the segments with a slope of -1 of length l indicate that one fixed vector is close to a sequence of l successive vectors elsewhere in the time series. A higher-embedding dimension will introduce different data values as the additional components of the higher-dimensional fixed vector. If these new data values are about the same magnitude as the original components (which is more likely if the data comes from a repeatedly sampled continuous signal), then we expect the feature to persist. If, however, some of these additional data values are sufficiently different from the component values of the original vector, then this fixed vector will not remain close to the same successive sequence of vectors and this feature will disappear.

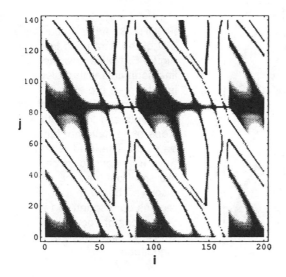

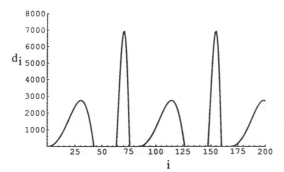

FIGURE 5 An $\Re^1$ recurrence plot of the relaxation oscillator $(\{100j\sin(s\pi j/(125-j))\}_{j=0}^{86},$ then repeat) whose first 200 values are shown below. As in Figure 1, we see how an $\Re^1$ recurrence plot tends to resemble the original time series. Notice that the vertical line segments and line segments with a slope of −1 correspond to the times when the time series assumes the median data value (0, in this case.) This demonstrates how a recurrence plot can act as a threshold indicator in lower embedding dimensions. This is another reason to concentrate our attention on higher embedding dimensions.

The line segments whose slope is positive or negative with a value other than −1 or −2 correspond to time dilations or contractions (with or without inversion) in the original time series. Since the additional data values being introduced are not coming from successive data values (even when the time series comes from a measured continuous signal), they are that much less likely to be about the same magnitude as the original components, and we see why these are the most likely segments to disappear upon using a higher embedding dimension. These phenomena can be seen in Figures 5 through 8.

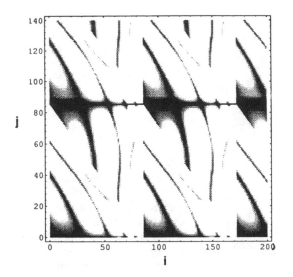

FIGURE 6   An $\Re^3$ recurrence plot of the data used in Figure 5. Notice how the line segments with a slope less than −2 are already disappearing. The vertical line segments and those with a slope of −1 have not faded as much. The filled regions have not changed much. We see from the time series displayed below that these regions correspond to two sequences of vectors, all of whose components are of small magnitude.

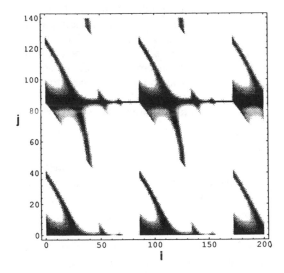

FIGURE 7   An $\Re^5$ recurrence plot of the data used in the preceding two figures. We can now clearly distinguish the horizontal lines which reflect the periodicity in the time series. In addition to the regions mentioned in Figure 6, we still see a prominent line segment with a slope of −1 which curves down through the filled region to become a line segment with a slope greater than negative two. This reflects the near periodic behavior of the time series when the amplitude is still small.

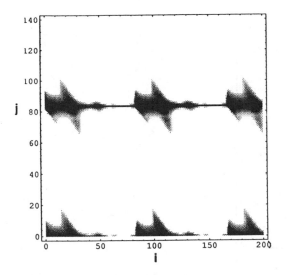

FIGURE 8 An $\Re^{10}$ recurrence plot of the same data used in Figures 5 through 7. By now, most of the previously discussed line-segment orientations have vanished leaving the horizontal lines, which will always be present, and the filled regions, though we can see that the size of these regions will be inversely proportional to the rate of amplitude increase.

EXPERIMENTAL DATA

Here we look at two different portions of a data set from human EEG described in Mayer-Kress et al.[7] The vectors used to create the recurrence plot shown in Figure 9 begin 240 msec after the initiation of the recording. There are several line segments appearing in this plot which have a slope of 2 and several with a slope less than −2, which indicate that portions of the time series are shortly re-appearing inverted in time. This is not discernible in the original data. Of greater physiological interest, however, is the fact that the horizontal line segments which are visible in this plot are relatively short, indicating a high local rate of divergence. This contrast with the horizontal line-segment length seen in Figure 10, which is a recurrence plot of a later portion of the same data file used in Figure 9. The plot shown in Figure 10 was created using the same embedding dimension, delay, and cutoff as used to create Figure 9, but we now see long, vertically repeating, horizontal line segments corresponding to a much lower local divergence rate and the development of a periodic structure within the time series. The frequency of this oscillation can be determined by the spacing between successive segments and in this case corresponds to the alpha waves of the brain (about 10 Hz)

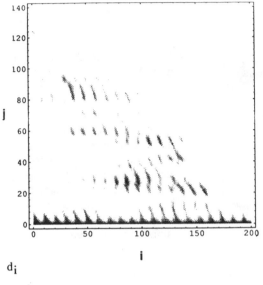

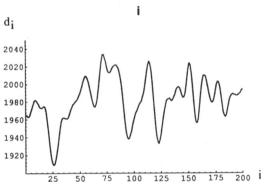

FIGURE 9   An $\Re^{10}$ recurrence plot of human EEG reading (resting, awake, eyes closed; see Fraser[2] for details) with a delay of 20 msec and a cutoff $c = .28d$. The 340 vectors used to create this recurrence plot begin 240 msec after the initiation of the recording corresponding to the figures in Mayer-Kress et al.[7] Notice the line segments with a slope of -1 and 0.

INVARIANTS

Lyapunov exponents and dimension are two of the most-studied diffeomorphic invariants of autonomous dynamical systems. When a non-autonomous system can be viewed as an autonomous system with slowly varying parameters, then much useful insight into the underlying dynamics can be gained by calculating the variation of these two characteristics with respect to time.

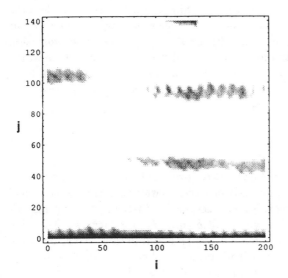

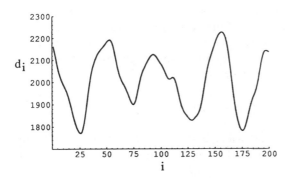

FIGURE 10 This is an $\Re^{10}$ recurrence plot of a later segment (10 sec after the beginning of the recording) of the same EEG file shown in Figure 9, also with a delay of 10 and a cutoff of $c = .28d$. Notice that horizontal line segments which are much longer than the ones appearing in Figure 9 predominate, indicating a periodicity in the data and a small local divergence rate. In this case, the frequency shown through the spacing between the lines is about 10 Hz, which corresponds to the alpha waves of the brain.

LYAPUNOV EXPONENT (LARGEST)

As we have seen, one of the more obvious, and most important, features in our plots are the horizontal line segments. Recall that such line segments correspond to a sequence of distances:

$$\{\|\vec{v}_i - \vec{v}_{(i+j)}\|, \|\vec{v}_{(i+1)} - \vec{v}_{(i+j+1)}\|, \|\vec{v}_{(i+2)} - \vec{v}_{(i+j+2)}\|, \ldots, \|\vec{v}_{(i+l-1)} - \vec{v}_{(i+j+l-1)}\|\}.$$

If we suppose that, at least locally, the system is described by a hyperbolic dynamic, then we know that all vectors are diverging in the average at the same rate of $e^{\lambda \cdot l}$, where λ is the largest positive Lyapunov exponent.[10] If we assume a constant rate of divergence, then l, the length of this horizontal line segment, must be inversely proportional to λ. Thus, observing the variation of horizontal line-segment lengths within a particular recurrence plot, or between recurrence plots taken from different portions of a large data set (see Figures 9 and 10) enables us to perceive the variation of the local rate of divergence within our reconstructed vector sequence. It is worth noting that relatively few calculations are needed to estimate the variation of local divergence in this way, and that the ability to observe this variation is present in data sets containing as few as 400 points.

DIMENSIONAL COMPLEXITY

In addition to local rate of divergence information provided by the lengths of horizontal line segments, a recurrence plot also provides a way to estimate the local dimension of the time series in a more precise sense.[4,5,9] Instead of mapping values of δ into $\mathcal{Z}_2$, generalize ρ to map values of δ into $\mathcal{Z}_k$, for large k ($k = 256$, for example). Consider the previously defined recurrence function, $\rho \circ \delta$, on a log scale. Then we have a direct connection from taking a slice at a fixed column of a particular recurrence plot to local gauge functions. To see how this is so, let us suppose that blue is the color which corresponds to the least recurrence function value C_{min}, i.e., the smallest distance, while red is the color which corresponds to the greatest recurrence function value, C_{max}. Consider a slice at the i_0th column of a recurrence plot defined on an $m \times n$ array. Starting at the bottom of the column, color the first p elements with the color corresponding to the closest recurrence to the vector $\vec{v}_{i_0}$, where p is the number of vectors such that

$$\rho(\|\vec{v}_{i_0} - \vec{v}_{i_0+j}\|) \leq C_{min}, j \in \mathcal{Z}_m.$$

Continuing in this fashion up the column will result in the top q elements being colored with the color corresponding to the largest separation from vector $\vec{v}_{i_0}$, where q is the number of vectors such that

$$\rho(\|\vec{v}_{i_0} - \vec{v}_{i_0+j}\|) \geq C_{max}, j \in \mathcal{Z}_m.$$

If we would now apply a log function to our color map, then the distance from the red region to the blue region is related to the dimension of the time series at time i_0. A gradual change in color from blue to red indicates a low local dimension, while a sharp transition occurring somewhere within the column corresponds to a steep line on a $\log(n)$ vs. $\log(r)$ plot, i.e., a high local dimension. A more direct representation would be a color coding of the change in color as a function of the log of the distance.

CONCLUSION

Many biological, physiological, and ecological processes have been characterized as autonomous systems with slowly varying parameters. Constructing a recurrence plot in the fashion we've described is an efficient means of recovering dynamical information from an orbit which has been reconstructed from a time series generated by such a dynamic.

Continuing the investigation outlined above, we can obtain a complete classification of line-segment orientations within a recurrence plot:

1. With a slope less than −2: These line-segment orientations indicate that a successive sequence of vectors reappears in the reconstructed vector sequence inverted and dilated with respect to time.
2. A slope of −2: These segment orientations indicate that a successive sequence of vectors reappears in the reconstructed vector sequence inverted with respect to time.
3. With a slope (strictly) between −2 and −1: This orientation indicates a sequence of vectors which reappears inverted and contracted in time.
4. Vertical and with a slope of −1: These line-segment orientations correspond to a particular vector which remains close to a successive sequence of vectors elsewhere in the reconstructed vector sequence.
5. With a slope (strictly) between −1 and 0: This orientation indicates that a sequence of vectors reappears contracted in time.
6. Horizontal: These line segments indicate a recurrent sequence of successive vectors.
7. Positive slope: This orientation indicates that a sequence of vectors reappears dilated in time.

In addition to the above, the variation of the local rate of divergence of the dynamic underlying the original time series is visually accessible. The definition of a recurrence plot can also be modified, as described in the preceding section, in order to make the variation of the local dimensional complexity visually accessible. This method can provide these estimates for data sets with only a few hundred values.

REFERENCES

1. Eckmann, J.-P., S. Oliffson Kamphorst, and D. Ruelle. "Recurrence Plots of Dynamical Systems." *Europhys. Lett.* **4(9)** (1987): 973–977.
2. Fraser, A. M. "Information and Entropy in Strange Attractors." Dissertation, University of Texas at Austin, 1988.

3. Layne, S. P., G. Mayer-Kress, and J. Holzfuss. "Problems Associated with Dimensional Analysis of Electro-Encephalogram Data." In *Dimensions and Entropies in Chaotic Systems*, edited by G. Mayer-Kress, 246–256. Berlin: Springer Verlag, 1986

4. Mayer-Kress, G., ed. *Dimensions and Entropies in Chaotic Systems—Quantification of Complex Behavior*. Berlin: Springer-Verlag, 1986

5. Mayer-Kress, G. "Application of Dimension Algorithms to Experimental Chaos." In *Directions in Chaos*, edited by Hao Bai-lin, 122–147. Singapore: World Scientific, 1987

6. Mayer-Kress, G., and S. P. Layne. "Dimensionality of the Human Electroencephalogram." *Annals of the New York Academy of Sciences* **504** (1987): 62–86.

7. Mayer-Kress, G., F. E. Yates, L. Benton, M. Keidel, W. Tirsch, S. J. Poppl, and K. Geist. "Dimensional Analysis of Nonlinear Oscillations in Brain, Heart, and Muscle." *Math. Biosci.* **90** (1988).

8. Mayer-Kress, G. "Monitoring Chaos of Cardiac Rhythms." In *Biotech USA*, Proceedings of the 6th annual industry conference and exhibition, Oct. 2-4, 1989, San Francisco, CA, 435–443.

9. Mayer-Kress, G., and A. Hubler. "Time Evolution of Local Complexity Measures and Aperiodic Perturbations of Nonlinear Dynamical Systems." In *Measures of Complexity and Chaos*, edited by N. B. Abraham, A. M. Albano, A. Passamante, and P. E. Rapp, 155–172. New York: Plenum, 1990.

10. Nicolis, J., G. Mayer-Kress, and G. Haubs. "Non-Uniform Chaotic Dynamics with Implications to Information Processing." *Z. Naturforsch.* **38a** (1983): 1157–1169.

11. Packard, N. H., J. P. Crutchfield, J. D. Farmer, and R. S. Shaw. "Geometry From a Time Series." *Phys. Rev. Lett.* **45** (1982): 712.

12. Takens, F. "Detecting Strange Attractors in Turbulence." In *Lecture Notes in Mathematics*, vol. 898, edited by D. A. Rand and L. S. Young, 366–381. New York: Springer-Verlag, 1980.

13. Zbilut, J., M. Koebbe, H. Loeb, and G. Mayer-Kress. "Use of Recurrence Plots in the Analysis of Heart Beat Intervals." *Proc. IEEE Conference on Computers in Cardiology*, Chicago, Sept. 1990.

Section IV: Applications

Blake LeBaron
Department of Economics, University of Wisconsin-Madison, Madison, WI 53706

Nonlinear Forecasts for the S&P Stock Index

1. INTRODUCTION

Few time series have more people involved in forecasting than stock prices. For almost a century various stock analysts and technical traders have attempted to convert supposed forecasting prowess into trading profits in the stock market. Recent studies have found evidence for nonlinearities in stock returns series.[7,13] These early results suggest that some forecast improvements may be possible using nonlinear time-series techniques. Several recent papers have proceeded to test for nonlinear forecast improvements in several economic series.[1] All these have found almost no evidence for any improvement using standard nonparametric estimation techniques. If there is no out-of-sample forecast improvement, then why do these tests indicate the presence of nonlinearities? These results could indicate that the previous tests are just picking up nonstationarities in the series tested, and there is little evidence for any nonlinearity which stays stable over time.

---

[1]This group includes Diebold and Nason,[4] Meese and Rose,[10] and Mizrach[11] for exchange rates, LeBaron[8] and White[15] for stock returns, and Prescott and Stengos[12] for gold.

This paper presents some evidence which suggests that these earlier conclusions may have been premature. Some out-of-sample forecast improvements are demonstrated for the weekly S&P 500 series. The evidence shows that these improvements are difficult to detect since they only occur during certain time periods. The forecasting model has some periods in which absolutely no forecast improvement is possible. Although forecast improvements appear significant, they are extremely small and occur only for a fraction of the weeks tested. Stock returns remain, as they should, a relatively difficult series to forecast.

The second section will introduce some of the basic facts about well-known properties of stock returns. Also, some graphical evidence indicating possible forecast improvements will be shown. Section 3 will do some initial forecasting tests. Section 4 will present the results on nonlinear forecasting, and section 5 will conclude.

2. STOCK RETURN FACTS

This section presents some facts about stock returns which will be used in the remaining sections. For all the tests performed in this paper, the S&P 500 index will be used from 1947–1985. This common stock-market index is a value-weighted index of 500 stocks. Unfortunately, dividend payments are not included in this index. Weekly stock returns are calculated from this index by taking the log first differences from Friday close to the following Friday close. This weekly returns series will be the focus of all forecasting experiments done in this paper.

The unconditional distribution of this returns series shares many common features with other asset returns. It is close to symmetric, but it is highly leptokurtic relative to a standard normal distribution. For this series the kurtosis is 5.63. Table 1 presents the autocorrelations at lags 1 through 10 for this series in the column labeled "returns." The returns series is very close to being uncorrelated. The final row gives the Bartlett standard errors for the estimated correlations. Only the correlation at lag 3 appears significant. This indicates very little possibility for linear forecastability.

While returns themselves are close to uncorrelated, their squares are not. This is shown in the second column of Table 1. The autocorrelations of the squared returns for the first 10 lags are all significantly different from zero, indicating large persistence. This implies that while returns are close to uncorrelated, large returns of either sign tend to follow large returns. Stock returns go through periods of high and low volatility. This fact has led to a new type of models, ARCH/GARCH,

TABLE 1 Autocorrelations: Returns and Squared
Returns

Lag	Returns	Squared Returns
1	0.0259	0.1766
2	0.0236	0.2342
3	0.0483	0.2234
4	-0.0218	0.2461
5	-0.0356	0.1325
6	0.0368	0.1334
7	-0.0018	0.1406
8	-0.0294	0.2154
9	0.0298	0.0894
10	-0.0263	0.1104
Bartlett Std.	0.0219	0.0219

designed to capture this effect.[2] In this paper a specific type of GARCH model
will be estimated and simulated. It takes the following form,

$$r_t = a + \gamma h_t + b\epsilon_{t-1} + \epsilon_t, \qquad \epsilon_t = h_t^{1/2} z_t,$$
$$h_t = \alpha_0 + \alpha_1 \epsilon_{t-1}^2 + \beta h_{t-1},$$
$$z_t \sim N(0,1).$$

The GARCH model allows the conditional variance of the disturbance terms, h_t,
to persist. It depends linearly on the past variance and disturbances squared. This
model also allows for some weak correlation at the one-week lag through an MA(1)
term, $b\epsilon_{t-1}$. For the simple GARCH model, the γ term above is not estimated. It is
used in what is known as the GARCH-M form. Here the conditional mean for r_t is
allowed to move linearly with the conditional variance. Estimates of this parameter
are generally positive, but sometimes they are not stable over time. This suggests
a simple form of risk-return tradeoff for stock returns. The parameter estimates for
these models are given in Table 2. Estimation is by maximum likelihood. These
models will be used to generate simulated stock returns. It is felt that they capture
some of the important volatility aspects of the series, and therefore are better than
simple random number simulations.

The results here show stock returns to have changing conditional variances
and very little correlation. Recent work in LeBaron[9] suggests that this is not all
that is going on and forms a basis for many of the forecast experiments done here.

[2] These are developed in Engle[5] and Bollerslev.[1]

TABLE 2 GARCH(1,1) Parameter Estimates[1]

$$r_t = a + \gamma h_t + b\epsilon_{t-1} + \epsilon_t \qquad \epsilon_t = h_t^{1/2} z_t$$
$$h_t = \alpha_0 + \alpha_1 \epsilon_{t-1}^2 + \beta h_{t-1}$$
$$z_t \sim N(0,1)$$

Period	α_0 $*10^5$	α_1	β	b	a $*10^3$	γ
Weekly S&P 1946–85	1.19 (0.29)	0.117 (0.016)	0.851 (0.018)	0.069 (0.024)	1.60 (0.35)	
Weekly S&P 1946–85	1.18 (0.296)	0.116 (0.016)	0.852 (0.019)	0.067 (0.024)	0.734 (0.733)	3.82 (2.43)

[1] Estimation is by maximum likelihood and numbers in parenthesis are asymptotic standard errors.

This evidence will be presented in Figures 1 and 2. These figures show that for stock returns there are changes in the autocorrelations measured conditionally on previous information. For all the tests run in this paper, this conditional information is a simple sum of squares over several previous periods,

$$\hat{\sigma}_t^2 = 1 over L \sum_{i=0}^{L-1} x_{t-i}^2.$$

For all experiments in this paper, L will be set to 10.

Figure 1 is inspired by a test of Tsay.[14] This procedure is performed in five steps.

1. Get an estimate of the conditional variance at time t, $\hat{\sigma}_t^2$.
2. Form the data into triplets of the form $\hat{\sigma}_t^2, x_t, x_{t+1}$.
3. Sort these by the estimated conditional variance, $\hat{\sigma}_t^2$.
4. Form a moving rectangular band of length J over ranges $\hat{\sigma}_t^2$ through $\hat{\sigma}_{t+J}^2$ for $t = 1$ through $t = N - J + 1$. (N is the length of the sorted triplet series.)
5. Estimate the serial correlation in each part. This vector of serial correlations will then be displayed. Serial correlations are estimated, removing the local means.

Figure 1 shows the conditional correlations using the above tests for the S&P 500 series. The conditional correlation is plotted on the y-axis and the volatility quantile order on the x-axis, where 0.50 corresponds to the median volatility

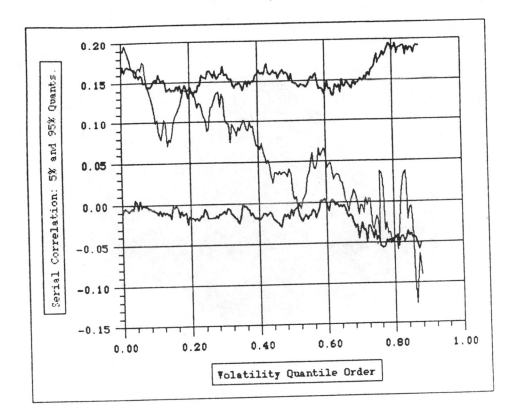

FIGURE 1   Sorted Local Correlation: Weekly S&P 1946–1985.

level. The band size used here is 250. The figure shows the conditional serial correlation dropping from a maximum of about 0.20 to a minimum of -0.10. To check the significance of these results, the GARCH model estimated in Table 2 is simulated. The model estimated allows for a constant serial correlation with an MA(1) term. This is then simulated 100 times and run through the above procedures. Generated 5% and 95% confidence bands are presented in Figure 1. They do not reveal any pattern similar to that from the actual data.

Figure 2 presents a second graphical test of this effect. Using a standard nonparametric regression with a Gaussian kernel, the expected value of the return at time $t + 1$, r_{t+1}, is estimated using r_t and $\hat{\sigma}_t^2$. Before estimation is done the two series are "normalized." First, the returns series is divided by its unconditional standard deviation. Second, the volatility series is mapped into its fractiles (i.e., the median volatility becomes 0.50). This uniform distribution is then divided by its standard deviation. The return at $t + 1$ is then estimated using independent

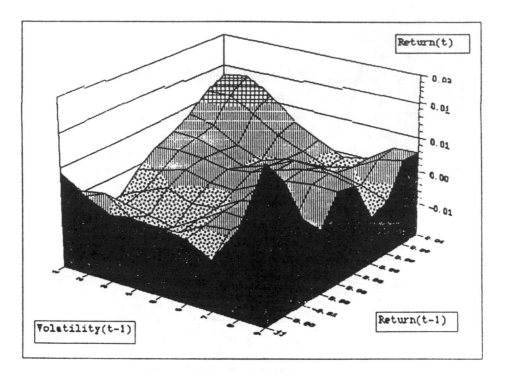

FIGURE 2   Weekly S&P 1946–1985 Kernel Estimation.

kernels over the two "normalized" series with the bandwidth set to 0.5.[3] These conditional expectations are then plotted against the two pieces of conditioning information. The front axis corresponds to conditional volatility and the right axis is lagged return. The vertical distance is the conditional return. From this plot we see a strong positive correlation between r_{t+1} and r_t for low levels of volatility, but this is seen to drop off quickly as volatility increases (moving to the right across the front axis). This figure agrees with the previous figure in the connection between correlations and volatility.

Both these figures point to possible increased predictability during certain time periods when volatility is low. They show high correlations during the quietest periods. This predictability may go away when volatility increases. This suggests two things for forecasting. First, forecast improvements should be evaluated for different volatility levels separately. Second, volatility may be a key piece of information for improving forecasts.

[3] If the statistical significance of Figure 2 were to be tested, some form of cross validation would be used to determine the optimal bandwidth. Since this is not done here, Figure 2 remains only interesting supporting graphical evidence.

3. INITIAL FORECASTS

This section will perform some initial forecasting experiments on the weekly stock returns series. All forecasting tests can suffer extreme data mining problems when forecast evaluation is done using the same set of data as is used in estimating parameters. This is why the use of a hold-out sample is useful. Here the 2000 weekly returns are split in two and the first half is used for estimation while the second half is used for forecast evaluation.

All forecasts will be evaluated using mean squared forecast error (MSE),

$$(1/N)\sum_{t=1}^{N} e_t^2,$$

where e_t is the forecast error at time t. Numbers presented will represent the percentage difference in MSE from one method to another. Table 3 begins some forecasting experiments with results from three linear models, an AR(1), AR(5), and an AR(10) of the form

$$\hat{r}_{t+1} = b + \sum_{j=0}^{p-1} \rho_j r_{t-j},$$

where $\hat{r}_{t+1}$ is the forecast for r_{t+1}. Forecast errors will be compared with those estimated using the sample mean as the optimal forecast. This makes the percentage improvement in MSE similar to measures used in Farmer and Sidorowich,[6] but for this measure, 1 indicates perfect predictability and 0 is no better than guessing the mean. Improvements can also be negative, indicating that the method is doing worse than guessing the mean.

The panel A of Table 3 in the column labeled "all" lists the percentage improvement in MSE over the sample mean from the various linear models for in-sample forecasting. The forecasting and estimation is done over the first half of the returns series only. These linear models are estimated using ordinary least squares. The numbers indicate that even in sample the improvements from the linear models are quite small with the improvement for the AR(10) at 1.4 percent. These improvements are further subdivided according to volatility level. Local volatility is measured as in the last section as the sum of lagged squares. Forecast errors at time $t + 1$ are classified according to the estimated volatility at time t, $\hat{\sigma}_t^2$. Volatility is divided into five quintiles. Table 3 reports the forecast improvements for each quintile. In each quintile the forecast improvements are still fairly small for an in-sample experiment.

Panel B of Table 3 repeats the experiments for out-of-sample forecasts. Here the models are estimated over the first half of the series and forecasts are made during the second half. The sample mean used for comparison is also estimated during the first half only. Moving out of sample reduces all the MSE improvement percentages, with most of the numbers turning negative, indicating that the linear models are

TABLE 3 Forecast Improvements: Linear Models[1]

AR Lag	Q1	Q2	Q3	Q4	Q5	All
			Panel A: In Sample			
1	0.0165	0.0086	-0.0009	-0.0068	0.0094	0.0039
5	0.0148	0.0229	-0.0019	0.0011	0.0224	0.0115
10	0.0105	0.0200	0.0057	0.0105	0.0211	0.0141
			Panel B: Out of Sample			
1	0.0167	0.0068	0.0150	-0.0095	-0.0113	-0.0031
5	0.0121	0.0147	-0.0046	-0.0011	-0.0356	-0.0151
10	0.0121	0.0066	-0.0074	-0.0086	-0.0401	-0.0201

[1] In-sample forecasts performed on the first half of the data set. Out-of-sample forecasts use the first half for estimation and the second half for forecasting. Qn's are percentage improvements in MSE for various quintile levels. All is the percentage improvement in MSE over all observations.

doing worse than just using the mean from the first half. These are the kinds of properties that would be expected when trying to forecast an unforecastable series.

4. NONLINEAR FORECASTING

For an uncorrelated series it should not be too surprising that linear techniques failed. However, the evidence suggests that there may be nonlinear forecastable structure. Nearest-neighbor local regression techniques are used to look for this.[4] These methods fit a linear regression to local neighborhoods of points. Close neighbors are determined using Euclidean distance over the lagged returns. The method used will choose the N closest neighbors for each point r_t. Table 4 reports in-sample results for the first half of the returns series. Panel A shows results using one lag and panel B shows results using five lags. The format is similar to Table 3 in that "All" is the standard MSE improvement, and "Qn" represents the improvement for

[4] These were developed in Cleveland.[3] Some closely related techniques are also used in Casdagli,[2] Diebold and Nason,[4] Farmer and Sidorowich,[6] Meese and Rose,[10] Mizrach,[11] and Prescott and Stengos.[12]

TABLE 4 In-Sample Forecast Improvements: Nearest Neighbors[1]

Neighbors	Q1	Q2	Q3	Q4	Q5	All
			Panel A: $p = 1$			
25	-5.7729	-18.8040	-75.0597	-27.4379	-29.5613	-33.6915
50	-1.8765	-3.3448	-3.7818	-2.0231	-5.7761	-3.6876
100	-0.0436	-0.1428	-0.2337	-0.2296	-0.3885	-0.2487
150	-0.0201	-0.0396	-0.1449	-0.0696	-0.1562	-0.1019
200	0.0053	-0.0360	-0.0582	-0.0402	-0.1004	-0.0570
250	0.0091	-0.0076	-0.0379	-0.0172	-0.0643	-0.0318
300	0.0002	-0.0064	-0.0142	-0.0188	-0.0323	-0.0185
350	-0.0045	-0.0052	-0.0210	-0.0109	-0.0327	-0.0183
400	0.0003	-0.0077	0.0087	-0.0167	-0.0294	-0.0128
450	0.0005	-0.0043	0.0085	-0.0195	-0.0175	-0.0094
500	0.0028	-0.0070	0.0134	-0.0092	-0.0196	-0.0065
			Panel B: $p = 5$			
25	-0.5923	-0.6028	-1.0608	-1.1200	-2.8996	-1.5411
50	-0.1374	-0.2708	-0.4275	-0.3911	-1.3931	-0.6680
100	0.0132	-0.1150	-0.2021	-0.1865	-0.8142	-0.3546
150	0.0143	-0.0743	-0.0961	-0.0863	-0.5223	-0.2119
200	-0.0060	-0.0207	-0.0885	-0.0681	-0.2312	-0.1101
250	0.0075	0.0040	-0.0605	-0.0444	-0.2008	-0.0845
300	-0.0138	0.0130	-0.0357	-0.0095	-0.1398	-0.0528
350	-0.0218	0.0125	-0.0382	-0.0037	-0.1144	-0.0448
400	-0.0210	0.0191	-0.0104	0.0079	-0.1192	-0.0371
450	-0.0072	-0.0020	-0.0151	0.0184	-0.0901	-0.0273
500	-0.0085	-0.0039	-0.0257	0.0198	-0.0566	-0.0188

[1] In-sample forecasts performed on the first half of the data set. Qn's are percentage improvements in MSE for various quintile levels. All is the percentage improvement in MSE over all observations. Local regressions are constructed for both $p = 1$ and $p = 5$ lags.

volatility quintile n. All the numbers again show the percentage improvement over a simple sample mean. These results strongly confirm previous authors' findings that nearest-neighbor forecasting yields no improvements for stock-return forecasting. In Table 4 most of the numbers are negative, indicating the forecasts are not doing as well as the simple mean. Even in sample the nearest-neighbor techniques do not perform very well for these series.

The results from Figures 1 and 2 suggest a different method. Conditioning the local forecasts on volatility levels should improve forecasts. In other words, fit local regressions using the volatility measure to determine local neighborhoods. The distance between two points i and j would be

$$d_{i,j} = |\hat{\sigma}_i^2 - \hat{\sigma}_j^2|.$$

This is done in sample in Table 5. The local regressions are run on only one lagged return, with no constant term, and the volatility measure does not enter the regression. All local regressions are fit over the first half of the series, and forecasts are evaluated using this data. The column "All" shows no improvements for any of the choices of local neighbor sizes. Remember that the figures showed little possibility for forecast improvements during the noisiest time periods. Looking at the quietest periods, $Q1$, shows an interesting result. The forecasts are starting to show some improvement with a maximum improvement for $N = 300$ of 3.8 percent.

Using $N = 300$ as the estimated optimal forecasting method, Table 6 takes the test out of sample. Forecasts are now done in the second half of the series using local regressions estimated from the first half only. The only information used from the second half to forecast r_{t+1} are r_t and $\hat{\sigma}_t^2$. Once again the comparison forecast is generated from the mean estimated over the first half of the series. For the first quintiles the forecast improvement is large, about six percent. The second quintile shows a smaller improvement of two percent. The statistical magnitude

TABLE 5 Volatility Index Forecasts: S&P 500 Weekly Returns: In Sample[1]

Neighbors	Q1	Q2	Q3	Q4	Q5	All
25	0.0273	-0.0602	-0.0357	-0.0135	-0.0951	-0.0448
50	0.0144	-0.0337	-0.0040	0.0131	-0.0510	-0.0161
100	0.0279	-0.0259	-0.0180	0.0162	-0.0336	-0.0100
150	0.0336	-0.0164	-0.0142	0.0122	-0.0270	-0.0064
200	0.0343	-0.0143	-0.0183	-0.0046	-0.0278	-0.0114
250	0.0371	-0.0155	-0.0235	-0.0066	-0.0253	-0.0120
300	0.0386	-0.0156	-0.0274	0.0037	-0.0195	-0.0081
350	0.0371	-0.0143	-0.0231	0.0065	-0.0206	-0.0069
400	0.0347	-0.0172	-0.0169	0.0039	-0.0224	-0.0076
450	0.0337	-0.0135	-0.0126	0.0030	-0.0215	-0.0064
500	0.0292	-0.0131	-0.0138	0.0044	-0.0175	-0.0055
750	0.0238	-0.0147	-0.0134	0.0028	-0.0150	-0.0058
1000	0.0208	-0.0151	-0.0170	-0.0041	-0.0128	-0.0079

[1] In-sample forecasts performed on the first half of the data set. Qn's are percentage improvements in MSE for various quintile levels. All is the percentage improvement in MSE over all observations.

TABLE 6 Out-of-Sample Tests: S&P 500 Weekly Returns[1]

Test	Q1	Q2	Q3	Q4	Q5	All
300	0.0623	0.0205	0.0136	-0.0029	-0.0115	0.0023
GARCH	(0.000)	(0.076)	(0.149)	(0.430)	(0.735)	(0.161)
GARCHM	(0.004)	(0.116)	(0.161)	(0.386)	(0.502)	(0.137)
Reverse	0.0273	-0.0071	0.0019	-0.0127	0.0017	0.0021
GARCH	(0.024)	(0.406)	(0.297)	(0.675)	(0.361)	(0.165)
GARCHM	(0.112)	(0.486)	(0.289)	(0.554)	(0.237)	(0.141)

[1] Out-of-sample forecasts performed on the second half of the data set using the first half to build local regressions. Qn's are percentage improvements in MSE for various quintile levels. All is the percentage improvement in MSE over all observations. Numbers in parenthesis are simulated "p-values" from GARCH and GARCH-M models. Reverse flips the estimation and forecast sets around.

of these numbers is still somewhat questionable. To get a better feel for their size, the previously estimated GARCH and GARCH-M models are simulated 250 times to generate returns series of the same length as the weekly returns series. These series are then sent through the same out-of-sample forecasting procedure as the original series and the forecast improvements are compared with those from the original series. The numbers in parentheses in Table 6 represent the fraction of simulations generating as large an improvement as the original series. For $Q1$ none of the GARCH simulations generated an improvement as large as 6.2 percent, and only 0.4 percent of the GARCH-M models generated an improvement that large.

To check the robustness of these results, the estimation and testing sections of the sample are reversed. The same technique is used, but the second half is used for local regression estimation while the first half is used for forecast evaluation. This reduces the forecast improvement to 2.7 percent. The simulated models once again show this to be significant, but the fraction for the GARCH-M model of 11 percent is the weakest result for the group. These simulations can only compare these forecasts to the simulated null models. They cannot guarantee that these results will be significant against all possible null models. However, for the moment these results look extremely interesting.

5. CONCLUSIONS

The results presented here suggest that some small out-of-sample forecast improvements are possible for weekly stock returns using nonlinear techniques. In comparison with other papers on nonlinear forecasting in other fields, the improvements here are small. In the best case there is a reduction in forecast error variance of only six percent. The economic significance of these measures must be viewed with some caution. From the numbers presented so far, there is no way to tell if they suggest profitable trading strategies. To accomplish this task careful attention needs to be paid to issues such as transactions costs and risk adjustment. It is interesting to note that the forecast improvements here suggest that improved forecasting is possible during only the quietest of periods. It is possible that stock-market movements during these periods are so small that the improved forecasts would not help much. When the market really is moving, forecast improvements are not possible.

These results suggest two ideas to others working in this area. First, measuring global forecast improvements may be misleading and not give a complete picture of how well a forecasting method is working. There may be periods in which forecast improvements would be impossible due to the way in which noise is entering the system. Forecasting tools which perform well during other periods should not be penalized for missing these extremely difficult periods. Second, for some systems forecasts may be improved by reducing the dimensionality of lagged information using a cleverly chosen function. Using previous knowledge of the system, some functions may be found which summarize the lagged information more concisely and help to average out some of the noise in this information.

These results must be viewed as preliminary since only the simplest of methods are used here. The nearest-neighbor forecasting techniques can be varied in a large number of ways, including adding more lags, using weighted local regressions, and bringing in other variables such as trading volume. Also, the volatility estimate used here is very primitive and other estimates of volatility could yield some different results. The numbers reported here suggest a possibility for improvement, but they are far from being tuned to yield the optimal forecast.

REFERENCES

1. Bollerslev, Tim. "Generalized Autoregressive Conditional Heteroskedasticity." *J. Econometrics* **21** (1986): 307–328.
2. Casdagli, Martin. "Nonlinear Prediction of Chaotic Time Series." *Physica D* **35** (1989): 335–356.
3. Cleveland, W. S. "Robust Locally Weighted Regression and Smoothing Scatterplots." *J. Am. Stat. Assoc.* **74(368)** (1979): 829–836.
4. Diebold, Francis X., and James M. Nason. "Nonparametric Exchange Rate Prediction?" *J. Int'l. Econ.* **28** (1990): 315–332.
5. Engle, Robert F. "Autoregressive Conditional Heteroskedasticity with Estimates of the Variance of United Kingdom Inflation." *Econometrica* **50** (1982): 987–1007.
6. Farmer, J. D., and J. J. Sidorowich. "Predicting Chaotic Time Series." *Phys. Rev. Lett.* **59(8)** (1987): 845–848.
7. Hinich, M. J., and D. M. Patterson. "Evidence of Nonlinearity in Daily Stock Returns." *J. Bus. & Econ. Stat.* **3(1)** (1985): 69–77.
8. LeBaron, B. "The Changing Structure of Stock Returns." In *Nonlinear Dynamics, Chaos and Instability*, edited by W. A. Brock, D. Hsieh and B. LeBaron. Cambridge, MA: MIT Press, in press.
9. LeBaron, B. "Some Relations Between Volatility and Serial Correlations in Stock Market Returns." Technical paper #9002, University of Wisconsin, SSRI, Madison, Wisconsin.
10. Meese, R. A., and A. K. Rose. "An Empirical Assessment of Nonlinearities in Models of Exchange Rate Determination." Technical paper, University of California, Berkeley, 1989.
11. Mizrach, Bruce. "Multivariate Nearest-Neighbor Forecasts of EMS Exchange Rates." Technical paper, Boston University, Chestnut Hill, MA, 1989.
12. Prescott, David M., and Thanasis Stengos. "Do Asset Markets Overlook Exploitable Nonlinearities? The Case of Gold." Technical paper, University of Guelph, Guelph, Ontario, 1988.
13. Scheinkman, José A., and Blake LeBaron. "Nonlinear Dynamics and Stock Returns." *J. Business* **62(3)** (1989): 311–338.
14. Tsay, Ruey S. "Testing and Modeling Threshold Autoregressive Processes." *J. Am. Stat. Soc.* **84(405)** (1989): 231–240.
15. White, H. "Economic Prediction Using Neural Networks: The Case of IBM Daily Stock Returns." Technical paper, University of California, San Diego, San Diego, CA, 1988.

Andreas S. Weigend,† Bernardo A. Huberman,‡ and David E. Rumelhart*

†Physics Department, Stanford University, Stanford, CA 94305; ‡Dynamics of Computation Group, Xerox PARC, Palo Alto, CA 94304; and *Psychology Department, Stanford University, Stanford, CA 94305

Predicting Sunspots and Exchange Rates with Connectionist Networks

We investigate the effectiveness of connectionist networks for predicting the future continuation of temporal sequences. The problem of overfitting, particularly serious for short records of noisy data, is addressed by the method of *weight-elimination*: a term penalizing network complexity is added to the usual cost function in back propagation.

The ultimate goal is prediction accuracy. We analyze two time series. On the benchmark sunspot series, the networks outperform traditional statistical approaches. We show that the network performance does not deteriorate when there are more input units than needed. Weight elimination also manages to extract some part of the dynamics of the notoriously noisy currency exchange rates and makes the network solution interpretable.

1. INTRODUCTION

In many instances, the desire to predict the future is the driving force behind the search for laws that explain the behavior of certain phenomena. Examples range from Newton's laws of motion to forecasting the weather and anticipating currency exchange rates.

The ability to forecast the behavior of a given system hinges on two types of knowledge. The first and most powerful one is knowledge of the laws underlying a given phenomenon. When this knowledge is expressed in the form of deterministic equations that can in principle be solved, the future outcome of an experiment can be predicted once the initial conditions are completely specified.[1]

A second, albeit less powerful, method for predicting the future relies on the discovery of strong empirical regularities in observations of the system. The motion of the planets, the small amplitude oscillations of a pendulum, or the rhythm of the seasons carry within them the potential for predicting their future behavior from knowledge of their cycles without resorting to knowledge of the underlying mechanism. There are problems, however, with the latter approach. Periodicities are not always evident, and they are often masked by noise. Even worse, there are phenomena—although recurrent in a generic sense—that seem random, without apparent periodicities.

We use feed-forward networks to predict the continuation of a temporal sequence by extracting knowledge from its past. In distinction to previous connectionist approaches that dealt only with noise-free, computer-generated time series,[2] we focus on noisy, real-world data of limited record length. In this case, the *problem of overfitting* can become very serious. *A priori*, it is not clear what network size is required to solve a given problem. If the network is chosen too small, it will not be flexible enough to emulate the dynamics of the system that produced the time series ("underfitting"). If it is chosen too large, the excess freedom will allow the network to fit not only the signal but also the noise ("overfitting"). Both too small and too large networks thus give poor predictions in the presence of noise.

The problem of overfitting is approached from two angles: by using internal validation and by the method of weight-elimination. The key idea of *internal validation* is to split the training data into two sets. The first set is used to estimate the parameters. The purpose of the second set is to estimate the performance on new data and help determine the stopping point in training. This procedure, where only one pair of such two sets is used, is a special case of the statistical method of cross-validation.

---

[1]For the moment, we ignore the principal limitations quantum mechanics imposes on the complete specification of the initial conditions. Uncertainties in the initial conditions are particularly important for chaotic systems where errors grow exponentially in time. We show in Section 2.1 that our approach is consistent with the requirements for predicting chaotic systems.

[2]Using noise-free, computer-generated data means shunning the problem of overfitting. For example, the feat of forecasting the dazzling displays of the chaotic characteristics of the quadratic map, $x_t = 4x_{t-1}(1 - x_{t-1})$, amounts to nothing but approximating a parabola! Given the theoretical results that networks can essentially fit any function,[2,10,17,56] there is not much content in showing that networks can, yes, also fit a parabola.

The key idea of *weight elimination* is to add a penalty term to the cost function that represents network complexity. The sum reflects the dynamical trade-off between performance and complexity: the sum has a u-shaped minimum between the extremes of having a too simple network that produces horrendous errors and a network with small errors on the training data which has enormous complexity. The sum of both terms is minimized through back propagation.

We analyze the time series of yearly sunspot averages from the year 1700 onward, a benchmark used by many time-series analysts. On iterated predictions into the future, the network performance turns out to be significantly better than most other models.[3] Comparable results were only obtained by multivariate adaptive regression splines (MARS, by Friedman,[15] applied to the sunspot series by Lewis and Stevens[30]). Furthermore, by increasing the number of inputs to four times the usual size, we also show that networks can ignore irrelevant information.

This feature encouraged us to apply networks to the prediction of the notoriously noisy foreign exchange rates. In our second example, we pick one day (Monday) and one currency (DM vs. US$). Given daily exchange rate information up to and including a Monday, the task is to predict the rate for the following Tuesday. Weight elimination manages to extract some part of the dynamics and makes the solution interpretable. The solution turns out to be fairly linear: the key to the solution for exchange rates thus lies in the selection of the relevant variables through weight elimination.

These two examples—medium noise and medium nonlinearity for the sunspots, very high noise and small nonlinearity for the exchange rates—complement the example of a chaotic time series with little noise for which the network solution was found to be highly nonlinear: in Weigend et al.,[57] we analyzed a time series from a computational ecosystem and found that connectionist networks can predict the fraction of agents choosing a certain strategy for hundreds of steps into the future.

In this paper, after introducing time-series models and their evaluation, we address the key issue of generalization and describe our method of weight elimination for networks. We then apply this method to the time series of sunspots and of foreign exchange rates, and close with some ideas for the future.

---

2. REPRESENTATION

2.1 EMBEDDING

We begin our analysis with a discussion of the representation of the time series $\{x_t\}$, a sequence of measurements of an observable x that were taken at equal time

---

[3] Specifically, we compare the network model to the threshold autoregressive model by Tong and Lim,[51] the bilinear model by Subba Rao and Gabr[45] and the weighted linear predictor model by Stokbro.[47] All models are globally nonlinear.

intervals. Following Yule,[61] we express the present value x_t as a function of the previous d values of the time series itself,

$$x_t = f(\text{past values}) = \begin{cases} \mathbf{R}^d \to \mathbf{R} \\ (x_{t-1}, x_{t-2}, \ldots, x_{t-d}) \mapsto x_t \end{cases}. \tag{1}$$

The vector $(x_{t-1}, x_{t-2}, \ldots, x_{t-d})$ lies in the d-dimensional time delay space or *lag space*. This standard approach is also called "state-space reconstruction" in the physics community and "tapped delay line" in the engineering community. Since it is a regression onto past values of the variable that is to be predicted, statisticians refer to it as *autoregressive*. For noise-free data, Takens[49] showed that this approach constitutes an embedding (in a mathematical sense) that preserves invariances.[4]

The optimal *window-size d*, however, is not supplied with the time series to be predicted. On the one hand, d has to be large enough to resolve ambiguities. On the other hand, the higher dimensional the space becomes, the more distributed the data will appear in that space. If the average distance ϵ to the next point is to remain constant, the number of points N to cover the space increases exponentially with the dimension of the space d,

$$N \propto \left(\frac{1}{\epsilon}\right)^d.$$

This is known as "the curse of dimensionality." It is desirable that a predictive model is robust under too large values of d. This issue will be discussed in Section 4.3.

The lag space approach can only deal with *stationary* time series: since f does not depend on time, the underlying process that generated the series is assumed to be time invariant. Actually, any model that does not explicitly include time assumes that the underlying process is invariant under time translation. Trends on a time scale longer than d will not be picked up but smeared over.

The problem of prediction, usually framed as *extrapolation* in time, is re-framed for time invariant systems as *interpolation* in lag space. This lag space representation is consistent with the short-term predictability of chaotic systems. It is to be contrasted to a stronger approach for non-chaotic systems with long-term predictability, where one usually builds a model from first principles, determines boundary conditions and initial values, and finds the solution x_t for all times t,

$$x_t = f(\text{time}) = \begin{cases} \mathbf{R} \to \mathbf{R} \\ t \mapsto x_t \end{cases}. \tag{2}$$

---

[4] A clear introduction is given by Gershenfeld,[18] new results on embedding are presented by Sauer, Casdagli and Yorke.[44] More general introductions to chaotic dynamics are Schuster[42] and Eubank and Farmer.[12] Refreshingly critical remarks on dimensions of chaotic systems are expressed by Ruelle.[40] More specifically on the problem of prediction, see Casdagli[8] and Farmer and Sidorowich.[16] Chaos (or its absence) in economic and financial time series is discussed by Brock.[6]

The latter model is global in time in the sense that information (such as the frequency distribution) is averaged over the whole record. It amounts to a shortcut in time, impossible for chaotic systems.

2.2 INGREDIENTS

Having chosen to represent the data in lag space, three ingredients are required to specify a predictive model.

1. *Primitives.* Past relationships $\{x_t(x_{t-1}, x_{t-2}, \ldots, x_{t-d})\}$ are approximated for deterministic systems by a smooth surface above the $(x_{t-1}, x_{t-2}, \ldots, x_{t-d})$-plane. Different approaches in time-series prediction mainly differ in the choice of primitives for that surface (polynomials, splines, sigmoids, radial basis functions,...) and also in the choice between one global fit in lag space vs. many local fits. (Locality refers here to proximity in lag space, not in time.) The architecture and the primitives for our networks are introduced in Section 2.3.

2. *Error model.* Assumptions about the underlying distribution of measurement noise have to be made. The error model determines how likely a deviation Δ_t of an observation from the true value is. Two common choices are:

   - likelihood $\propto \exp(-(1/2)\Delta_t^2/\sigma^2)$ (GAUSS) $\rightarrow$ minimize squared differences, $\sum_t \Delta_t^2$
   - likelihood $\propto \exp(-|\Delta_t|/\sigma)$ (LAPLACE) $\rightarrow$ minimize absolute differences, $\sum_t |\Delta_t|$

   As Vapnik[53] points out, a Gaussian model is appropriate if the measurements are carried out under fixed conditions, whereas a Laplacian model reflects "maximally varying" experimental conditions.[5] We here assume the errors to be Gaussian distributed.

3. *Complexity model.* The third ingredient reflects assumptions about the prior probability of the set of parameters. A reasonable assumption might be that simple models with fewer parameters are more likely than complex models with many parameters. This issue of model complexity becomes absolutely central for relatively short and noisy data sets. It will be discussed in detail in Sections 3.2 and 3.3.

---

[5] An intermediate model that interpolates between quadratic errors (for small deviations) and linear errors (for large deviations) can be obtained with a logistic distribution for the cumulative probability of the errors. All of these error models only deal with noise of the dependent variable (output) and assume that the independent variables (inputs) are noise free. For autoregression, this assumption is poor since the same observation first serves as target output, then as input. Robust error models specific for time-series prediction are discussed by Martin.[32]

Then, given the lag space, primitives, error model, and complexity model, all that is left to do is to estimate the parameters of the surface minimizing the sum of the second ingredient (error cost) and the third ingredient (complexity cost). In back-propagation learning,[41] this minimization is simply done by gradient descent. Once the surface has been determined, the prediction for the value following a point in lag space is given by the value of the surface above that point in the $(x_{t-1}, x_{t-2}, \ldots, x_{t-d})$-plane.

Viewed from the perspective of statistics, the prediction f estimates the *conditional mean*,

$$f(x_{t-1}, x_{t-2}, \ldots, x_{t-d}, \text{parameters}) = \mathbf{E}\left[p(x_t | x_{t-1}, x_{t-2}, \ldots, x_{t-d}, \text{parameters})\right] . \tag{3}$$

The symbol $\mathbf{E}[\cdot]$ denotes the mean. p is the probability distribution over the output values x_t for each input vector. Note that this probability distribution p of the model, given inputs and parameters, is not to be confused with the probability distribution of the observed output given the predicted output, i.e., the error model. Whereas the error model, usually of relatively simple form, reflects our prior assumptions about the noise, the conditional mean depends on the data and can be fairly complicated.

2.2 EVALUATION

Whereas the choice of the error model is part of the model, we want a fairly general measure to evaluate and compare the predictive power of different algorithms. Given a set S of pairs of the actual values (or targets, x_k) and predicted values ($\hat{x}_k$), we define the *average relative variance* of a set S as

$$\mathbf{arv}(S) = \frac{\sum_{k \in S} (\text{target}_k - \text{prediction}_k)^2}{\sum_{k \in S} (\text{target}_k - \text{mean})^2}$$
$$= \frac{1}{\hat{\sigma}^2} \frac{1}{N} \sum_{k \in S} (x_k - \hat{x}_k)^2 . \tag{4}$$

The averaging (division by N, the number of data points in a set S) makes the measure independent of the size of the set. The normalization (division by $\hat{\sigma}^2$, the estimated variance of the data) removes the dependence on the dynamic range of the data. This normalization implies that if the estimated mean of the data is used as predictor, $\mathbf{arv} = 1.0$ is obtained.[6]

This quantity corresponds to the fraction of the variance of the data that is not "explained" by the model. It is also called the *relative mean squared error*. It is

---

[6] If the variances of the prediction, training and possibly validation sets differ, a choice has to be made. For merely historical reasons, we use the variance of the entire record for sunspots, but normalize each set by its own variance for the exchange rates.

related to the *correlation coefficient*, ρ, measuring the (linear) dependence between pairs of desired values and predictions by

$$\mathtt{arv} = 1 - \rho^2 . \tag{5}$$

This is exact if the predictions have the same mean and variance as the original data.

The symbol S denotes which part of the data the variance is computed for:

- *Training set*, with a certain *fitting error* or *approximation error* or *in-sample performance*. This part of the data is used to estimate the parameters. If the model also needs to be determined, this set is further split into two sets. The first set is still called *training set*. It is used for direct parameter estimation. The second set is referred to as *internal validation set* and can in network training be used to determine the stopping point of the training process.
- *Prediction set*. A certain part of the available data is strictly kept apart and only used to quote the expected performance in the future as *prediction error* or *out-of-sample performance*. In contrast to errors derived from the training set (see below), this error is sometime called "honest" prediction error.

Ultimately, we are interested in good performance for future predictions. Can we simply use the performance on the training set as an estimate of the predictive performance? Do we really need to set some data apart as prediction set?

It is well known that the in-sample performance can be a poor estimate of the out-of-sample performance, particularly in the presence of noise. For linear regression, some heuristics have evolved to "correct" for the usually over-optimistic estimate. We believe that these heuristics cannot be blindly applied to connectionist networks. An example of such a heuristic is to multiply the fitting error with $(N + k)/(N - k)$, where N is the number of data points and k is the number of parameters of the model.[1] It is not at all clear to what degree such approximations break down for nonlinear models, such as connectionist networks.

Now, even if we decided to ignore the issue of nonlinearities completely, what value should we use for k? Although the the number of *available* parameters of the network is fixed, the number of *effective* parameters increases during training, as shown by Weigend and Rumelhart. Let us consider the network at the very beginning of the training process when the parameters are just initialized with random values. Although all parameters are already present, the number of parameters that are effective for solving the task is essentially zero since they have not learned to be useful for anything yet. The focus on the learning process is very different from the typical assumption in statistics that the parameters are fully estimated at the time of model selection.

In our work, we always use a separate prediction set to determine the prediction error. We now turn to the specific choices for the primitives of our connectionist networks. In Section 3, we describe how the values of the parameters are determined.

2.3 ARCHITECTURE

Figure 1 shows a feed-forward network with one hidden layer. (For the time series we analyzed, it turned out that one hidden layer sufficed.) The abbreviation d-n-1 denotes the following network:

- The d *input units* are given the values $x_{t-1}, x_{t-2}, \ldots, x_{t-d}$.
- The n *nonlinear hidden units* are fully connected to the input units.
- The linear *output unit* is fully connected to the hidden units, producing the prediction $\hat{x}_t$ as the weighted sum of the activations of the hidden units.
- Output and hidden units have adjustable *biases, b*.
- The *weights* can be positive, negative, or zero.

The nonlinearities are located in the *activation function* (or nonlinear transfer function) of the hidden units. We use hyperbolic tangent (tanh) and sigmoid activations.[7]

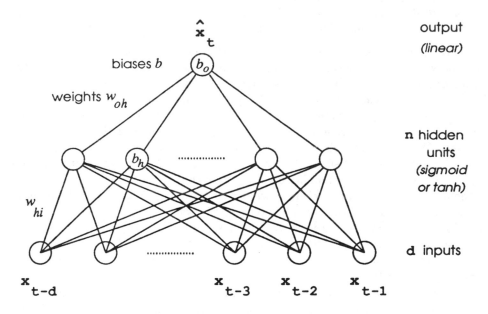

FIGURE 1 Architecture of a simple feed-forward network.

[7]For merely historical reasons, we use sigmoids in the sunspot example and tanh-units for the exchange rates. Tanh-units train faster than sigmoids under conditions given by Le Cun et al.[29] A comparison to radial basis functions is given in Weigend et al.[57]

The output or the response of a *hidden unit* is called its *activation*. It is a composition of two operators: an affine mapping, followed by a nonlinear transformation.

1. First, the inputs into a hidden unit h are linearly combined, and an offset or bias b_h is added. The sum is the *network input* ξ_h,

$$\xi_h = \sum_{i=1}^{d} w_{hi} x_i + b_h = \vec{w}_h \cdot \vec{x} + b_h.$$

x_i stands for x_{t-i}, the value of input i, and w_{hi} is the weight between input unit i and hidden unit h.

Before turning to the second step, some interpretation is in order. A hidden unit only responds to $\vec{w}_h \cdot \vec{x}$, the projection of the input vector $\vec{x} = (x_1, x_2, \dots, x_d)$ onto the weight vector $\vec{w}_h = (w_{h1}, w_{h2}, \dots, w_{hd})$. Changes in the input that are orthogonal to the direction of the weight vector have no effect on the activation of the hidden unit. The "equi-activation surfaces" (on which a hidden unit's activation is constant) are hyperplanes orthogonal to the direction of $\vec{w}_h$. (Perceptrons with a simple threshold activation function can thus be viewed as partitioning the input space with such a hyperplane orthogonal to the weight vector.)

The network input into a hidden unit h can be characterized by

- a direction, $\vec{w}_h / \|\vec{w}_h\|$,
- a scale parameter, $\|\vec{w}_h\|$, and
- a location parameter, $b_h / \|\vec{w}_h\|$.

The symbol $\| \cdot \|$ denotes the length (Euclidean norm) of the vector.

2. The second step can be viewed as "piping" ξ_h through a nonlinear *activation function*. Two popular choices for monotonically increasing activation functions are hyperbolic tangents and sigmoids.

- The activation T_h of a *tanh-unit* is given by

$$T_h = T(\xi_h) = \tanh a\,\xi_h = \tanh a\,(\vec{w}_h \cdot \vec{x} + b_h).$$

The tanh performs a smooth mapping $(-\infty, +\infty) \rightarrow (-1, 1)$. The gain a can be absorbed into weights and biases without loss of generality and is set to one.

- The activation S_h of a *sigmoid* (or logistic) unit is given by

$$S_h = S(\xi_h) = \frac{1}{1 + e^{-a\xi_h}} = \frac{1}{2}\left(1 + \tanh \frac{a}{2}\xi_h\right).$$

The sigmoid performs a smooth mapping $(-\infty, +\infty) \rightarrow (0, 1)$. We here also set $a = 1$.

To summarize, our connectionist networks globally superimpose nonlinear functions to produce an output that can be viewed as a surface above the $(x_1, x_2, \ldots, x_d)$ plane of the inputs. Up to now, we have ignored the question of how to determine the values of the weights and biases. In the next section, we turn to this question of parameter estimation. Particular attention is devoted to the problem of model selection for noisy data.

---

3. LEARNING

3.1 BACK PROPAGATION

We use the error back-propagation algorithm of Rumelhart et al.[41] to train the network: the parameters are changed by gradient descent on the cost surface over the weights and biases. This is numerically very simple. This paper does not address issues of how the learning speed can be improved; we do not use any acceleration scheme such as a momentum term, and choose learning rates small enough to avoid artifacts. The learning rate is 0.1 for the sunspot example and 2.5×10^{-4} for the exchange rates. The noisier the data, the smaller the learning rate should be for convergence of the solution. Too large learning rates for the exchange rates lead to too large weight changes, corresponding to a too big effect of the last pattern (if updated after each pattern), or to too large overall weight changes (if updated at the end of each epoch).

On the whole, the problem of building a network that readily memorizes a set of training data has proven easier than expected. However, the problem of good generalization has proven more difficult. Thus, we focus on the question: exactly what cost function is it that we want to minimize in order to get good predictions?

3.2 GENERALIZATION

Learning procedures for connectionist networks are essentially statistical devices for performing inductive inference. There is a trade-off between two goals: on the one hand, we want such devices to be as general as possible so that they are able to learn a broad range of problems. This recommends large and flexible networks. On the other hand, the true measure of an inductive device is not how well it performs on the examples it has been shown, but how it performs on cases it has not yet seen, i.e., its out-of-sample performance.

Too many weights of high precision make it easy for a network to fit the noise of the training data. In this case, when the network picks out the idiosyncrasies of the training sample, the generalization to new cases is poor. This *overfitting problem* is familiar in inductive inference, such as polynomial curve fitting. There are a number of potential solutions to this problem. We focus here on the so-called minimal network strategy. The underlying hypothesis is: if several networks fit the data equally well, the simplest one will on average provide the best generalization. Evaluating this hypothesis requires (i) some way of measuring simplicity and (ii) a search procedure for finding the desired network.

The complexity of an algorithm can be measured by the length of its minimal description in some language. The old but vague intuition of Occam's razor—or dream—can be formalized as the information theoretic *Minimum Description Length Criterion*[8]: Given some data, the most probable model is the model that minimizes

$$\underbrace{\text{description length}}_{\text{cost}} =$$

$$\underbrace{\text{description length}(\text{data given model})}_{\text{error}} + \underbrace{\text{description length}(\text{model})}_{\text{complexity}} .$$

This sum represents the trade-off between residual error and model complexity. The goal is to find a network that has the lowest complexity while fitting the data adequately. The complexity is dominated by the number of bits needed to encode the weights. It is roughly proportional to the number of weights times the number of bits per weight. We focus here on the procedure of weight elimination that tries to find a network with the smallest *number of weights.*

In Section 4.1.1, we compare weight elimination to the approach of *internal validation*: in that case, the cost function only consists of an error term, and overfitting is prevented by stopping the training before the minimum is reached, thereby having an effective number of parameters less than the total number of weights and biases. In Weigend et al.,[58] we also compare weight elimination with yet another approach that tries to minimize the *number of bits per weight*, thereby creating a network that is not too dependent on the precise values of its weights.

3.3 WEIGHT ELIMINATION

In 1987, Rumelhart proposed several methods for finding minimal networks within the framework of back propagation learning. A natural choice for describing the complexity of a connectionist network uses quantities such as the the size of the weights, the number of connections, the number of hidden units, the number of layers of hidden units, or the symmetries of the network. We here explain the method of weight elimination that considers the size of the weights and the number of weights, and interpret the complexity term as a prior distribution of the weights.

---

[8] Wallace and Boulton[54,55] introduce the term Minimum Message Length (MML), Rissanen[38,39] uses the term Minimum Description Length (MDL). For our discussion, differences between these approaches are not important. An introduction is given by Cheeseman.[9]

3.3.1 METHOD The idea is indeed simple in conception: add to the performance term a term which counts the number of parameters. In weight elimination, the new term is

$$\text{complexity cost} = \sum_{i \in \mathcal{C}} \frac{w_i^2/w_0^2}{1 + w_i^2/w_0^2}. \tag{10}$$

w_0 is the scale for the weights. The subscript i in w_i simply enumerates the weights. The sum extends over all connections $\mathcal{C}$. Note that the biases do not enter the cost function: all offsets are *a priori* equally probable. (In the framework developed below, this corresponds a non-informative prior: the probability density for the location parameter is flat.)

The performance term depends on the error model; see Section 2.2. Since we assume in this paper the errors to be Gaussian distributed, the complete cost function to be minimized over the set of training examples $\mathcal{T}$ is given by

$$\sum_{k \in \mathcal{T}} (\text{target}_k - \text{output}_k)^2 + \lambda \sum_{i \in \mathcal{C}} \frac{w_i^2/w_0^2}{1 + w_i^2/w_0^2}. \tag{11}$$

The first term measures the performance of the network. The second term measures the size of the network. λ represents the relative importance of the complexity term with respect to the performance term.

Given these two terms, the learning rule is then to change the weights according to the gradient of the *entire* cost function, continuously doing justice to the trade-off between error and complexity. This differs from the methods mentioned in Section 2.2 that consider a set of fixed models, estimate the parameters for each of them, and then compare between the models.

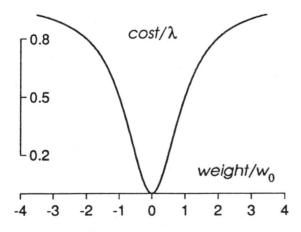

FIGURE 2   Complexity cost (in units of λ) of a weight as a function of the size of the weight (in units of w_0).

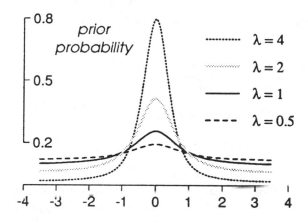

FIGURE 3 Prior probability of a weight as a function of the size of the weight (in units of w_0), plotted for different values of λ.

The complexity cost is shown in Figure 2 as function of w_i/w_0. The regions where the absolute value of the weight $|w_i|$ is either very large or very small are easily interpreted. For $|w_i| \gg w_0$, the cost of a weight approaches unity (times λ). This justifies the interpretation of the complexity term as a counter of significantly sized weights. For $|w_i| \ll w_0$, the cost is close to zero. "Large" and "small" are defined with respect to the scale w_0. It is a free parameter of the weight-elimination procedure. In our experience, choosing w_0 of order unity is good for activations of order unity. It is discussed further at the end of Section 3.3.3.

λ is dynamically adjusted in training. This dynamic increase, described in detail in the Appendix, is not a quick fix but is closely related to the concept of iterated training as opposed to one-shot parameter estimation. At the beginning of the training, the weights are not useful yet, since they were just initialized randomly. Any significant cost for complexity would just devour the whole network. So, λ has to start at zero. The usual subsequent increase corresponds to attaching more importance to the complexity term or, from the perspective developed in the next section, to sharpening the peak around zero of the prior distribution of the probability density function of the weights.

3.3.2 INTERPRETATION AS PRIOR PROBABILITY Starting with the relation $p(\mathcal{M}, \mathcal{D}) = p(\mathcal{M}|\mathcal{D})p(\mathcal{D})$ between

- $p(\mathcal{M}, \mathcal{D})$, the joint probability that both $\mathcal{M}$ and $\mathcal{D}$ occur,
- $p(\mathcal{M}|\mathcal{D})$, the conditional probability that $\mathcal{M}$ occurs given that $\mathcal{D}$ is the case, and
- $p(\mathcal{D})$, the probability that $\mathcal{D}$ occurs,

and using the symmetry of the joint probability, $p(\mathcal{M}, \mathcal{D}) = p(\mathcal{D}, \mathcal{M})$, we get

$$p(\mathcal{M}|\mathcal{D})p(\mathcal{D}) = p(\mathcal{D}|\mathcal{M})p(\mathcal{M}). \tag{12}$$

Interpreting $\mathcal{D}$ as a set of experimental data and $\mathcal{M}$ as the set of parameters that describe the model, we obtain *Bayes' learning equation*,

$$\underbrace{p(\mathcal{M}|\mathcal{D})}_{\text{posterior}} = \underbrace{p(\mathcal{D}|\mathcal{M})}_{\text{likelihood}} \times \underbrace{p(\mathcal{M})}_{\text{prior}} / p(\mathcal{D}).$$

Turning the products into sums by taking logarithms yields

$$\log p(\mathcal{M}|\mathcal{D}) = \log p(\mathcal{D}|\mathcal{M}) + \log p(\mathcal{M}) + c.$$

Since the probability $p(\mathcal{D})$ that the data were taken is not affected by the specific model that we are trying to fit, we set $\log p(\mathcal{D}) = c$, a constant.[9] The ultimate goal is to find the model that has the highest probability given the data. This corresponds to maximizing $p(\mathcal{M}|\mathcal{D})$ with respect to the parameters (weights and biases). Since the logarithm is monotonic, this is equivalent to maximizing $\log p(\mathcal{M}|\mathcal{D})$. We can interpret this logarithm of the posterior probability as the negative of a total cost that is to be minimized:

▷ total cost $= - \log p$ (posterior)
  $= - \log$ (probability of network $\mathcal{M}$, given data $\mathcal{D}$).

Identifying these terms with Eq. 9, we find

▷ performance cost $= - \log p$ (likelihood) $= - \log p(\mathcal{D}|\mathcal{M})$, and
▷ complexity cost $= - \log p$ (prior) $= - \log p(\mathcal{M})$.

The complexity cost is the negative logarithm of the prior probability of the weights.

In Fig. 3, we show the prior probability density function from which single weights of size w_i are drawn,

$$\text{prior} \propto \left[\exp\left(-\frac{w_i^2/w_0^2}{1 + w_i^2/w_0^2} \right) \right]^{\lambda}. \tag{14}$$

As can be seen from the figure, this prior can be interpreted as a "mixture" of a flat distribution and a bump around zero. Relevant weights are drawn from the flat distribution. Weights that are merely the result of "noise" are drawn from the bump centered on zero; they are expected to be small. If we wish to approximate the bump around zero by a Gaussian, its variance is given by $\sigma^2 = w_0^2/\lambda$. This means that its width scales with w_0.

Due to the exponentiation, the weighting factor λ now influences the width: the variance of the noise is inversely proportional to λ. The larger λ is, the closer to zero a weight must be to have a reasonable probability of being a member of the noise distribution. Also, the larger λ is, the more "pressure" small weights feel to become even smaller.

---

[9] This assumption is justified since weight elimination works within a given class of models. If the goal is to compare different classes of models such as sigmoid networks and radial basis function networks, $p(\mathcal{D})$ cannot be treated as constant any more. This more general case is addressed by MacKay.[31]

So far, we have only described our choice of the prior for a single weight. How do we get to the whole network? Assuming that the weights can be treated as independent, we sum over the connections in Eq. (11).

3.3.3 ALTERNATIVES We here only discuss the relationship of our method of weight *elimination* to weight *decay*, proposed by Hinton and by Le Cun in 1987.[10] In weight decay, a small percentage of the weight is subtracted at each weight update,

$$\Delta w_i = \text{(weight change due to back propagation of error)} - \alpha w_i.$$

This can be viewed as an exponential decay of the weight. It corresponds to a quadratic complexity cost (α w_i^2), known in the statistics community as *ridge regression*. It is contained in the weight elimination scheme as the special case of large w_0. Weight decay always prefers networks with many small weights. Weight elimination prefers few large weights over many medium-sized weights in the region where it acts as a counter. The scale parameter w_0 allows us to express a preference for many small weights (w_0 large) vs. a few large weights (w_0 small). Depending on the dynamic range and the number of the units of the preceding layer, w_0 might be given different values for different layers of the network.

4. SUNSPOTS

Sunspots, often larger in diameter than the earth, are dark blotches on the sun. They were first observed around 1610, shortly after the invention of the telescope.[14] Yearly averages have been recorded since 1700. The sunspot numbers are defined as $k(10g + f)$, where g is the number of sunspot groups, f is the number of individual sunspots, and k is used to reduce different telescopes to the same scale.[33] The observations are shown as black squares in Figure 4. The average time between maxima is 11 years. Note, however, that the time between maxima ranges from 7 to 15 years.[11]

The underlying mechanism for sunspot appearances is not exactly known. No first-principles theory exists, although it is known that sunspots are related to other

[10] Variations and alternatives have been developed by Hinton,[21] Scalettar and Zee,[48] Hanson and Pratt,[22] Mozer and Smolensky,[35] Schürmann,[43] Ishikawa,[24] Le Cun, Denker and Solla,[26] Ji, Snapp and Psaltis,[25] and others.

[11] Given the periodic appearance of the series, one might want to try to approximate the series by sinusoids. Such a Fourier interpolation captures the *global* structure in time over the entire fitting range. Bretthorst [5] carries out a Bayesian Fourier Analysis of the sunspot data. Extending his solution into the future gives poor predictions: almost immediately after the end of the fitting region, the extrapolation yields negative numbers for the sunspots.

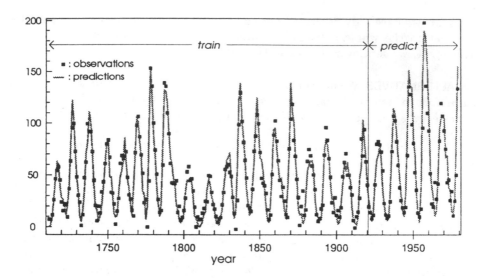

FIGURE 4 The sunspot data (black squares) and the one-year-ahead prediction (connected by the gray line) of the network.

solar activities. For example, the magnetic field of the sun changes with an average period of 22 years. Sunspots usually appear in pairs, corresponding to magnetic dipoles. Sunspot pairs reverse their polarity from one cycle to the next, reflecting the underlying magnetic cycle.

The sunspot series has served as a benchmark in the statistics literature. Within the lag space paradigm, different models differ in the specific choice of the primitives for the surface above the input space. In the simplest case, a single hyperplane approximates the data points. Such a *linear autoregressive model* is a linear superposition of past values of the observable.

Our evaluation of the network model, however, is carried out by comparison to a *nonlinear* model, the *threshold autoregressive model* (TAR) by Tong and Lim.[51,52] It has served as a benchmark for Priestley,[36,37] for Subba Rao and Gabr,[20,45] for Lewis and Stevens,[30] for Stokbro[47] and others.

The TAR model is globally nonlinear: it consists of two local linear autoregressive models. Tong and Lim found optimal performance for input dimension $d = 12$. They used yearly sunspot data from 1700 through 1920 for training, and the data from 1921 to 1979 for evaluation of the prediction.

To make the comparison between network and TAR performance as close as possible, we use their exact data for training and evaluation, their choice for the input dimension, i.e., 12 input units, their error model, and their evaluation criterion. The only difference between the models lies in the choice of the primitives used for the fitting of the surface.

4.1 LEARNING THE TIME SERIES

In this section we analyze the in-sample learning behavior of the networks, first with an internal validation set (needed to determine the stopping point when there is no complexity term in the cost function), and then with weight elimination. Subsequently, we turn to the out-of-sample performance.

4.1.1 WITH INTERNAL VALIDATION (EARLY STOPPING)

The learning of the sunspot series of a 12-8-1 network is shown in Figure 5 as a function of epochs: in one epoch the network sees each point from the training set exactly once. The *average relative variance*, as defined in Eq. (4), is normalized by the variance of the entire record, $\hat{\sigma}^2 = \hat{\sigma}^2_{\text{all}} = 1535$. Thus, in any of the three sets, a value of $\mathbf{arv} = 0.1$ corresponds to an average *absolute* quadratic error of $\mathbf{arv} \times \hat{\sigma}^2_{\text{all}} = 0.1 \times 1535 = 153.5 \approx (12.4)^2$.

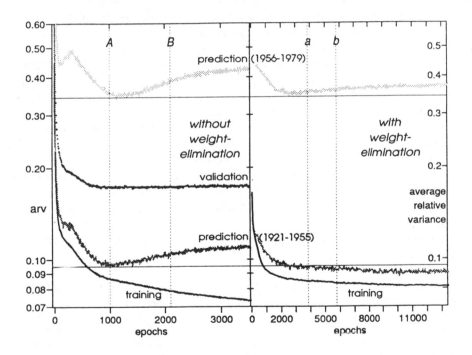

FIGURE 5 Learning curves of a 12-8-1 network as function of training time in epochs. The average relative single-step prediction variances are given for the training sets, and early and late prediction sets (as well as for the internal validation set for the network trained without weight elimination on the left side). The vertical lines (A, B, a, b) indicate different stopping points.

In Figure 5, the success in mastering the training set is indicated by the monotonic decrease of the lowest curve, indicating the *in-sample performance (fitting error* or *approximation error)*. Standard error back propagation without weight elimination is shown in the left panel of Figure 5. To get a feeling for the non-stationarity of the time series, the prediction set was split in two parts, 1921–1955 and 1956–1979. On both prediction sets, the error first decreases, but then starts to increase: the network begins to use its resources to fit the noise of the training set, i.e., it starts to pick out properties that are specific to the training set, but not present in the prediction sets. This overfitting leads to deteriorated generalization.[12]

The question to be addressed is when the training should be stopped. Since prediction sets must not be used for this decision, a validation set is required for a statistically proper determination of the end of the training process. To get a feeling for the effect of the sampling error by picking a specific training set/validation set combination, we investigated several training set–validation set pairs.

The validation sets consisted of 22 years chosen at random from the time before 1920. Those points were removed in the corresponding training sets, reducing their size by 10%. For the validation set of the example shown, the average relative variance approaches an asymptotic value; it happens not to increase. For this specific choice, the fitting of the noise of this training set happens to have no effect on the error of this validation set. Because the sunspot data set is rather small, different pairs of training and validation sets lead to results differing by factors of up to two. These variations are larger than the variations due to different random initial weights and biases.[13] In the given random choice for the validation set, it is not entirely clear from the error of the validation set when the training process should be terminated. In the evaluation of the performance in Section 4.2, we compare the performance for two stopping points, *A* after 1000 epochs and *B* after about 2100 epochs.

To summarize, some of the problems with early stopping through internal validation are that

[12] A motivation for early stopping can be given by observing the learning behavior of the network. In Weigend and Rumelhart,[60] we show how the effective number of parameters increases during training by analyzing the eigenvalue spectra of the covariance matrix of hidden unit *activations* and of the matrix of *weights* between inputs and hidden units. We show that the effective ranks of these matrices are equal to each other when a solution is reached, and interestingly also equal to the number of hidden units of the minimal network obtained with weight-elimination.

[13] Since the validation set is used to estimate only one quantity, as opposed to the training set estimating all of the weights and biases, an uneven split seems appropriate. Some improvement might be gained by only considering random splits where the first and second moments (mean and variance) of the validation set match the training set. Another idea is to first train, stop, and save several networks on different training-validation pairs, and then linearly superimpose their individual predictions. The combination is done by freezing the weights and biases of the sub-nets, and only letting the few new combination weights adapt to the entire training set.

- a part of the available training data cannot be used directly for parameter estimation,
- the monitored validation set error often shows multiple minima as a function of training time (even in the simple linear case analyzed by Baldi and Chauvin[3]),
- the specific solution at the stopping points depends strongly on the specific pair of training set and validation set, and
- the results are also sensitive to the specific initialization of the parameters.

We now present training with weight elimination as an alternative to simple error minimization with an internal validation set.

4.1.2 WITH WEIGHT ELIMINATION As in the case of back propagation without weight elimination, we start with a network large enough to guarantee a decrease of the error with training. The training curve for back propagation with weight elimination is shown in the right panel of Figure 5. With the same learning parameters as without weight elimination (zero momentum and a learning rate of 0.1), significant overfitting is avoided, even for training times four times as long. Since the entire training set is used, we are relieved from the uncertainty of a specific choice for a validation set. A decision, however, has to be made as to when the network reaches its asymptotic state. The performance of two solutions (a after 3900 epochs, b after about 5800 epochs) is compared in Section 4.2. It turns out that the exact stopping point is not important. In the first 5000 epochs, the procedure eliminated the weights between the output unit and five of the eight hidden units: only three hidden units survive.

We analyzed the specific solution of the network that was stopped at point b and subsequently trained with a very small learning rate for a few epochs. (The parameters of the network can be found in Weigend et al.[57]) The main contribution to the first hidden unit comes from x_{t-9}, to the second hidden unit from x_{t-2}, and to the third hidden unit from x_{t-1}. In contrast to the output weights, only very few of the weights from the input units to the active hidden units were eliminated. The fact that the remaining weights are of moderately small size points to a moderate use of the available nonlinearities.

Predictions are obtained by adding the values of these three hidden units to the bias of the output unit. The solution of the network can thus be interpreted as a nonlinear transformation from the twelve-dimensional input lag space to the three-dimensional space of hidden units.

4.2 PREDICTIONS AND COMPARISONS

So far, we have concentrated on the *learning* behavior of the network. Just obtaining a small network, however, is not an end in itself: the ultimate goal is to *predict* future values of the time series. In this section, we assess the predictive power of the network and compare it to other approaches. We first analyze single-step predictions and then turn to multi-step predictions.

4.2.1 SINGLE-STEP PREDICTION The term *single-step prediction* (or *one-step-ahead prediction*) is used when all input units are given the actual values of the observed time series. To assess the single-step prediction performance, we use the average relative variance, **arv**. It is independent of the dynamic range of the data and of the record length of the series.

The solution of the weight-eliminated network gives

$$\mathbf{arv}(\text{train}) = 0.082\,,$$
$$\mathbf{arv}(\text{predict})_{1921-1955} = 0.086\,,$$
$$\mathbf{arv}(\text{predict})_{1956-1979} = 0.35\,.$$

The corresponding values for the TAR model are

$$\mathbf{arv}(\text{train}) = 0.097\,,$$
$$\mathbf{arv}(\text{predict})_{1921-1955} = 0.097\,,$$
$$\mathbf{arv}(\text{predict})_{1956-1979} = 0.28\,.$$

As can be seen by comparing this measure for the network with the TAR model, the *single-step* prediction qualities of the network and the benchmark model are *comparable*. The numeric differences are not significant. Despite this similarity, however, significant differences will appear for predictions further than one step into the future.

4.2.2 MULTI-STEP PREDICTION There are two ways to predict further than one step into the future. We first present the results of iterated single-step predictions and subsequently turn to direct multi-step predictions. In *iterated single-step* predictions, the predicted output is fed back as input for the next prediction and all other input units are shifted back one unit. Hence, the inputs consist of *predicted* values as opposed to actual observations of the original time series. The predicted value for time t, obtained after I iterations, is denoted by $\widehat{x}_{t,I}$.

The prediction error will not only depend on I but also on the time $(t - I)$ when the iteration was started. We wish to obtain a performance measure as a function of the number of iterations I that averages over the starting times. Since we want to fully exploit the standard prediction set range for the sunspot data from $t_{\text{BEGIN}} = 1921$ to $t_{\text{END}} = 1955$, we compute for each I the average

$$\mathbf{arv}(I) = \frac{1}{\sigma^2} \frac{1}{t_{\text{END}} - (t_{\text{BEGIN}} - 1 + I)} \sum_{t_{\text{BEGIN}} - 1 + I}^{t_{\text{END}}} (x_t - \widehat{x}_{t,I})^2\,. \tag{15}$$

This (average relative) prediction variance after I iterations is shown in Figure 6. Only to indicate the spread of network performances, we give several network solutions. The letters A, B, a, b refer to the different stopping points, shown in Figure 5. The differences between the different network solutions are *not* significant.

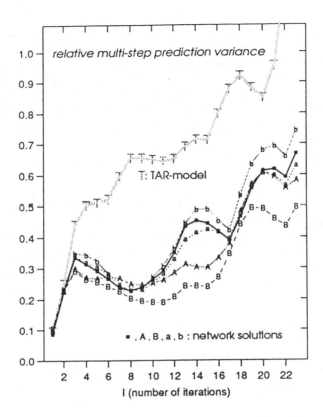

FIGURE 6 Relative prediction error after I iterations for the sunspot series. Gray T's indicate the performance of the TAR model. Black squares show the performance of the weight-eliminated network with three hidden units. The other curves indicate the variation of network solutions.

An alternative to this multi-step prediction by *iterated single-step* prediction is *direct multi-step* prediction: the network is trained to predict directly several steps ahead. On the sunspot data set, the prediction error for direct multi-step prediction was significantly worse than the error for iterated single-step prediction.

In summary, although we took extreme care not to gain any unfair advantage over Tong and Lim[51] (by taking the same input dimension, using identical data sets, minimizing the same sum of squared errors, etc.), the *multi-step* predictions were found to be *significantly better*: on average, the iterated prediction variances of the network were about half the iterated prediction variances of the threshold autoregressive model. This concludes the comparison with the benchmark model.

Gabr and Subba Rao[20,45] apply a *bilinear model* [14] to the sunspot data and find an improvement of about 15% over the TAR model, both for single-step and iterated predictions. On predictions further than one step into the future, our networks outperform the bilinear model on average by 35% in mean squared error.

Stokbro[47] uses a *weighted linear predictor* (WLP). In a WLP, each primitive is the product of a first-order polynomial and a normalized Gaussian radial basis function. The predictor is the linear superposition of these primitives. Stokbro compares WLP with the network solution on the 1921 to 1946 prediction set that is given in Weigend et al.[57] For one and two iterations, both methods perform comparably. When iterated more than twice, however, the network outperforms the WLP model.

Recently, Lewis and Stevens[30] applied *multivariate adaptive regression splines* (MARS, Friedman[15]) to the sunspot series. We find that the performance of MARS is very similar to the network performance. Given that the primitives of both schemes (sigmoids and splines) are biased towards smoothness, and given that both approaches employ a regularization scheme that penalizes complexity, the similar performance is not astonishing but rather encouraging.

4.3 VARYING THE EMBEDDING DIMENSION

In the previous sections, the predictions were always based on information of the previous 12 years. We now vary the number of inputs units from 1 to 41. The prediction error for iterated single-step predictions (for the standard 1921-1955 set) is shown in Figure 7 as a surface above the number of input units of the network and the prediction time into the future.

Networks with one input unit already manage to capture two thirds of the single-step variance, reducing it to $\mathbf{arv}(\text{predict})_{1921-1955} = 0.33$. The solution is practically linear with an offset. Networks with two input units reduce the relative variance to 0.17; they begin to use the available nonlinearities.

With increasing number of input units, the error reaches a roughly constant value. The performance does not degrade with input dimension several times larger than necessary: the network ignores irrelevant information. This important insensitivity to the input dimension is an advantage over other prediction methods such as the simplex algorithm employed by Sugihara and May.[46]

---

[14]In addition to linear autoregression (terms proportional to x_{t-i}), Subba Rao and Gabr allow terms proportional to the forecasting errors ϵ_{t-j} as well as terms proportional to the product $x_{t-k}\epsilon_{t-l}$ ("bilinear interactions"). On the one hand, a bilinear model is less general than our network approach in that it only considers linear dependence on past values of the series. On the other hand, it is more general than the lag space approach in that it uses information about past prediction errors, viewed as outside shocks that drive the system.

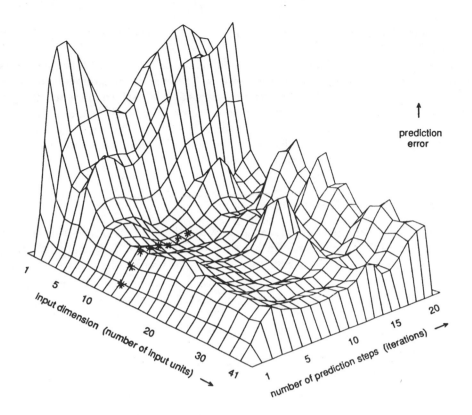

FIGURE 7 The prediction error for sunspots as a function of the number of inputs and forecasting time. Asterisks indicate the cross-section shown in Figure 6.

5. CURRENCY EXCHANGE RATES

The foreign exchange market is a prime example of a system that is believed to be very noisy. No equations from first principles (such as in some areas of physics) are known. This makes it an excellent testbed for any algorithm that is supposed to be able to deal with noisy data.

The data are from the Monetary Yearbook of the Chicago Mercantile Exchange. They are daily closing bids for five currencies (German Mark (DM), Japanese Yen,

Swiss Franc, Pound Sterling and Canadian Dollar) with respect to the U.S. Dollar.[15]

It is crucial to be absolutely clear about in-sample and out-of-sample errors, see Section 2.2. Any simple predictor can easily reduce the in-sample error by 20 percent—immediately yielding disastrous out-of-sample performance, much worse than the random walk prediction that tomorrow's rate is the same as today's. Diebold and Nason[11] review attempts to forecast foreign exchange rates and conclude that none of them succeeded out of sample: the *efficient market hypothesis* implies that the vicissitudes of financial markets are unpredictable.

5.1 TASK AND ARCHITECTURE

If the daily rates really performed a random walk, then the difference from one day to the next would be pure noise. Rather than predicting the daily rates (or *prices*, p_t), we actually turn to the harder task and try to predict the *returns* at day t, defined as

$$r_t := \ln \frac{p_t}{p_{t-1}} = \ln \left(1 + \frac{p_t - p_{t-1}}{p_{t-1}} \right) \approx \frac{p_t - p_{t-1}}{p_{t-1}}. \tag{16}$$

For small changes, the return is the difference to the previous day normalized by the price p_{t-1}.

It is well known[23] that there is a strong "day-of-the-week effect": the dynamics of the market differ for different days of the week. Since we do not want to average out the dynamics over different days, we define the prediction task to be

> learn the *Monday DM dynamics:* given exchange rate information up to and including a Monday, predict the DM–US$ rate for the following Tuesday.

The available data are split into three sets. The boundaries of the sets were determined randomly; no other combinations were tried. The training set is used to estimate the parameters. The two test sets monitor the performance; they can be viewed as an internal validation set and a prediction set.

The dates in Table 1 correspond to the first and last Mondays in that set. N is the number of Mondays. σ_r^2 denotes the relative variances of DM returns of the set in comparison to the variance over the entire period. Variations of a factor of two in σ_r^2 point to non-stationarity. We later use the σ_r^2 of the respective sets for the normalizations of the average relative variances.

We add some potentially useful inputs to the network. We hasten to emphasize that in all of the results presented here, these extra inputs were solely derived from the exchange rates themselves.

[15] We thank Blake LeBaron for giving us the data, for the friendly discussions at Santa Fe, for sharing his knowledge about the problem domain, and for sending us a draft of a forthcoming paper.[27]

- The *k-day trend* at day t is the running mean of the returns of the k last days,

$$\text{trend}_{k,t} := \frac{1}{k} \sum_{t-k+1}^{t} r_t = p_t - p_{t-k+1}. \tag{17}$$

- Similarly, the *k-day volatility* is the running standard deviation of the returns of the k last days,

$$\sqrt{\frac{1}{k-1} \sum_{t-k+1}^{t} (r_t - \text{trend}_{k,t})^2}. \tag{18}$$

Although the network could configure itself to compute these quantities from the raw returns, we offer them explicitly since we would like to find out whether the network can retrieve some of the rules used in technical trading.

The network is shown in Figure 8. It has 45 inputs for past daily DM returns, 5 inputs for the present Monday's returns of all available currencies, and 11 extra inputs. Both volatilities and trends are computed for various window sizes between one week ($k = 5$) and three months ($k = 65$). Just in case the prices themselves play a role in the dynamics (as opposed to the returns only), we also offer the present Monday DM price and the average of the prices of the remaining four currencies. All input values are *normalized* to zero mean and unit standard deviation (over the period of all three sets).

The input units are fully connected to the *five hidden tanh-units* with range $(-1, 1)$. The hidden units in turn are fully connected to *two output units*. The first one is to predict the next day *return*, r_{t+1}. This is a linear unit, trained with quadratic error, reflecting the Gaussian error model.

TABLE 1 Data sets for currency exchange rate predictions

name	from		to	N	σ_r^2
early test	September 3, 1973	–	April 28, 1975	87	0.96
train	May 5, 1975	–	December 3, 1984	501	0.69
late test	December 10, 1984	–	May 18, 1987	128	1.50

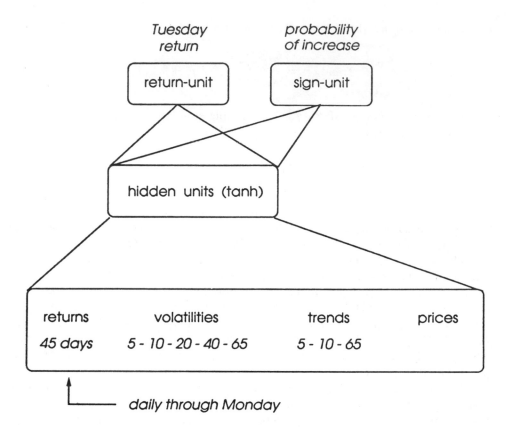

FIGURE 8 Architecture of the network for currency exchange rate prediction.

The second output unit focuses on the *sign* of the change. Its target value is one when the price goes up and zero otherwise. Since we want this unit to predict the probability that the return is positive, we choose a sigmoidal unit with range $(0,1)$ and minimize the cross-entropy error,[7,19]

$$\sum_{i=1}^{N} [t_i \ln o_i \ + \ (1 - t_i) \ln(1 - o_i)]. \tag{19}$$

TABLE 2 Output units for currency exchange rate predictions

	return-unit	sign-unit
predict	Tuesday DM return	probability of increase
activation	linear	(0,1)–sigmoid
error	squared difference	cross-entropy

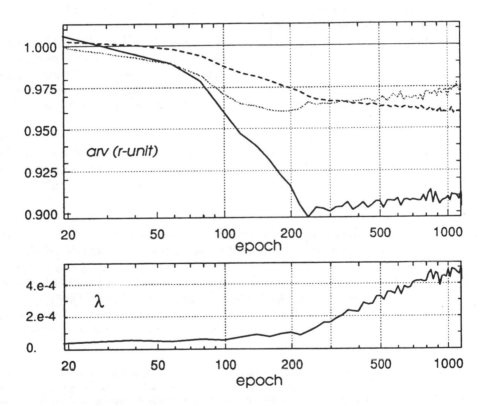

FIGURE 9 Learning curves of currency exchange rates. The average relative variance of the unit predicting the return (r-unit) is shown as a function of training time. The in-sample error is shown as solid line, the out-of-sample errors in gray (early test) and dashed (late test). The bottom panel depicts the increasing importance of the complexity term during training.

t_i denotes the target value and o_i the actual output value for day i. Information about the output units is summarized in Table 2.

Having described the task and the architecture, we now turn to the learning and performance of the network.

5.2 LEARNING

The central question is whether the network is able to extract any signal from the training set that generalizes to the test sets or not. In Figure 9, the performance is plotted as function of training time in epochs. As usual, it is expressed in terms of the average relative variance, defined in Eq. (4). Since the mean of the returns is close to zero, the random-walk hypothesis corresponds to **arv** = 1.0. Will it be possible to predict better than chance?

The bottom panel of Figure 9 shows the dynamics of weight elimination, mentioned in Section 3.3.1 and described in detail in the Appendix. λ first grows very slowly. Then, around epoch 230, the in-sample error reaches the externally given performance criterion. The network is starting to focus on the elimination of weights (indicated by growing λ) without further reducing its in-sample errors, since that might lead to overfitting.

The resulting out-of-sample prediction is *better than chance*. Weight-elimination reliably extracts a signal that accounts for between 2.5 and 4.0 percent of the variance. The latter value corresponds to a linear-correlation coefficient (between pairs of predictions and actual values) of 0.2 for both test sets taken together (number of out-of-sample Tuesdays = 87 + 128 = 215) .

5.2.1 CONFIDENCE INTERVALS

To determine the significance of the result, we would like to know not only the expected value of the out-of-sample correlation coefficient but also how accurate that estimate is: ideally, we would like to obtain the entire probability distribution of ρ. We obtain an estimate of the distribution of ρ by straightforwardly applying the bootstrap method by Efron.[13] Our use of the bootstrap idea should be contrasted with Brock et al.[4] and LeBaron[27]: whereas we bootstrap the correlation coefficient, they bootstrap the time series. [16] We take

---

[16] In more detail: Our simple goal is to obtain the probability distribution of the correlation coefficient for *one* specific solution on the *given* data sets. We exactly follow the procedure given by Efron[13] to obtain this distribution and then simply read off the probabilities of certain values of ρ for that time series and that solution.

Brock et al.[4] and LeBaron[27] use the bootstrap idea to generate *new* time series that embody the null hypothesis. The simple null hypothesis of a random walk, for example, is implemented by resampling the returns of the original series. Tests against arbitrarily complicated null hypothesis can be carried out by generating data from arbitrarily complicated models (see Theiler et al[62]).

We analyzed both time-reversed and randomized versions of the original series. In neither case could the networks reduce the out-of-sample error. Randomization, however, also destroys the "day-of-the-week effect." For the present task (using daily information up to and including Mondays to forecast Tuesdays), one would ideally like to use a method that randomizes the series without averaging over the days of the weeks.

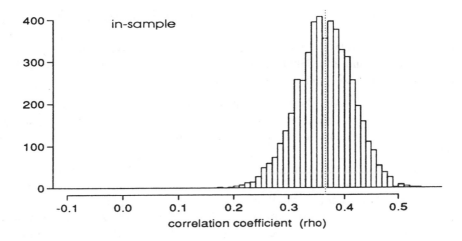

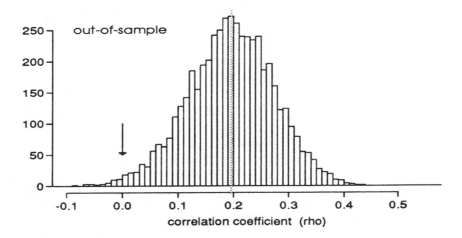

FIGURE 10 Histogram of 5000 bootstrap samples of the correlation coefficient between target and prediction. Top: training set (501 Tuesdays). Bottom: prediction sets (215 Tuesdays). The means are indicated by gray lines. The arrow points to $\rho = 0$.

the network solution from Figure 9 after 500 epochs, and closely follow Efron[13] (p. 465):

1. Use a random number generator to draw 215 pairs independently and with replacement from the set of 215 out-of-sample target-output pairs. Such a new

Yet another variation is to average the ρ distributions of many network solutions. As described in the main text, we here only pick one typical solution.

sample, called a bootstrap sample, is a subset of the original 215 out-of-sample pairs.

2. Compute the correlation coefficient $\widehat{\rho}^*$ for that bootstrap sample.
3. Repeat steps 1 and 2 for 5000 times, each time using an independent set of new random numbers to generate the new bootstrap sample.

In the lower panel of Figure 10, we histogram the distribution of the 5000 values of $\widehat{\rho}^*$ for the prediction set. For comparison, we show in the upper panel the distribution of the correlation coefficient for the training set, obtained by the same procedure.

The means of the distributions are 0.365 (in-sample) and 0.196 (out-of-sample). 5000 samples should determine them to about 1%. The standard deviations of the $\widehat{\rho}^*$ distributions are 0.050 (in-sample) and 0.077 (out-of-sample).

Let us interpret the out-of-sample correlation coefficient by giving a central confidence interval and some one-tail confidence limits. The central 98% confidence interval is $(0.007, 0.37)$. With probability 5%, the true correlation coefficient is less than 0.064, with 2% below 0.031, and with 1% below 0.007. 38 of the 5000 bootstrap samples, or 0.76%, had a negative correlation coefficient.

To summarize, the true *correlation* coefficient is *zero* or less with probability below *one percent*. In contrast, networks without precautions against overfitting show hopeless out-of-sample performance almost before the training has started.

5.2.2 ANALYSIS OF THE NETWORK Finally, we analyze the weight eliminated network solution. The weights from the hidden units to the outputs are in a region where the complexity term acts as a counter. In fact, only one or two hidden units remain. All the other hidden-to-output weights go to zero (within machine precision). Subsequently, the weights from the inputs to the dead hidden units also get eliminated, since there is no reason for the network to pay a price for weights to these deceased hidden units since they have no effect on the output. For time series prediction, weight elimination acts as hidden unit elimination.

Since all inputs are scaled to zero mean and unit standard deviation, we can gauge the importance of different inputs directly by the size of the weights. With weight elimination, it becomes fairly clear which quantities are important, since connections that do not manage to reduce the error are not worth their price. In Table 3, we give the importance of the units that were added to the network in addition to the 45-day window of DM returns. Important quantities are boldfaced, medium ones are in normal font, and insignificant ones are italicized. For example, the Canadian Dollar returns are less important than the returns of the other four currencies. This finding, consistent with LeBaron,[27] could be due to the strong coupling between the Canadian and U.S. economies. The sign in parentheses indicates the sign of the weight. The signs of the three previous days' DM return are $(+)$, $(-)$, and $(+)$. The magnitude of their weights is similar to the present-day return.

TABLE 3 Importance of the extra inputs. "Medium" (0.1 < absolute value of weight < 0.04; shown in roman), "**large**" (> 0.04; shown in boldface), and "*small*" (< 0.01; shown in italics).

Returns	Trends	Volatilities	Prices
DM (+)	**5-day DM (+)**	5-day DM (+)	**DM (+)**
Japanese Yen (+)	*10-day DM (+)*	10-day DM (+)	*rest (+)*
Swiss Franc (+)	*65-day DM (+)*	*20-day DM (−)*	
Pound Sterling (+)	65-day rest (−)	40-day DM (−)	
Canadian Dollar (−)		*65-day DM (+)*	

The weights between inputs and remaining hidden units are fairly small. Weight elimination is in its quadratic region and prevents them from growing too large. Consequently, the activation of the hidden units lies usually in $(-0.4, 0.4)$. This prompted us to try a linear network where our procedure also works surprisingly well, yielding comparable performance to tanh-nonlinearities. For exchange rates, the key to the success thus lies in variable selection by weight elimination.

Our results suggest some further steps.

- Offer the returns of the other currencies of the recent past in addition to the present day.
- Combine five such networks, one for each day, to make predictions for every day of the week possible.
- Estimate also the error of each forecast, as outlined in Buntine and Weigend.[7]
- Add connections that link the inputs directly with the outputs, skipping the hidden layer. These weights can encode only a linear part of the solution.
- Add input units that are given previous forecasting errors, allowing the network to model the response to exogenous shocks or innovations. In the language of statistics, this means to extend the nonlinear autoregressive model to a nonlinear autoregressive moving average (ARMA) model; see Weigend.[59]
- Finally, leaving the turf of mere academic comparison of forecasting methods,
- Offer additional information that cannot be derived from the exchange rates, such as interest rates.
- Change the cost function to reflect the actual profit. Back propagation is used to estimate the parameters maximizing this generalized cost function that also incorporates transaction costs.

This concludes our study of connectionist networks for foreign exchange rates.

6. SUMMARY

We investigated connectionist networks for short-term prediction of time series. We first applied the networks to the sunspot series. On this noisy real-world time series, our networks outperformed the threshold autoregressive model by Tong and Lim,[51] the bilinear model by Gabr and Subba Rao,[20,45] and the weighted linear predictors by Stokbro.[47] We found the network performance to be very comparable to the performance obtained with multivariate adaptive regression splines (MARS) by Friedman,[15] applied to the sunspot series by Lewis and Stevens.[30]

In the second example, the notoriously noisy foreign exchange rates series, we defined a sub-task by picking one day (Monday) and one currency (DM vs. U.S.$). Given exchange rate information up to and including a Monday, the networks managed to predict the return for the following Tuesday with an out-of-sample correlation coefficient of 0.2, corresponding to an "explained" variance of four percent.

Both results rely heavily on a procedure called weight elimination that dynamically eliminates weights during training, addressing the related problems of network size and overfitting. In both examples, weight elimination drastically reduced the number of hidden units, thus making the solutions more interpretable. Possible further applications of these methods are in medicine (electrocardiograms and electroencephalograms), seismic data, speech (nonlinear predictive coding), language, and music.

ACKNOWLEDGMENTS

We thank Martin Casdagli for the invitation and support to attend this workshop; Blake LeBaron, Tali Tishby, Buz Brock and Peter Grassberger for discussions at Santa Fe; and Art Owen and Jerry Friedman for discussions back at Stanford.

APPENDIX

We here describe how we change the weighting factor of the complexity term, λ, during training. λ is defined in Eq. (11) in Section 3.3.1.

Although the basic form of the weight-elimination procedure is simple, it is sensitive to the choice of λ. If λ is too small, it will have no effect. If λ is too large, all of the weights will be driven to zero. Worse, a value of λ which is useful for a problem that is easily learned may be too large for a hard problem, and a problem which is difficult in one region (at the start, for example) may require a larger value of λ later on. We have developed some rules that make the performance relatively insensitive to the exact values of the parameters.

We start with $\lambda = 0$ so that the network can initially use all of its resources. λ is changed after each epoch. It is usually gently incremented, sometimes decremented, and, in emergencies, cut down. The choice among these three actions depends on the value of the *error on the training set* $\mathcal{E}_n$.

The subscript n denotes the number of the epoch that has just finished. (Note that $\mathcal{E}_n$ is only the first term of the cost function (Eq. (11)). Since gradient descent minimizes the sum of both terms, $\mathcal{E}_n$ by itself can decrease or increase.) $\mathcal{E}_n$ is compared to three quantities, the first two derived from previous values of that error itself and the last one given externally:

- $\mathcal{E}_{n-1}$: *Previous* error.
- $\mathcal{A}_n$: *Average* error (exponentially weighted over the past). It is defined as $\mathcal{A}_n = \gamma \mathcal{A}_{n-1} + (1 - \gamma)\mathcal{E}_n$ (with γ close to 1).
- $\mathcal{D}$: *Desired* error, the externally provided performance criterion. The strategy for choosing $\mathcal{D}$ depends on the specific problem. For example, "solutions" with an error larger than $\mathcal{D}$ might not be acceptable. Or, we may have observed (by monitoring the out-of-sample performance during training) that overfitting starts when a certain in-sample error is reached. Or, we may have some other estimate of the amount of noise in the training data. For toy problems, derived from approximating analytically defined functions (where perfect performance on the training data can be expected), a good choice is $\mathcal{D} = 0$. For hard problems, such as the prediction of currency exchange rates, $\mathcal{D}$ is set just below the error that corresponds to chance performance, since overfitting would occur if the error was reduced further.

After each epoch in training, we evaluate whether $\mathcal{E}_n$ is above or below each of these quantities. This gives eight possibilities. Three actions are possible:

- $\lambda \leftarrow \lambda + \Delta\lambda$: In six cases, we increment λ slightly. These are the situations in which things are going well: the error is already below the criterion ($\mathcal{E}_n < \mathcal{D}$) and/or is still falling ($\mathcal{E}_n < \mathcal{E}_{n-1}$). Incrementing λ means attaching more importance to the complexity term and making the Gaussian a little sharper. Note that the primary parameter is actually $\Delta\lambda$. It is fairly small, of order 10^{-6}.

In the remaining two cases, the error is worse than the criterion and it has grown compared to just before $(\mathcal{E}_n \geq \mathcal{E}_{n-1})$. The action depends on its relation to its long-term average $\mathcal{A}_n$.

- $\lambda \leftarrow \lambda - \Delta\lambda$ [if $\mathcal{E}_n \geq \mathcal{E}_{n-1} \wedge \mathcal{E}_n < \mathcal{A}_n \wedge \mathcal{E}_n \geq \mathcal{D}$]: In the less severe of those two cases, the performance is still improving with respect to the long term average $(\mathcal{E}_n < \mathcal{A})$. Since the error can have grown only slightly, we reduce λ slightly. (If the new value of λ would be negative, we set λ to zero.)
- $\lambda \leftarrow 0.9\,\lambda$ [if $\mathcal{E}_n \geq \mathcal{E}_{n-1} \wedge \mathcal{E}_n \geq \mathcal{A}_n \wedge \mathcal{E}_n \geq \mathcal{D}$]: In this last case, the error has increased and exceeds its long-term average. This can happen for two reasons. The error might have grown a lot in the last iteration. Or, it might not have improved by much in the whole period covered by the long-term average, i.e., the network might be trapped somewhere before reaching the performance criterion. The value of λ is cut, hopefully preventing weight elimination from devouring the whole net.

In our experience, this set of heuristics for finding a minimal network while achieving a desired level of performance on the training data works well.

REFERENCES

1. Akaike, Hirotugo. "Statistical Predictor Identification." *Ann. Inst. of Stat. Math.* **22** (1970): 203–217.
2. Barron, Andrew R. "Universal Approximation Bounds for Superpositions of a Sigmoidal Function." Technical Report 58, Statistics Department, University of Illinois, Urbana, February 1991.
3. Baldi, Pierre, and Yves Chauvin. "A Study of Generalization in Simple Linear Networks." Submitted to *Neural Computation*, 1991.
4. Brock, William A., Josef Lakonishok, and Blake LeDaron. "Simple Technical Trading Rules and the Stochastic Properties of Stock Returns." Technical Report 9022, Social Systems Research Institute, University of Wisconsin, Madison, 1990.
5. Bretthorst, G. Larry. *Bayesian Spectrum Analysis and Parameter Estimation.* Lecture Notes in Statistics, volume 48. New York: Springer, 1988.
6. Brock, William A. "Causality, Chaos, Explanation and Prediction in Economics and Finance." In *Beyond Belief: Randomness, Prediction, and Explanation in Science*, edited by J. Casti and A. Karlqvist, 230–279. Boca Raton, FL: CRC Press, 1991.
7. Buntine, Wray L., and Andreas S. Weigend. "Bayesian Back-Propagation." Submitted to *Complex Systems*. Technical Report, NASA/Ames, 1991.
8. Casdagli, Martin. "Nonlinear Prediction of Chaotic Time Series." *Physica D* **35** (1989): 335–356.
9. Cheeseman, Peter C. "On Finding the Most Probable Model." In *Computational Models of Scientific Discovery and Theory Formation*, edited by Jeff Shrager and Pat Langley, 73–95. Morgan Kaufmann, 1990.
10. Cybenko, George. "Approximation by Superpositions of a Sigmoidal Function." *Mathematics of Control, Signals, and Systems* **2** (1989): 303–314.
11. Diebold, Francis X., and James A. Nason. "Nonparametric Exchange Rate Prediction?" *J. Int'l. Econ.* **28** (1990): 315–332.
12. Eubank, Stephen G., and J. Doyne Farmer. "An Introduction to Chaos and Randomness." In *1989 Lectures in Complex Systems*, edited by Erica Jen. Santa Fe Institute Studies in the Sciences of Complexity, Lect. Vol. II, 75–190. Redwood City: Addison-Wesley, 1990.
13. Efron, Bradley. "Computers and the Theory of Statistics: Thinking the Unthinkable." *SIAM Review (Society for Industrial and Applied Mathematics)* **21** (1979): 460–480.
14. Foukal, Peter V. "The Variable Sun." *Sci. Am.* **262** (February 1990): 34.
15. Friedman, Jerome H. "Multivariate Adaptive Regression Splines (with discussion)." *Annals of Statistics* **19** (1991): 1–141.
16. Farmer, J. Doyne, and John J. Sidorowich. "Exploiting Chaos to Predict the Future and Reduce Noise." In *Evolution, Learning and Cognition*, edited by Y. C. Lee. New York: World Scientific, 1989.

17. Funahashi, Ken-ichi. "On the Approximate Realization of Continuous Mappings by Neural Networks." *Neural Networks* **2** (1989): 183–192.

18. Gershenfeld, Neil A. "An Experimentalist's Introduction to the Observation of Dynamical Systems." In *Directions in Chaos*, edited by Bai-Lin Hao, volume 2, 310–384. New York: World Scientific, 1989.

19. Golden, Richard M. "A Unified Framework for Connectionist Systems." *Biological Cybernetics* **59** (1988): 109–120.

20. Gabr, M. M., and T. Subba Rao. "The Estimation and Prediction of Subset Bilinear Time Series Models with Applications." *J. Time Series Analysis* **2** (1981): 155–171.

21. Hinton, Geoffrey E. "Connectionist Learning Procedures." Technical Report CMU-CS-87-115 (version 2), Carnegie-Mellon University, 1987.

22. Hanson, Stephen José, and Lorien Y. Pratt. "Comparing Biases for Minimal Network Construction with Back-Propagation." In *Advances in Neural Information Processing Systems 1 (NIPS*88)*, edited by D. S. Touretzky, 177–185. Morgan Kaufmann, 1989.

23. Hsieh, David. "Testing for Nonlinear Dependence in Daily Foreign Exchange Rates." *J. Business* **62** (1989): 339–369.

24. Ishikawa, Masumi. "A Structural Learning Algorithm with Forgetting of Link Weights." Technical Report TR-90-7, Electrotechnical Laboratory, Life Electronics Research Center, Tokyo, 1990. Modified version of a paper presented at *IJCNN*, Washington, D.C., 1989.

25. Ji, Chuanyi, Robert R. Snapp, and Demetri Psaltis. "Generalizing Smoothness Constraints from Discrete Samples." *Neural Computation* **2** (1990): 188–197.

26. Le Cun, Yann, John S. Denker, and Sara A. Solla. "Optimal Brain Damage." In *Advances in Neural Information Processing Systems 2 (NIPS*89)*, edited by D. S. Touretzky, 598–605. Morgan Kaufmann, 1990.

27. LeBaron, Blake. "Technical Trading Rules and Simulated Processes for Foreign Exchange Rates." Technical report, Department of Economics, University of Wisconsin, Madison, 1991.

28. Lapedes, Alan S., and Robert M. Farber. "Nonlinear Signal Processing using Neural Networks: Prediction and System Modelling." Technical Report LA-UR-87-2662, Los Alamos National Laboratory, 1987.

29. Le Cun, Yann, Ido Kanter, and Sara A. Solla. "Second-Order Properties of Error Surfaces: Learning Time and Generalization." In *Advances in Neural Information Processing Systems 3 (NIPS*90)*, edited by R. P. Lippmann, J. Moody, and D. S. Touretzky, 918–924. Morgan Kaufmann, 1991.

30. Lewis, Peter A. W., and J. G. Stevens. "Nonlinear Modeling of Time Series using Multivariate Adaptive Regression Splines (MARS)." Submitted to *J. Am. Stat. Assoc.*, 1991.

31. MacKay, David J. C. "*1.* Bayesian Interpolation *and 2.* A Practical Bayesian Framework for Backprop Networks." To be submitted to *Neural Computation*, 1991.

32. Martin, R. Douglas. "Robust Methods for Time Series." In *Applied Time Series Analysis II*, edited by D. F. Findley, 683–759. New York: Academic Press, 1981.

33. Marple, S. Lawrence. *Digital Spectral Analysis with Applications*. New York: Prentice-Hall, 1987.

34. Moody, John, and Christian J. Darken. "Fast Learning in Networks of Locally Tuned Processing Units." *Neural Computation* **1** (1989): 281–294.

35. Mozer, Michael C., and Paul Smolensky. "Using Relevance to Reduce Network Size Automatically." *Connection Science* **1** (1989): 3–16.

36. Priestly, Maurice B. *Spectral Analysis and Time Series*. New York: Academic Press, 1981.

37. Priestly, Maurice B. *Non-linear and Non-stationary Time Series Analysis*. New York: Academic Press, 1988.

38. Rissanen, Jorma. "Stochastic Complexity." *J. Royal Stat. Soc. B* **49** (1987): 223–239 (with discussion: 252–265).

39. Rissanen, Jorma. *Stochastic Complexity in Statistical Inquiry*. Singapore: World Scientific, 1989.

40. Ruelle, David. "Deterministic Chaos: The Science and the Fiction." *Proc. Royal Soc. London A* **427** (1990): 241–248.

41. Rumelhart, David E., Geoffrey E. Hinton, and Ronald J. Williams. "Learning Internal Representations by Error Propagation." In *Parallel Distributed Processing*, edited by David E. Rumelhart, James L. McClelland, and the PDP Research Group, 318–362. Cambridge: MIT Press, 1986.

42. Schuster, Heinz-Georg. *Deterministic Chaos*. VCH Verlagsgesellschaft, 1988.

43. Schürmann, Bernd. "Stability and Adaptation in Artificial Neural Systems." *Phys. Rev. A* **40** (1989): 2681–2688.

44. Sauer, Tim, Martin Casdagli, and James A. Yorke. "Embedology." *J. Stat. Phys.*, to appear.

45. Subba Rao, T., and M. M. Gabr. *An Introduction to Bispectral Analysis and Bilinear Time Series Models*. Lecture Notes in Statistics, volume 24. New York: Springer, 1984.

46. Sugihara, George, and Robert M. May. "Nonlinear Forecasting as a Way of Distinguishing Chaos from Measurement Error in Time Series." *Nature* **344** (April 1990): 734–741.

47. Stokbro, Kurt. "Predicting Chaos with Weighted Maps." This volume.

48. Scalettar, R., and A. Zee. "Emergence of Grandmother Memory in Feed-Forward Networks: Learning with Noise and Forgetfulness." In *Connectionist Models and their Implications*, edited by D. Waltz and J. Feldman, 309–327. Ablex, 1988.

49. Takens, Floris. "Detecting Strange Attractors in Turbulence." In *Dynamical Systems and Turbulence*, edited by D. A. Rand and L.-S. Young. Lecture Notes in Mathematics, volume 898, 366–381. New York: Springer, 1981.

50. Theiler, J., B. Galsrikian, A. Longtin, S. Eubank, an J. D. Farmer. "Using Surrogate Data to Detect Nonlinearity in Time Series." This volume.

51. Tong, Howell, and K. S. Lim. "Threshold Autoregression, Limit Cycles and Cyclical Data." *J. Royal Stat. Soc. B* **42** (1980): 245–292.

52. Tong, Howell. *Non-linear Time Series: A Dynamical System Approach.* Oxford: Oxford University Press, 1990.

53. Vapnik, Vladimir N. *Estimation of Dependencies Based on Empirical Data.* New York: Springer, 1982.

54. Wallace, Chris S., and D. M. Boulton. "An Information Measure for Classification." *Comp. J.* **11** (1968): 185–195.

55. Wallace, Chris S., and P. R. Freeman. "Estimation and Inference by Compact Coding." *J. Royal Stat. Soc. B* **49** (1987): 240–265.

56. White, Halbert. "Connectionist Nonparametric Regression: Multilayer Feedforward Networks Can Learn Arbitrary Mappings." *Neural Networks* **3** (1990): 535–549.

57. Weigend, Andreas S., Bernardo A. Huberman, and David E. Rumelhart. "Predicting the Future: A Connectionist Approach." *Int'l. J. Neural Systems* **1** (1990): 193–209.

58. Weigend, Andreas S., David E. Rumelhart, and Bernardo A. Huberman. "Generalization by Weight-Elimination with Application to Forecasting." In *Advances in Neural Information Processing Systems 3 (NIPS*90)*, edited by R. P. Lippmann, J. Moody, and D. S. Touretzky, 875–882. Morgan Kaufmann, 1991.

59. Weigend, Andreas S. "Connectionist Architectures for Time Series Prediction." Ph.D. thesis, Stanford University, 1991 (in preparation).

60. Weigend, Andreas S., and David E. Rumelhart. *Proceedings of INTERFACE'91— Computational Science and Statistics*, edited by Elaine Keramidas. Springer-Verlag, to appear.

61. Yule, G. U. "On a Method of Investigating Periodicities in Disturbed Series with Special Reference to Wolfer's Sunspot Numbers." *Philosophical Transactions Royal Soc. London Ser. A* **226** (1927): 267–298.

Brent Townshend
TCT, 230 Sherbrooke East, Suite 502, Montréal, Québec, Canada, H2X 1E1

Nonlinear Prediction of Speech Signals

Measurements were made of the correlation dimension of normally spoken speech from a single speaker and the results reveal that most of the points in the state space of the signal lie very close to a manifold of surprisingly low dimensionality, approximately 3.3. To validate this measurement, a nonparametric predictor was constructed using local approximation techniques. By using this predictor in tandem with a linear predictor, the residual error was reduced by 25% compared with the linear predictor alone. This type of nonlinear predictor has applications in many areas of speech analysis and processing.

MODELING SPEECH AS A CHAOTIC TIME SERIES

In speech coding, recognition, and synthesis, there is a need to be able to reduce the amount of data necessary to describe a speech signal. In coding, this allows a signal to be transmitted transparently with the minimum number of bits; in recognition, reduced descriptions may lead to better front-end processors; in synthesis, such descriptions can reduce the number of control variables necessary to produce high-quality speech. One avenue to finding better representations of speech is to model the processes which a human speaker uses in speech production. This type of modeling can be carried out on a number of levels ranging from physiological or physical to purely phenomenological. Most models, however, lie somewhere between these two extremes; they are usually driven by data but are also based, sometimes loosely, on the underlying physics. For example, the vocal tract, with the exception of the nasal cavity, can be modeled as an all-pole filter corresponding to its resonances. By using linear prediction on the acoustic waveform, a model of the speech can be inferred and the effect of any poles in the vocal tract or in the excitation can be removed from the signal. The set of linear predictive coding (LPC) parameters plus the residual forms a more compact description of the speech signal than the acoustic waveform. However, linear prediction has its limitations—it models only the poles of the vocal tract and possibly some of the excitation.

It is the purpose of this work to investigate the possibility of using nonlinear models to further improve speech modeling. In particular, the goal is to find a method which is able to predict the acoustic waveform of a speech signal with greater accuracy than linear prediction. Although the models under study are more in the realm of phenomenological models, they do have a basis in the physics of the vocal apparatus. In essence, one can extract the underlying state of the system and use observations of state transitions to estimate the next state of the system. This state should capture the position of the articulators as well as that of the vocal cords. There is, however, no attempt to relate the state variables used to their physical counterparts.

Many of the ideas used in this work come from the field of nonlinear dynamics and modeling of chaotic processes and parallel those expounded by Farmer and Sidorowich[9] and by Casdagli.[5] Although speech production is not strictly a low-dimensional chaotic process, it may still be possible to produce very good forecasts of speech using only a small subset of its state space. We can then think of speech as the output of a deterministic generator of low dimension, much as chaotic processes are. In modeling speech in this way, there are two implicit assumptions:

- The state of the human speech production system at any instant of time can be captured by a low-dimensional state vector.
- There is a deterministic rule that can be used to compute the state at one time from the state at a prior time.

Neither of these assumptions are strictly true. It is likely that the speech production mechanism is very high dimensional. In addition, the system is not simply a generator; it has external inputs. However, it is not unreasonable to assume that for a first-order approximation, a low-dimensional state vector may suffice. This is similar to modeling the vocal tract using only 10–12 LPC coefficients. The justification for such an assumption will come from the empirical results—the ability to perform prediction using a small number of state variables as well as from direct measurements of the dimension of the state space.

The second assumption is also questionable. In the long term it is ridiculous to postulate that it might be possible to predict everything that a person will say based on the current state of the system. Again, however, the short-term picture is much simpler. Predicting the state a very short time later, less than one millisecond, than a given state with limited accuracy can be a boon to many areas of speech analysis.

RECONSTRUCTION OF STATE SPACE

To model the dynamics of a time series such as speech, it is first necessary to reconstruct the (hidden) state at each time. Since the only information available are the one-dimensional samples of the series, the state must be reconstructed by looking over a time window of non-zero extent. A simple way to create a state representation is simply to form vectors of the n most recent samples of the time-series. That is, the delay vector, d_t, at time t is given by:

$$d_t \equiv \begin{pmatrix} x_t \\ x_{t-1} \\ x_{t-2} \\ \vdots \\ x_{t-n+1} \end{pmatrix} \tag{1}$$

and the state at time t is just $s_t = d_t$.

This choice of state representation is commonly known as the *method of delays*.[14] If the choice of n is large enough, then these vectors will form a sufficient state representation. It can be shown that to model the state of a p-dimensional system, we need only choose the length of the delay vector (the embedding dimension), n, such that[20]:

$$n \geq 2p + 1 \tag{2}$$

to guarantee that the representation is sufficient. In practice, it appears that a more relaxed bound of $n \geq p$ is often adequate.[7]

Although mathematically equivalent, different choices of the sampling rate, τ, and of n may result in very different numerical stability. The method of delays suffers from the problem of choosing the right values for these parameters. Another reconstruction which does not have this problem was suggested by Broomhead and King.[3] Their method applies a singular value decomposition to the delay coordinates. By using only the components whose eigenvalues are above a noise floor, the resulting state space can be made virtually independent of the choices of n and τ. Specifically, using some training material, a matrix, D, is formed whose rows are the delay vectors:

$$D \equiv \begin{pmatrix} d_1' \\ d_2' \\ d_3' \\ \vdots \end{pmatrix} . \tag{3}$$

Applying a singular value decomposition to D produces two orthonormal matrices, U and V, and the diagonal matrix w such that:

$$D = UwV' . \tag{4}$$

The state vectors are then given by:

$$s_t = \tilde{V}'d_t \tag{5}$$

where $\tilde{V}$ consists of the columns of V which correspond to singular values greater that some threshold, $w \geq w_T$. A further improvement to this method, suggested by Farmer and Sidorowich, is to weight the prior delay samples by a decaying exponential before performing the decomposition.[9] The specific embedding used in this work was

$$d_t = \begin{pmatrix} x_t \\ e^{-0.005}x_{t-1} \\ e^{-0.010}x_{t-2} \\ e^{-0.015}x_{t-3} \\ \vdots \\ e^{-0.120}x_{t-24} \end{pmatrix} \tag{6}$$

and the state vectors were

$$s_t = \tilde{V}'d_t \tag{7}$$

where $\tilde{V}$ is a 12×25 matrix, each column of which is an eigenvector of the decomposition. This set of embedding parameters was found, by trial and error, to provide the best prediction performance for the speech signals examined.

In applying these methods to speech signals, it is important to recognize that the data available is seldom tied to an absolute physical scale. Usually speech is scaled during recording or later processing to utilize the entire dynamic range possible. Thus, it would be difficult to use state mappings which depend on absolute levels. To remedy this problem, state vectors can be scaled to a constant length. Then scaled versions of a time series map onto the same state sequence. That is, the dynamics of the system can be modeled by a vector function, $f(\cdot)$, such that

$$s_{t+1} = \|s_t\| f\left(\frac{s_t}{\|s_t\|}\right) . \tag{8}$$

Scaling has the added advantage of collapsing the domain of f, and reducing its complexity. It is, however, imposing a partial linearity constraint on the model of the dynamics. Another problem due to the scaling is that during silent intervals, background noise will be scaled up to a fixed amplitude. If such state sequences are used to infer $f(\cdot)$, the resulting predictions will be adversely affected. To alleviate this, a noise threshold can be determined. Then, during training, rather than scaling state vectors below the noise threshold, they can be zeroed.

It should be emphasized that although many different choices of the state representation are mathematically equivalent, in practice the choice can have a great effect on the results of a predictor. Improper choice can result in amplification of any noise present. Furthermore, choosing a state space which makes use of some of the known properties of the speech signal can result in an embedding of lower dimension than that given in Eq. (2). For example, we will show that if the speech signal is first processed by a linear predictor, removing the effect of the vocal tract resonances, an embedding of the residual will give different results than simply embedding the raw time series.

MEASURES OF SPEECH DIMENSIONALITY

Once a state representation is formed, one might ask if the system under study fully occupies the state space or whether it spends most of its time near a subset of the space, known as an attractor. To answer this, one can measure the dimensionality of the attractor and see how it compares with that of the entire state space. In fact, there is not only one dimension which characterizes an attractor embedded in state space, but many. For an excellent review of some of these, see Farmer et al.[8] One dimension which is relatively easy to compute is the correlation dimension, ν.[10,11] Also, it can be shown that the correlation dimension establishes a lower bound on the other dimensions.[11] If we define a correlation integral, $C(l)$, by

$$C(l) = \lim_{N \to \infty} \frac{1}{N^2} \left\{ \begin{array}{c} \text{number of pairs of points, } (S_i, S_j), \text{on the attractor} \\ \text{with separation } < l \end{array} \right\}, \quad (9)$$

then one observes that as $l \to 0$,

$$C(l) \propto l^{\nu}, \quad (10)$$

giving the working definition of the correlation dimension, ν. Thus, one needs only to count the number of pairs of points separated by a distance of less than l. Plotting the value of $C(l)$ versus l on a log-log plot gives a slope of ν for small l. Actually, when observing an attractor with noisy measurements, the plot will have three regions: for very small l the slope will equal the dimension of the embedding space since uncorrelated noise is space filling; for l larger than the noise components, the slope will be ν; while for large l the edge of the attractor will reduce the slope below l.

The non-zero autocorrelation of speech required that a correction be made to the correlation integral computation to avoid bias. Only vector pairs which originated from speech material separated by more that 50 ms. were used in Eq. (9). This helps prevent the short-term, linear-deterministic dynamics of the signal from interfering with the strictly geometric correlation-dimension measurement.[17] Recall also that the state vectors are scaled to lie on the unit sphere. This should not affect the correlation integral other than reducing the effective embedding dimension by one. Thresholding was necessary to correct for the undue weight that would have been attributed to small magnitude signal segments, which are mostly background noise, while investigating the correlation integral at small radii. The value of ν was measured using the the prism-assisted algorithm suggested by Theiler[17] that requires time of order $N \log N$; a substantial savings over the $\mathcal{O}(N^2)$ direct implementation.

Using over 30 seconds, or 250,000 samples, of speech from a single speaker, the correlation integral for values of l varying from 0.0001 to the maximum possible, 2, was measured. The integral as a function of l is shown in Figure 1 for state vector lengths varying from 4 to 12. This corresponds to embedding dimensions varying from 3 to 11 after consideration of the scaling of the state vectors. The figure shows the characteristic signs of a low-dimensional system. For medium values of l, where $C(l)$ is between 10^{-6} and 10^{-2}, scaling behavior is clearly exhibited. The slope of the correlation integral, ν, is relatively constant at 3.2–3.3 independent of the

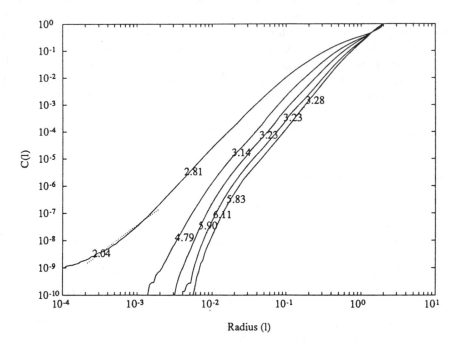

FIGURE 1 Correlation integral for scaled speech. The probability of two randomly selected segments of speech being within a distance, l, is plotted as a function of l for state vector lengths of 4, 6, 8, 10, and 12 (left to right). The dotted lines represent linear fits to the model $C(l) = kl^{\nu}$, with the values of ν superimposed.

embedding dimension, as long as it is sufficiently large. This is in agreement with the results reported for voiced speech over a number of speakers of between 3 and 5.[18] At very large values of l, the scaling breaks down due to edge effects. For values of l below 0.03, higher-dimensional behavior is evident for the larger embedding dimensions and scaling is lost. One possible interpretation of this knee is that the signal under study contains a non-deterministic component with energy 30 dB below that of the overall signal. This noise may be due to either actual background noise recorded along with the speech signal or a more complex component of the speech signal, which appears only at smaller scales. Measurements of the correlation integral using unscaled state vectors reveal similar behavior, with nearly identical slopes, but the regions are not as distinct and statistical errors are larger.

The lower slopes evident in the smaller embedding dimensions for small l, near the lower left of Figure 1, are due to the quantization of the speech signal. This saturation occurs when $C(l)$ becomes small compared to the number of state vectors with exactly the same (quantized) value. The value of $C(l)$ at which this saturation occurs, C_q, can be bounded by considering the case of an uncorrelated signal.

$$C_q \geq \left[2 \sum_{x=T}^{\infty} p(x)^2 \right]^m \tag{11}$$

where $p(x)$ is the probability of a sample having value x, m is the embedding dimension, and T is the threshold magnitude that was used to discard the vectors with magnitude less than $\sqrt{m}T$. For the statistics, $p(x)$, of the given speech sample, C_q was found to be greater than 10^{-9} for an embedding dimension of 3. This is in close agreement with Figure 1. Note that for embedding dimensions of 3 or 5, increasing the amount of data used to compute the correlation integral would not affect the results significantly—the main limitation is the resolution of the data itself. This assumes, of course, that the test data is typical of speech in general. For higher embedding dimensions, utilization of more data could extend the correlation integral to values of $C(l)$ below 10^{-9}, but would not affect the linear regions as was verified using cross-validation.

LINEAR MODELS

A great deal of research has been done on the study of the speech signal and it would be unwise to treat this signal as simply another time series of unknown origin and analyze it blindly. In particular, there exists a vast body of literature concerning linear analysis of speech. It is, thus, important to ascertain whether the above results are simply a new manifestation of old knowledge.

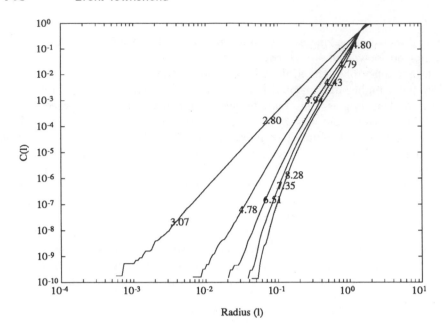

FIGURE 2 Correlation integral for the LPC residual of speech. The probability of two randomly selected, scaled segments of speech being within a distance, l, is plotted as a function of l for state vector lengths of 4, 6, 8, 10, and 12 (left to right). The dotted lines represent linear fits to the the model $C(l) = kl^{\nu}$.

Linear prediction models speech using the relation

$$\hat{x}(t) = \sum_{i=1}^{d} \alpha_i x(t - i) \, . \tag{12}$$

One possible approach to using linear prediction is to find a single set of parameters, α_i, for a body of speech and to use them for all subsequent processing. Using this approach no side information needs to be provided to a predictor. However, the prediction gain is obtained by simply removing any tilt in the spectrum of the speech signal, limiting its utility.

Another approach is to use time-varying linear prediction. Since linear prediction is usually used to model the vocal tract resonances, which are not fixed but change slowly, the predictor parameters can be updated at a commensurate rate; typically every 10–20 ms. The vocal tract contains three or four major resonances and to adequately model these requires 10–12 parameters. Using more does not significantly improve the prediction gain.

As the linear models are quite well matched to speech, one might postulate that the low correlation dimensions observed above are due simply to the linear-deterministic components of this signal. To address this point, the correlation dimension of the residual after applying a linear predictor to speech was examined.

The LPC analysis was performed every 10 ms on a frame of length 12.5 ms which was windowed with a Hamming window. The autocorrelation method with bandwidth expansion and high-frequency compensation was used to compute the first twelve predictor coefficients.[1] The correlation integral for the residual of the speech signal after filtering by this predictor is shown in Figure 2. In this case the scaling behavior is less clear, but there appears to be a saturation at $\nu \approx$ 5 for embedding dimensions greater than 6, although such high values of ν are questionable.[15]

This result is, unfortunately, inconclusive. One possible explanation for the increase in ν is that the low original correlation dimension was indeed due to the linear correlations in the signal. Once the linear determinism is removed, the remnant appears more like space-filling noise.

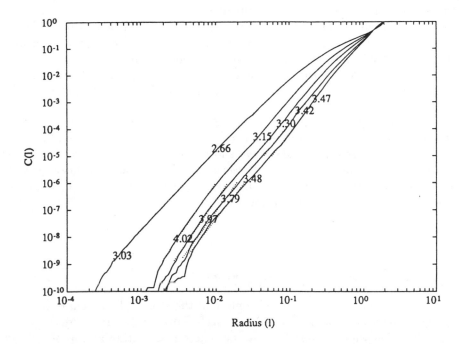

FIGURE 3   Correlation integral for an LPC filter excited by random noise. The probability of two randomly selected segments being within a distance, l, is plotted as a function of l for state vector lengths of 4, 6, 8, 10, and 12 (left to right). The dotted lines represent linear fits to the the model $C(l) = kl^{\nu}$.

However, this is not the only explanation. Equally plausible is the possibility that the linear model is a good, but not perfect, model of the signal. Rather than simplify the system, we have complicated it by adding an additional stage which also contains internal state variables, the LPC parameters. We would then expect to see dimensions measurements 12 higher than previously, well beyond the capability of these techniques to resolve.

To distinguish these two possibilities and determine the relative contributions of the linear-predictable part of speech to the low observed correlation dimension, the following experiment was performed. An LPC analysis was carried out on the speech and the residual computed for each frame was replaced with Gaussian white noise with the same energy. This noise was used to excite the LPC filter to produce a synthetic speech. This new signal has the feature that its determinism is limited to the linear components; that is, the best possible short-term predictor of this signal is a time-varying, linear one.[17] By comparing the correlation dimension of this signal with that obtained for the original speech, properties of the LPC residual can be investigated. If the correlation dimension of the noise-excited LPC filter is higher than that of the original speech, then we can conclude that the LPC residual contains a nonlinear, deterministic component.

The correlation integral for the noise-excited LPC filter is shown in Figure 3. From the figure, it is evident that this time series has a correlation dimension, ν, of about 3.4–3.5. Although this is slightly higher than that obtained for the original speech, the difference is less than the experimental error. Thus, it appears that the linear determinism may account, at least partially, for the low correlation dimension.

One other point of interest in Figure 3 is the wide range over which linear scaling persists. Since the excitation was replaced by completely space-filling noise, the radius at which scaling breaks down should be at least as high as that for the raw speech, which was at $r = 0.03$. However, it appears that the linear region is maintained down to the smallest radii measured, less than 0.005. Thus far, this apparent paradox lacks an explanation.

IMPLEMENTATION OF PREDICTOR

The state of a dynamical system tells only half the story. The actual dynamics lie in the mapping which carries the current state into the future. In discrete time, this is a function which maps the state at time t to the state at time $t + 1$. This mapping could be any linear or nonlinear map. Linear prediction, consisting of a time-varying 10- to 12-order linear map has been widely used for speech analysis. Other methods that have been used include layered neural networks,[18] and radial basis functions.[12]

In this work a non-parametric map was formed using *local approximation*.[6,9] The idea of local approximation is to break the mapping into a set of small neighborhoods and apply some parametric model to each one. If the neighborhoods are small enough, the mapping will be smooth in each one making the fit to a local model much easier.

There are a number of ways of using local approximation to model a function, but we will restrict the discussion to that used in this work. The procedure to find the value of a function, $f(x)$, at some point, $x = a$, is as follows:

1. Store a set of training data in a table as pairs of the form $(x, f(x))$.
2. Find the k nearest neighbors of a in the table. That is, find the set of entries such that $\|x - a\|$ is minimized.
3. Compute the parameters of a local model which approximates $f(x)$ over the neighborhood found above. The fit is weighted by the distance between each x and a.
4. Evaluate the local model at a to get $\hat{f}(a)$.

This can be used to model the dynamics of the speech by choosing x to be the state of the system at time t and $f(x)$ to be the state predictor of Eq. (8). The local approximation model then allows us to find s_{t+1} from s_t. In fact, since we are currently interested only in predicting the time waveform of speech, a further simplification can be be made. Since the state captures the system completely at each instant of time, there exists another function, $g(s_t)$, which maps the state to the actual scalar level of the waveform, x_t. The two mappings $s_t \rightarrow s_{t+1}$ and $s_{t+1} \rightarrow x_{t+1}$ can be composed and, instead of storing the state at time $t+1$ in the table, x_{t+1} can be stored instead. This reduces the size and computation involved in the mapping process.

The complexity of the above procedure is dominated by step 2 above, the nearest-neighbor search. A direct implementation would require N distance computations to perform the search, where N is the training set size. Fortunately, there are algorithms which allow this step to be completed in $\mathcal{O}(\log N)$ time. One of these, the $k - d$ tree, was used in this work.[2,13,16]

ORDER OF THE LOCAL MAP

There is a trade-off possible between the number of neighbors considered, k, and the order of the local model used to approximate the mapping over each neighborhood. The order, in this case, refers to the number of derivatives which are matched by the model, i.e., linear, quadratic, etc. If very small neighborhoods are used, then the same accuracy of the approximation can be obtained with simpler local models than required if the neighborhoods are larger. In the limit of a zeroth-order local model and $k = 1$, this approach reduces to vector quantization.[4] For a given value of a, the table entry which minimizes $\|x - a\|$ is chosen and the corresponding value of $f(x)$ is used for $\hat{f}(a)$. It is interesting to note that vector quantization has been used with great success for coding of speech signals, although the motivation for use is much different than that presented here. However, with a small increase in neighborhood size, local linear models were found to produce superior performance.

LOCAL FITS USING SVD

The local maps are meant to model an attractor of dimension D embedded in a higher-dimensional space. A knowledge of the value of D can be used to improve the quality of the local fits. For example, if a linear model is being used, then the local fit can be performed used a singular value decomposition. In each neighborhood, the first D singular values should represent the behavior of the attractor while the remaining values are influenced by noise and curvature due to the non-zero size of the neighborhood. By zeroing all but the D largest singular values when performing the fit, the effect of the noise and curvature can be reduced while retaining the desired behavior of the mapping. A more commonly used method, zeroing the singular values less than some threshold, requires an estimate of the noise component and is not well suited to handling the curvature problem.

CONTROLLING POOR PREDICTIONS

Another issue which arises when the amount of training data is finite is the handling of regions of the attractor for which little or no training data is available. Fortunately this situation is easily identified by examining the radius of the neighborhood found while querying the mapping table. If this radius is larger than some threshold value, then the prediction can be marked as invalid. Depending on the application of the predictor, this can handled as is appropriate, such as by generating a zero prediction.

   With a finite amount of training data, or if the state representation does not completely capture the state of the system under study, then there will also be cases such that the data points within a single neighborhood will map to very different function values. Fitting a local model under such conditions is guaranteed to generate incorrect results by interpolating two or more disjoint function values. This problem can be alleviated by finding neighborhoods larger than the minimum size required to fit the local model and then clustering the neighbors based on their function values. If there are multiple clusters, then it is not clear which mapping to use for the current point, but the nearest cluster to the current point can be used to pick the correct cluster much of the time. The contents of the cluster are then used to fit the local model and interpolate $f(x)$ at the current point. As well as giving correct results some of the time, this technique allows problem areas to be readily identified.

SPEECH PREDICTION RESULTS

DATA SET

The data used were from a single speaker recorded with a sampling rate of 8 kHz and 14-bit resolution. The speech was filtered to a range extending from 100 Hz to 3.6 kHz. Three disjoint bodies of speech were used. A nine-second segment were used to compute the basis vectors of the embedding space, $\tilde{V}$. Using this basis set, 30 seconds of speech was used to populate the local approximation table. This body of speech was also used to find the parameters for the fixed linear predictors that were tested. The third body of speech, a two-second sentence, was used to test the performance of the various predictors.

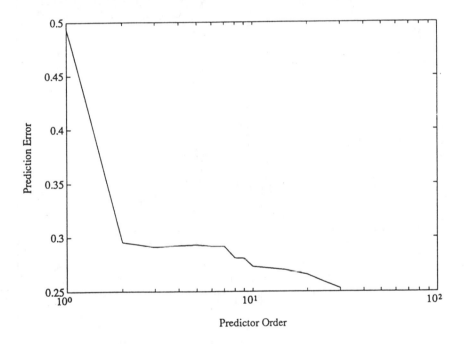

FIGURE 4   Prediction gain for a fixed linear predictor. Relative error after prediction using a fixed linear predictor is shown as a function of the number of parameters of the predictor.

LINEAR PREDICTION

Before examining the nonlinear prediction results, a baseline can be established using linear prediction. A plot of prediction error versus predictor order, d, for a fixed linear predictor is shown in Figure 4. The error is defined as the energy of the residual divided by the original signal energy:

$$E^2 = \frac{\langle(\hat{x}_t - x_t)^2\rangle}{\langle(x_t - \langle x_t\rangle)^2\rangle} .$$

(13)

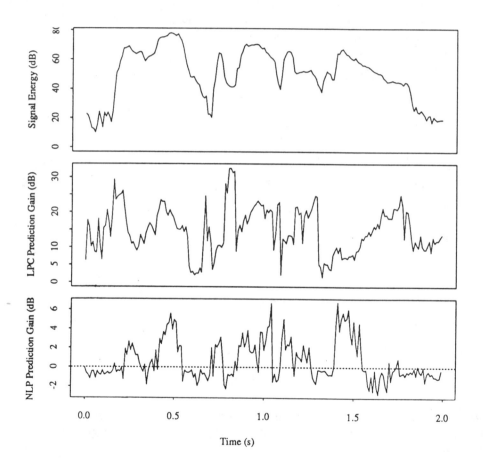

FIGURE 5 Prediction gain as a function of time for the test sentence, "We lost the golden chain." Upper frame is signal energy; middle is LPC prediction gain using 10ms frames and a 12th-order predictor; lower frame is additional gain obtained with local approximation.

The speech community more commonly uses prediction gain, defined as

$$G = -20\log_{10}(E)\text{dB} \tag{14}$$

as a figure of merit. The minimum error obtained using a fixed linear predictor for the test material was approximately 0.25, a prediction gain of 12 dB. Note that the test data is disjoint from that used to choose the linear predictor parameters, though both sets are from the same speaker. Much lower prediction gains would be obtained if the testing and training data were obtained from different speakers. Using a time-varying linear predictor, as described earlier, the prediction error was 0.155 (16.2 dB), but the error in each frame was highly variable, ranging from less than 0.03 (30 dB) to 0.9 (1 dB). The signal energy and the resulting prediction gain as a function of time are shown in the top two frames of Figure 5.

NONLINEAR PREDICTION OF SPEECH TIME SERIES

A nonlinear predictor was trained on the residual of the same time-varying LPC predictor used previously. As well as short-term correlations, speech time series show a high correlation between samples separated by one pitch period during voiced speech. To allow a fair comparison between linear prediction and nonlinear prediction, the number of past samples was limited to 25, preventing the possibility of making use of the pitch periodicity. The additional gain that can be achieved by using pitch prediction can then be investigated separately. The embedding shown in Eqs. (6) and (7) was found to maximize the prediction gain.

A linear local model was used to fit the data in each neighborhood. The model was of the form

$$f(s) = a_0 + \sum_{i=1}^{12} a_i s_i . \tag{15}$$

The values of the a_i were determined using neighborhoods containing 26 pairs, $(s, f(s))$, and a weighted regression. The weights used were inversely proportional to the distance between each s and the point at which the value of the function was computed. To improve the conditioning of the fitting procedure, the singular-value decomposition regression procedure described above was used. All but the two largest singular values of the decomposition were zeroed during the fitting procedure. The result of this is that the resulting parameters model the function values along the plane which best fits the s values. Since the measured dimensionality was between three and four, one might expect that retaining this many components should perform better. However, due to increased noise contributions, using more components produced marginally more prediction error than using just two.

Figure 5 illustrates the linear and nonlinear prediction gains as a function of time. The upper panel shows the test signal energy over a period of two seconds. The middle frame shows the LPC prediction gain obtained, while the lower frame shows the additional prediction gain obtained by using local approximation. The overall prediction gains obtained were 16.2 dB from LPC and an additional 2.9 dB from the nonlinear mapping. It is clear from the figure that the most prediction gain was obtained in areas of high signal energy, usually voiced vowels. To further illustrate the effect of the nonlinear mapping on the LPC residual, Figure 6 shows the speech waveform, the LPC residual, and the residual after the nonlinear mapping for a few pitch periods during a vowel. It is clear from the figure that the nonlinear predictor is able to make the residual look more like a sequence of impulses corresponding to the excitation from the glottis. The height of the pitch impulses are reduced and the ringing following each spike is removed by the nonlinear predictor. This ringing is probably due to an all-pass component of the vocal-tract which cannot be removed by the linear analysis. The nonlinear predictor is, however, able to model this component more closely.

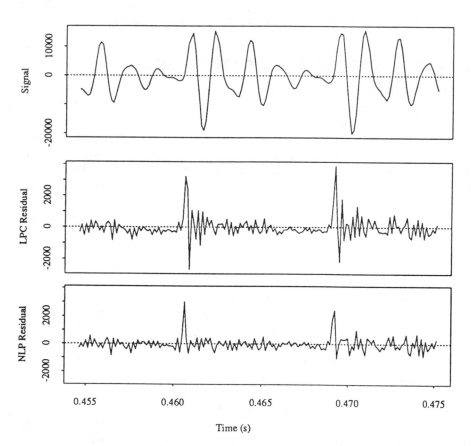

FIGURE 6 Waveforms for a 12.5 millisecond segment of speech. The upper panel shows the original speech signal as a function of time, the middle is the residual after linear prediction. The lower panel shows the remainder after subtracting the nonlinear prediction. The two residuals are on the same scale. This segment is from the same analysis as shown in Figure 5.

A predictor was also designed to operate on the raw speech samples. After optimization of the embedding parameters, this predictor was only able to obtain 9.2 dB of prediction gain. This was not as good as an equivalent fixed linear predictor, which obtains 11.1 dB of gain. One reason for the failure may be that the nonlinear predictor is over-learning the mapping and ends up amplifying noise. Another possibility is that the optimal model has a large linear component. The compact, parametric model is much better suited to learn this component than the nonparametric system.

SCALING OF PREDICTION ERROR

Figure 7 shows the prediction error for the nonlinear predictor following a linear predictor as a function of the training set size. Also shown in the figure is a least-squares fit to the model:

$$E = kN^{-\frac{q}{D}} + e_0$$

where q is the order of the local model and D is the dimension of the attractor. This is similar to the error scaling relationship derived by Farmer and Sidorowich[9] with one important distinction: the addition of the e_0 term. This term reflects the fact that we are really dealing with only a partial model of the system. Only the gross dynamics of the system can be adequately modeled by a low-dimensional representation. The e_0 term accounts for an additional component of the prediction error due to the components of the output which are not captured by the model.

The fitted values of q/D and e_0 were 0.34 and 0.69, respectively. Since the order, q, of the local approximation models was 2, this allows us to compute another measure of the dimension of the attractor, D, of 5.8. This is significantly different from the correlation dimension measurements shown earlier. However, the value of q found using this model is very sensitive to the data points. In fact, a fit using $D = 3$ also passes within the spread of the data points, only slightly increasing the mean square fitting error. Thus it seems that for the narrow range of prediction errors encountered here, this approach does not provide a very precise method for estimating the dimension.

The large value of e_0 found, 0.69, is also somewhat disconcerting. It indicates that we are obtaining nearly the maximum prediction gain possible using a low-dimensional model. It is unclear whether this limit will be affected by the choice of embedding and preprocessing strategies.

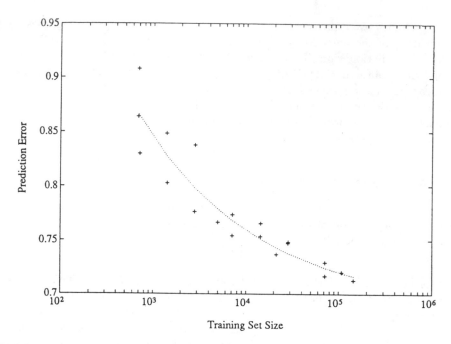

FIGURE 7   Prediction error as a function of training set size for nonlinear predictor operating on an LPC residual. The dotted line is a fit to the model discussed in the text.

CONCLUSIONS

Measurements of the correlation dimension of speech revealed that most of speech lies on an attractor with dimension 3.3. It is, however, possible that this observation is due solely to the well-known linear-predictable component of speech. Measurements of the correlation dimension of the LPC residual as well as that of linear-deterministic synthetic speech were unable to rule out this possibility. Better evidence of a nonlinear, deterministic component of speech is that a nonlinear predictor was able to provide additional gain over a time-varying linear system. Prediction results are summarized in Table 1.

There are innumerable avenues for continued work in this area. Aside from the implementation problems; optimization and simplification, there exist a number of more fundamental issues. One is the determination of an optimal set of parameters for the nonlinear predictor. The values measured for ν give us valuable insight

TABLE 1 Predictor Results

Condition	Prediction Gain (dB)	Relative RMS Error
Linear Fixed	11.0	0.28
Local Approximation	12.2	0.24
Linear 10ms Frames	16.2	0.15
Linear 10ms + Local Approx.	19.0	0.11

here, but further work in this area would be beneficial. Given an underlying dimensionality, the parameters of the embedding procedure might also be improved. These are the length of the delay vectors, the weighting of them, and the choice of the basis.

More experimentation is also required in local approximation. In this work, only first-order local models have been used. There is considerable evidence that higher-order models may drastically improve the prediction gain.[5] Another area is the details of the local modeling—the number of principal components to retain and criteria for rejecting poor predictions.

Of course, the speaker dependency issue must be addressed. Much of the prediction gain obtained so far may vanish when a speaker-independent mode is adopted. This will also impact the training set issues above and a pruning technique will almost certainly be required to keep the number of computations under control.

Although this work has only scratched the surface of the possibilities in using nonlinear prediction in speech analysis, the results are encouraging. A nonlinear predictor is able to add about 3 dB of prediction gain to that obtained using a linear predictor. Combined with measurements of the underlying dimensionality of the speech generation mechanism, these results indicate that it is feasible to model speech as the output of a chaotic generator. The techniques have potential applications in speech coding,[19] speech enhancement, noise reduction, and speech recognition.

ACKNOWLEDGMENTS

A large portion of this work was performed while the author was with AT&T Bell Laboratories, Murray Hill, NJ. The author would like to thank Dr. Naftali Tishby both for suggesting the idea of modeling speech as the output of a chaotic system and for his useful discussions in this area. I would also like to thank Dr. Bishnu Atal for his assistance and guidance.

REFERENCES

1. Atal, B. S., and M. R. Schroeder. "Predictive Coding of Speech Signals and Subjective Error Criteria." *IEEE Trans. Acoust., Speech, Signal Processing* **27(3)** (1979).
2. Bentley, J. L. "Multidimensional Binary Search Trees Used For Associative Searching." *Communications of the ACM* **18(9)** (1975): 509–517.
3. Broomhead, D. S., and G. P. King. "Extracting Qualitative Dynamics From Experimental Data." *Physica D* **20** (1986): 217–236.
4. Buzo, A., A. H. Gray Jr., R. M. Gray, and J. D. Markel. "Speech Coding Based Upon Vector Quantization." *IEEE Trans. Acoust., Speech, Signal Processing* (1980): 562–574.
5. Casdagli, M. "Nonlinear Prediction of Chaotic Time Series." *Physica D* **35** (1989): 335–356.
6. Cleveland, W. S., and S. J. Devlin. "Locally Weighted Regression: An Approach to Regression Analysis by Local Fitting." *J. Am. Statistical Assoc.* **83(403)** (1988): 596–610.
7. Eckmann, J. P., and D. Ruelle. "Ergodic Theory of Chaos and Strange Attractors." *Rev. Mod. Phys.* **57** (1985): 617.
8. Farmer, J. D., E. Ott, and J. A. Yorke. "The Dimension of Chaotic Attractors." *Physica D* **7** (1983): 153–180.
9. Farmer, J. D., and J. J. Sidorowich. "Exploiting Chaos to Predict the Future and Reduce Noise." In *Evolution, Learning, and Cognition*, edited by Y. C. Lee, 277. Singapore: World Scientific, 1988.
10. Grassberger, P., and I. Procaccia. "Characterization of Strange Attractors." *Phys. Rev. Lett.* **50(5)** (1983): 346–349.
11. Grassberger, P., and I. Procaccia. "Measuring the Strangeness of Strange Attractors." *Physica D* **9** (1983): 189–208.
12. Lowe, D., and A. Webb. "Adaptive Networks, Dynamical Systems, and the Predictive Analysis of Time Series." In *IEEE Int. Conf. on Neural Networks*, 1989.
13. Murphy, O. J., and S. M. Selkow. "The Efficiency of Using k-d Trees for Finding Nearest Neighbors in Discrete Space." *Information Proc. Lett.* **23** (1986): 215–218.
14. Packard, N. H., J. P. Crutchfield, J. D. Farmer, and R. S. Shaw. "Geometry From a Time Series." *Phys. Rev. Lett.* **45** (1980): 712–716.
15. Procaccia, I. "Weather Systems: Complex or Just Complicated?" *Nature* **333** (1988): 498–499.
16. Ramasubramanian, V., and K. K. Paliwal. "A Generalized Optimization of the K-d Tree for Fast Nearest-Neighbour Search." *Int'l Conference on Acoustics, Speech, and Signal Processing*, 1989.
17. Theiler, J. "Quantifying Chaos: Practical Estimation of the Correlation Dimension." Ph.D thesis, California Institute of Technology, 1988.

18. Tishby, N. Z. "A Dynamical Systems Approach to Speech Processing." Paper presented at the *Int'l Conference on Acoustics, Speech, and Signal Processing*, 1990.
19. Townshend, B. "Nonlinear Prediction of Speech." *Int'l Conference on Acoustics, Speech, and Signal Processing*, May 1991.
20. Whitney, H. "Differentiable Manifolds." *Ann. Math.* **37(3)** (1936): 645–680.

William W. Taylor
The RTA Corporation, 5613 Artesian Drive, Rockville, MD 20855

Application of Nonlinear Prediction to Signal Separation

INTRODUCTION

As we gain a clearer understanding of the nature of chaotic behavior in nonlinear dynamical systems, we must look more closely at how we can use this understanding to our advantage. The prospect of capitalizing on short-term predictability to analyze data generated by physical systems whose behavior was previously regarded as completely random has significant appeal and deserves aggressive pursuit. One approach to such an effort is to re-examine our definition of random behavior to determine whether an alternative mathematical model is available for the required data analysis. This approach is suggested by the work of A. N. Kolmogorov, who developed the theory of *random processes* which is crucial to current signal processing methodology, for which he and Norbert Wiener originally provided the mathematical foundations. Kolmogorov had a lifelong fascination with the concept of random behavior, and the following quote was attributed to him in 1982:

Nonlinear Modeling and Forecasting, SFI Studies in the Sciences of Complexity,
Proc. Vol. XII, Eds. M. Casdagli & S. Eubank, Addison-Wesley, 1992

In everyday language we call random those phenomena where we cannot find the regularity allowing us to predict precisely their results. Generally speaking, there are no grounds to believe that a random phenomenon should possess any definite probability [distribution]. Therefore we should distinguish between randomness proper (as the absence of any regularity) and stochastic randomness (which is the subject of probability theory).[22]

In the 1960s he had identified this "absence of regularity" with algorithmic complexity and also exhibited the central role played by *entropy* by relating the combinatorial, algorithmic, and probabilistic approaches to its definition. He also pursued the relationship of these notions to the foundations of information theory and probability theory, respectively, in an effort to quantify randomness.[20,21,22,23] A similar approach to random behavior was pursued independently by Chaitin.[10] An approach based on algorithmic complexity has not proved readily applicable to signal extraction applications so that useful techniques still universally assume that any randomness encountered in a data set results from definite probability distributions. The quote, however, does propose an alternative approach to quantifying randomness, since it suggests the option of bypassing the search for regularity (or lack thereof) in data sets obtained from certain phenomena and turning directly to an effort "to predict precisely their results."

This alternative approach is also motivated by the wealth of recent work concentrating on applying nonlinear prediction methods[1,2,8,11,13,14,15,16,24,27] and qualitative dynamic analysis[4,5,6] to actual data sets generated by dynamical systems. These techniques exhibit the potential to dramatically extend the state-space reconstruction techniques originally developed to quantify chaotic behavior[12,17,18,28,29] so that dynamic behavior as well as geometric structure becomes accessible to the analyst. The third essential ingredient in the alternative approach has been the parallel development of *ergodic theory* which enables us to consider random processes and dynamical systems within the same mathematical framework.[19,30] This suggests that we can shift the emphasis in signal processing applications from the requirement to know (or be able to estimate from the data) all of the underlying probability distributions (plus the finite joint distributions) of the random variables which define the random process to the requirement to know (or be able to estimate from the data) the underlying equations of motion of the dynamical system which generates the data set. This means that as we analyze an actual data set produced by a real-world physical process, we should have the option to view any observed random behavior either in terms of chaotic dynamics produced in accordance with physical laws or in terms of random processes defined by their underlying probability distributions. It is perhaps worthwhile to note that the first step in charting this remarkable equivalence between the statistical and dynamical properties of an actual data set was again taken by Kolmogorov. Subsequent results have been the work of Sinai, Anosov, Bowen, Ornstein, and others.[19,30]

If we call this new alternative the *dynamical processing approach*, as opposed to the traditional *statistical processing approach*, our purpose in this paper is to document some encouraging initial results obtained using dynamical processing methods, while recognizing that we are rapidly obscuring the traditional distinctions between statistics and dynamics. In order to test our theory that dynamical processing methods should be used to extract a physically generated signal process from a contamination process which was also physically generated, we were given two data sets containing acoustical data composed of a voice signal contaminated by noise from an air-conditioning system. Both data sets were contaminated and did not contain either the signal or the noise separately. We made independent estimates of the signal strengths in the data. One data set (the easy one) contained the voice signal buried by 30 decibels (dB) of noise for a signal-to-noise ratio of minus 30 dB or an amplitude ratio between the air-conditioning acoustic pressure and that of the voice signal of slightly over 30 (actually the square root of 1000). In the hard data set, the signal was buried by 40 dB of noise for a signal-to-noise ratio of minus 40 dB and an amplitude ratio of 100. We measured our signal processing success in terms of *intelligibility*, which is the proportion of words in the data sample which are understandable to a panel of listeners. The easy data set had an original intelligibility of less than 50%, and the original intelligibility of the hard data set was less than 25%. After applying the dynamical processing methods for signal extraction, the resulting intelligibility for both samples was above 95%, and the results were quite noticeable on audio tape. The remainder of the paper will be devoted to describing in some detail these dynamical processing methods and their underlying theory. Since a primary objective in developing these techniques is to exploit short-term predictability, we will begin by discussing the elements required to quantify the predictability of a process modeled by nonlinear prediction methods.

QUANTIFYING PREDICTABILITY
UNDERLYING STRUCTURE

Our development, which adapts the work of Kravtsov,[25,26] assumes that we are working with a fixed nonlinear prediction model of the sort referenced above. Our purpose will be to quantify the predictability of a fixed (inaccessible) physical process which generates a time-dependent data set (in time-series form). The quantification will be accomplished in terms of the given nonlinear model. Though this means that the predictability (and thus the randomness) of the process becomes model dependent, the approach will still apply in fairly general circumstances.

Let $x(t)$ denote the state at time t of an actual physical system that we are studying (time may be either discrete or continuous). This yields an idealized process which evolves in accordance with dynamic equations which are unknown and not directly accessible to an observer. Let $y(t)$ denote an observation of this process (at time t). These observations, of course, differ significantly from the ideal states in several respects. First of all, $x(t)$ is typically multi-dimensional, indicating many degrees of freedom for a physical system, whereas most observations are scalar valued (multiple sensors can provide vector valued data, but intuition is perhaps best served by regarding $y(t)$ as a scalar-valued observation process). Also, our observations are almost always distorted in some manner, yielding contaminated data. Next, let $w(t)$ denote a model process reconstructed from the observed data process $y(t)$ and designed to replicate within certain limits the physical process $x(t)$. Thus, $w(t)$ must be reconstructed in a multi-dimensional state space, although its dimension will typically be less than that of the original physical process, which may not reveal all of its degrees of freedom through our available data observations.

The equations of motion which govern the evolutionary behavior of $w(t)$ are also estimated from the data based on a nonlinear dynamical model. We further assume that we can observe the model process to obtain a scalar-valued process $z(t)$ which can be compared directly to the observed data $y(t)$. Thus we have identified four processes essential to our estimation problem. They are:

$$\begin{aligned}
\text{The original physical process,} && x(t)\,, \\
\text{The observed data process,} && y(t)\,, \\
\text{The reconstructed model process,} && w(t)\,, \\
\text{The estimated, or predicted, data process,} && z(t)\,.
\end{aligned}$$

The traditional measure of model accuracy involves the *error term* at time t, defined by

$$e(t) = y(t) - z(t)\,, \tag{1}$$

which gives the error when the estimated, or predicted, data process resulting from the reconstructed model is compared to the observed data. Many applications use the variance of this error process, given by (presuming that we have an unbiased estimate)

$$\text{var}_e = \langle\, |\, y(t) - z(t)\, |^2 \,\rangle\,, \tag{2}$$

where the angle brackets, $\langle \ldots \rangle$, denote an averaging process. Statistical processing methods would require this to be an ensemble average, though we will use time averages. Still following Kravtsov we will introduce an alternative performance measure which incorporates in a natural manner the time period during which behavior can be accurately predicted.

DEGREE OF PREDICTABILITY

First of all, observe that the modeling process will normally be based only on data preceding a specified time, designated by t_0, and our forecast, or prediction, will continue until a future time, denoted by $t = t_0 + s$. This establishes an initial condition that can always be successfully met by constructing the model process to ensure that

$$z(t_0) = y(t_0)\,. \tag{3}$$

Then the prediction at time $t = t_0 + s$ is given by

$$z(t) = z(t_0 + s) = u_s(t), \qquad (4)$$

where Eq. (4) defines the term on the far right, which may be interpreted as the prediction over s time units into the future of the data point at time t based on data taken through time t_0. Thus s simply denotes the *time lag* associated with the prediction process. Now for a fixed time lag s, the prediction error becomes a function of the time lag as well as the time of prediction, so that Eq. (1) becomes

$$e(t, s) = y(t) - u_s(t), \qquad (5)$$

which can be interpreted as the prediction error at time $t = t_0 + s$ based on a projection s time units into the future from time t_0. The error variance also becomes a function of s with

$$\text{var}_e(s) = \langle |\, y(t) - u_s(t)\, |^2 \rangle. \qquad (6)$$

A more useful measure of prediction accuracy, however, is obtained from the normalized correlation of the predicted data process with the observed data process. The result is termed the *degree of predictability*, denoted $D(s)$,

$$D(s) = \frac{\langle y(t)u_s(t)\rangle}{(\langle y(t)^2\rangle\langle u_s(t)^2\rangle)^{1/2}}, \qquad (7)$$

where the angle brackets now denote the average over the admissible values of t (computed as a sum for discrete time and as an integral for continuous time). This measure is used to define time periods for which accurate predictions can be made. It is clear (from the Cauchy-Schwartz inequality) that the absolute value of $D(s)$ is bounded by unity; and for those values of s for which $D(s)$ remains close to one (recall that Eq. (3) ensures that $D(0) = 1$), we can expect to be able to make very accurate predictions with our model, while those values of s which drive $D(s)$ close to zero can be interpreted as defining an extremely *unpredictable*, or random, region. Values of s for which D lies in between define time scales for which the process can be regarded as *partially predictable*. Figure 1 depicts a nominally shaped $D(s)$ curve and illustrates these three regions of interest which this interpretation quantifies for any observed process $y(t)$ by selecting an appropriate value of $D(s)$ (0.5 in this example) and solving Eq. (7) for the corresponding value of s, denoted by s_{det}, a neighborhood which separates the time axis into the three regions of interest.

   This approach enables us to quantify predictability by specifying those time scales for which the data process can be accurately predicted using a given model. It is this *short-term predictability* which we would like to exploit for dynamical signal separation. Statistical methods are definitely preferred in the unpredictable region, and it should be clear that the degree of predictability is a function of the modeling procedure as well as the data process. It is worth noting that the (normalized) autocorrelation, often employed in traditional signal processing applications and defined by

$$R(s) = \frac{\langle y(t)y(t-s)\rangle}{\langle y(t)^2\rangle},\qquad(8)$$

provides a lower bound to $D(s)$, since it can be obtained from Eq. (7) by setting

$$u_s(t) = y(t-s) = y(t_0),\qquad\text{since }t-s = t_0\,.\qquad(9)$$

Thus the autocorrelation quantifies the degree of predictability we get using the primitive model which employs the current data value $y(t_0)$ as the predicted value to estimate $y(t) = y(t_0 + s)$ s time units in the future. We now turn directly to signal separation.

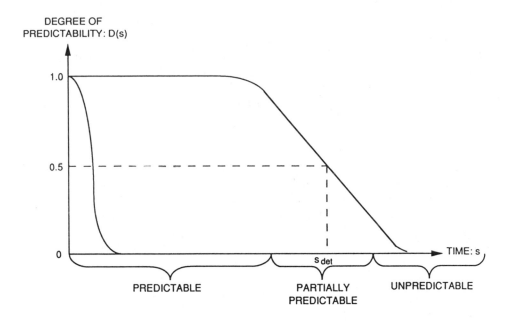

FIGURE 1   Source: Kravtsov, Yu. A. "Randomness, Determinism, Predictability." *Soviet Physics-Uspekhi* **32(5)** (1989): 434–449.

SIGNAL SEPARATION
PROCESSING METHODOLOGY

GENERAL DISCUSSION. In an actual situation the observed process $y(t)$ will typically contain information from a signal of interest which has been contaminated or obscured by one or more competing processes. The objective is to separate the signal of interest from the effects of these contaminating processes. Since these contaminating effects are present in the observed data, they will remain present in the reconstruction process required to implement any of the nonlinear prediction models, and the resulting degree of predictability computations to determine the $D(s)$ curves will incorporate components from both the signal of interest process and the contamination process (where all contaminating effects are lumped in the latter, and we assume no data preprocessing for their removal). The situation of interest in the present context is where these competing processes exhibit sufficiently distinct dynamic behavior to react differently to the prediction modeling procedure.

One of the component processes may be more faithfully replicated in the reconstruction methodology of the modeling procedure than the other. A process displaying such *dominant dynamic* behavior can be predicted over a longer period of time than can the other process. If we could display the $D(s)$ curves separately for these two processes, we would expect them to resemble the two curves in Figure 1 where one process is accurately predicted by the model over much longer time scales than the other. This circumstance is then exploited by using the prediction model to generate an estimated data process $\{u_s(t)\}$ for values of s which place $D(s)$ near unity for the dominant process, while $D(s)$ remains near zero for the less predictable process. The effect of the dominant process can then be eliminated from the data by taking the difference indicated in Eq. (5), since this dominant process has been accurately predicted by our model and its effect is present in the predicted data process. This difference still contains the effect of the less predictable process in the error term.

SPECIFIC RESULTS. In our data samples the air-conditioning noise exhibited the dominant dynamic behavior, so we were able to replicate the effect of this process more accurately than the voice signal in the modeling procedure. This allowed us to recapture the voice signal by taking the difference indicated in Eq. (5). We used our prediction model to generate estimated data processes over a large range of time scales, and all of our listeners identified improved results when hearing the processed signals resulting from Eq. (5) for s values from one time unit up to six time units (where the time units were defined by the sampling rate at which the data sets were originally digitized). The best results occurred when we used s values of two, three, or four time units, since all of the listeners selected one of these values as providing the best intelligibility for both data sets, and they all indicated that the selection among these values in terms of intelligibility was difficult to make.

The method described here is analogous to the traditional technique used to separate narrow band signals of interest from wide band noise using the characteristic distinction between their respective autocorrelation functions. The distinction in dynamical processing, though, is that either process can dominate dynamically, and the method remains applicable as long as the nonlinear prediction model can distinguish between the dynamics.

No preprocessing of the data was attempted, and, indeed, no traditional signal-processing methods were employed in our efforts. At the request of our client, the only processing accomplished was to determine estimated data processes $\{u_s(t)\}$ for various values of s and then require our listeners to evaluate their ability to understand the resulting $\{e(t,s)\}$ error process given by Eq. (5) for a fixed value of s. We feel that more traditional signal-processing methods might have helped exploit the fact that we were able to generate several distinct processes (for different values of s) which were definitely superior in intelligibility to the original data series, since each of these processes was obscured by background noise of an audibly different frequency.

THE NONLINEAR PREDICTION MODEL

The prediction model which was used in our analysis incorporated techniques from a variety of results.[1,2,3,4,5,6,8,11,15,16] The state-space reconstruction for the model incorporated the traditional delay methods, and the embedding dimension was eventually estimated with the Arbarbanel/Kadtke technique.[3] The dynamic equations of motion were estimated with the radial basis functions suggested by Casdagli,[8] using global prediction techniques which were updated periodically to afford better localization properties. The model parameters were optimized for a one-step prediction match to the data. The equations were then iterated to obtain predictions over time scales further in the future.

We will not dwell on the details of the model used to obtain these results because we feel that it represents a fairly crude and unsophisticated first attempt to apply dynamic processing techniques to signal separation. We are already looking at model improvements which should be incorporated into subsequent efforts. We would prefer, for example, to implement localized techniques as well as recursive or adaptive methods in place of the periodic updating procedure which was actually used. It also seems clear that we should look at bringing the degree of predictability analysis directly into determining the model parameters by instituting an optimization process which chooses model parameters which force $D(s)$ to remain close to unity for a specified range of values for s (instead of optimizing only for $s = 1$). This could mean optimizing s-step predictions over a predetermined range of values of s. Additional details and further recommendations are presented in the final report,[7] which also documents several alternative approaches which proved less successful in our signal separation effort.

CONCLUDING COMMENTS

It should be stressed that the nonlinear dynamic techniques which we advocate here have the potential to be most effective in dealing with problems where it is necessary to separate physically generated signal processes from physically generated contamination processes where little is known about the statistical behavior of either process. It is also important to emphasize that the nonlinear prediction model must be able to distinguish the dynamic behavior of the two processes. There appear to be a number of circumstances in acoustics, communications, remote sensing, and other applications where the above conditions prevail, however, and traditional statistical methods requiring linearized dynamic equations perturbed by random processes with known probability distributions are inadequate. These are precisely the circumstances for which these new methods should be exploited. We fully acknowledge that the results presented in this paper represent only an initial step toward developing the required techniques, but we also feel that these results are sufficiently encouraging to argue that further effort toward the full development of dynamic processing methods certainly seems warranted.

REFERENCES

1. Abarbanel, Henry D. I., Reggie Brown, and James B. Kadtke. "Prediction and System Identification in Chaotic Nonlinear Systems: Time Series with Broadband Spectra." Technical Report, University of California, San Diego, 1989.
2. Abarbanel, Henry D. L, Reggie Brown, and James B. Kadtke. "Prediction in Chaotic Nonlinear Systems: Methods for Time Series with Broadband Fourier Spectra." Technical Report, University of California, San Diego, 1989.
3. Abarbanel, Henry D. I., and James B. Kadtke. "Information Theoretic Methods for Determining Minimum Embedding Dimensions for Strange Attractors." Technical Report, University of California, San Diego, 1989.
4. Broomhead, D. S., and Gregory P. King. "Extracting Qualitative Dynamics from Experimental Data." *Physica* **20D** (1986): 217–236.
5. Broomhead, D. S., and Gregory P. King. "On the Qualitative Analysis of Experimental Dynamical Systems." In *Nonlinear Phenomena and Chaos*, edited by S. Sarkar, 113–144. Bristol: Adam Hilger, 1986.
6. Broomhead, D. S., R. Jones, G. P. King, and E. R. Pike. "Singular System Analysis with Application to Dynamical Systems." In *Chaos, Noise, and Fractals*, edited by E. R. Pike and L. A. Lugiato, 15–27. Bristol: Adam Hilger, 1987.

7. Brush, J. S., W. W. Taylor, and R. K. White. "Nonlinear Dynamics for Acoustic Signal Processing." RTA Technical Report 4–90, Rekenthaler Technology Associates, 1990.

8. Casdagli, Martin. "Nonlinear Prediction of Chaotic Time Series." *Physica D* **35** (1989): 335–356.

9. Catlin, Donald E. *Estimation, Control, and the Discrete Kalman Filter.* Applied Mathematical Sciences, Vol. 71. New York: Springer-Verlag, 1989.

10. Chaitin, Gregory J. *Information, Randomness & Incompleteness.* Singapore: World Scientific, 1987.

11. Crutchfield, James P., and Bruce S. McNamara. "Equations of Motion from a Data Series." *Complex Systems* **1** (1987): 417–452.

12. Eckmann, J. P., and D. Ruelle. "Ergodic Theory of Chaos and Strange Attractors." *Rev. Mod. Phys.* **57(3)** (1985): 617–656.

13. Eubank, Stephen, and J. Doyne Farmer. "An Introduction to Chaos and Randomness." In *1989 Lectures in Complex Systems*, edited by Erica Jen, 75–190. Santa Fe Institute Studies in the Sciences of Complexity, Lect. Vol. II. Redwood City CA: Addison-Wesley, 1990.

14. Farmer, J. Doyne, and J. J. Sidorowich. "Predicting Chaotic Time Series." *Phys. Rev. Lett.* **59(8)** (1987): 845–848.

15. Farmer, J. Doyne, and J. J. Sidorowich. "Exploiting Chaos to Predict the Future and Reduce Noise." In *Evolution, Learning and Cognition*, edited by Y. C. Lee. Singapore: World Scientific, 1988.

16. Farmer, J. Doyne, and J. J. Sidorowich. "Predicting Chaotic Dynamics." In *Dynamic Patterns in Complex Systems*, edited by J. A. S. Kelso, A. J. Mandell, and M. F. Shlesinger. Singapore: World Scientific, 1988.

17. Grassberger, P., and I. Procaccia. "Measuring the Strangeness of Strange Attractors." *Physica* **9D** (1983): 189–208.

18. Grassberger, P., and I. Procaccia. "Dimensions and Entropies of Strange Attractors from a Fluctuating Dynamics Approach." *Physica* **13D** (1984): 34–54.

19. Gray, Robert M. *Entropy and Information Theory.* New York: Springer-Verlag, 1990.

20. Kolmogorov, A. N. "Three Approaches to the Quantitative Definition of Information." *Problems in Information Transmission* **1(1)** (1965): 1–7.

21. Kolmogorov, A. N. "Logical Basis for Information Theory and Probability Theory." *IEEE Transactions on Information Theory* **IT-14(5)** (1968): 662–664. An independent translation of this paper appeared as "On the Logical Foundations of Information Theory and Probability Theory. " In *Problems in Information Transmission* **5(3)** (1969): 1–4.

22. Kolmogorov, A. N. "On Logical Foundations of Probability Theory." In *Probability Theory and Mathematical Statistics*, edited by K. Ito and J. V. Prokhorov. Lect. Notes in Mathematics, Vol. 1021. Berlin: Springer-Verlag, 1983.

23. Kolmogorov, A. N., and V. A. Uspenskii. "Algorithms and Randomness." *Theory of Probability and Its Applications* **32(3)** (1987): 389–412.

24. Kostelich, Eric J., and Jarnes A. Yorke. "Noise Reduction in Dynamical Systems." *Phys. Rev. A* **38(3)** (1988): 1649–1652.

25. Kravtsov, Yu. A. "Randomness and Predictability in Dynamic Chaos." In *Nonlinear Waves 2*, edited by A. V. Gaponov-Grekhov, M. I. Rabinovich, and J. K. Engelbrecht, 44–56. Research Reports in Physics. Berlin: Springer-Verlag, 1989.

26. Kravtsov, Yu. A. "Randomness, Determinism, Predictability." *Soviet Physics Uspekhi* **32(5)** (1989): 434–449.

27. Lapedes, Alan, and Robert Farber. "Nonlinear Signal Processing Using Neural Networks: Prediction and System Modelling." Los Alamos National Laboratory Report LA-UR-87-2662, 1987.

28. Mane, R. "On the Dimension of the Compact Invariant Sets of Certain Nonlinear Maps." In *Dynamical Systems and Turbulence*, edited by D. A. Rand and L-S Young, 230–242. Lect. Notes in Mathematics, Vol. 898. Berlin: Springer-Verlag, 1981.

29. Mayer-Kress, G., ed. *Dimensions and Entropies in Chaotic Systems.* Springer Series in Synergetics, Vol. 32. Berlin: Springer-Verlag, 1986.

30. Ornstein, D. S. and B. Weiss. "Statistical Properties of Chaotic Systems." *Bulletin (New Series) of the American Mathematical Society* **24(1)** (1991): 116.

31. Takens, F. "Detecting Strange Attractors in Fluid Turbulence." In *Dynamical Systems and Turbulence*, edited by D. A. Rand and L-S Young, 366–381. Lect. Notes in Mathematics, Vol. 898. Berlin: Springer-Verlag, 1981.

Norman F. Hunter, Jr.
WX-11, Los Alamos National Laboratory, Los Alamos, NM 87545

Application of Nonlinear Time-Series Models to Driven Systems

BACKGROUND

Techniques for the experimental modeling of linear systems have been extensively used for most of this century. Classical experimental techniques contrast with theoretical methods based on first principles in that experimental approaches move from analysis of measured data to a description of the physical process being measured. In contrast, theoretical approaches move from first principles to construct a model of a system. Science and engineering move forward through a unified application of both approaches. Here, our primary interest is focused on the development of system models based on experimental data. Basic theoretical knowledge of the system guides the model-building process.

Experimental system modeling may be performed in either the time or the frequency domains. Through the theorems of Fourier analysis, these approaches are equivalent. For linear systems, characterization through computation of the transfer function, or frequency response function, lends considerable insight into system behavior. Each pole of the transfer function signifies a resonant mode of the system. For a linear system the response is the sum of these orthogonal modes. Reduction of a system response into orthogonal modes offers a fruitful way of viewing the behavior of a complex system as the sum of many simpler features, the orthogonal modes.

The response of a linear system may also be expressed as a time-series model. Such a model expresses the time response of the system as a linear combination of past responses and current and past system inputs. Such models are often referred to as autoregressive moving average (ARMA) models. A close relationship exists between the modes of a linear system, the time-series model which describes evolution of the system responses, and the state-space model of the system. Given any of these model forms, the evolution of the system response can, in principle, be predicted for an indefinite time into the future.

For nonlinear systems, each of the model forms noted in the case of linear systems also exist. Based on the early work of Volterra and Weiner,[24] the frequency response function was generalized into the higher-order frequency response functions. Conceptually, higher-order frequency response functions of any order may exist for a nonlinear system, but most current work centers around computation of the second-order and third-order functions (bispectrum and trispectrum). In the bispectrum, system response at a frequency f depends on the input at two frequencies which sum to f, while in the trispectrum, the system response at frequency f depends on three frequencies which sum to f. Progress has been made in computing the bispectrum and trispectrum using both fourier transform and parametric methods.[22,24]

In our laboratory, we have been engaged in an effort to model nonlinear systems using time-series methods. Our objectives have been, first, to understand how the time-series response of a nonlinear system unfolds as a function of the underlying state variables; second, to model the evolution of the state variables; and finally, to predict nonlinear system responses. We hope to address the relationship between model and system parameters in the near future. Control of nonlinear systems based on experimentally derived parameters is a planned topic for future research.

MODEL FORMULATION

For autonomous nonlinear systems, models may be formulated based on the time series of the system response. Models of this type, which are referred to as response-based models or response-based embeddings, are often formulated in terms of delay coordinates. In a response-based model using delay coordinates, we embed the data in a state space by using delayed response values as coordinates. The future $y(t)$ is a function of the past as:

$$y(t) = f\{y(t - \tau), y(t - 2\tau), y(t - 3\tau), \ldots, y(t - j\tau)\}. \tag{1}$$

Here τ is usually taken as the time delay between samples. Proper selection of the number of lags, j, is made based on the dimension of the system. We refer to this type of embedding as a delay coordinate embedding. Delay coordinate embeddings of dynamic systems were originally formulated by Packard et al.,[19] and put on a firm theoretical basis by Takens.[27] Takens' theorem guarantees that, at least in the noise-free case, a system of state dimension s may be embedded using a maximum of d lags where $d \leq 2s + 1$.

Numerous methods of approximating the function in Eq. (1) exist. Local linear models[9] or local splines[10] form a "local" fit to the function in the neighborhood of the point for which a response prediction is desired. Global functional forms, such as polynomials,[2] rational polynomials, or radial basis functions[3] have also been used to approximate the functional form in Eq. (1). Local functional forms have considerable flexibility in fitting functions of arbitrary complexity. In contrast, global functional forms, by making use of the full data set, have potential for noise reduction for systems in which limited data is available. To our knowledge no functional form has, at this time, been clearly shown to be superior in all cases. Our experience does indicate, however, that local linear functional forms are very useful in approximating Eq. (1) in a number of practical applications.

The model formulated by Eq. (1) is quite useful. For certain specific applications, however, other model forms may be appropriate. There are three specific cases where other model formulations may be appropriate, including:

1. Input–Output systems (driven systems).
2. Systems with multiple responses.
3. Multiple-input, multiple-response systems.

The work described here concentrates on the application of nonlinear time-series analysis methods to input-output, or driven, systems. Conceptually, a typical single input-single output system is shown in Figure 1.

In a manner analogous to that used in Eq. (1) for the autonomous system, we formulate the response of a driven system based on a delay coordinate embedding of response and input values. For causal systems the fundamental equation describing the evolution of the time-series response $y(t)$ (as a function of past responses and inputs $u(t)$) using delay embeddings is:

$$y(t) = f\{y(t-\tau), y(t-2\tau), \ldots, y(t-j\tau), u(t), u(t-\tau), u(t-2\tau), \ldots, u(t-k\tau)\} . \quad (2)$$

The embedding now explicitly includes k delays of the input to the dynamic system and j delays of the response of the system in the model formulation. Casdagli[4] has shown that a straightforward extension of Takens' theorem assures that a diffeomorphism exists between Eq. (2) and the state evolution of a driven system of state dimension s using a maximum $2s + 1$ lags of both the input and response. The model formulation (2) allows application of time-series models to driven systems with complex drives when the drive signal is known. For such systems the driven model formulation in Eq. (2) often leads to a more universally applicable formulation of the state evolution than the autonomous model form of Eq. (1).

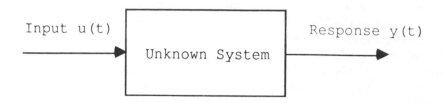

FIGURE 1   Conceptual model of a single input–single output system.

The input-output or driven formulation of Eq. (2) has both advantages and disadvantages in comparison to the autonomous form of Eq. (1). Knowledge of the input provides additional information which may simplify the model-building process. However, complex input forms (inputs with large dimensions) such as random noise may be encountered. These may range from inputs of low dimension, such as a simple constant-level pressure in a flow experiment or a sinusoidal voltage applied to a circuit, to random inputs of conceptually infinite dimension. On the one hand, knowledge of the input gives us more information to use in our model, simplifying the prediction process. On the other hand, complex inputs may cause the system to explore broader regions of the state space, making prediction more difficult. With any nonlinear system, it is imperative that extrapolation to unexplored regions of the state space be limited in any attempts at prediction. If knowledge of the system is limited to experimental measurements, no intelligent extrapolation to regions of the state space that have not been explored can, in general, be made. Clearly, linking of experimental measurement to analytical modeling is necessary for extrapolation to have any validity. If, however, we wish to explore system behavior over a broad range of states, with control of the input we can excite the system into the regions of interest, successively exploring the many complex attractive regions typically exhibited by nonlinear systems. As in the case of autonomous systems, we rely on the tendency of very complex dynamic systems with a potentially infinite number of degrees of freedom to collapse onto an attractor of relatively low dimension.

The functional forms used in the input-output models of Eq. (2) are essentially the same as those used in the autonomous case of Eq. (1).[2,9,11,14] Our best results in the input-output case have been obtained with local linear models. We plan to investigate the local spline approach of Friedman in the near future. In our experience, global polynomial model forms, while they lead directly to means of computing higher-order frequency response functions,[2] have limited utility for iterated time-series predictions as they tend to become unstable under repeated iteration.

Optimal embedding techniques for models of the form of Eqs. (1) and (2) have been investigated extensively.[2,5] More compact formulations of the system state than a strict embedding based on input and response delays may be realized through the use of singular value decomposition on local formulations of Eq. (2) or through the use of Canonical Variate Analysis.[16] Currently our approaches concentrate in three areas: (1) the direct application of Eq. (2) using delay embeddings; (2) use of singular value decomposition (principal component analysis) of local delay embeddings; and (3) canonical variate analysis of local delay embeddings. At this time canonical variate analysis based on local embeddings is giving an optimal estimate of local state dimension.

Possibilities other than the delay embeddings used in the formulation of Eq. (2) exist. For some systems, it is probable that embeddings based on approximations to the integral of the response may be more appropriate. In fact, time-series modeling based directly on the differential equations governing the system can lead to considerable insight. The nonlinear nature of the forces in some classes of vibrating systems has been revealed using direct modeling of the differential equations governing the system.[7,8]

Special consideration must be given to the potential predictability of the time-series response $y(t)$ in Eq. (2), or, in fact, in any model formulation which attempts to predict the time-series response of nonlinear, driven systems. It is well known that nonlinear autonomous systems (as described by Eq. (1)) potentially possess chaotic behavior. Driven systems (as described in Eq. (2)) also may behave in a chaotic fashion.[4] Consequently the specific time-series response of some nonlinear, driven, chaotic systems is unpredictable as errors grow exponentially. Both Lyapunov exponent calculations and time-series predictions for the driven Duffing equation, which we discuss later, clearly indicate chaotic behavior, especially in systems where the nonlinear terms are dominant.

APPLICATIONS

Here, we concentrate on the application of Eq. (2) to experimental data for driven systems. Numerous potential applications exist. The input-response formulation is appropriate when describing numerous physical, chemical, environmental, biological, climatic, and economic systems. Linear system models of driven or autonomous forms see extensive application in control theory, mechanical engineering (mechanical oscillators), electrical engineering (circuit models, finite impulse response and infinite impulse response filters, control system models), and statistics (economic time-series modeling). Application of time-series models to nonlinear systems is expected to provide greater insight into the behavior of nonlinear systems in all of the above fields.

To illustrate applications of nonlinear time-series modeling and to emphasize its advantages over linear modeling techniques, several applications are illustrated. These, of course, represent only a small set of the possible range of applications. We apply nonlinear time-series modeling of driven systems to experimental structural dynamics (a driven Duffing oscillator and a beam moving in two potential wells), and to modeling of climatic time-series data (prediction of ice ages). Additional examples of application to a heat exchanger system,[2] ship rolling,[15] a Van Der Pol oscillator,[15] and to nonlinear prediction of speech signals[28] have been reported in the literature.

In each of our examples, the application of Eq. (2) to the problem of system modeling and prediction follows a fixed series of steps, as follows:

1. If the option to use a controlled excitation is available, the system is excited with a representative input and a set of sampled input values $u(t)$ and output values $y(t)$ are obtained. The input should be representative of those inputs for which response prediction and modeling are desired. In particular, the input should cause the system to explore regions of interest in the state space. If, as is often the case, control over the system excitation is not achievable, a representative training set of measured input and output values is chosen from measured data. The data is then embedded using delay coordinates. For direct application of delay coordinates, we used a sampling interval which met, but did not substantially exceed, the Nyquist criterion. When the initial delay embedding is followed by the use of singular value decomposition, the use of sampling rates which exceed the Nyquist criterion by an order of magnitude is desirable. Selection of the singular values and eigenvectors in the system modeling process effectively re-embeds the data.

2. Nearest neighbors are selected from the training data set. These neighbors are embedded in a state space either directly using delays or through the use of local singular value decomposition or local canonical variate analysis. A linear function is chosen to relate the past values of input and response to the future response in the neighborhood chosen. Each prediction involves selection of an appropriate neighborhood from the training data. Euclidian distance is used as a metric in the selection of nearest neighbors.

3. Equation 2 is used to predict the system response a single step ahead.

4. The output of Eq. (2), the "one-step-ahead" response, is used as a lagged response $y(t - \tau)$ term and this value, together with the known input, forms a new, estimated, vector of the past. An appropriate neighborhood is chosen from the training data set and steps 2 and 3 above are repeated to obtain the response prediction two steps ahead.

5. Continued iteration of Eq. (2) through steps 2, 3, and 4 above is used to predict the system response many steps in the future.

Most of the examples which we cite use the iterated prediction as a measure of model quality. Single-step prediction is often of limited utility as successive increases in model complexity reduce the single-step error as more parameters are added to the model. In the examples shown for driven systems, we have emphasized the use of iterated prediction error as a model validation criterion. Relatively low, but not minimal, single-step prediction error is achieved in each application that we illustrate. We have found, especially in the case of noisy systems, that the model which produces minimal single-step prediction error does not lead to minimal iterated prediction error.

To obtain a valid model, it is essential that the training data and prediction data form disjoint sets. In the examples we illustrate, the prediction data is not used as part of the training dataset, so in every case, we are performing out-of-sample iterated forecasts.

APPLICATIONS TO EXPERIMENTAL STRUCTURAL DYNAMICS

In experimental structural dynamics, the input-output system of Figure 1 and Eq. (2) is applied to the specific case shown in Figure 2, where a test structure is excited by an input force or acceleration from a vibration machine. The input time series may be a transient (such as a simulated earthquake pulse), a sine wave (whose frequency is perhaps swept slowly with time), or random noise with a specified power spectrum. In Figure 2, the structure is fixed to the vibration exciter and the input waveform is generally under the control of the experimenter. In a more general case, the structure may be in service, as in the case of an item undergoing truck transport, a building subjected to the input acceleration of an earthquake, or

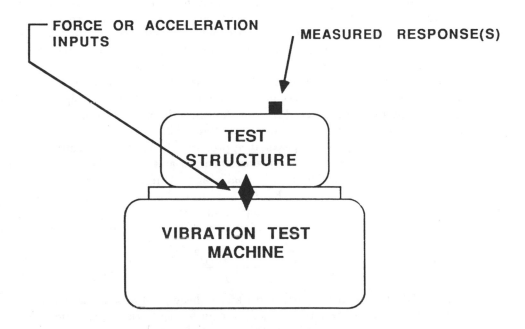

FIGURE 2 Experimental setup for structural characterization.

a satellite undergoing launch accelerations. In each case, the measured-input time series and the measured-response time series would be used to model the system.

DRIVEN DUFFING OSCILLATOR

Consider the driven oscillator described by Eq. (3).

$$y'' + 2\zeta\omega_n(y' - y_0') + \omega_n^2(y - y_0) + \alpha\omega_n^2(y - y_0)^2 + \beta\omega_n^2(y - y_0)|y - y_0| = 0. \quad (3)$$

This equation describes a base excited oscillator with linear and nonlinear restoring forces. The symmetric and antisymmetric nonlinear restoring forces, represented by the α and β terms in Eq. (3), provide Duffing-like behavior, though a cubic term is not strictly present. y'' is the response acceleration and y_0'' is the acceleration of the base of the system. This is a typical, though simple, example of a driven structure. The degree of nonlinearity is controlled by the coefficients α and β, and by the input waveform y_0''. The behavior exhibited by a driven Duffing system is characteristic of many of the structural nonlinearities encountered in practice.

For our first example we consider a moderately nonlinear case of the Duffing oscillator where $\alpha = 1500, \beta = 1500, \zeta = 0.04$, and $\omega_n = 2\pi(11.5)$. To provide a known and controllable degree of nonlinearity while simultaneously allowing for the necessity of making experimental measurements, the equation is simulated on an analog computer. The input y_0'' is band-limited random noise generated by a Genrad 2514 vibration control system. The noise input level is 0.300 volts rms. The band-limited input spectrum is centered around 10 Hz and contains significant energy in the 5 Hz to 15 Hz range. Figure 3 shows the input and response power spectra for this moderately nonlinear driven Duffing oscillator.

The response power spectrum peaks at approximately 16 Hz. Substantial second and third harmonic responses are present near 30 and 45 Hz. This harmonic response is characteristic of a significantly nonlinear system. In fact, the nonlinear response terms (the α and β terms in Eq. (3) have a combined rms value, averaged over long time periods, of approximately 40 percent of the linear ω_n^2 term in Eq. (3).

To evaluate model prediction accuracy, an average iterated prediction error, e_r, is defined.

$$e_r = \frac{\sum_{i=1}^{n}(y(i) - y_{e(i)})^2}{\sum_{i=1}^{n} y(i)^2}. \quad (4)$$

Here y_e is the iterated prediction, iterated i steps ahead, and y is the actual measured response value. The error is averaged over an n-step iterated prediction and normalized by the mean square value of the measured response.

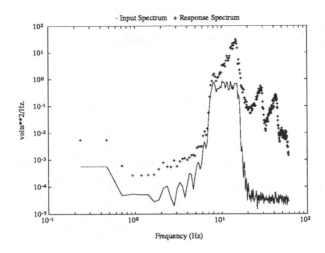

FIGURE 3 Input and response power spectra of a driven, moderately nonlinear, duffing oscillator.

Typical measured and predicted response time histories for the moderately nonlinear Duffing oscillator are shown in Figure 4.

Starting at the beginning of the interval shown linear (Figure 4(a)) and nonlinear (Figure 4(b)), iterated predictions are made for a 300-point interval. These 300 points represent a time interval of 2.5 seconds or about 15 cycles of response at 16 Hz. Delay coordinate embeddings are used in both the linear and nonlinear models. The average iterated prediction error e_r over the 300-point interval for the linear model is 0.39, for the nonlinear model, 0.18. The graphs in Figures 4(a) and 4(b) clearly show the superiority of the nonlinear model predictions. Table 1 gives some comparative iterated prediction errors for linear and nonlinear models for various cases of the Duffing oscillator.

In our second example we consider the driven Duffing oscillator of Eq. (3) with increased nonlinearity. The significance of the nonlinear terms is increased by setting $\alpha = 3500$ and $\beta = 3500$. The random noise is increased to an rms drive level of 0.405 volts. As in the previous example, the system is driven by Gaussian, band-limited random noise. The input and response power spectra for the system are shown in Figure 5.

Substantial harmonic response again indicates that this is a nonlinear system. The nonlinear terms, whose coefficients are α and β in Eq. (3), fluctuate over a very wide range as is typical of strongly nonlinear systems. The long-term average of the rms value of the nonlinear terms in Eq. (3) substantially exceeds that of the linear term (the ω_n^2 term in Eq. (3)).

Figure 6 compares the iterated predictions and measured responses for the case of the strongly nonlinear driven Duffing oscillator. The superiority of the local linear model is even more evident than in the case of the moderately nonlinear Duffing oscillator. For the strongly nonlinear system, the average error for the linear predictor is 0.62, and for the local linear model, 0.32.

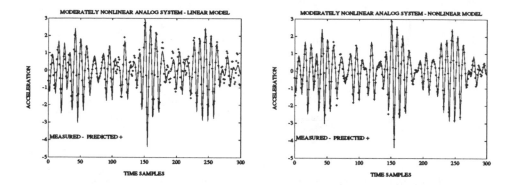

FIGURE 4 Measured and predicted response time histories for a moderately nonlinear duffing oscillator. (a) Linear model. (b) Nonlinear model.

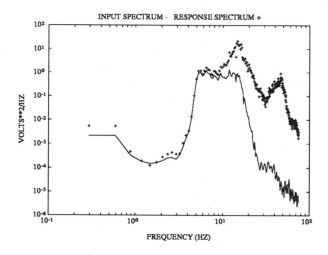

FIGURE 5   Input and response power spectra of a driven, strongly nonlinear, duffing oscillator.

TABLE 1 Comparative iterated prediction errors for the driven Duffing's equation

Model	Embedding Lags	Iterated Prediction Range	Iterated Prediction Error (Typical)
Moderately Nonlinear Duffing's Oscillator			
Linear	Delays 4 input, 4 response	300 points	0.39
Nonlinear	Delays 4 input, 4 response	300 points	0.18
Linear	SVD 4 input, 4 response	200 points[1]	0.56
Nonlinear	SVD 4 input, 4 response	200 points[1]	0.23
Strongly nonlinear Duffing's Oscillator			
Linear	Delays 4 input, 4 response	300 points	0.62
Nonlinear	Delays 4 input, 4 response	300 points	0.32
Linear	SVD 5 input, 5 response	200 points[1]	0.62
Nonlinear	SVD 6 input, 6 response	200 points[1]	0.34

[1] averaged over 125 different initial conditions, 1000 training points.

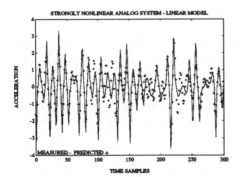
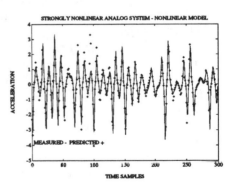

FIGURE 6 Measured and predicted response time histories for a strongly nonlinear duffing oscillator. (A) Linear model. (b) Nonlinear model.

Comparative iterated prediction errors are shown in Table 1 for the moderately and strongly nonlinear cases of the Duffing oscillator. Delay embeddings and delay embeddings, followed by local singular value decomposition, are compared for linear and local linear models. The advantage of the nonlinear (local linear) modeling technique is evident in each case. Typical single-step prediction errors also clearly show the advantage of the local linear model. For example, for the strongly nonlinear Duffing oscillator average, one-step prediction error for the linear model is 0.023, for the nonlinear model, 0.014. For the case of the Duffing oscillator, the local singular value decomposition does not provide a significant advantage, perhaps because the noise levels are relatively low.

Application of nonlinear models to the time series for this simulated Duffing oscillator is clearly advantageous. To what degree errors in the iterated predictions are due to chaotic behavior exhibited by the system, as opposed to errors in the model, remains to be addressed. Present indications are that the strongly nonlinear Duffing oscillator is exhibiting marginally chaotic behavior.

Prediction of the time-series responses of the Duffing oscillator clearly indicates the validity of the nonlinear (local linear) modeling techniques used. We are currently analyzing the local linear model and comparing the model form with the known nonlinearity present in the Duffing system.

BEAM MOVING IN A DOUBLE POTENTIAL WELL

The driven Duffing oscillator is one typical nonlinear system encountered in structural dynamics. Characteristically, the Duffing system exhibits a continuum of nonlinear behavior as the magnitudes of the coefficients of the nonlinear terms are varied. The nonlinearity is strongly dependent on the magnitude of the excitation. For low-drive levels, Eq. (3) behaves in a nearly linear fashion, and for increased drive levels, behavior is chaotic.

As our next illustration of the application of nonlinear time-series analysis to driven systems, we choose a system which is clearly exhibiting chaotic behavior, the "chaos beam" described by Moon.[17] A sketch of the experimental setup for this system is shown in Figure 7.

In this experiment, the upper horizontal cantilever beam is excited into nearly sinusoidal motion by a small motor driving an eccentric weight. The oscillations of this horizontal beam excite the lower vertical steel vane, which vibrates in the fields of the two magnets as illustrated. Stable oscillations may occur in the potential well of either magnet. At certain excitation levels, erratic transitions in the position of the vane occur as the vane transits between the two potential wells.

System input is the acceleration at the end of the upper horizontal cantilever beam near the motor. System response is the strain at the root of the vertical beam. Equation (5) gives an approximate description of the motion of the vane.

$$y'' + 2\zeta\omega_n(y' - y_0') - \omega_n^2(y - y_0) + \gamma\omega_n^2(y - y_0)^3 = 0 . \tag{5}$$

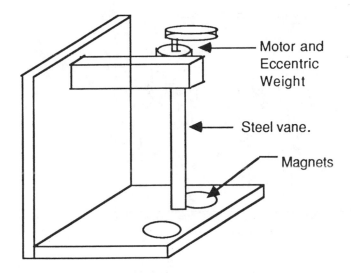

FIGURE 7 Chaos beam experimental setup.

Here y_0 is the motion of the base of the horizontal cantilever beam, typically measured as acceleration y_0'', and y is the strain at the root of the vertical vane. Equation 5 is a modified form of Duffings equation, where the linear stiffness term is negative and the cubic term is positive. Generically, Eq. (5) describes systems where oscillation about either of two stable states may occur. In addition to the case of the chaos beam illustrated here, Eq. (5) describes motion of a buckled beam. While buckling is not as typical of nonlinear structural behavior as the earlier example of Duffing's equation, it is clearly applicable to numerous structural systems.

Rather than simulate the behavior of the chaos beam using Eq. (5), we chose to construct an experimental model of the beam shown in Figure 7. Vibration excitation was provided by a small DC motor, input acceleration was measured using an Endevco 2250 accelerometer, and strain at the root of the steel vane was measured using a strain gage. With the proper excitation frequency and amplitude, the system behaves chaotically. In our experiment, 8000 samples of the input (top acceleration) and output (vertical beam strain) were digitized. A local linear model was used with 4000 training points, 4 input lags, 4 response lags, and delay coordinate embedding. Single-step prediction errors less than 2% were readily achieved. Iterated prediction error was used as a criterion for model validity. Iterated predictions were made over a 1000-point range of time samples. Measured and predicted strain time series are shown in Figure 8.

Due to the chaotic behavior of the system, the predicted and measured strain response time series rapidly diverge. The iterated prediction error is essentially unity. Clearly for this system, or for any strongly chaotic system, long-term iterated forecast error provides a poor measure of model validity. Characterization of an invariant measure of the system is required. In this case we chose to compare

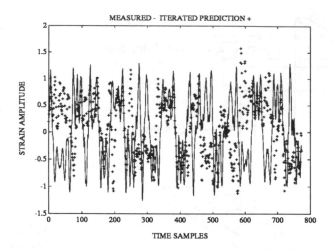

FIGURE 8 Measured response and iterated model predictions for the chaos beam.

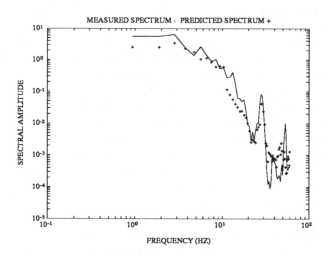

FIGURE 9 Power spectra of the measured and iterated model predictions for the chaos beam.

the power spectra of the measured and the predicted strains in Figure 8. Figure 9 shows the power spectra of the measured and predicted response strain signals.

The similarity of the power spectra indicates that, as might be suspected from an examination of the time series in Figure 8, the model is capturing essential features of the response of this strongly chaotic system. Alternately, iterated prediction error over a much shorter time interval could be used as a measure of model validity for strongly chaotic systems.

Application of nonlinear models to structural systems is in the early stages of development. Future tasks include application of nonlinear time-series modeling techniques to a wider range of structural systems, including multiple-degree-of-freedom systems. Extension of these techniques to multiple-input, multiple-response driven systems is anticipated. Examination of the model form and its relation to types and degrees of structural nonlinearity will be a major focus of future effort.

APPLICATION TO CLIMATIC DATA TIME SERIES

As a final example of nonlinear time-series modeling, we consider an application involving long-term climatic data. Modeling measured climatic data, which is based on geologic records, represents a considerable increase in the difficulty and complexity of the modeling process since climatic data series are often derived from geologic data whose time scale is often poorly defined and whose time series are quite noisy.

BACKGROUND

Numerous complex oscillations exist in the record of global climate. In recent geologic time, the dominant long-term variation is a complex cyclical variation of global ice volume. While the nature of the global ice volume variations is quite complex, cyclical changes in global ice volume show a dominant period of about 100,000 years. Glacial periods of relatively high ice volume are separated by periods of relatively low global ice volume, termed interglacials. Additional cyclic variations with periods of about 2500, 80, 100, 11, and 350 years have been noted in the climatic record.[6] Here we focus on modeling the long-term variations in global climate. Many recent studies of long-term climatic variations have focused on the approximately 100,000 year ice-age cycle with emphasis on the Milkanovich theory of the ice ages.

In the Milkanovich theory the driving force for the ice ages is the variation in solar insolation ($cal/cm^2/day$ of solar energy incident on the atmosphere) based on long-term changes in the earth's orbital parameters. The total amount of sunlight received at any point on the earth's surface during any given month varies cyclically depending on the tilt of the earth's axis (obliquity), whether summer in the northern hemisphere falls at the point when the north pole of the earth is tilted toward or away from the sun (precession), and the degree of ellipticity of the orbit (eccentricity). Obliquity, precession, and eccentricity each have a characteristic period. The dominant period for obliquity is 41,000 years, precession has a period of 20,000 years, and eccentricity has dominant periods of 100,000 years and 400,000 years. The existence of a relationship between orbital parameters and long-term global climate was postulated by Milkanovich and more recently shown to hold by Imbrie and Shackleton.[13,14] A central feature of the theory is the hypothesis that summer insolation in the northern hemisphere is a dominant factor in the evolution of global climate. Winter temperatures at high northern latitudes are low enough that snow always accumulates. Minimal summer temperatures and corresponding minimal summer insolation are necessary for limiting summer melting, which enhances the growth of ice sheets. The dominant effect of the northern hemisphere climate on the ice ages is attributed to the current geometry of the continents and to the effect of this geometry on global ocean circulation. Substantially different continental geometry could alter the current dominant effect of the Northern hemisphere on global ice volume changes, so our model is limited to relatively recent geologic periods when the continental geometry was similar to the present.

Global ice volume is typically inferred from measurements of the oxygen isotope ratios for O^{16}–O^{18} in the shells of bottom dwelling Formanifera. During an ice age, the ocean is relatively depleted of the lighter O^{16} isotope which, having evaporated preferentially from the oceans, is precipitated on land and locked up in global ice sheets. Oxygen isotope ratios also vary with water temperature. However, based on an assumption of relatively constant bottom water temperature, the $O^{16} - O^{18}$ isotope ratio for benthonic (bottom dwelling) organisms is primarily dependent on variations in global ice volume.[25] Measurements of oxygen isotope ratios for bottom dwelling organisms have been made for a number of ocean cores taken from both the Atlantic and Pacific oceans. For the purposes of this analysis, we will concentrate on data obtained from two Equatorial Pacific cores, V28-238 and V28-239. Emphasis is placed on data from these cores because they provide ice volume data with reasonable fidelity over the entire span of the Pliestocene. Most analysis of global ice volume have used data from core V28-238. In all of these ocean cores, the first assumption is that sedimentation rate is approximately constant, so depth in the core is representative of time in the past. A number of time-scale corrections, used to "tune up" the time scale to compensate for a variable sedimentation rate, have been proposed. We will discuss these methods as we proceed with analysis of the oxygen isotope data. Our two basic data sources are as follows:

1. Orbital fluctuations based on data from Berger,[1] giving eccentricity, precession, and obliquity for the past million years. This orbital data also extends 100,000 years into the future. Knowledge of eccentricity, precession, and obliquity allows calculation of the incident solar radiation (insolation) at any latitude for past or future time periods. For the purposes of this analysis, we use July insolation at 60 degrees north latitude as our model input.
2. Global ice volume based on oxygen isotope measurements for benthonic organisms from equatorial pacific cores V28-238 and V29-239.

The quality of the orbital data is quite high as it is based on celestial mechanics. In contrast, the quality of the ice volume data is lower, as the oxygen isotope ratio is very likely contaminated by effects other than ice volume and, further, the time scale is based on sedimentation rate which leads to uncertainties in the precise timing of events.

The complex cyclical variation of July insolation at 60 degrees north latitude is shown in Figure 10. Global ice volume (derived from oxygen isotope analysis of core V28-239) is shown in Figure 11. Spectral analysis of the insolation and ice volume data reveals that common frequency components with periods of approximately 21,000 years and 41,000 years occur in both time series.[14] The dominant frequency present in the global ice volume time series has a period of about 100,000 years. The most significant problem in the construction of a model relating the ice volume time series and the solar insolation time series is the origin of the major 100,000-year variation in global ice volume.

For at least the most recent ice ages, regions of high ice volume in the isotope record correspond to regions where the variance of the solar insolation is low. These regions of low variance in the solar insolation result from the modulating effect of the eccentricity on precession. Variations in the earth's orbital eccentricity have a dominant period of approximately 100,000 years, though in fact the eccentricity variation is complex with sinusoidal variations of other (predominantly longer) periods present. Ice volume and eccentricity are compared in Figure 12.

A relationship appears to exist between eccentricity and ice volume, particularly for the more recent ice ages. From Figure 12, the ice volume is typically high at times when eccentricity is low. This relationship is in accord with that observed in the solar insolation signal, since, when eccentricity is low, its modulating effect on the precession causes the variance of the solar insolation signal to be minimized. From a linear dynamic modeling standpoint, difficulties are encountered in producing the 100,000-year variation in the ice volume, since the 100,000-year variation in eccentricity is not explicitly present as a dominant frequency component in the solar insolation. The modulating effect of eccentricity on precession causes the approximately 20,000-year variation in precession to split into sinusoids with periods of 19,000 and 21,000 years. In a linear model translation of energy between frequencies cannot occur; hence, the 100,000-year variation of precession cannot be coupled to the approximately 100,000-year variation in global ice volume indicated by the ocean core data. In contrast, since variations with periods of 41,000, 19,000, and 21,000 years jointly exist in the ice volume and insolation data, it is possible that a linear model could explain variations with these periods

Numerous climatic models have been proposed to explain the complex variation in ice volume. These include models by Imbrie,[13,14] Suarez and Held,[26] and Pollard.[20,21] In all of these models the driving force behind the global ice volume changes is the variation in solar insolation over geologic time based on celestial mechanics. All of the models also use physical principles underlying ice sheet growth to develop hypothetical models for ice sheet dynamics. Typically these models postulate slow growth of ice sheets and rapid decay, and use nonlinear differential

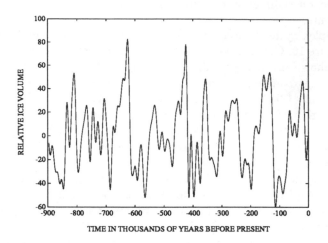

FIGURE 10   July solar insolation at 60 degrees north latitude.

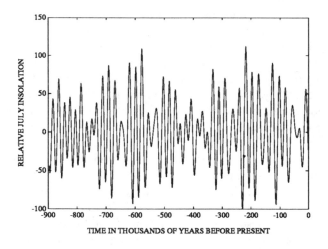

FIGURE 11   Relative global ice volume based on V28-239.

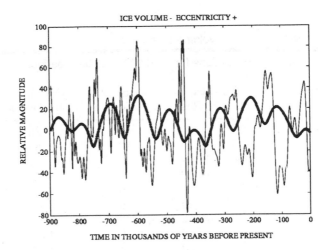

FIGURE 12   Relative global ice volume compared to eccentricity.

equations to model the ice volume variations. These models have met with varying degrees of success. In particular a model by Imbrie, based on a nonlinear first-order differential equation,[14] using solar insolation as a driving term, gives a reasonable simulation of the two most recent ice ages.

Here we focus on the relationship between insolation at high northern latitudes and the global ice volume. Our goal is to use solar insolation and ice volume data to show that the long-term evolution of climate is a nonlinear deterministic system, and to utilize past data for prediction of future global ice volume changes. In contrast to previous models, we do not assume an *a priori* model form but use a general, nonlinear, delay coordinate model which, when trained on the measured data, is used to define the relationships between past ice volume, solar insolation, and future ice volume.

CLIMATIC MODEL FORMULATION

A major problem which must be addressed prior to application of time-series modeling is the assignment of a time scale to the core data. The cyclic variations in global ice volume extend back approximately 800,000 years. Prior to this time the global ice volume record does not support the existence of relatively regular ice volume changes characteristic of the ice ages.

TABLE 2 Prediction error for ice volume time series

| | | | | | | Iterated |
| Core | Time-Scale Correction | Delays | | Singular Values | One-Step Error (Overall) | Error (Recent) |
		Input	Resp.			
		Iterated predictions for a 90,000-year time period.				
239	none	20	20	8	0.86	0.83
239	eccentricity	20	20	12	0.85	0.89
239	eccentricity	14	4	6	0.76	0.70
239	Hays	20	20	8	0.81	0.71
239	Hays	14	2	6	0.84	0.66
238	none	20	20	6	0.82	0.73
238	none	4	2	6	0.79	0.64
238	eccentricity	18	4	6	0.84	0.62
238	Hays	20	20	8	0.66	0.66
238	Hays	14	1	6	0.81	0.82

Correction of the ice volume time scale is made in two ways. An initial correlation of depth with time is made based on the depth of occurrence of magnetic reversals, whose time of occurrence is approximately known. These magnetic reversals are observed in core material and provide time markers. Shackleton[25] has developed a time scale based on magnetic reversal data. The Shackleton time scale is used in Figures 11 and 12. Further refinements of the time scale are based on the "phase locking" of common frequency components in the insolation and ice volume data. Phase locking is based on the assumption that phase relationships observed in the past 200,000 years are preserved in data from the more distant past. Hays[11] has developed a time scale based on phase locking the 21,000-year (precession) and 41,000-year (obliquity) signals. We chose to phase lock the eccentricity and the corresponding component (100,000-year period) of ice volume. Time-scale corrections brought about by phase locking are generally relatively small in magnitude (less than 10%) of a given time value, but may have significant effects on the model. Fitting models to the ice volume and solar insolation data requires consideration of a number of model parameters. These parameters and the model fitting results are summarized in Table 2.

In general, several types of models may be fitted to the solar insolation, ice volume time-series data. Linear or nonlinear time-series models may be applied. For either linear or nonlinear models, reconstruction of the ice volume time series may be attempted based on driven systems models (where the solar insolation is used as the driving time series) or autonomous system models (where the future ice volume is based only on past ice volume values). In general, we achieved poor results using autonomous system models. Iterated prediction errors were no better than a random guess. Linear driven system models performed better than autonomous models, giving a reasonable fit for data from core V28-238 where the Hays time-scale correction was used. No linear model gave a reasonable fit to the data where the input was emphasized in the embedding process. In contrast to the Duffing oscillator results, the use of local singular value decomposition consistently gave results superior to delay coordinate embeddings alone. This may be due to noise present on the ice volume data.

Nonlinear model results are summarized in Table 2. Iterated prediction errors are shown for cores V28-238 and V28-239 for no time-scale corrections, eccentricity corrected data, and Hays corrected data. Iterated prediction error was calculated over a time period of 90,000 years in all cases. Iterated predictions for recent data were made using training data from 250,000 years before the present to 600,000 years before the present. Average iterated predictions were made using ten separate ranges of data spaced throughout the data sets for prediction and data outside the iterated prediction range for training. A clear resemblance between the measured ice volume and the iterated prediction is generally noted for iterated prediction errors less than 0.85. By this criterion reasonable nonlinear model fits were obtained for

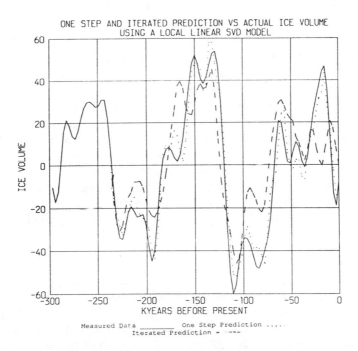

FIGURE 13 Iterated and one-step predictions of global ice volume compared to measured data for the most recent glacial periods.

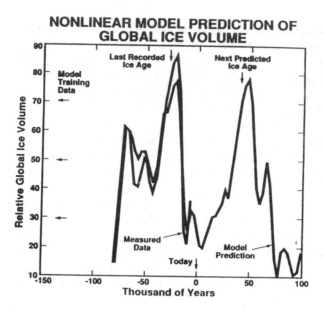

FIGURE 14   Prediction of the next ice age using a nonlinear model.

both cores V28-238 and V28-239. The best fits were obtained using models where both input (solar insolation) and output (past ice volume) values were used in the embedding process. Reasonably good models were also obtained in several cases with models which emphasized the input solar insolation in the model process by using 14 lagged input (solar insolation) values and 1 or 2 lagged response (ice volume) values. This is especially significant as it implies that a nonlinear relationship is present between solar insolation and ice volume, even in the case of data with minimal time-scale corrections. In general, the more recent ice volume data is more predictable than data more than 250,000 years before the present. The shorter the prediction range and the more data that is used for training, the better the predictions. Iterated predictions, one-step predictions, and measured ice volume data are compared in Figure 13 for a model based on 20 input lags, 20 response lags, and eccentricity corrected data. This type of forecast is typical for iterated prediction errors in the 0.60 to 0.75 range.

Overall, nonlinear input-output models gave superior results in comparison to nonlinear autonomous models or linear models of any form.

Since solar insolation data are available for the future, nonlinear forecasting techniques can be applied to predict the next ice age. For this prediction eccentricity corrected data is used from core V28-239. Except for the last 60,000 years, the entire past insolation and ice volume data sets were used for training. Future ice volume predictions for the next 100,000 years using a nonlinear model, local singular value decomposition, and 20 input and 20 response lags are shown in Figure 14. For the last 50,000 years the measured ice volume data and the iterated model prediction are compared. Then the model is used to predict a future peak in global ice volume occurring in approximately 50,000 years, a future ice age.

CONCLUSION

The use of nonlinear models for driven systems has been illustrated using examples from structural dynamics and climatic time series. These examples represent only a small range of the potential applications of nonlinear time-series analysis. Further applications may be expected wherever nonlinear system modeling is advantageous. These methods are especially valuable for detecting and modeling nonlinear relationships in experimental data. Current techniques using delay coordinate embeddings of both system input and system response have given reasonable results for nonlinear single-degree-of-freedom systems. Local singular value decomposition techniques simplify the selection of the data sample interval and are useful for noisy data, as illustrated in the example of the ice volume time series.

Either one-step or iterated prediction error may be used as a criterion for determining model validity. In our examples we emphasized the use of iterated prediction error over a moderate range of iterated predictions. The range of iterated predictions must be kept reasonably short as the positive Lyapunov exponents of chaotic systems may cause divergence of the predicted and measured time series, even with a valid model. In these cases the response power spectra may be used to validate the nonlinear model.

A number of future enhancements to nonlinear model construction are expected. In particular, the direct use of local singular value decomposition on the delay vectors used in the embedding process corresponds to finding the principal components of the past as used to predict a future value of the time series. This approach, while optimal for autonomous systems, is probably not optimal for driven systems. For a driven system, optimal selection of the system state involves simultaneous diagonalization of correlation matrices involving the past, the future, and the cross-correlation of the past and future, a procedure known as canonical variate analysis.[16] Use of canonical variate analysis is expected to provide improved predictions and should lead to improved model stability. The work we have done has emphasized the use of local linear models. Use of local splines as suggested by Friedman[10] should also lead to improved modeling.

Careful analysis of the variation in model coefficients is very desirable as a means of detecting both the degree and type of nonlinearity present in experimental data.

Concurrent with the improvements in technique, we plan to apply nonlinear modeling to a greater number and variety of nonlinear systems including multiple-input, multiple-response structural systems, geologic systems, and climatic systems.

REFERENCES

1. Berger, A. L. "Obliquity and Precession for the Last 5,000,000 Years." *Astronomy and Astrophysics* **51(1)** (1976): 127–135.
2. Billings, S. A., K. M. Tsang, and G. R. Tomlinson. "Application of the Narmax Method to Nonlinear Frequency Response Estimation." Proceedings of the 6th International IMAC Conference, 1987, 1433–1438.
3. Casdagli, Martin. "Nonlinear Prediction of Chaotic Time Series." *Physica D* **35** (1989).
4. Casdagi, Martin. "A Dynamical System Approach to Modelling Driven Systems." This volume.
5. Casdagli, Martin, John Gibson, Stephen Eubank, and J. Doyne Farmer. "State Space Reconstruction in the Presence of Noise." *Physica D* **51** (1991): 52–98.
6. Crowley, "The Geologic Record of Climatic Change." *Rev. Geophys. and Space Phys.* **21(4)** (1983): 828–877.
7. Endebrock, Elton. "Characteristics of Dynamic Systems From Test Data." Proceedings of the 61st Shock and Vibration Symposium, Pasadena, CA, October 16–18, 1990.
8. Endebrock, Elton G. "Restoring Force Method and Response Estimation." 10th Annual Modal Analysis Conference Proceedings, February, 1990.
9. Farmer, J. Doyne, and John J. Sidorowich. "Exploiting Chaos to Predict the Future and Reduce Noise." In *Evolution, Games, and Cognition*, edited by Y. C. Lee, 277. Singapore: World Scientific, 1988.
10. Friedman, Jerome H. "Multivariate Adaptive Regression Splines." Technical Report No. 102, Department of Statistics, Stanford, University, 1989.
11. Hays, James D., and Joseph J. Morley. "Towards a High-Resolution, Global, Deep Sea Chronology for the Last 750,000 Years." *Earth and Planetary Science Letters* **53** (1981): 279–295.
12. Hunter, Norman F. "Analysis of Nonlinear Systems Using Delay Coordinate Models." 10th Annual Modal Analysis Conference Proceedings, February, 1990.
13. Imbrie, John, and N. J. Shackelton. "Variations in the Earth's Orbit: Pacemaker of the Ice Ages." *Science* **194** (1976): 1121–1132.
14. Imbrie, John, and John Z. Imbrie. "Modeling the Climatic Response to Orbital Variations." *Science* **207(29)** (1980): 943–952.
15. Lai, Hsin-Yi, and Su-Hua Hsieh. "Threshold Modeling of Nonlinear Oscillatory Systems." Proceedings of the 5th International Modal Analysis Conference, 1987, 1487.
16. Larimore, W. E. "System Identification, Reduced Order Filtering, and Modeling Via Canonical Variate Analysis." *Proceedings of 1983 American Control Converence*, edited by H. S. Rao and T. Dorato, 445–451. New York: IEEE 1984,.
17. Moon, *Chaotic Vibrations*. New York: Wiley, 1987.

18. Nikias, Chrysostomos, L. "Bispectral Estimation: A Digital Signal Processing Framework." Proceedings of the IEEE, Vol. 75, No. 7, July 1987.

19. Packard, N. H., J. P. Crutchfield, J. D. Farmer, and R. S. Shaw. "Geometry from a Time Series." *Phys. Rev. Letts.* **45(9)** 712.

20. Pollard, D., A. P. Ingersoll, and J. G. Lockwood. "Response of a Seasonal Colmate—Ice Model to the Orbital Perturbations During the Quaternary Ice Ages." *Tellus* **32** (1980): 301–319.

21. Pollard, D. "A Simple Ice Sheet Model Yields 100 kyr Glacial Cycles." *Nature* **296** (1982): 334–338.

22. Powers, E. J., J. Y. Hong, and Y. C. Kim. "On Modeling the Nonlinear Relationship Between Fluctuations with Nonlinear Transfer Functons." Proceedings of the IEEE, Vol. 68, No. 8, Aug. 1980.

23. Priestley, M. B., V. Haggan, and S. M. Heravi. "A Study of the Applicaton of State-Dependent Models in Nonlinear Time Series Analysis." *J. Time Series Analysis* **5(2)**: 69.

24. Schetzen, M. *The Volterra and Weiner Theories of Non-linear Systems.* New York: Wiley, 1980.

25. Shackleton, N. J. and N. D. Opdyde. "Oxygen Isotope and Paleomagentic Stratigraphy of Equatorial Pacific Core V28-238: Oxygen Isotope Temperatures and Ice Volumes on a 105 and 106 Year Scale." *Quaternary Resarch* **3** (1973): 30–55.

26. Suarez and Held. "Note on Modeling Climate Response to Orbital Parameter Variations." *Nature* **263** (1976): 46–47.

27. Takens, F. "Detecting Strange Attractors in Fluid Turbulence." In *Dynamical Systems and Turbulance*, edited by D. A. Rand and L. S. Young. Berlin: Springer-Verlag, 1981.

28. Townsend, Brent. "Nonlinear Prediction of Speech Signals." This volume.

Daniel P. Lathrop[†] and Eric J. Kostelich[‡]
†Center for Nonlinear Dynamics, Department of Physics, University of Texas, Austin, Texas 78712. and ‡Department of Mathematics, Arizona State University, Tempe, Arizona 85287

Periodic Saddle Orbits in Experimental Strange Attractors

The metric and topological invariants associated with periodic saddle orbits can be used to estimate Lyapunov exponents and deduce the global structure of low-dimensional chaotic attractors. Moreover, the periodic orbits and their eigenvalues can be used as detailed descriptions of experimental and numerical chaotic dynamics. An example is given using data from a driven-diode oscillator experiment.

INTRODUCTION

Chaotic dynamical systems are characterized by sensitive dependence on initial conditions. In many situations of physical interest, such behavior appears to be governed by relatively few degrees of freedom. The sensitivity to initial conditions can be quantified in various ways. The metric entropy and Lyapunov exponents are measures of the rates of information and predictability loss.[7] Several notions of dimension have been developed to quantify in some sense the number of independent variables governing the dynamics and the fractal (geometric) structure of the attractor.[17]

It has recently been shown that the Lyapunov exponents, entropy, and fractal dimension can be determined by the unstable periodic points in the attractor, at least in cases where the attractor is "Axiom A."[2,13] By this we mean two things: (1) at each periodic point P in the attractor, no eigenvalue of the the Jacobian matrix of the map evaluated at P has absolute value 1; and (2) the periodic points are dense in the attractor.[3] (A periodic point that satisfies (1) *is hyperbolic*.) Although the so-called strange attractors that one finds in physical experiments typically are not Axiom A, it is conjectured that many of the Axiom A results carry over in certain cases where the attractor is not uniformly hyperbolic.[13] Indeed, the periodic points can be thought of as a skeleton from which we can decompose the attractor into simpler parts.[5]

In this paper, we show that it is relatively easy to extract quantitative information about the periodic saddle orbits (and, hence, the Lyapunov exponents and related quantities) in an attractor obtained from a suitable set of experimental observations, as long as the dimension of the attractor is not too large. (A saddle fixed point is a hyperbolic fixed point with at least one eigenvalue greater than 1 in absolute value and at least one eigenvalue less than 1 in absolute value.)

A typical experimental data set consists of a time series of measurements taken at one or more points in the apparatus. If one assumes that the behavior is governed by a low-dimensional attractor (not directly observable), then one must reconstruct the attractor from the measurements. A common procedure is the time-delay embedding method.[22] Given a sequence of scalar measurements $\{S_i\}$, a set of points $x_i = (S_i, S_{i+\tau}, S_{i+2\tau} \ldots S_{i+(d-1)\tau})$ is formed, where τ is a constant called the time delay. Under suitable hypotheses it can be shown that the points $\{x_i\}$ are equivalent to the original attractor for many choices of τ if the embedding dimension d is large enough.[22] Several criteria have been suggested for determining optimum values of d and τ from experimental data sets,[4,10] and alternative procedures like singular value decompositions have been used to estimate dimension.[1] Any good embedding preserves the topological structure of the attractor, including the periodic points.[20,21]

FINDING PERIODIC POINTS

The periodic saddle points in chaotic attractors are unstable because the associated Jacobian matrix has an eigenvalue larger than 1 in absolute value; in other words, typical nearby initial conditions eventually leave the neighborhood of the saddle point. However, in low-dimensional cases, this often occurs sufficiently slowly that the location of the unstable periodic point can be determined accurately.

The basic strategy is to locate unstable periodic points by looking for orbits (sequences of points) on the reconstructed attractor that are nearly recurrent. That is, we try to find points x_i such that $\| x_{i+m} - x_i \| < \epsilon$, i.e., x_i returns to within ϵ of its original location after m iterations. Such points are said to be (m, ϵ) recurrent (Figure 1).

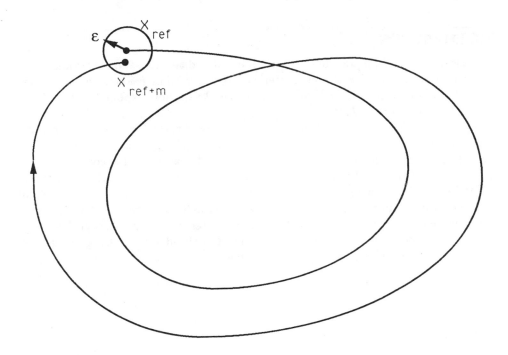

FIGURE 1 Recurrent point returns within ε.

This procedure is easy to program on a computer. We find it helpful to generate a histogram of the recurrence times to see which periods are observed and the number of associated recurrent points. In the simplest case, the periods are all integer multiples of some basic frequency, as in the Belousov-Zhabotinskii chemical reaction.[15] If the noise level is low and the periodic points are not too unstable, then one may find that 80-90% of the points on the attractor are recurrent with ϵ equal to 1-2% of the attractor extent.

The dynamics in a neighborhood of each periodic point x is approximated by the Jacobian matrix of partial derivatives of the map evaluated near the periodic point. Although the map f is not known explicitly when the attractor is reconstructed from experimental data, the Jacobian matrix of partial derivatives $Df(x)$ can be estimated using linear least squares.[8,9,14,19] In this procedure, we approximate $f(x) = Ax + b$ for some matrix A and translation vector b. By collecting all the points in a suitably small neighborhood of the periodic point, together with their observed images, a least squares estimate of A and b can be found in a straightforward way.[14] Here A is an approximation to the Jacobian $Df(x)$ at the periodic point x.

EIGENVALUES

The eigenvalues of the Jacobian matrices evaluated at the saddle periodic points are useful invariants for the characterization of low-dimensional chaotic attractors. They can also be used to compute the Lyapunov exponents. Let $D_n = Df^{n-1}(x)Df^{n-2}(x)\ldots Df(x)$. Then, whenever the limit $\lim_{n\to\infty}(D_n^T D_n)^{1/2n} = \Lambda_x$ exists and is independent of almost every initial point x, the logarithms of the eigenvalues of Λ_x are called the Lyapunov exponents.[18] The periodic saddle orbits are exceptional: the logarithms of the eigenvalues of the associated Jacobians are usually different from the Lyapunov exponents. For example, in the Hénon attractor with the usual parameters $\alpha = 1.4$ and $\beta = 0.3$, the expansion rate of the saddle fixed point in the attractor differs from the numerically obtained positive Lyapunov exponent by about 40%. In cases where one is interested in prediction, knowledge of the variation in the local expansion rates is helpful for assessing the relative predictability of different portions of the time series.

ENTROPY

The number of periodic orbits of a given period p grows exponentially in p for chaotic attractors.[18] The growth rate of the number N_p of periodic orbits of period p is called the topological entropy h_t, where $N_p = 2^{h_t p}$ for large p. Although one can estimate the topological entropy by counting the number of observed periodic orbits (or symbol sequences), we often use a method involving transfer matrices. This method requires that one find a partition of the attractor which approximates a generating partition. The partition elements must map completely across other partition elements along the unstable direction. If we label each partition element $i \in (A, B, C, \ldots)$, then, we can form a binary transition matrix $M_{ij} = (i \to j)$ whose entries are 0 for disallowed transitions and 1 for allowed transitions. The topological entropy can be found from the logarithm of the largest eigenvalue of the matrix M, $h_t = \log \lambda_{\max}$.[6] For example, consider the logistic map $f(x_n) = \lambda x_n(1 - x_n)$ at the parameter value $\lambda = 1.745\ldots$ immediately before the period 3 window.[11,12] At this parameter value, the chaotic attractor has a convenient generating partition. The interval can be partitioned into two maximal subsegments, A and B. The interval A maps onto B, and the interval B maps onto both A and B. This is a good generating partition because $f(A) = B$ and $f(B) = A \cup B$. From our knowledge of the images of the intervals, we construct the binary transition matrix

$$A = \begin{array}{c} \\ A \\ B \end{array} \begin{array}{cc} A & B \\ \left(\begin{array}{cc} 0 & 1 \\ 1 & 1 \end{array} \right) \end{array}.$$

The largest eigenvalue is $\lambda_{\max} = (\sqrt{5} - 1)/2$, so the topological entropy is $h_t = \log \lambda_{\max}$.

THE DIODE OSCILLATOR

It is well known that driven diode oscillators exhibit low-dimensional chaotic behavior.[23] We implement a circuit with a resistor, an inductor, and a bridge diode for the nonlinear element (Figure 2). Because the circuit is driven with a sinusoidal signal, it is possible to sample on a constant phase, thus giving a Poincaré section. The voltage drop across the diode is measured using a 12-bit analog to digital card from Metrabyte, sampling at the driving frequency 104.1 kHz. This circuit exhibits period doubling from a steady oscillation, leading to chaos separated by windows of periods 7, 5, and 3. The period 3 window in this circuit has a subcritical transition, and leads to a dramatic change in the shape and size of the attractor. We illustrate the technique here using a time series from a parameter immediately before the tangent bifurcation leading to the period 5 window. This parameter is not arbitrary: it is chosen close to the periodic window to facilitate our analysis. Although the first return map (Figure 3) is not a simple unimodal form, the dynamics of this attractor are easily understood. The chaotic motion in this system arises through a period-doubling cascade. The attractor appears locally to be one dimensional in this section. The two neighboring lines in the central portion of the attractor are examples of stable foliation: when two portions of the attractor are folded together, the contraction rates are finite and leave gaps in the attractor. In the absence of noise, a Cantor structure in this direction would be visible. Since we can see part of that structure, it may be possible that some of the negative Lyapunov exponents can be measured from the data and used to estimate the Lyapunov dimension of the attractor.[16] Since the attractor is near the period 5 window in parameter space, trajectories spend a lot of time near a period 5 saddle orbit (that soon gains stability).

These features make convenient landmarks for understanding the dynamics of the attractor. Let us label the locations of the ensuing period 5 orbit by I, II, III, IV and V along the unstable manifold. The order of iteration of the period 5 orbit is $I \to V \to III \to II \to IV \to I$. This information illustrates the dynamics on all portions of the attractor. Consider the "intervals" whose endpoints are on the period 5 saddle: $A = [I, II], B = [II, III], C = [III, IV]$ and $D = [I, II]$. Figure 4 is a schematic diagram showing how points on these intervals are mapped. The segments A to D form a good generating partition, and a binary transition matrix can be formed from the allowed transitions between them:

$$M = \begin{array}{c} \\ A \\ B \\ C \\ D \end{array} \begin{array}{c} A \quad B \quad C \quad D \\ \left(\begin{array}{cccc} 0 & 0 & 1 & 1 \\ 0 & 1 & 0 & 1 \\ 0 & 1 & 0 & 0 \\ 1 & 0 & 0 & 0 \end{array} \right) \end{array}.$$

From this transition matrix, and its largest eigenvalue $\lambda_{\max} = 1.513$, we estimate the topological entropy for the entire period 5 window as $h_t = 0.5972$ bits/orbit.

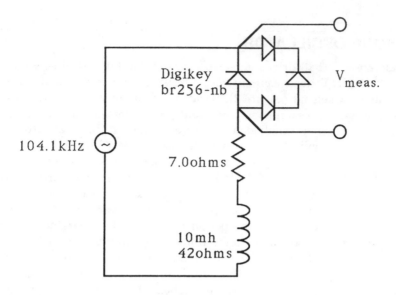

FIGURE 2   Diode oscillator circuit diagram.

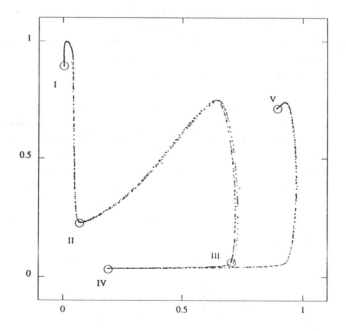

FIGURE 3   Return map for diode oscillator just prior to period 5 window.

The topological entropy h_t is an upper bound to the metric entropy h_μ. The experimental data suggest that this attractor has a smooth invariant measure along unstable directions and that there is only one positive Lyapunov exponent. Under these assumptions, the metric entropy is given by the positive Lyapunov exponent,[18] so that $h_t > h_\mu = \lambda_1$. The Lyapunov exponents are measured using an algorithm suggested by Eckmann and Ruelle[7]; the estimated value for the positive one is $\lambda_1 = 0.471$ bits/orbit. Although this value of λ_l is an average expansion rate on the attractor, the local expansion rates can fluctuate from this; i.e., the expansion rates in a neighborhood of a periodic saddle point are different. We have located low-order periodic orbits from the data using the (m, ϵ) procedure. Using the linearization method described above, there appears to be a period 1 orbit whose largest eigenvalue is $\lambda = 1.702 \pm 0.07$; a period 2 point with $\lambda = 0.446 \pm 0.014$; and a period 4 orbit with $\lambda = 0.660 \pm 0.04$. The fluctuations in the expansion rates near periodic points need to be taken into account when evaluating predictive methods: one can make more accurate predictions for a longer time into the future near a weakly repelling periodic orbit.

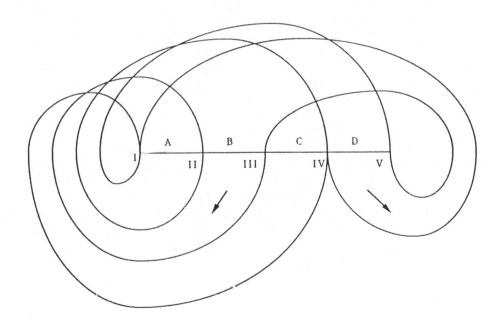

FIGURE 4 The period 5 points (I, II, III, IV, and V) and their sequence define a branched manifold for the attractor.

PRACTICAL CONSIDERATIONS

Observational noise must be taken into account when choosing a value of ϵ for locating recurrent points. If the noise level is η, then one must have $\epsilon > \eta$. The size of the ball also influences the number of points observed with a particular recurrence time. If ϵ is too large, then nonlinearities may be important and the algorithm finds extraneous points; conversely, a smaller ϵ gives a better estimate of the location of the periodic saddle orbit, but there are fewer observations to consider. If one is interested in period p orbits and λ_l is the largest Lyapunov exponent, then ϵ should be a multiple of $2^{\lambda_l p}$, because in general, the expansion rate is greater for orbits of higher period. For example, if we want points within 0.1% of a period 6 orbit, and $\lambda_l = 1$ bit/orbit, then ϵ should be approximately 0.001×26 or 6% of the attractor extent. The periodic saddle orbits that can be located accurately with this procedure are limited by the available data, the expansion rate associated with each orbit, and the dimension of the attractor.

There are several problems that arise in locating periodic orbits in higher-dimensional data sets. The primary one is that more samples are necessary to get a reasonable number of observations in suitably small neighborhoods on the attractor. In general, the number of points in a typical ϵ neighborhood of a D-dimensional attractor decreases at a rate proportional to ϵ^D. In addition, the topological entropy of the attractor increases with dimension (so there are more periodic saddle orbits), as does the number of unstable directions associated with typical saddle orbits. If the topological entropy is large, then points of high period are not observable from laboratory data sets, because they tend to be more strongly repelling. In many cases, only a period 1 orbit (if one exists) is seen.

Difficulties also arise when there is a separation of time scales between different parts of the motion. If the ratio of the largest to the smallest positive Lyapunov exponent is large, then it is difficult to measure the small Lyapunov exponents accurately, because the statistical problem for finding the Jacobian matrix is ill conditioned. If there is a separation of the amplitudes in the motion, then the small-scale motions can often be mistaken for noise. In fact, this is often what is meant by noise in experimental observations: that part of the signal which is not obviously deterministic but is smaller than the large-scale deterministic motions.

ZERO-NOISE DOMAINS

If we know that a period-m point exists, we may work out the domain of nearby points which appear as (m, ϵ) points. Here we do it for the case of a saddle point with a one-dimensional unstable manifold with eigenvalue Λ_u and a one-dimensional stable manifold with eigenvalue Λ_x. These locally linear manifolds which might have an angle θ between them are transformed to a new coordinate system where the manifolds are perpendicular and the periodic point is at the origin. The distances in this coordinate system are denoted by (δ_s, δ_u) as the distances along the stable and unstable directions, respectively. The action of the iterate of the map is thus:

$$f(\delta_s, \delta_u) \rightarrow (\Lambda_s \delta_s, \Lambda_u \delta_u).$$

The condition that a point at coordinates (δ_s, δ_u) be a (m, ϵ) point is that the distance between it and its image under the map be less than epsilon:

$$\| ((\Lambda_s - 1)\delta_s, (\Lambda_u - 1)\delta_u \| < \epsilon.$$

Using a L^2 norm, this leads to domains which are ellipses with major axis $\epsilon/(\Lambda_s - 1)$ and minor axis $\epsilon/(\Lambda_u - 1)$. In the L^1 norm which we used in the diode example, the domains are diamonds, and the L^∞ (max) norm has box shaped domains, each with similar dimensions to the ellipsoidal case.

REFERENCES

1. Albano, A. M., J. Meunch, C. Schwartz, A. I. Mees, and P. E. Rapp. *Phys. Rev. A* **38** (1988): 3017.
2. Auerbach, D., P. Cvitanovic, J.-P. Eckmann, G. H. Gunaratne, and I. Procaccia. *Phys. Rev. Lett.* **58** (1987): 2387.
3. Bowen, R. *On Axiom A Diffeomorphisms.* CBMS, Vol. 35. Providence, RI: American Mathematical Association, 1978.
4. Broomhead, D. S., and G. P. King. *Physica D* **20** (1986): 217.
5. Cvitanovic, P. *Phys. Rev. Lett.* **61** (1988): 2729.
6. Devaney, R. L. *An Introduction to Chaotic Dynamical Systems.* Menlo Park, CA: Benjamin/Cummings, 1986.
7. Eckmann, J.-P., and D. Ruelle. *Rev. Mod. Phys.* **57** (1985): 617.
8. Eckmann, J.-P., S. 0. Kamphorst, D. Ruelle, and S. Ciliberto. *Phys. Rev. A* **34** (1986): 4971.
9. Farmer, J. D., and J. J. Sidorowich. *Phys. Rev. Lett.* **59** (1987): 845.
10. Fraser, A. M., and H. L. Swinney. *Phys. Rev. A* **33** (1986): 1134.
11. Grebogi, C., E. Ott, and J. A. Yorke. *Phys. Rev. Lett.* **48** (1982).
12. Grebogi, C., E. Ott, and J. A. Yorke. *Physica D* **7** (1983).
13. Grebogi, C., E. Ott, and J. A. Yorke. *Phys. Rev. A* **36** (1988): 3522.
14. Kostelich, E. J., and J. A. Yorke. *Physica D* **41** (1990): 183.
15. Lathrop, D. P., and E. J. Kostelich. *Phys. Rev. A* **40** (1989): 431.
16. Lathrop, D. P., and E. J. Kostelich. In *Quantitative Measures of Complex Dynamical Systems*, edited by N. D. Abraham and A. Albano. NATO Advanced Science Institute series. New York: Plenum, 1990.
17. Mayer-Kress, G. "Dimensions and Entropies." In *Chaotic Systems*, edited by G. Mayer-Kress. Berlin: Springer-Verlag, 1986.

18. Ruelle, D. *Chaotic Evolution and Strange Attractors*. Cambridge: Cambridge University Press, 1989.
19. Sano, M., and Y. Sawada. , *Phys. Rev. Lett.* **55** (1985): 1082.
20. Sauer, T., M. Casdagli, and J. A. Yorke. "Embedology." Preprint, University of Maryland, 1991.
21. Solari, H. G., and R. Gilmore. *Phys. Rev. A* **37** (1988): 3096.
22. Takens, F. In *Dynamical Systems and Turbulence*, edited by D. A. Rand and L.-S. Young, 366. Springer Lecture Notes in Mathematics, Vol. 898. New York: Springer-Verlag, 1981.
23. Testa, J., J. Perez and C. Jeffries. *Phys. Rev. Lett.* **48** (1982): 714.

Christopher G. Atkeson
The Artificial Intelligence Laboratory and the Brain and Cognitive Sciences Department, Massachusetts Institute of Technology, NE43-771, 545 Technology Square, Cambridge, MA 02139. email: cga@ai.mit.edu

Memory-Based Approaches to Approximating Continuous Functions

This paper explores the use of locally weighted regression in memory-based robot learning. A local model is formed to answer each query, using a weighted regression in which close points (similar experiences) are weighted more than distant points (less relevant experiences). This approach implements a philosophy of modeling a complex function with many simple local models. The paper explains how an appropriate distance metric, or measure of similarity, can be found, and how the distance metric is used. It also explains how irrelevant input variables and terms in the local model are detected. An example from the control of a robot arm is used to compare the performance of this approach with that of other approaches that have been used for robot control and learning.

1. INTRODUCTION

A common problem in motor learning is approximating a continuous function from samples of the function's inputs and outputs. This paper explores a memory-based algorithm that simply remembers experiences (samples) and builds a local model to answer any particular query (an input for which the function's output is desired). Our approach is to model complex functions using simple local models. This approach avoids the difficult problem of finding an appropriate structure for a global model. A key idea is to form a training set for the local model after a query to be answered is known. This approach allows us to include in the training set only relevant experiences (nearby samples) and to weight the experiences according to their relevance to the query. We form a local model of the portion of the function near the query point, much as a Taylor series models a function in a neighborhood of a point. This local model is then used to predict the output of the function for that query. After answering the query, the local model is discarded and a new local model is created to answer the next query.

The memory-based architecture can represent nonlinear functions which are smooth, yet has simple training rules with a single global optimum for building a local model in response to a query. This allows complex nonlinear models to be identified (trained) quickly. Currently we are using polynomials as the local models. Since the polynomial local models are linear in the unknown parameters, we can estimate these parameters using a linear regression. We use cross validation to choose an appropriate distance metric and weighting function, and to help find irrelevant input variables and terms in the local model. The local modeling approach minimizes interference between old and new data, and allows the range of generalization to depend on the density of the samples.

In order to explore the feasibility of the memory-based approach and identify research issues and problems, we used the approach to model and control a simulated, planar, two-joint robot arm moving in a horizontal plane. The memory-based approach performed surprisingly well in these simple evaluations. The simulated arm was able to follow a desired trajectory after only a few practice movements. The cross validation identified irrelevant input variables and terms in the local model in the arm dynamics problem.

We also compared the performance of memory-based motor learning with that of other approaches such as CMAC[2] and a three-layer feedforward neural network. Our memory-based approach generalizes more effectively than CMAC, and has much less interference between old and new data than the feedforward neural network model structure.

2. ROLES OF FUNCTION APPROXIMATION IN LEARNING CONTROL

Memory-based modeling is one approach to approximating functions. There are many others. In order to understand why memory-based modeling might be useful, I will outline some potential roles of function approximation in learning control. In this outline I will talk about general "learning." Learning problems that have been studied extensively include refining kinematic models, refining dynamic models, learning a feedforward command to follow a particular trajectory more accurately, and learning models of particular tasks. Function approximation can play a role in all of these forms of learning. I will use the following notation that can be applied to all of these types of learning problems. A command $\mathbf{c}$ is given to a plant in state $\mathbf{x}$ which results in output $\mathbf{p}$:

$$\mathbf{p} = \mathbf{f}(\mathbf{x}, \mathbf{c}). \tag{1}$$

A unique inverse for the true plant may exist:

$$\mathbf{c} = \mathbf{f}^{-1}(\mathbf{x}, \mathbf{p}). \tag{2}$$

2.1 FORMING AN INVERSE MODEL OF THE PLANT

One application of learning control is to directly learn an inverse model of the plant

$$\mathbf{c} = \widehat{\mathbf{f}}^{-1}(\mathbf{x}, \mathbf{p}) \tag{3}$$

from samples $(\mathbf{x}_i, \mathbf{c}_i, \mathbf{p}_i)$. This approach assumes a unique inverse exists. Once the inverse model is formed, a command to achieve a particular goal can be found with a single query. It is reasonable to expect that there will be as many data points to learn from as there will be queries when using an inverse model in on-line learning. Each query leads to an actual command applied and a new data point to be learned. This suggests that function approximation methods should be efficient to train, and that expensive iterative searches for parameters in nonlinear global models may not be appropriate in this context, although the global model may be inexpensive to evaluate for any particular query. Another important issue with this approach is generating sufficiently rich training data so that the inverse model makes accurate predictions and the learning process converges. There are a wide range of teaching strategies to generate the training data, including random experimentation and using a crude feedback controller for initial control.

2.2 FORMING A FORWARD MODEL OF THE PLANT

In cases where a unique inverse for the plant does not exist a forward model can be used in an iterative search for the appropriate command to apply to achieve a particular goal. For example, the following three-step process can be used to find a command that achieves a goal $\mathbf{p}_d$:

1. Guess an initial command $\mathbf{c}_0$. An approximate inverse model could be used:

$$\mathbf{c}_0 = \widehat{\mathbf{f}}^{-1}(\mathbf{x}, \mathbf{p}_d). \tag{4}$$

2. Use the forward model to predict the performance that will result from applying a command (mental simulation):

$$\mathbf{p}_i = \widehat{\mathbf{f}}(\mathbf{x}, \mathbf{c}_i). \tag{5}$$

3. Use derivatives of the forward model (the Jacobian $\mathbf{J} = \partial \mathbf{p}/\partial \mathbf{c}$) to refine the command. Gradient descent is one approach[36]:

$$\mathbf{c}_{i+1} = \mathbf{c}_i - \delta \mathbf{c}_i \tag{6}$$

$$\delta \mathbf{c}_i = \widehat{\mathbf{J}}^T (\mathbf{p}_i - \mathbf{p}_d). \tag{7}$$

More aggressive nonlinear equation-solving (root finding) techniques can be used, such as variants of the Newton Raphson method, which involve using pseudoinverse techniques to solve

$$\widehat{\mathbf{J}} \delta \mathbf{c}_i = (\mathbf{p}_i - \mathbf{p}_d)$$

for $\delta \mathbf{c}_i$. If $\widehat{\mathbf{J}}$ is almost singular, these corrections may be invalid because they are too large. Various schemes can be used to limit the size of the command corrections.

Any number of iterations of steps 2 and 3 can be performed, until the predicted command converges or the forward model is queried outside the range of the training data. At this point the command should be applied, new training data obtained, and the above process repeated. In this approach the forward model is likely to be evaluated many times relative to the number of training points, which suggests that a function approximation method that is cheap to evaluate and to find derivatives should be used, even if it is expensive to train.

The approach of using a forward model and mental simulation may give this method a wider range of convergence than methods that form inverse models. We are currently exploring this issue. In cases where more than one command can achieve the desired output, the solution chosen by these methods depends on the starting point of the learning process. Additional constraints, or a criteria to be optimized can be added to force a unique choice of command independent of the starting point of the search. Generating initial training data and active exploration during the learning process are important issues that have not been carefully examined. Random or deliberate experiments, based on the observed performance improvement rate, could be used.

2.3 FUNCTION OPTIMIZATION

Often a goal in learning is to optimize a particular criteria, rather than achieve a particular output. Function approximation techniques can be used to represent the cost function directly, and to speed the search for extreme points.[16,52,55] A linear local model can be used to estimate the first derivatives (gradient) and a quadratic local model can be used to estimate the second derivatives (Hessian) of the function at the current point in the optimization procedure. These estimates can be used in gradient descent, or in a Gauss Newton method which uses estimates of second derivatives. This optimization process may be constrained in that some particular constraints on the output must be satisfied as well as a criteria optimized, a combination of root finding and optimization.

2.4 OPTIMIZATION OVER TIME: DYNAMIC PROGRAMMING

In situations where commands must be optimized over time, dynamic programming provides a framework to perform the optimization.[35] Let us consider a discrete time problem although the continuous-time problem is closely related. A forward model of the plant $\mathbf{x}_{k+1} = \mathbf{f}(\mathbf{x}_k, \mathbf{u}_k)$ is learned as well as the single-step cost function $L(\mathbf{x}_k, \mathbf{u}_k)$. The cost to go for a state $\mathbf{x}_k$ at any time step is the sum of single-step costs incurred in following an optimal policy from that state. Assume we know the cost to go for all states at time $k+1$: $V_{k+1}(\mathbf{x}_{k+1})$. Bellman's principle of optimality gives us a rule for finding the cost to go V_k and the optimal control $\mathbf{u}_k$ at time step k, given an initial state $\mathbf{x}_k$:

$$V_k(\mathbf{x}_k) = \min_{\mathbf{u}_k}[L_k(\mathbf{x}_k, \mathbf{u}_k) + V_{k+1}(\mathbf{f}(\mathbf{x}_k, \mathbf{u}_k))]. \tag{9}$$

For smooth problems this minimization can be performed by continuous optimization techniques. As described in the previous section, local linear forward models of the plant and local quadratic models of the one-step cost and cost to go can be used to optimize the control.[35]

Thus, goals for function approximation in learning control go beyond being able to represent the training set and generalize appropriately. Other desired criteria include:

- Fast training based on a continuous stream of incoming data, rather than a fixed training set.
- Ability to find first (and potentially second) derivatives of the learned function.
- Ability to tell where in the input space the function is accurately approximated. This is typically based on the local density of samples, and an estimate of the local variance of the outputs. This ability is used in iterative use of the model to determine when to terminate search and collect more data.
- Ability to find good initial guesses for search or optimization procedures. Memory-based approaches typically scan the entire data set. If the data set is represented as a set of parameters, this is more difficult.

3. RELATED WORK

Two forms of modeling previous experience are those approaches that represent the experiences directly, as in this study, and those that represent the experiences using parameters or weights, as in many table-based schemes, perceptrons, and multilayer connectionist or model neural networks. For a more extensive review of related work, see Atkeson.[5]

3.1 DIRECT STORAGE OF EXPERIENCE

Memory-based modeling has a long history. Approaches which represent previous experiences directly and use a similar experience or similar experiences to form a local model are often referred to as nearest-neighbor or k-nearest-neighbor approaches. Local models (often polynomials) have been used for many years to smooth time series,[43,58,59,73] and interpolate and extrapolate from limited data. Barnhill[6] and Sabin[54] survey the use of nearest-neighbor interpolators to fit surfaces to arbitrarily spaced points. Eubank[19] surveys the use of nearest-neighbor estimators in nonparametric regression. Lancaster and Šalkauskas[39] refer to nearest-neighbor approaches as "moving least squares" and survey their use in fitting surfaces to data. Farmer and Sidorowich[22,23] survey the use of nearest-neighbor and local model approaches in modeling chaotic dynamic systems.

An early use of direct storage of experience was in pattern recognition. Fix and Hodges[25,26] suggested that a new pattern could be classified by searching for similar patterns among a set of stored patterns, and using the categories of the similar patterns to classify the new pattern. Steinbuch and Taylor proposed a neural network implementation of the direct storage of experience and nearest-neighbor search process for pattern recognition,[61,67] and pointed out that this approach could be used for control.[62] Stanfill and Waltz[60] proposed using directly stored experience to learn pronunciation, using a Connection Machine and parallel search to find relevant experience. They have also applied their approach to medical diagnosis[71] and protein structure prediction.

Nearest-neighbor approaches have also been used in nonparametric regression and fitting surfaces to data. Often, a group of similar experiences, or nearest neighbors, is used to form a local model, and then that model is used to predict the desired value for a new point. Local models are formed for each new access to the memory. Watson,[72] Royall,[53] Crain and Bhattacharyya,[14] Cover,[13] and Shepard[57] proposed using a weighted average of a set of nearest neighbors. Gordon and Wixom[31] and Barnhill, Dube, and Little[7] analyze such weighted-average schemes. Crain and Bhattacharyya,[14] Falconer,[20] and McLain[45] suggested using a weighted regression to fit a local polynomial model at each point a function evaluation was desired. All of the available data points were used. Each data point was weighted by a function of its distance to the desired point in the regression. McIntyre, Pollard, and Smith,[44] Pelto, Elkins, and Boyd,[51] Legg and Brent,[40] Palmer,[50] Walters,[70]

Lodwick and Whittle,[42] Stone,[64] and Franke and Nielson[28] suggested fitting a polynomial surface to a set of nearest neighbors, also using distance weighted regression. Stone scaled the values in each dimension when the experiences where stored. The standard deviations of each dimension of previous experiences were used as the scaling factors, so that the range of values in each dimension were approximately equal. This affects the distance metric used to measure closeness of points. Cleveland[10] proposed using robust regression procedures to eliminate outlying or erroneous points in the regression process. A program implementing a refined version of this approach (LOESS) is available by sending electronic mail containing the single line, *send dloess from a*, to the address netlib@research.att.com.[32] Cleveland, Devlin, and Grosse[12] analyze the statistical properties of the LOESS algorithm and Cleveland and Devlin[11] show examples of its use. Stone,[65,66] Devroye,[17] Lancaster,[37] Lancaster and Šalkauskas,[38] Cheng,[9] Li,[41] Farwig,[24] and Müller[49] provide analyses of nearest-neighbor approaches. Franke[27] compares the performance of nearest-neighbor approaches with other methods for fitting surfaces to data.

4. HOW DOES LOCALLY WEIGHTED REGRESSION WORK?

There are several stages in the version of locally weighted regression developed in this research.

1. Initially, experiences are simply stored in memory.
2. To answer any particular query, a weighted linear regression is performed. The weights in the regression depend on parameters used to calculate a distance metric and the weighting function, and stabilize the solution to the regression. I will refer to these parameters as "fit parameters."
3. The fit parameters can be optimized using cross validation.

4.1 STORING THE EXPERIENCES

The choice of the method of storing experiences depends on what fraction of the experiences are used in each locally weighted regression and what computational technology is available. If all of the experiences are used in each locally weighted regression, then simply maintaining a list or array of experiences is sufficient. If only nearby experiences are included in the locally weighted regression, then an efficient method of finding nearest neighbors is required.

K-d trees[29] are a tree data structure used to speed up the search for a best match, and are an appropriate technique for finding nearest neighbors on a serial computer. However, the number of data points that must be examined to find the best match grows exponentially with the number of dimensions used in the distance metric, in the worst case. The actual performance of k-d trees depends on the distribution of the data, the less randomly distributed the better. Farmer and Sidorowich[21,22,23] used k-d trees and a variety of generalization functions to model chaotic dynamic systems. Cleveland and Devlin[11] and Cleveland, Devlin, and Grosse[12] used k-d trees in the LOESS program. Moore[47] used k-d trees to model a robot manipulator.

On a parallel computer, such as the Connection Machine,[33] exhaustive search is often faster than using k-d trees, due to the limited number of experiences allocated to each processor. The Connection Machine can have up to 2^{16} (65,536) processors, and can simulate a parallel computer with many more processors. Experiences are stored in the local memory associated with each processor. An experience can be compared to the desired experience in each processor, with the processors running in parallel, and then a hardwired global-OR bus can be used to find the closest match in constant time independent of the number of stored experiences. This approach is similar to many Connection Machine algorithms that find a best match or set of nearest neighbors.[71] The search time depends linearly on the number of dimensions in the distance metric, and the distance metric can be easily changed or made to depend on the current query point.

We have implemented both of these forms of storing and accessing the experiences, using standard workstations for the implementation of k-d trees, and using the Connection Machine to do parallel exhaustive search.

4.2 PERFORMING THE REGRESSION

As an example of modeling a function using locally weighted regression, we will consider a problem from motor control and robotics, two-joint arm inverse dynamics. We will predict torques at the shoulder, τ_1, and elbow, τ_2, on the basis of joint positions, θ_1 and θ_2, joint velocities, $\dot{\theta}_1$ and $\dot{\theta}_2$, and joint accelerations, $\ddot{\theta}_1$ and $\ddot{\theta}_2$. We use this example because we already know the idealized function, and will be able to assess how well the locally weighted regression procedure is doing and interpret the parameters used to improve the fit. In an actual application a structured model based on *a priori* knowledge would be used to fit the dynamics data and the locally weighted regression would be used to fit the errors (residuals) of the parametric model. The structured model will generalize most effectively from the limited training data. An, Atkeson, and Hollergach[4] describe an example of using a structured model to fit data from an existing robot arm. The residuals were typically small, but repeatable.

We have a query point $(\theta_1^*, \theta_2^*, \dot{\theta}_1^*, \dot{\theta}_2^*, \ddot{\theta}_1^*, \ddot{\theta}_2^*)$ for which we want to predict the shoulder and elbow torques. We will first show how an unweighted regression can be used to form a global model. Then we will show how a weighted regression can be used to form a local model appropriate to answer this particular query. For the purposes of this example, we will assume a quadratic model is used in the regression.

In this dynamics example there are 28 terms in the quadratic model:

$$
\begin{array}{cccccc}
1 & \theta_1, & \theta_2, & \dot{\theta}_1, & \dot{\theta}_2, & \ddot{\theta}_1, & \ddot{\theta}_2, \\
\theta_1 * \theta_1, & \theta_1 * \theta_2, & \theta_1 * \dot{\theta}_1, & \theta_1 * \dot{\theta}_2, & \theta_1 * \ddot{\theta}_1, & \theta_1 * \ddot{\theta}_2, \\
\theta_2 * \theta_2, & \theta_2 * \dot{\theta}_1, & \theta_2 * \dot{\theta}_2, & \theta_2 * \ddot{\theta}_1, & \theta_2 * \ddot{\theta}_2, \\
\dot{\theta}_1 * \dot{\theta}_1, & \dot{\theta}_1 * \dot{\theta}_2, & \dot{\theta}_1 * \ddot{\theta}_1, & \dot{\theta}_1 * \ddot{\theta}_2, \\
\dot{\theta}_2 * \dot{\theta}_2, & \dot{\theta}_2 * \ddot{\theta}_1, & \dot{\theta}_2 * \ddot{\theta}_2, \\
\ddot{\theta}_1 * \ddot{\theta}_1, & \ddot{\theta}_1 * \ddot{\theta}_2, \\
\ddot{\theta}_2 * \ddot{\theta}_2
\end{array}
$$

where 1 represents the constant term in the model.

Let us assume we have 1000 samples of the two-joint arm dynamics function. The equation to be solved is

$$\mathbf{X}\beta = \mathbf{y} \tag{10}$$

where $\mathbf{X}$ is a 1000 × 28 data matrix, in which each row has the 28 terms of the quadratic model corresponding to a point (sample of the function), and each column corresponds to a particular term in the quadratic model. β is the vector of 28 estimated parameters of the quadratic model, and $\mathbf{y}$ is the vector of 1000 torques from the 1000 points included in the regression.

An unweighted regression finds the solution to the normal equations:

$$\left(\mathbf{X}^T \mathbf{X}\right)\beta = \mathbf{X}^T \mathbf{y} . \tag{11}$$

The estimated parameters are used, with the query point, to predict the torques for the query point.

However, we assume the global quadratic model is not the correct model structure for predicting the torques. These structural modeling errors imply that different sets of parameters are estimated by the regression, given different data sets. The data set can be tailored to the query point by emphasizing nearby points in the regression. The origin of the input data is first shifted by subtracting the query point from each data point. Then each data point is given a weight.

Unweighted regression gives distant points equal influence with nearby points on the ultimate answer to the query. To weight similar points more, a weighted regression is used. Each row of $\mathbf{X}$ and $\mathbf{y}$ is multiplied by a weight:

$$w_i = (d_i^2)^{-p} \tag{12}$$

where w_i is the weight for row i (corresponding to the ith point), d_i is the Euclidean distance of point i to the query point, and p is a parameter that determines how local the regression will be (the rate of drop-off of the weights with distance). d_i^2 is calculated for each point in the following way (in this example):

$$
\begin{aligned}
d^2 = &m_1^2(\theta_1 - \theta_1^*)^2 + m_2^2(\theta_2 - \theta_2^*)^2 + m_3^2(\dot{\theta}_1 - \dot{\theta}_1^*)^2 \\
&+ m_4^2(\dot{\theta}_2 - \dot{\theta}_2^*)^2 + m_5^2(\ddot{\theta}_1 - \ddot{\theta}_1^*)^2 + m_6^2(\ddot{\theta}_2 - \ddot{\theta}_2^*)^2 .
\end{aligned}
\tag{13}
$$

The superscript * indicates the query point, and the m_i are the components of the distance metric.

A potential problem is that the data points may be distributed in such a way as to make the regression matrix $\mathbf{X}$ nearly singular. A technique known as ridge regression[18] is used to prevent problems due to a singular data matrix. The following equation, with $\mathbf{X}$ and $\mathbf{y}$ already weighted, is solved for β:

$$\left(\mathbf{X}^T\mathbf{X} + \Lambda\right)\beta = \mathbf{X}^T\mathbf{y} \tag{14}$$

where Λ is a diagonal matrix with small positive diagonal elements λ_i^2. This is equivalent to adding i extra rows to $\mathbf{X}$, each having a single non-zero element, λ_i, in the ith column. Adding additional rows can be viewed as adding "fake" data, which, in the absence of sufficient real data, biases the parameter estimates to zero.[18] Another view of ridge regression parameters is that they are the Bayesian assumptions about the *a priori* distributions of the estimated parameters.[56]

4.3 OPTIMIZING THE FIT PARAMETERS

For the example problem, we have introduced 34 free parameters into the local regression process: the weighting function drop-off parameter p, the 6 elements of the distance metric m_i, and the 27 variable diagonal elements of Λ (the ridge regression parameters λ_i). The element of Λ corresponding to the constant term, λ_1, is held fixed.

A cross-validation approach is used to choose values for these fit parameters. The sum of the squared cross-validation error is minimized using a nonlinear parameter estimation procedure (MINPACK[48] or NL2SOL,[15] for example). We can take the derivative of the cross-validation error with respect to the parameters to be estimated, which greatly speeds up the search process (see Appendix).

The cross validation to optimize the fit parameters may be done globally, using all the experiences in the memory to produce one set of fit parameters. Different fit parameters can be used for different outputs. The cross validation may also be done locally, either with each query, or separately for different regions of the input space, producing different sets of fit parameters specialized for particular queries. We have only experimented with global cross validation so far.

We can use the optimized distance metric to find which input variables are irrelevant to the function being represented. In the horizontal, two-joint arm, inverse dynamics problem, m_1, the weight on the distances in the θ_1 direction typically drops to zero, indicating that the input variable θ_1 is irrelevant to predicting τ_1 and τ_2. This is actually the case, as θ_1 does not appear in the true dynamics equations for an arm operating in a horizontal plane.

We can also interpret the ridge regression parameters, λ_i. Since the arm dynamics are linear in acceleration, the terms in the local model that are quadratic in acceleration ($\ddot{\theta}_1^2$, $\ddot{\theta}_1 * \ddot{\theta}_2$, $\ddot{\theta}_2^2$) are not relevant to predicting torques. Similarly the products of velocity and acceleration ($\dot{\theta}_1 * \ddot{\theta}_1$, $\dot{\theta}_1 * \ddot{\theta}_2$, $\dot{\theta}_2 * \ddot{\theta}_1$, $\dot{\theta}_2 * \ddot{\theta}_2$) are also not relevant to the dynamics. The ridge regression parameter for each of these terms becomes very large in the parameter optimization. The effect of this is to force the estimated parameter β_i for these terms to be zero and the terms to have no effect on the regression.

We have also explored stepwise regression procedures to determine which terms of the local model are useful[18] with similar results.

5. PERFORMANCE COMPARISONS

Two methods, CMAC[2,3] and sigmoidal feedforward neural networks, were compared to the approach explored in this paper. The parameters for the CMAC approach were taken from Miller, Glanz, and Kraft[46] who used the CMAC to model arm dynamics. The architecture for the sigmoidal feedforward neural network was taken from Goldberg and Pearlmutter[30] (section 6) who also modeled arm dynamics.

The ability of each of these methods to predict the torques of the simulated two-joint arm at 1000 random points was compared. Figure 1 plots the normalized RMS prediction error. The points were sampled uniformly using ranges comparable to those used in Miller et al.[46] Initially, each method was trained on a training set of 1000 random samples of the two-joint arm dynamics function, and then the predictions of the torques on a separate test set of 1000 random samples of the two-joint arm dynamics function were assessed (points 1, 3, and 5). Each method was

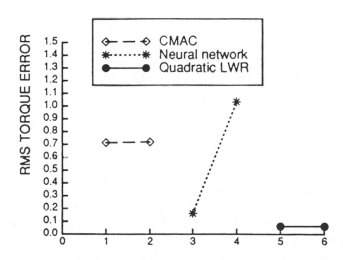

FIGURE 1   Performance of various methods on two-joint arm dynamics.

then trained on ten attempts to make a particular desired movement. Each method successfully learned the desired movement. After this second round of training, performance on the random test set was again measured (points 2, 4, and 6).

The data indicate that the locally weighted regression approach (filled in circles) and the sigmoidal feedforward network approach (asterisks) both generalize well on this problem (points 3 and 5 have low error). The CMAC (diamonds) did not generalize well on this problem (point 1 has a large error), although it represented the original training set with a normalized RMS error of 0.000001. A variety of CMAC resolutions were explored, ranging from a basic CMAC cell size covering the entire range of data to a cell size covering a fifth of the data range in each dimension. A cell size covering one half the data ranges in each dimension generalized best (the data shown here).

After training on a different training set (the attempts to make a particular desired movement), the sigmoidal feedforward neural network lost its memory of the full dynamics (point 4), and represented only the dynamics of the particular movements being learned in the second training set. This interference between new and previously learned data was not prevented by increasing the number of hidden units in the single-layer network from 10 up to 100. The other methods explored did not show this interference effect (points 2 and 6).

6. CONCLUSIONS

It will be necessary to try out this memory-based approach to modeling on real data from several problems in order to more fully assess the potential of the approach. We are currently attempting to use these methods for both learning control (see An, Atkeson, and Hollerbach[4] for further references on learning control) and task level learning.[1] Several points can be made at this early stage of the research.

This approach to modeling is appropriate for situations where new data is being collected continuously. Storing new data is cheap: the new data point is simply stored in a memory. However, making a prediction using this approach is much more expensive than just evaluating a global model at a point, as a new local model is fit for each query. If there is a fixed training set followed by many queries, other approaches will be preferable. If there is a continuously growing training set intermixed with queries, a memory-based approach seems worth exploring. The local modeling approach, because of its retention of all the data in memory, reduces interference between old and new data, and allows the range of generalization to depend on the density of the samples.

The cross-validation approach to optimizing the fit parameters reduces the number of arbitrary choices that need to be made before the training data is collected. However, like other modeling approaches, the choice of representation of the data (number and selection of dimensions to be measured, etc.) play a large role in determining the success of the approach.

One concern with memory-based approaches is that more data will eventually be collected than will fit in the memory. In many problems and applications, this is not an issue. However, algorithms for forgetting data and algorithms that determine whether or not new data is worth storing remain to be explored.

APPENDIX: OPTIMIZING THE FIT PARAMETERS

The cross-validation error for point i is the difference of y_i and the prediction of y_i generated by removing point i from the memory and performing a locally weighted regression using the current set of fit parameters with point i as the query point. The derivative of the cross-validation error for point i is given by the derivative of the prediction for point i with respect to the fit parameters, since y_i itself is a known fixed quantity.

The dependence on the distance metric occurs indirectly through the weights:

$$\frac{\partial \hat{y}_i}{\partial m_k} = \sum_j \left(\frac{\partial \hat{y}_i}{\partial w_j} \frac{\partial w_j}{\partial m_k} \right) . \tag{15}$$

$\hat{y}_i$ is the prediction for the ith cross-validated point, m_k is the weight for the kth dimension in the distance metric, and w_j is the weight for the jth point or row in $\mathbf{X}$.

$$\frac{\partial \hat{y}_i}{\partial w_j} = \frac{2 e_j \left[\left(\mathbf{X}^T \mathbf{X} \right)^{-1} \mathbf{x}_j^T \right]_l}{w_j} \tag{16}$$

where $\mathbf{X}^T \mathbf{X}$ is computed from the weighted regression matrix leaving point i out and includes the ridge regression parameters, $\mathbf{x}_j$ is the jth row of $\mathbf{X}$, and e_j is the jth residual for the current weighted regression.[8] $\mathbf{X}$, $\mathbf{x}_j$, and e_j are all evaluated with the current set of fit parameters. The l index indicates that the element of the vector in brackets corresponding to the constant term is selected (since the origin is moved to the query point, $\hat{y}_i$ is the constant term estimated by the ith regression). Thus, I is a fixed quantity.

$$\frac{\partial w_j}{\partial m_k} = -2 w_j m_k p \frac{\delta_{jk}^2}{d_j^2}$$

where p is the power term, w_j is the weight for the jth point, m_k is the metric for the kth dimension, δ_{jk} is the difference between the jth point and the desired point in the kth dimension, and d_j^2 is the squared total distance between the jth point and the desired point.

The dependence on the power parameter in the weighting function also occurs indirectly through the weights:

$$\frac{\partial \hat{y}_i}{\partial p} = \sum_j \left(\frac{\partial \hat{y}_i}{\partial w_j} \frac{\partial w_j}{\partial p} \right) , \tag{18}$$

$$\frac{\partial w_j}{\partial p} = -w_j \ln(d_j^2) . \tag{19}$$

The ridge regression parameters can be treated as weights directly:

$$\frac{\partial \hat{y}_i}{\partial \lambda_k} = \frac{2 e_k \left[\left(\mathbf{X}^T \mathbf{X} \right)^{-1} \mathbf{x}_k^T \right]_l}{\lambda_k} .$$

$\mathbf{x}_k$ is a row with a single non-zero element λ_k in the kth position (the kth fake data point), and e_k is the residual for that row of $\mathbf{X}$.

ACKNOWLEDGMENTS

Support was provided under Office of Naval Research contract N00014-88-K-0321 and under Air Force Office of Scientific Research grant AFOSR-89-0500. Support for Christopher G. Atkeson was provided by a National Science Foundation Engineering Initiation Award and Presidential Young Investigator Award, an Alfred P. Sloan Research Fellowship, the W. M. Keck Foundation Assistant Professorship in Biomedical Engineering, and a Whitaker Health Sciences Fund MIT Faculty Research Grant.

REFERENCES

1. Aboaf, E. W., S. M. Drucker, and C. G. Atkeson. "Task-Level Robot Learning: Juggling a Tennis Ball More Accurately." Proceedings, IEEE International Conference on Robotics and Automation, May 14-19, 1989, Scottsdale, Arizona.
2. Albus, J. S. "A New Approach to Manipulator Control: The Cerebellar Model Articulation Controller (CMAC)." *ASME J. Dynamic Systems, Meas., Control* **97** (1975): 220–227.
3. Albus, J. S. "Data Storage in the Cerebellar Model Articulation Controller (CMAC)." *ASME J. Dynamic Systems, Meas., Control* **97** (1975): 228–233.
4. An, C. H., C. G. Atkeson, and J. M. Hollerbach. *Model-Based Control of a Robot Manipulator.* Cambridge, MA: MIT Press, 1988.
5. Atkeson, C. G. "Learning Arm Kinematics and Dynamics." *Ann. Rev. Neurosci.* **12** (1989): 157–183.
6. Barnhill, R. E. "Representation and Approximation of Surfaces." In *Mathematical Software III,* edited by J. R. Rice, 69–120. New York: Academic Press, 1977.
7. Barnhill, R. E., R. P. Dube, and F. F. Little. "Properties of Shepard's Surfaces." *Rocky Mountain J. Math.* **13(2)** (1983): 365–382.
8. Belsley, D. A., E. Kuh, and R. E. Welsch. *Regression Diagnostics: Identifying Influential Data and Sources of Collinearity.* New York: John Wiley, 1980.
9. Cheng, P. E. "Strong Consistency of Nearest Neighbor Regression Function Estimators." *J. Multivariate Analysis* **15** (1984): 63–72.
10. Cleveland, W. S. "Robust Locally Weighted Regression and Smoothing Scatterplots." *J. Amer. Stat. Assoc.* **74** (1979): 829–836.
11. Cleveland, W. S., and S. J. Devlin. "Locally Weighted Regression: An Approach to Regression Analysis by Local Fitting." *J. Amer. Stat. Assoc.* **83** (1988): 596–610.

12. Cleveland, W. S., S. J. Devlin, and E. Grosse. "Regression by Local Fitting: Methods, Properties, and Computational Algorithms." *J. Econ.* **37** (1988): 87–114.

13. Cover, T. M. "Estimation by the Nearest Neighbor Rule." *IEEE Transactions on Information Theory* **IT-14** (1968): 50–55.

14. Crain, I. K., and B. K. Bhattacharyya. "Treatment of Nonequispaced Two-Dimensional Data with a Digital Computer." *Geoexploration* **5** (1967): 173–194.

15. Dennis, J. E., D. M. Gay, and R. E. Welsch. "An Adaptive Nonlinear Least-Squares Algorithm." *ACM Transactions on Mathematical Software* **7(3)** (1981).

16. Devroye, L. P. "The Uniform Convergence of Nearest Neighbor Regression Function Estimators and Their Application in Optimization." *IEEE Transactions on Information Theory* **IT-24** (1978): 142–151.

17. Devroye, L. P. "On the Almost Everywhere Convergence of Nonparametric Regression Function Estimates." Ann. of Stat. **9(6)** (1981): 1310–1319.

18. Draper, N. R., and H. Smith. *Applied Regression Analysis*, 2nd ed. New York: John Wiley, 1981.

19. Eubank, R. L. *Spline Smoothing and Nonparametric Regression*, 384–387. New York: Marcel Dekker, 1988.

20. Falconer, K. J. "A General Purpose Algorithm for Contouring over Scattered Data Points." Nat. Phys. Lab. Report NAC 6, 1971.

21. Farmer, J. D., and J. J. Sidorowich. "Predicting Chaotic Time Series." *Phys. Rev. Lett.* **59(8)** (1987): 845–848.

22. Farmer, J. D., and J. J. Sidorowich. "Exploiting Chaos to Predict the Future and Reduce Noise." In *Evolution, Learning, and Cognition*, edited by Y. C. Lee, 277. Singapore: World Scientific, 1988.

23. Farmer, J. D., and J. J. Sidorowich. "Predicting Chaotic Dynamics." In *Dynamic Patterns in Complex Systems*, edited by J. A. S. Kelso, A. J. Mandell, and M. F. Schlesinger, 265–292. New Jersey: World Scientific, 1988.

24. Farwig, R. "Multivariate Interpolation of Scattered Data by Moving Least Squares Methods." *Algorithms for Approximation*, edited by J. C. Mason and M. G. Cox, 193–211. Oxford: Clarendon Press, 1987.

25. Fix, E., and J. L. Hodges, Jr. "Discriminatory Analysis, Nonparametric Regression: Consistency Properties." Project 21-49-004, Report No. 4. USAF School of Aviation Medicine Randolph Field, Texas. Contract AF-41-(128)-31, February 1951

26. Fix, E., and J. L. Hodges, Jr. "Discriminatory Analysis: Small Sample Performance." Project 21-49-004, Rep. 11 USAF School of Aviation Medicine Randolph Field, Texas. August 1952.

27. Franke, R. "Scattered Data Interpolation: Tests of Some Methods." *Mathematics of Computation* **38(157)** (1982): 181–200.

28. Franke, R., and G. Nielson. "Smooth Interpolation of Large Sets of Scattered Data." *International Journal Numerical Methods Engineering* **15** (1980): 1691-1704.

29. Friedman, J. H., J. L. Bentley, and R. A. Finkel "An Algorithm for Finding Best Matches in Logarithmic Expected Time." *ACM Trans. on Mathematical Software* **3(3)** (1977): 209–226.

30. Goldberg, K. Y., and B. Pearlmutter. "Using a Neural Network to Learn the Dynamics of the CMU Direct-Drive Arm II." Technical Report CMU-CS-88-160, Carnegie-Mellon University, Pittsburgh, PA, August 1988.

31. Gordon, W. J., and J. A. Wixom. "Shepard's Method of Metric Interpolation to Bivariate and Multivariate Interpolation." *Mathematics of Computation* **32(141)** (1978): 253–264.

32. Grosse, E. "LOESS: Multivariate Smoothing by Moving Least Squares." In *Approximation Theory VI*, edited by C. K. Chui, L. L. Schumaker, and J. D. Ward, 1–4. Boston: Academic Press, 1989.

33. Hillis, D. *The Connection Machine.* Cambridge, MA: MIT Press, 1985.

34. Kazmierczak, H., and K. Steinbuch. "Adaptive Systems in Pattern Recognition." *IEEE Trans. on Electronic Computers* **EC-12** (1963): 822–835.

35. Jacobson, D. H., and D. Q. Mayne. *Differential Dynamic Programming.* New York: American Elsevier, 1970.

36. Jordan, M. I., and D. A. Rosenbaum. "Action." Tech. Rep. 88-26, Computer and Information Science, University of Massachusetts, Amherst, 1988.

37. Lancaster, P. "Moving Weighted Least-Squares Methods." In *Polynomial and Spline Approximation,* edited by B. N. Sahney, 103–120. Boston: D. Reidel, 1979.

38. Lancaster, P., and K. Šalkauskas. "Surfaces Generated by Moving Least Squares Methods." *Mathematics of Computation* **37(155)** (1981): 141–158.

39. Lancaster, P., and K. Šalkauskas. *Curve And Surface Fitting.* New York: Academic Press, 1986.

40. Legg M. P. C., and R. P. Brent. "Automatic Contouring." Proc. 4th Australian Computer Conference, 1969, 467–468.

41. Li, K. C. "Consistency for Cross-Validated Nearest-Neighbor Estimates in Nonparametric Regression." *The Annals of Statistics* **12** (1984): 230–240.

42. Lodwick, G. D., and J. Whittle. "A Technique for Automatic Contouring Field Survey Data." *Australian Computer Journal* **2** (1970): 104–109.

43. Macauley, F. R. *The Smoothing of Time Series.* National Bureau of Economic Research, New York, 1931.

44. McIntyre, D. B., D. D. Pollard, and R. Smith. "Computer Programs For Automatic Contouring." Kansas Geological Survey Computer Contributions 23, University of Kansas, Lawrence, Kansas, 1968.

45. McLain, D. H. "Drawing Contours From Arbitrary Data Points." *The Computer Journal* **17(4)** (1974): 318–324.

46. Miller, W. T., F. H. Glanz, and L. G. Kraft. "Application of a General Learning Algorithm to the Control of Robotic Manipulators." *International Journal of Robotics Research* **6** (1987): 84–98.

47. Moore, A. W. "Acquisition of Dynamic Control Knowledge for a Robotic Manipulator." In *Machine Learning: Proceedings of the Seventh International*

Conference, edited by B. W. Porter, and R. J. Mooney, 244–252. San Mateo, CA: Morgan Kaufmann, 1990.

48. More, J. J., B. S. Garbow, and K. E. Hillstrom. "User Guide for MINPACK-1." ANL-80-74, Argonne National Laboratory, Argonne, Illinois, 1980.

49. Müller, H. G. "Weighted Local Regression and Kernel Methods for Nonparametric Curve Fitting." *J. Amer. Stat. Assoc.* **82** (1987): 231–238.

50. Palmer, J. A. B. "Automated Mapping." Proc. 4th Australian Computer Conference, 1969, 463–466.

51. Pelto, C. R., T. A. Elkins, and H. A. Boyd. "Automatic Contouring of Irregularly Spaced Data." *Geophysics* **33** (1968): 424–430.

52. Powell, M. J. D. "Radial Basis Functions for Multivariable Interpolation: A Review." In *Algorithms for Approximation*, edited by J. C. Mason and M. G. Cox, 143-167. Oxford: Clarendon Press, 1987.

53. Royall, R. M. "A Class of Nonparametric Estimators of a Smooth Regression Function." Ph. D. Dissertation and Tech Report No. 14, Public Health Service Grant USPHS-5T1 GM 25-09, Department of Statistics, Stanford University, 1966.

54. Sabin, M. A. "Contouring—A Review of Methods for Scattered Data." In *Mathematical Methods in Computer Graphics and Design*, edited by K. W. Brodlie, 63–86. New York: Academic Press, 1980.

55. Schagen, I. P. "Sequential Exploration of Unknown Multi-dimensional Functions as an Aid to Optimization." *IMA Journal of Numerical Analysis* **4** (1984): 337–347.

56. Seber, G. A. F. *Linear Regression Analysis.* New York: John Wiley, 1977.

57. Shepard, D. "A Two-Dimensional Function for Irregularly Spaced Data." Proceedings of 23rd ACM National Conference, 1968, 517–524.

58. Sheppard, W. F. "Reduction of Errors by Means of Negligible Differences." *Proceedings of the Fifth International Congress of Mathematicians*, vol. II, edited by E. W. Hobson and A. E. H. Love, 348–384. Cambridge, MA: Cambridge University Press, 1912.

59. Sherriff, C. W. M. "On a Class of Graduation Formulae." *Proceedings of the Royal Society of Edinburgh* **XL** (1920): 112–128.

60. Stanfill, C., and D. Waltz. "Toward Memory-Based Reasoning." *Communications of the ACM* **29(12)** (1986): 1213–1228.

61. Steinbuch, K. "Die lernmatrix." *Kybernetik* **1** (1961): 36–45.

62. Steinbuch, K., and U. A. W. Piske. "Learning Matrices and Their Applications." *IEEE Trans. on Electronic Computers* **EC-12** (1963): 846–862.

63. Steinbuch, K., and B. Widrow. "A Critical Comparison of Two Kinds of Adaptive Classification Networks." *IEEE Trans. on Electronic Computers* **EC-14** (1965): 737–740.

64. Stone, C. J. "Nearest Neighbor Estimators of a Nonlinear Regression Function." Proc. of Computer Science and Statistics: 8th Annual Symposium on the Interface, 1975, 413–418.

65. Stone, C. J. "Consistent Nonparametric Regression." *The Annals of Statistics* **5** (1977): 595–645.

66. Stone, C. J. "Optimal Global Rates of Convergence for Nonparametric Regression." *The Annals of Statistics* **10(4)** (1982): 1040–1053.
67. Taylor, W. K. "Pattern Recognition By Means Of Automatic Analogue Apparatus." *Proceedings of The Institution of Electrical Engineers* **106B** (1959): 198–209.
68. Taylor, W. K. "A Parallel Analogue Reading Machine." *Control* **3** (1960): 95–99.
69. Taylor, W. K. "Cortico-Thalamic Organization and Memory." *Proc. Royal Society B* **159** (1964): 466–478.
70. Walters, R. F. "Contouring by Machine: A User's Guide." *Amer. Assoc. Petroleum Geologists Bull.* **53(11)** (1969): 2324–2340.
71. Waltz, D. L. "Applications of the Connection Machine." *Computer* **20(1)** (1987): 85–97.
72. Watson, G. S. "Smooth Regression Analysis." *Sankhyā: The Indian Journal of Statistics, Series A* **26** (1964): 359–372.
73. Whittaker, E., and G. Robinson. *The Calculus of Observations.* London: Blackie & Son, 1924.

Index

A

Abarbanel, H. D., 244
activation, 403
 function, 402-403
actual forecast, 306
affine functions, 11
algorithm
 back-propagation, 404
 computational, 283
 Grassberger-Procaccia, 46
 Grassberger-Procaccia-Takens, 170
 learning, 41-42, 44, 45, 66
 off-line, 41
 on-line, 41
 QR, 269
almost smooth approximation, 12
An, C. H., 510
analysis
 canonical factor, 208
 canonical variate, 283, 489
approximation, 13
 almost smooth, 12
 annealed, 34
 error, 412
 of continuous functions, 503
 smooth, 13
Arbarbanel/Kadtke technique, 462
ARCH/GARCH, 382
architecture, 47
ARIMA model, 144
arm dynamics, 513
ARMA, 139
 model, 144, 149
 process, 139
Aronszagn, N., 99
asymptotic exactness, 13
asymptotic sampling distribution
theory, 156
Atkeson, C. G., 510
Autoregressive Moving Average
 see ARMA
average relative variance, 400

B

β-instruments, 320
back propagation, 39, 308-309, 395
 algorithm, 404
 learning, 400
 networks, 40

Baek, E., 154, 156
Barnhill, R. E., 508
barycentric coordinates, 10-11
basis function centers, 39, 55
Bayes
 learning equation, 408
 rule, 325
BDS, 149
 statistics, 148-149, 152, 155, 170
 test, 150, 168
Bellman's principle of optimality, 507
Benton, L., 373
Bertero, M., 98
beta-binomial distribution, 127
BetaMultiplier parameters, 50
Bhattacharyya, B. K., 508
bias-variance trade-off, 133
bilinear models, 199, 205, 208, 220, 223, 416
Billings, S. A., 266
binomial
 model, 125
 procedure, 127
Bischol, R., 106
bispectrum, 199, 201, 204, 207, 209, 212
bivariate normal distribution, 128
bleaching, 175
Boolean function, 33
bootstrapping approach, 157, 163, 165, 422
Bowyer, A., 6
Bowyer's method, 6
Box and Jenkins time-series analysis, 144
Box-Ljung test, 145
Boyd, H. A., 508
brain, 97
branched manifold, 8-9
branches, 3
Brent, R. P., 508
Brock, William, 145-147, 149, 154, 157-158, 164, 422
Brockett, P. L., 205
Brockwell, P., 144-145, 155
Broomhead, D. S., 435
Brown, R., 244
Brown, T. C., 117
Bryant, D., 244
Buntine, Wray L., 425

C

Canadian lynx, 205

canonical factor analysis, 208
canonical variate analysis, 283, 489
    independent of, 283
Casdagli, Martin, 46, 74, 80, 96, 171, 265, 267, 274, 434, 462
causality, 335
Čenys, A., 170
Chaitin, Gregory J., 456
Chaitin-Kolmogorov complexity, 334
chaos, 138, 140, 145, 157-158, 208, 265, 267, 279, 355
    beam, 478-479
    period, 244
chaotic
    strange attractor, 46
    behavior, 471
    time series, 39
Cheng, P. E., 509
Chomsky hierarchy, 317
Ciliberto, S., 171
circumsphere, 14
Cleveland, W. S., 508-509
climate
    data, 481
    models, 483
CMAC, 513
CNLS-net, 39-41, 47-49, 51, 69
CNLSTOOL, 48, 64
coherence timescale, 55
communication channel, 324
complexity, 319, 332-333, 355
    cost, 408
    finitary, 332
compression, 313
computation, 319
    theory, 317
    time, 90
computational algorithms, 283
computational irreversibility, 341, 353
computational mechanics, 319
computational structure, 332
conditional correlations, 384
confidence intervals, 422
connection machine, 510
connection matrix, 333
Connectionist Normalized Local Spline network
    see CNLS-net
connectionist networks, 395
conservation of information, 332
Constable, C., 106

context, 343
continuous function, 503
contraction, 370
control, 356
    theory, 273
controlled Markov process, 291
convergence, 337
convex hull, 5, 16
Corcoran, E., 97
correlation coefficient, 20-21
correlation dimension, 115-116, 118, 123, 128, 163, 170, 433
correlation integral, 140, 145-146, 149, 156
cost function, 327
Cover, T. M., 508
Cox, D., 100
Crain, I. K., 508
critical phenomena, 353
cross-validation error, 512
Crutchfield, J. P., 263, 468
currency exchange rates, 395
curse of dimensionality, 398
CVA, 288-289, 292-294
    independent, 293
Cybenko, G., 96-97

D

da Silva, M. G. A., 208
Darken, C. J., 40, 74
data compression, 354
Davis, R., 144-145, 155
Dechert, W. D., 145-148, 164
degree of meaning, 343
Delaunay triangulation, 5-6
delay
    embeddings, 471
    vector, 273
Denker, M., 140, 146
design, 356
    matrix, 92
desktop computers, 164
determinism, 20, 173
    map, 122
    modeling, 273
Devlin, S. J., 509
Devroye, L. P., 509
Diebold, Francis X., 418
dilation, 370
dimensional complexity, 376
diode oscillators, 497

direct multi-step prediction, 415
Dirichlet tesselation, 5, 10, 21
discriminating statistic, 165, 170
divergence, 271
dominant dynamic behavior, 461
driven Duffing oscillator, 474-475, 478
driven systems, 467
    models, 486
Dube, R. P., 508
Duchon, J., 103
Duffing equation, 10
dwell time, 322
dynamic programming, 507
dynamical noise, 173
dynamical processing, 462
    approach, 456
dynamical systems, 3, 356
dynamics, 14

E

ϵ-machines, 331, 336
Eckmann, J.-P., 171, 268-269, 362-363, 499
EEG data, 17-18
effective parameters, 401
efficient market hypothesis, 418
Efron, B., 165, 422
eigenvalues, 496
electroencephalogram, 179
Elkins, T. A., 508
Ellner, S., 165
embedding, 4, 39, 47, 52, 54, 279, 397-399
    dimension, 5, 416
    theorem, 116
energy, 348
entropy, 25, 29, 31-32, 37, 319, 328, 350
    352, 493, 496, 499-500
    excess, 341, 354
    Kolmogorov-Sinai, 231, 334
    rate, 332
    relative, 25
    Renyi, 309, 354
    Shannon's, 351
epidemic data, 18
equations
    Duffing, 10
    of motion, 320, 458
    stochastic evolution, 320
equilibrium, 350
error
    bars, 165

error (cont'd.)
    model, 399-400
    term, 458
    tolerance, 30
    estimation, 152
Eubank, Stephen, 137-138, 151, 508
evolutionary systems, 355
excess entropy, 341, 354
exchange rates, 397, 417-418
explanation, 327
explanatory channel, 325
extrinsic noise, 325

F

falsifiable predictions, 14
Farber, Rob, 46-47, 64, 74
Farmer, J. Doyne, 14, 16, 46, 73, 81, 137-
138, 151, 178, 265, 267, 276, 387, 434, 449,
468, 508-509
Farwig, R., 509
fat baker's transformation, 122
feed-forward neural network, 27, 74
finitary complexity, 332
Fix, E., 508
Flannery, B. P., 92
fluctuation spectra, 353
Folconer, K. J., 508
folds, 3, 9
forecasting, 356
    actual forecasts, 306
    errors, 163, 170-171
foreign exchange, 417
forward model, 505
Fourier model, 318
Franke, R., 508-509
free energy, 349
free information, 354
Freeman, F. P., 327
Friedman, J. H., 191, 397
Fujisaka, H., 244
function
    activation, 402-403
    affine, 11
    approximation, 507
    boolean, 33
    estimation, 95, 98
    masking, 30
    multivariate, 95-96
    of vector valued, 95, 106
    optimization, 507

function (cont'd.)
    weighting, 74
functional diversity, 29
fuzzy logic, 322
fuzzy partition, 323

G

Gabr, M., 200, 205, 208, 416
GARCH, 391
    model, 383-385
Gaussianity, 205
GCV, 104
    test, 100
gedanken experiment, 333
Geist, K., 373
Gelfand, I. M., 295
generalization, 26, 33, 35-36, 404
    ability, 33
generalized baker's transformation, 121-
    122, 124, 128, 132
geometric approach, 3
geometry, 5
Gibbs formulation, 37
Gibbsian statistical mechanics, 319
Girosi, F., 98, 106
Glanz, F. H., 513
global forcasting
    improvements, 392
    method, 90
global Lyapunov exponents, 229
Godfrey, M. D., 200
Goldberg, K. Y., 513
Golomb, M., 97
Golosov, J., 107
Gordon, W. J., 508
gradient descent, 404
    learning, 43
Granger, C. W. J., 189
Grassberger, P., 116, 168, 244
Grassberger-Procaccia
    algorithm, 46
    formalism, 245
    method, 116, 119
    procedure, 115, 117, 124
Grassberger-Procaccia-Takens
    algorithm, 170
    dimension plots, 148
Green, M., 157
Green, P. J., 5
Grosse, E., 509

Gu, C., 104

H

Hannan, E., 153
Hasselman, K., 200
Heller, W., 189
Hénon, M., 173
Hénon attractor, 128
Hénon map, 129, 132, 212, 239
hidden units, 402
higher-order cumulant spectra, 207
Hilbert space, 99, 104
Hinich, M. J., 205
Hinich's test, 202
Hinton, Geoffrey E., 404
Hodges, Jr., J. L., 508
Holland, K. N., 200
Hollergach, J. H., 510
Hsieh, Davis, 158
Hsieh, E., 149
Hubel, D., 97
Huber, P. J., 191
human EEG, 373
Hunter, N. F., 266
Hutchinson, M., 106

I

Ikeda map, 235, 269, 271, 277
Imbrie, John, 482
in-sample performance, 401, 412
independent and identically distributed
process, 138, 146
independent canonical variate analysis, 293
inductive inference, 327
inference methodology, 320
information, 317
    acquisition rate, 323
    dimension, 276
input-output, 279
    maps, 25
    models, 269, 470
    systems, 265-267, 269,
    273-274, 276
instrument, 317
    temperature, 322
intelligibility, 457
interaction splines, 105
internal validation set, 396, 401, 411
intrinsic computation, 319
invariants, 374

inverse
methods, 96
model, 505
isotropic reproducing kernel, 101
iterated single-step, 414

J

Jacobian matrices, 232
of partial derivatives, 495
Jaynes, E. T., 319

K

k-d trees, 509
k-means clustering, 106
Kalman-Bucy, 16
Kalman Noise Filtering, 40, 54
Kamphorst, S. Oliffson, 362-363
Kanter, M., 153
Kaplan-Yorke conjecture, 46
Keidel, M., 373
Keller, G., 140, 146
Kemeny, J. G., 327
kernel, 145
density estimation, 77
Kimeldorf, G., 97
King, G. P., 435
knowledge relaxation, 335, 338
Koh, E., 100
Kolmogorov, A. N., 324, 455-456
Kolmogorov-Sinai entropy, 231, 334
Kolmogorov's theorem, 329
Kraft, L. G., 513
Kravtsov, Yu. A., 457
Kullback information, 296

L

Labonishok, Josef, 422
lacunarity, 115-116, 121, 124, 133
lag, 4
lag space, 398
Lakonishok, J., 157-158
Lancaster, H. O., 294
Lancaster, P., 508-509
Lapedes, Alan, 46-47, 64, 74
Law of Iterated Expectations, 144
Law of Large Numbers, 153, 356
learning, 25-26, 28-33, 96, 356, 404, 422,
503-504
algorithm, 41-42, 44-45, 66

learning (cont'd.)
behavior, 59
control, 507
curves, prediction of, 35
error, 31
rate, 49, 60
rate parameter, 49
supervised, 25, 30, 33
LeBaron, Blake, 149, 157-158, 167, 383, 422
Legendre transform, 354
Legg, M. P. C., 508
Lempel, A., 306
Lempel-Ziv codes, 307
Lempel-Ziv complexity, 334
Lempel-Ziv-type codes, 306
Lewis, Peter A. W., 416
Li, K. C., 172, 509
Liebert, W., 170
Lii, K. A. M., 200
Lim, K. S., 410
Lindgren, K., 312
linear prediction, 434, 440
coding (LPC), 434
linear representations, 91
linear stochastic process, 183
linearity, 205
testing, 149, 155
linearization method, 499
linearly correlated noise, 168
Little, F. F., 508
Liu, T., 189
local linear map predictors, 80, 82-84, 86,
89
local quadratic map predictor, 80
local approximation, 433, 443
schemes, 73
local coordinates, 10-11
local linear (or quadratic) predictors, 40
local Lyapunov exponents, 229, 244
local maps, 74
local models, 503
locally weighted regression, 158, 503, 509-
510, 514
Lodwick, G. D., 508
LOESS, 509
logistic map, 208, 339
long-range dependency, 223
Lorenz attractor, 8, 233
chaotic, 283

Lyapunov exponents, 46, 163, 170-171, 229-230, 268, 375, 493-494, 496-497, 499-500
lynx pelt data, 20

M

Macdonald, G., 200
machine reconstruction, 331
machine thermodynamics, 347
Mackey-Glass (M-G)
    equation, 39-40, 46-47, 62, 64
    delay-differential equation, 78
    time-delay differential equation, 39
    time-delay equation, 235
MacQueen, J., 106
manifolds, 6, 21
    identification, 6
map reconstruction, 3
Markov chains, 324
Markov processes, 290
masking function, 30
maximal correlation, 288
    serial, 191
maximum likelihood estimation, 119, 125
May, R. M., 14, 18, 156
Mayer-Kress, Gottfried, 373
McIntyre, D. B., 508
McLain, D. H., 508
mean generalization ability, 35
mean squared forecast error (MSE), 119-120, 387
meaning, 343-344, 346-347
    absolute, 347
measles epidemic data, 18
measurements, 317, 343
    channel, 324
    error, 151
    partition, 322
    semantics, 343
    space, 320
    of speech dimensionality, 437
memory, 332
memory-based architecture, 504
memory-based modeling, 508
memory-based robot learning, 503
method of delays, 435
Micchelli, C., 99, 101
microstates, 347
Milkanovich theory of the ice ages, 481

Miller, W. T., 513
minimal state rank, 292
Minimum Description Length Criterion, 405
models
    ARMA, 144, 149
    bilinear, 199, 205, 208, 220 223, 416
    binomial, 125
    forward, 505
    Fourier, 318
    input-output, 269
    input-output, 470
    identification, 299
    nonhomogeneous binomial, 125
    nonlinear, 318, 324
    nonlinear dynamical, 458
    parametric, 191, 195
    predictive, 398
    response based, 468
    selection for noisy data, 404
    stochastic, 189
    threshold autoregressive, 410
model space, 327
modeler, 323
modeling, 317
    limits to, 324
Monte Carlo experiment, 152
Monte Carlo results, 101
Moody, J., 40, 74
Moon, F., 478
Moore, A. W., 509
morph, 330
Moser, R. D., 244
moveable centers, 106
multi-step predictions, 414
multilayer networks, 97
multivariate
    adaptive regression splines (MARS), 397, 416
    function, 95-96
    nonlinear CVA, 286
    nonparametric function, 97
    splines, 95
Munk, W., 200
mutual information, 295, 341
    expected relative, 296
Müller, H. G., 509

N

Nakamura, Y., 200
Nason, James A., 418
nearest-neighbor, 389, 392, 508
   approaches, 508
   distances, 116
   distributions, 117
   local regression techniques, 388
neighbors, 6
Ness, Jon M., 244
network complexity, 395
network ensemble, 29
   prediction of, 37
network size, 52
networks
   connectionist, 395
   feed-forward neural, 27, 74
   multilayer, 97
   network, 513
   sigmoidal feedforward, 514
neural nets, 311
neural networks, 25-27, 37, 305
Newton Raphson method, 506
Nielson, G., 508
noise, 4, 207, 272
   extrinsic, 325
   linearly correlated, 168
   propagated,151
   temporally uncorrelated, 167
   white, 192
nonhomogeneous binomial model, 125
nonhomogeneous binomial procedure, 127
nonlinear, 139-140, 145, 151, 157
   dynamical model, 458
   forecasting, 388
   long-range dependence, 199
   mapping, 448
   modeling, 318, 324
   parameter estimation procedure, 512
   predictor, 433
   prediction, 220, 447, 455
      model, 457, 462
   relationships, 189, 191
   static transformation, 183
   structure,detection of, 164
   time-series modeling, 471
nonlinearity, 139, 163, 165, 170, 172-173,
181, 318
nonparametric, 381, 385
   models, 191

nonstationarity, 158
normalization, 283
normalizing transformations, 294
null hypothesis, 163, 165-169, 181
Nychka, D., 96, 110

O

O'Sullivan, F., 106
observer, 317, 320
   function, 137
Occam's razor, 405
off-line algorithms, 41
Oliffson Kamphorst, D., 171
on-line algorithms, 41
one-step-ahead forecast errors, 151-152
operator estimation, 95
optimization, 507
   forecast, 306
   instruments, 328
   of fit parameters, 516
   projection, 286
optimality, 328
order, 355
   of the local map, 443
origami, 10
oscillating saw-tooth wave, 368
Oseledec, V. I., 231
Oseledec matrix, 234
Oseledec Multiplicative Ergodic Theorem,
231, 238
out-of-sample
   forecasts, 381-382, 387, 391
   performance, 401
overfitting problem, 404

P

Packard, Norman, 330, 468
Palmer, J. A. B., 508
parametric models, 191, 195
Parker, R., 106
parse tree, 330, 333
partial prediction, 459
partition function, 348
Pearlmutter, B., 513
Pelto, C. R., 508
Penrose, R., 97
performance
   in-sample, 401, 412
periodic orbits, 493, 496, 499-500
   saddle orbits, 500

periodic saddle points, 494
Peterson, D., 205
phase locking, 486
phase transitions, 355
philosophy of nonlinear prediction, 14
piecewise linear approximation, 11
piecewise linear fitting, 4
Poggio, T., 98, 106
Polland, D. D., 508
polynomial map, 232
Poppl, S. J., 373
Potter, S., 143
pre-whitening, 175
predictability, 229
predictions, 3, 14, 330
    accuracy, 395
    error, 79, 276, 449
    partial, 459
    set, 401
predictive model, 398
predictor, 10, 75, 442
Press, W. H., 92
prevalent, 274
Priestly, M. B., 140, 202
principal component analysis, 208
problem of overfitting, 397
Procaccia, I., 116, 244
projection
    method, 146
    operators, 288
    pursuit, 191
    pursuit experiment, 193
propagated noise, 151
Pyragas, K., 170

Q

QR algorithm, 269
quantification of predictability, 457
quantization, 439

R

radial basis functions, 39-40, 80, 95, 105
random behavior, 455
randomness, 328
Rank Theorem, 292
ratio test, 100
reconstruction, 330, 435
    hierarchy, 332
recurrence plot, 361-362, 373
recursive filtering, 16

recursive forecasting, 16
recursive prediction, 16
relative entropy, 25, 32
relative mean squared error, 400
relaxation oscillator, 370
Renyi, A., 288-289
Renyi entropy, 309, 354
representation, 397
reproducing kernels (rk's), 95, 98, 105, 108
response-based embeddings, 468
response-based models, 468
ridge regression, 512
Rissanen, J., 327
robot control, 503
robustness, 52, 62
root finding, 507
Rosenblatt, M., 143, 200
Royall, R. M., 508
Ruelle, D., 171, 362-363, 499
Rumelhart, David E., 401, 404-405
Rössler attractor, 15

S

S&P 500 index, 384
S&P 500 series, 386
Sabin, M. A., 508
saddle, 493-494, 496-497, 499-500
Sakai, H., 209
Šalkauskas, K., 508-509
Sampling Theorem, 55
Sano, M., 171
Sasaki, K., 200
Sato, T., 200
Sauer, T., 267, 274-275
Savit, R., 157
Sawada, Y., 171
scalar time series, 273
scaling
    laws, 276
    of prediction error, 449
Scheinkman, José, 145-147, 149, 167
Schuster, H. G., 170
Science, 354
semantics, 317, 343
sensitivity tests, 49
sequential selection, 288
Serfling, R., 145-146
Shackleton, John Z., 482
Shannon
    information, 346

Shannon (cont'd.)
    coding theorems, 324
    entropy, 351
    theorem for a channel with
    noise, 325
    theory, 327
Shaw, R. S., 468
Shepard, W. F., 508
short-term predictability, 459
Sibson, R., 5, 13
Sidorowich, J. J., 14, 16, 73, 81, 178, 267,
    276, 387, 434, 449, 508-509
Sierpinski gasket, 128-129
Sierpinski map, 132
sigmoid, 403
sigmoidal feedforward network approach,
514
sigmoidal feedforward neural networks, 513
signal separation, 455, 461-462
Silverman, B. W., 117
simplex, 5
single-step prediction, 414
singular value decomposition, 444
Smith, L. A., 121
Smith, R., 508
smooth approximations, 13
smoothing spline, 100
Solla, S. A., 309
speech, 433
    modeling, 434
    signals, 433
square wave, 364
stable foliation, 497
Stanfill, C., 508
state, 330
    reconstruction, 299
    representation, 435
state-space reconstruction, 265, 273-274,
    276, 398
static nonlinear filter, 168
statistical complexity, 334-335
statistical estimation, 352
statistical mechanics, 319
statistical processing, 456
Steinbuch, K., 508
Stevens, J. G., 416
stochastic   complexity, 334
    connection matrix, 331
    evolution equations, 320
    linear process, 140
    linearity, 149

stochastic (cont'd.)
    machine, 331
    modeling, 273
    models, 189
    nonlinear process, 158
    process, 140
stock prices, 381
stock returns series, 381
Stokbro, K., 77, 74, 78, 416
Stone, C. J., 508-509
strange attractors, 493
strictly Sofic system, 337
Stuetzle, W., 191
Subba Rao, T., 200, 205, 208, 220, 416
subshift of finite type, 337
subtile weights, 13
Sugihara, G., 14, 156, 18
Sugihara-May testing, 157
sunspot series, 395
sunspots, 207, 221, 180, 397, 409
superfluid convection cell, 178
supervised learning, 25, 30, 33
surrogate data, 166, 163, 165, 169, 173, 183
symbolic dynamics, 337
syntax, 317

T

Takens, F., 115-116, 119, 124-125, 127, 170,
    267, 274, 398, 468
Takens' embedding theorem, 46
tangent
    manifold, 5
    space, 8
tanh-unit, 403
Taylor, D., 508
Tempelman, A., 107
temporally uncorrelated noise, 167
tent map, 209
Teräsvirta, T., 190
tesselations, 3, 5-6, 10, 16, 21
test set, 79
Teukolsky, S. A., 92
Theiler, J., 123-124
theories
    asymptotic sampling distribution, 156
    of regularized estimates of nonlinear
    operators, 95
    of state-space reconstruction, 276
    of the estimation of nonlinear
    operators, 96

theories (cont'd.)
    Osedelec theorem, 238
    Rank theorem, 292
    Sampling Theorem, 055
    Shannon's coding theorem, 324
    universal coding, 327
    Wold decomposition theorem, 172, 175
thermodynamic
    description, 32
    entropy, 348
    potential, 352
    relation, 33
thermodynamics, 317, 319, 347-348
Thieler, J., 122
thin plate splines, 102-103
threshold autoregressive model, 410
tile, 5
time
    delay, 494
    inversion, 370
    series, 39
Tirsch, W., 373
Tokomaru, H., 209
Tong, H., 20, 172, 265, 410
topological
    complexities, 333
    entropy, 333, 496
    machine, 331
Torre, V., 98
torus, 8
traditional statistical approach, 172
training, 14, 47
    data set, 57-58
    procedures, 52
    set, 14, 78, 401
    set size, 57
transducer, 320
triangle wave, 367
triangulation, 5-10
Tsay, Ruey S., 384

U

U-statistics, 139, 145-147
unit, 403
universal coding theory, 327
universal Turing machine, 319
unpredictable, 459

V

V-statistics, 140, 145-147
validation set, 413
Van Atta, C., 200
Vastano, J. A., 244
vector-valued functions, 95, 106
versatility, 52, 62
Vetterling, W. T, 92
vibration testing, 266, 268
visual system, 97
volatility, 382-383, 385-387, 390, 392
Volterra expansion, 142-143
Voronoi polygon, 5

W

Wahba, G., 98-99, 104, 110
Wallace, C. S., 327
Walters, R. F., 508
Waltz, D., 508
Watson, G. S., 508
Weigend, Andreas S., 397, 401, 425
weight decay, 409
weight elimination, 397, 405
weighted constant map predictors, 76, 80-81
weighted linear map predictors, 76, 80-84, 86, 89, 416
weighted map, 73
    schemes, 90
weighted superposition, 74
weighting functions, 74
Weinberger, H. J., 97
Weinert, H., 99
white noise, 192
Whittle, J., 508
Widrow, B., 42
Wiener, Norbert, 455
Williams, Ronald J., 404
window, 4
Wixom, J. A., 508
Wold decomposition theorem, 172, 175

Y

Yaglom, A. M., 295
Yandell, B., 100
Yates, F. E., 373
Yorke, J. A., 267, 274

Z

Ziv, J., 306